Numerical Methods
Classical and Advanced Topics

This book presents a pedagogical treatment of a wide range of numerical methods to suit the needs of undergraduate and postgraduate students, and teachers and researchers in physics, mathematics and engineering. For each method, the derivation of the formula/algorithm, error analysis, case studies, applications in science and engineering and the special features are covered. A detailed presentation of solving time-dependent Schrödinger equation and nonlinear wave equations, along with the Monte Carlo techniques (to mention a few), will aid in students' understanding of several physical phenomena including tunnelling, elastic collision of nonlinear waves, electronic distribution in atoms and diffusion of neutrons through a simulation study.

The book covers advanced topics such as symplectic integrators and random number generators for desired distributions and Monte Carlo techniques, which are usually overlooked in other numerical methods textbooks. Interesting updates on classical topics include curve fitting to a sigmoid and Gaussian functions and product of certain two functions, solving of differential equations in the presence of noise and solving the time-independent Schrödinger equation.

Solutions are presented in the form of tables and graphs to provide visual aid and encourage a deeper comprehension of the topic. The step-by-step computations presented for most of the problems can be verifiable using a scientific calculator and are therefore appropriate for classroom teaching. The readers of the book will benefit from acquiring an acquittance, knowledge, experience and realization of the significance of the numerical methods covered, their applicability to physical and engineering problems and the advantages of applying numerical methods over theoretical methods for specific problems.

Shanmuganathan Rajasekar was born in Thoothukudi, Tamil Nadu, India, in 1962. He was awarded Ph.D. from Bharathidasan University in 1992 under the supervision of Prof. M. Lakshmanan. In 1993, he joined as a Lecturer at the Department of Physics, Manonmaniam Sundaranar University, Tirunelveli. In 2005, he joined as a Professor at the School of Physics, Bharathidasan University, Tiruchirapalli. His recent research focuses on nonlinear dynamics with a special emphasis on nonlinear resonances. He has authored or co-authored more than 120 research papers in nonlinear dynamics.

Numerical Methods
Classical and Advanced Topics

Shanmuganathan Rajasekar

CRC Press
Taylor & Francis Group
Boca Raton London New York

CRC Press is an imprint of the
Taylor & Francis Group, an **informa** business

Designed cover image: S. Rajasekar

First edition published 2024
by CRC Press
2385 NW Executive Center Drive, Suite 320, Boca Raton FL 33431

and by CRC Press
4 Park Square, Milton Park, Abingdon, Oxon, OX14 4RN

CRC Press is an imprint of Taylor & Francis Group, LLC

© 2024 S. Rajasekar

ISBN: 978-1-032-63894-2 (hbk)
ISBN: 978-1-032-64991-7 (pbk)
ISBN: 978-1-032-64993-1 (ebk)

DOI: 10.1201/9781032649931

Typeset in CMR10
by KnowledgeWorks Global Ltd.

Publisher's note: This book has been prepared from camera-ready copy provided by the authors.

To
Professor M. Lakshmanan
Professor K.P.N. Murthy

Contents

Preface

The widespread use of computers, rapid increase of speed of computation in computers and the growth of sophisticated utility tools for computation led to a great deal of focus on numerical methods by the scientists and engineers. For a certain class of mathematical problems even if the construction of exact analytical solutions are feasible, numerical methods are often preferred to extract first-hand information about the solutions. Simplified user-friendly computer graphics made it easy to visualize and understand complex numerical solutions. Nowadays, a course on numerical methods is a part of undergraduate and postgraduate programmes in physics, mathematics, computer science, biophysics, bioinformatics and in various engineering subjects.

This book is developed as a textbook on numerical methods. In addition to covering standard numerical methods, the book deals with basic aspects of some novel advanced topics that have applications in various branches of applied science. For each numerical method the book thoroughly presents basic idea, derivation of formulas, systematic implementation of procedures and illustrative examples. In addition, wherever possible, the error analysis, convergence of iterated solutions, efficiency and stability of the methods are addressed.

The pedagogic features of the present book, which are not usually found in the textbooks on numerical methods, are the presentation of symplectic integrators for Hamiltonian systems, methods for solving time-independent and time-dependent Schrödinger equations, multi-symplectic integrators for nonlinear partial differential equations, fractional order ordinary and partial differential equations, fast Fourier transform, generators for random numbers with desired distributions and Monte Carlo technique. There are numerous collection of problems at the end of each chapter. Most of the problems belong to physical systems. Solutions to a large number of problems are listed at the end of the book.

The book has 20 chapters. In Chapter 1, certain preliminaries of numerical methods are provided. Computation of solutions of polynomial and reciprocal equations are treated in Chapter 2. The next chapter presents various methods for determining a root of general nonlinear equations. Comparison of various methods is made. Chapter 4 is concerned with the solution of system of linear equations. Direct and iterative methods such as Gauss elimination, Gauss–Jordan, triangular factorization, Jacobi and Gauss–Seidel are described. Solving tridiagonal systems, Vandermonde matrix systems, least-squares and singular-value problems are also addressed. The least-squares method to fit straight-line, power-law and exponential functions, polynomials, Gaussian functions and trigonometric polynomials are illustrated in Chapter 5. Interpolation is discussed in Chapter 6. Newton, Gregory–Newton and Lagrange interpolating polynomials are derived. Cubic spline and rational function approximation are also covered. Chapter 7 is devoted to the computation of eigenvalues and eigenvectors of arbitrary, symmetric and symmetric tridiagonal matrices. The power, Jacobi and QL methods are presented.

Numerical differentiation of a function is considered in Chapter 8. Various formulas for first-, second- and third-order derivatives are derived. The case of fractional order derivative is also dealt. The topic of minimization of one-dimensional and two-dimensional functions is treated in Chapter 9. In Chapter 10 certain methods for numerical integration of a function including fractional order integration are covered. A section on double integration is also included.

Chapters 11 through 15 are concerned with the numerical integration of differential equations with integer order derivatives. Chapter 11 is exclusively for the initial-value problem of ordinary differential equations. The Euler, Runge–Kutta, Runge–Kutta Fehlberg, Adams–Bashforth–Moulton methods are discussed in depth. A section on stiff equations is included. Solving differential equations with Gaussian white noise and coloured noise are described. Chapter 12 deals with classical ordinary differential equations of Hamiltonian systems, that is, systems in which energy remains a constant. The basic idea of symplectic methods is elucidated. Symplectic Euler, Störmer–Verlet and Runge–Kutta type methods are considered in detail. Chapter 13 is reserved for solving ordinary differential equations with appropriate boundary conditions. The shooting method is illustrated. Solving of time-independent Schrödinger equation is presented. The next chapter is devoted for linear partial differential equations. Particularly, finite-difference schemes for wave, heat, Laplace and Poisson equations are derived and discussed. Solving of time-dependent Schrödinger equation for certain interesting potentials is also considered. In Chapter 15, first the multi-symplectic method of constructing numerical solution for nonlinear partial differential equations is discussed. Then, the schemes for the ubiquitous Korteweg–de Vries and sine-Gordon equations are derived and generation of different types of localized solutions is presented.

Numerical integration methods for fractional order differential equations are considered in Chapters 16 and 17. In Chapter 16 for the fractional order ordinary differential equations the backward-difference, fractional Euler, Adams–Bashforth–Moulton and a two-step Adams–Bashforth methods are developed and analyzed for linear and nonlinear systems. The next chapter is devoted for fractional order partial differential equations. Particularly, schemes for the time-fractional diffusion and advection-diffusion equations, wave and damped wave equations, Fisher equation are set up and the validity of them are discussed. A method for the space-fractional diffusion equation is also presented. The accuracies of the methods are tested by considering the cases of the systems with exact analytical solutions. A way of constructing stable equilibrium solutions is also described.

Chapter 18 is on Fourier analysis and power spectrum. Computation of Fourier coefficients is illustrated. A simplified procedure for obtaining fast Fourier transform is described and power spectra for a certain types of functions and solutions of differential equations are computed and discussed. In Chapter 19 generation of random numbers with specific emphasize on random numbers obeying uniform, Gaussian, exponential, Cauchy, dichotomous and Lévy distributions are described in detail. Tests for randomness are also presented. The final chapter is devoted for Monte Carlo technique. The basic idea of Monte Carlo technique is illustrated through the evaluation of definite integrals and nth root of a real positive number and estimation of values of π and e. Applications of Monte Carlo technique to the simulation of electronic distribution in an atom, radioactive decay, diffusion of neutrons in a material slab and percolation are described in detail.

The book is prepared with great supports from many colleagues, students and friends. In particular, the author is grateful to K.P.N. Murthy, M. Marudai, A. Tamilselvan, P. Philominathan, A. Venkatesan and V. Chinnathambi for critical reading of some of the chapters, their suggestions and encouragement. The author thanks his family members for their unflinching support, cooperation and encouragement during the course of preparation of this work.

Tiruchirapalli, Tamilnadu, India

S. Rajasekar
October, 2023

About the Author

Shanmuganathan Rajasekar was born in Thoothukudi, Tamilnadu, India in 1962. He received his B.Sc. and M.Sc. in Physics both from the St. Joseph's College, Tiruchirapalli. He was awarded Ph.D. degree from Bharathidasan University in 1992 under the supervision of Prof. M. Lakshmanan. In 1993, he joined as a Lecturer at the Department of Physics, Manonmaniam Sundaranar University, Tirunelveli. In 2003, the book on *Nonlinear Dynamics: Integrability, Chaos and Patterns* written by Prof. M. Lakshmanan and the author was published by Springer. In 2005, he joined as a Professor at the School of Physics, Bharathidasan University. In 2016, Springer has published the book on *Nonlinear Resonances* written by the author and Prof. Miguel A.F. Sanjúan. Professors U.E. Vincent, P.V.E. McClintock, I.A. Khovanov and the author compiled and edited two issues of Philosophical Transactions of the Royal Society A on the theme *Vibrational and Stochastic Resonances in Driven Nonlinear Systems* in 2021. He has also edited a book on *Recent Trends in Chaotic, Nonlinear and Complex Dynamics* with Professors Jan Awrejecewicz and Minvydas Ragulskis published by World Scientific in 2022. In 2023, the second revised edition of the books on *Quantum Mechanics I: The Fundamentals* and *Quantum Mechanics II: Advanced Topics* written by the author and Prof. R. Velusamy were published by CRC Press. In the same year, World Scientific has published the book on *Understanding the Physics of Toys: Principles, Theory and Exercises* written by the author, Prof. Miguel A.F. Sanjúan and Prof. R. Velusamy. His recent research focuses on nonlinear dynamics with a special emphasize on nonlinear resonances. He has authored or coauthored more than 120 research papers in nonlinear dynamics.

1

Preliminaries

1.1 Introduction

Even long before the invention of computer, numerical methods have been developed and used in many scientific and engineering problems. However, the invention of computer and rapid increase in the speed of computation gave a new life to numerical methods. As a result, a great deal of interest has been focused on the formulation of very effective and accurate new numerical procedures for mathematical problems and modification of existing techniques for higher accuracy. Numerical simulations of dynamics of complex systems have also become possible. It is noteworthy to mention that many phenomena exhibited by dynamical systems have been first realized in the numerical study of the dynamics of the systems, and then necessary theories and characteristic measures have been developed. Nowadays, time-consuming analytical procedures are often replaced by appropriate numerical methods. This book is concerned with the various numerical methods useful for mathematicians, physicists and engineers. Before starting the presentation of numerical methods, this chapter defines certain basic terms in numerical computation and mathematics which are essential to follow the various methods presented in this book.

1.2 Binary Number System

In everyday life the decimal number system is used. In the decimal system, the base is 10. That is, the number 382 can be expressed as

$$(382)_{10} = 3 \times 10^2 + 8 \times 10^1 + 2 \times 10^0. \tag{1.1}$$

The number 0.382 can be written as

$$(0.382)_{10} = 3 \times 10^{-1} + 8 \times 10^{-2} + 2 \times 10^{-3}. \tag{1.2}$$

Computers, however, perform arithmetic calculations, using the binary number system. The decimal numbers given as input to a computer are converted to machine numbers with digits 0 and 1 only. The base of the binary number system is 2, and in this system, a number is represented in terms of only two digits 0 and 1 called *bits*. The numbers 29 and 0.90625 are represented in the binary system as

$$
\begin{aligned}
29 &= 16 + 8 + 4 + 1 \\
&= 1 \times 2^4 + 1 \times 2^3 + 1 \times 2^2 + 0 \times 2^1 + 1 \times 2^0 \\
&= (11101)_2
\end{aligned}
\tag{1.3}
$$

DOI: 10.1201/9781032649931-1

and

$$0.90625 = 1 \times 2^{-1} + 1 \times 2^{-2} + 1 \times 2^{-3} + 0 \times 2^{-4} + 1 \times 2^{-5}$$
$$= (0.11101)_2 . \tag{1.4}$$

In general, a number M can be written in the binary form as

$$M = b_{n-1} \times 2^{n-1} + b_{n-2} \times 2^{n-2} + \cdots + b_0 \times 2^0$$
$$+ b_{-1} 2^{-1} + \cdots + b_{-m} \times 2^{-m}$$
$$= b_{n-1} b_{n-2} \ldots b_0 . b_{-1} \ldots b_{-m} , \tag{1.5}$$

where b_i's take the values 0 or 1. Note the binary point '·' between b_0 and b_{-1}.

The octal system of representation of numbers uses the eight digits 0, 1, 2, 3, 4, 5, 6 and 7 while the hexadecimal system uses the digits 0, 1, 2, 3, 4, 5, 6, 7, 8, 9 and the alphabets A, B, C, D, E, F for the numbers 10, 11, 12, 13, 14, 15, respectively.

1.3   Floating Point Arithmetic and Significant Digits

Computers have both an integer mode and a floating-point mode for representing numbers. The integer mode is used for performing calculations that are integer values. Computers usually perform scientific calculations in floating point arithmetic, and a real number is represented in a *normalized* floating point binary form. That is, computers store a real number x in the binary form as

$$x = \pm (\cdot d_1 d_2 \ldots d_n)_\beta \, \beta^e , \tag{1.6}$$

where $(\cdot d_1 d_2 \ldots d_n)_\beta$ is called *mantissa* and e is an integer called the *exponent*. In most of the computers $\beta = 2$. Other values of β used in some computers are 10 and 16. The term 'normalized' means that the leading digit d_1 is always nonzero unless the number represented is 0. That is, $d_1 \neq 0$ or $d_1 = d_2 = \ldots = d_n = 0$. The representation 0.00123 is not in normalized form. In the normalized form it reads as 0.123×10^{-2}. In the form $x = p.d_1 d_2 \ldots d_n$, p is called *characteristic*. The terms *characteristic* and *mantissa* were suggested by Henry Briggs in 1624. Mantissa is a Latin word of Etruscan origin, meaning a makeweight, a small weight added to a scale to bring the weight to a desired value [1].

Computers using 32 bits to represent real numbers in single-precision use 8 bits for the exponent and 24 bits for the mantissa. The magnitude of the numbers represented in this way has the range

$2.938736\mathrm{E} - 39 \, (2^{-128})$ to $1.701412\mathrm{E} + 38 \, (2^{127})$.

Note that in the above the numbers are represented with six decimal digits of numerical precision. A 48-bit computer uses 8 bits for the exponent and 40 bits for the mantissa. In this case, the range of real numbers is

$2.9387358771\mathrm{E} - 39 \, (2^{-128})$ to $1.7014118346\mathrm{E} + 38 \, (2^{127})$

with 11 decimal digits precision. If a computer has 64 bits double-precision real numbers, it might use 11 bits for the exponent and 53 bits for the mantissa. They can represent real numbers in the range

$5.562684646268003 \times 10^{-309} \, (2^{1024})$ to $8.988465674311580 \times 10^{307} \, (2^{1023})$

with about 16 decimal digits of precision. Calculation in double-precision usually requires more memory space and double the running time (that is, calculation slows down) as compared to single-precision.

All the digits beginning with a nonzero digit are called *significant digits*. For example, the underlined digits in

$$n_1 = 0.0\underline{72451} \text{ and } n_2 = 0.006\underline{500}$$

are the significant digits of n_1 and n_2, respectively. A significant digit is said to be *correct* if the error $n - n'$, where n' is an approximation of n, does not exceed unity of the order corresponding to that digit. When

$$n_1 = 0.0\underline{72451} \text{ and } n_1 - n_1' = 0.000004$$

then the underlined digits are correct. If $n_1 - n_1' = 0.0003000$, then n_1' is correct only up to the digits 2, that is $n_1 = 0.0\underline{72}451$. A number is said to be correct to m decimal places if there are m decimal digits up to the last correct digit. For

$$n_1 = 0.072451 \text{ and } n_1 - n_1' = 0.000300$$

the number of correct digits is 3.

A loss in significant digits occurs often when a number is divided by a small number. For example, the actual value of $f(x) = 1/(1 - x^3)$ for $x = 0.80005$ is 2.04958. If x is approximated as 0.8, then $f = 2.04918$. That is, rounding-off a number in 5th decimal causes a change in the 4th decimal. The observed loss of significant digits propagates when the value of f is used in further calculations.

1.4   Type of Errors

All experimental measurements and numerical computations, however, carefully performed are subject to some kind of errors. Here, the term *error* does not mean a *mistake* or a *blunder*. An error in a computation means the presence of an unavoidable deviation from the exact value. One cannot avoid errors; however, one can make the errors as small as possible.

The outcomes of measurements and numerical computations are often not absolutely precise due to various limitations of the apparatus/experimenter/adopted method. A knowledge of possible errors associated with a quantity is of great importance in analysing approximation methods, and it can often tell us whether a proposed theory or numerical algorithm or experiment needs refinements [2]. In the following let us discuss the errors occurring in numerical computations.

1.4.1   Round-Off Error

A given real number x can be converted into a floating point number denoted as $\mathrm{fl}(x)$ by

1.   rounding or
2.   chopping.

Suppose, a computer uses three-decimal digit floating point arithmetic. If the value of a real number extends beyond three-decimal places, then in *rounding* only the decimal values up to the third are retained. Further, if the fourth decimal value is > 5, then 0.001 is added to the number. On the other hand, in chopping, digits appearing after three-decimal places are dropped and 0.001 is not added even if the fourth decimal value is > 5.

Example:

Values of some $\mathrm{fl}(x)$ in rounding and chopping in three-decimal digit floating point arithmetic are given below:

x	$\mathrm{fl}(x)$	
	rounding	chopping
2473	0.247×10^4	0.247×10^4
2478	0.248×10^4	0.247×10^4
$1/3 = 0.3333\ldots$	0.333×10^0	0.333×10^0
$2/3 = 0.6666\ldots$	0.667×10^0	0.666×10^0

The difference between x and $\mathrm{fl}(x)$ is called *round-off error*. It depends on the size of x and is measured relative to x. Cases with $|x| \geq \beta^m$ and $0 < |x| \leq \beta^{m-n}$ are referred as *overflow* and *underflow*, respectively. Here, m and n are bounds on the exponents. In these cases either $\mathrm{fl}(x)$ is not defined and computation is stopped or $\mathrm{fl}(x)$ is represented by a specified number that is not subject to the rules of arithmetic. Round-off error plays an important role in numerical computations.

Consider

$$a = -1.0\,\mathrm{e} + 20,\ b = 1.0\,\mathrm{e} + 20,\ c = a + b + 1.0,\ d = a + (b + 1.0) \text{ and } e = 1.0 + b + a.$$

What are the values of c, d and e if they are calculated using a computer in double precision mode? The result is

$$c = 1.0,\ d = 0.0 \text{ and } e = 1.0.$$

What is happening? Note that the computation of the values of c, d and e involve only three numbers. There is no mysterious round-off error occurrence in the calculations. In computing the value of c as $a + b = 0$ the value of $c = a + b + 1.0$ becomes 1.0 and is correct. In the case of computing the value of d, the value of b and the number 1.0 are in highly different magnitudes. Compared to 1.0 the value of b is 20 orders high. To correctly represent the value of $b + 1.0$ the number of significant figures needed is 21 which is beyond the scope of double precision. The point is that in this example, $b + 1.0$ is simply b, in the calculation done by the computer. Consequently,

$$d = a + (b + 1.0) = a + b = 0.0.$$

Similarly, the value of e becomes 1.0. As pointed out in [3] the rounding errors are not random. They are correlated. Sometimes the way computation is performed can lead to an incorrect value of a quantity.

1.4.2   Truncation Error

Truncation error occurs in the computation of values of certain mathematical functions such as $\cos x$, e^x, $(1 + x)^{-1}$ from their series expansion. For example, consider the series expansion of e^x:

$$\mathrm{e}^x = 1 + x + \frac{1}{2!}x^2 + \cdots .\tag{1.7}$$

In the computation of e^x with $x \ll 1$ one may use only the first few terms in Eq. (1.7) and neglect the remaining higher powers of x. That is, the series is truncated. This truncation of an infinite series into a finite number of terms gives rise to an error which is called *truncation error*.

Consider the evaluation of $\int_0^{0.2} e^x \mathrm{d}x$. Its true value is $0.2214027\ldots$. Suppose, approximate e^x as $1 + x + x^2/2!$. Use of this in the integral gives the value of it as $0.2213333\ldots$. The error

$$0.2214027\ldots - 0.2213333\ldots = 6.94\ldots \times 10^{-5}$$

is called the *truncation error*. Next, the number $2/3 = 0.6666\ldots$ cannot be represented by finite number of digits. Consequently, an error occurs in representing this number. Numbers like π, $\sqrt{3}$, $\sqrt{2}$ and $1/3$ require strictly infinite digits. However, in a computer they are approximated by finite digit numbers leading to an unavoidable error.

1.4.3 Discretization Error

In the numerical integration of a function $(\int_a^b f(x)\mathrm{d}x)$ the interval $[a, b]$ is divided into number of subintervals with width h. Similarly, in the numerical integration of an ordinary differential equation the interval between the starting and the end values, a and b, respectively, of the independent variable, say t, is discretized with step size h. This discretization leads to an error and is called *discretization error* [4-6]. Consider the case of an ordinary differential equation. Denote X_{ode} and X_{dis} as the exact solution of the given differential equation and the solution of the discretized equation, respectively. Then, the discretized error, E_{dis}, is given by

$$E_{\text{dis}} = X_{\text{ode}} - X_{\text{dis}}. \tag{1.8}$$

The discretization error can be greatly reduced by the Richardson extrapolation. For details of the Richardson extrapolation see Subsection 13.4.2.

1.4.4 Noise or Random Error

Slight variations in a measurable quantity can be observed when an experiment is repeated in apparently identical conditions. These errors are called *random errors*. The errors that cannot be revealed by repeating the measurements are called *systematic errors*. Random errors may be caused by small fluctuations in certain quantities which are assumed to remain constant. Examples include small changes in room temperature, pressure, resistivity of a resistor and applied electric and magnetic fields. The distribution of random errors is approximately Gaussian. By statistically analyzing the spread, that is, standard deviation in results, a very reliable estimate of random error can be obtained.

1.4.5 Measurement of Error

To describe the accuracy of a result, one may estimate either absolute error or relative error. Suppose, x is an approximation to the true value x^*. The quantity $|x - x^*|$ is called an *absolute error* while $|x - x^*|/|x^*|$ is referred as a *relative error*. The absolute error is simply the absolute difference between the true and approximate values whereas the relative error is a portion of the true value. In any error analysis measuring relative error is often desirable, particularly, when $|x^*| \gg 1$ or $|x^*| \ll 1$. This is clear from the following examples.

Example 1:

Let us calculate the absolute and relative errors for $x^* = 10^5$ with $x = 100005$. By observation, one can conclude that x is a good approximation to x^*. Let us see what one may conclude by calculating the errors:

$$\text{Absolute error} \quad = \quad 5.$$
$$\text{Relative error} \quad = \quad 5 \times 10^{-5}.$$

Based on absolute error one can conclude that x is far away from x^* whereas the smallness of relative error implies x is close to x^*.

Example 2:

Consider the numbers $x^* = 1 \times 10^{-5}$ and $x = 6 \times 10^{-5}$. Now,

$$\text{Absolute error} \quad = \quad 5 \times 10^{-5}.$$
$$\text{Relative error} \quad = \quad 5.$$

In this example, x is 6 times higher than x^* and hence it is not at all a good approximation to x^*. But the absolute error says x is a good approximation to x^*. The relative error correctly predicts our conclusion. In the above two examples, disagreement is with the conclusions drawn based on absolute error while agreement is with that of relative error. Therefore, in any numerical error analysis, relative error is often desirable over absolute error. The relative error is frequently given in percentage.

Let us list some of the properties associated with relative error.

1. If a given number is correct up to n significant figures and k is the first significant digit of the number then the relative error is less than $k \times 10^{(1-n)}$.

2. The total relative error in the product of n numbers is equal to the sum of the individual relative errors.

3. The total relative error in x/y is equal to the sum of the relative errors of x and y.

4. The relative error of x^m is m times the relative error of x.

5. The relative error of $x^{1/m}$ is $1/m$ times the relative error of x.

1.5   Periodic and Differentiable Functions and Series

A function $f(x)$ is said to be *periodic* with period T if

$$f(x + T) = f(x) \tag{1.9}$$

for all x. The trigonometric functions $\sin x$ and $\cos x$ are periodic functions with period $T = 2\pi$. A function $f(x)$ with period T not equal to 2π can always be transformed into a function $g(x)$ of period 2π by the change of variable $Tx/(2\pi)$. This can be easily verified:

$$
\begin{aligned}
g(x + 2\pi) &= f\left(\frac{Tx}{2\pi} + T\right) \\
&= f\left(\frac{Tx}{2\pi}\right) \\
&= g(x).
\end{aligned}
\tag{1.10}
$$

A function $f(x)$ is said to be *differentiable* at a point x_0 if

$$
f'(x_0) = \lim_{x \to x_0} \frac{f(x) - f(x_0)}{x - x_0}
\tag{1.11}
$$

exists. Here, f' is called *first derivative* of f at $x = x_0$.

A most useful series expansion of a function $f(x)$ is the *Taylor series*. If the first $n + 1$ derivatives of $f(x)$ are continuous in the interval $[a, b]$ then $f(x)$ can be expressed near $x_0 \in [a, b]$ in series as

$$
f(x + x_0) = f(x_0) + (x - x_0) f'(x_0) + \frac{1}{2!}(x - x_0)^2 f''(x_0) + \cdots.
\tag{1.12}
$$

In Eq. (1.12) $f'(x_0)$, $f''(x_0)$, ... are the successive derivatives of $f(x)$ evaluated at x_0. When $x_0 = 0$, the series is called *Maclaurin series*.

1.6   Rolle's, Intermediate-Value and Extreme-Value Theorems

The *Rolle's theorem* states that if $f(x)$ is continuous and differentiable in the interval $x \in [a, b]$ with $f(a) = f(b) = 0$ then $f'(c) = 0$ for some c with $a < c < b$. The function $f(x) = x^2 - 5x + 6$ has zeros at $x = 2$ and 3, that is $f(2) = f(3) = 0$. According to the Rolle's theorem, there exists a point c between 2 and 3 with $f'(c) = 0$. For the above function, this point is computed as 2.5.

Suppose, $f(x)$ is continuous in the interval $[a, b]$ and m is any number between $f(a)$ and $f(b)$. Then, there exists a value of x, say c, with $a < x = c < b$ such that $f(x = c) = m$. This is the *intermediate-value theorem* for a function. As an example, let $f(x) = \sin x$, $x \in [-\pi/2, \pi/2]$. Here, $f(-\pi/2) = -1$ and $f(\pi/2) = 1$. Choose m as 0.5 then $c = 0.52359\ldots$ which falls between $-\pi/2$ and $\pi/2$.

If $g(x)$ is continuous and does not change its sign on an interval $[a, b]$ then, there exists a number c with $a < c < b$ such that

$$
\int_a^b f(x)g(x)\,\mathrm{d}x = f(c)\int_a^b g(x)\,\mathrm{d}x.
\tag{1.13}
$$

This is known as the *intermediate-value theorem* or *mean-value theorem* for an integral. To verify this theorem, consider the integral $\int_1^2 x(x^2 - 2)\mathrm{d}x$ with $g(x) = x$. The actual value of the integral is 3/4. By the above theorem, the integral is equivalent to $I = (c^2 - 2)\int_1^2 x\mathrm{d}x$ and $I = 3(c^2 - 2)/2$. The value of I is 3/4 if $c = \sqrt{2.5}$. This value of c lies in the interval $[1, 2]$.

The *extreme value theorem* is the following. For a continuous function $f(x)$ defined in the interval $[a, b]$ there exists a lower bound and an upper bound (denoted as m_{min} and m_{max}, respectively) at two numbers c_1 and c_2 with $c_1, c_2 \in [a, b]$ such that

$$m_{min} = f(c_1) \leq f(x) \leq f(c_2) = m_{max}. \tag{1.14}$$

m_{min} and m_{max} can also be expressed as

$$m_{min} = f(c_1) = \min_{a \leq x \leq b} [f(x)], \tag{1.15a}$$

$$m_{max} = f(c_2) = \max_{a \leq x \leq b} [f(x)]. \tag{1.15b}$$

For $f(x) = \sin x$, $x \in [0, 2\pi]$ $m_{min} = -1$ at $x = c_1 = -\pi/2$ and $m_{max} = +1$ at $x = c_2 = \pi/2$. In this example

$$m_{min} = f(-\pi/2) \leq f(x) \leq f(\pi/2) = -1. \tag{1.16}$$

When $f(x) = \cos x$, $x \in [0, 2\pi]$ then

$$m_{min} = f(\pi) \leq f(x) \leq f(0) = -1. \tag{1.17}$$

1.7 Iterations and a Root of an Equation

Numerical methods of finding the roots of nonlinear equations are iterative methods. Therefore, first define an iteration.

1.7.1 Iterations

Consider the difference equation $x_{n+1} = f(x_n)$, where $n = 0, 1, 2, \ldots$, $f(x_n)$ is a known function of x_n and x_n can take any value in the interval $[-\infty, \infty]$. Assume that x_0 is given. Starting with the given x_0 calculate x_1 as $f(x_0)$. Using the obtained x_1 the value of x_2 can be computed as $f(x_1)$ and so on. This process of obtaining the sequence $x_1, x_2, \ldots$ is known as an *iteration*. When

$$\lim_{n \to \infty} x_n = x^* \tag{1.18}$$

then x^* is called the *limit* of the sequence. When x^* is finite, the sequence is a convergent series. If x_n is assumed as an approximation of the solution of a problem, for example, a root of the equation $f(x) = 0$, then the sequence $x_0, x_1, \ldots$ is assumed to be converging to an actual root. When a method of finding the solution of a problem involves iteration as described above, then it is called an *iterative method*. There are iterative methods for finding a root of a function, solution of a system of equations, eigenvalues of a square matrix and so on.

1.7.2 Root of an Equation

Consider the function $f(x)$ with one variable x. Assume that $f(x)$ is a continuous function of x. A value of x, say x^*, for which $f(x^*) = 0$ is called a *root of the equation* $f(x) = 0$ or *zero of the function* $f(x)$. For example, the roots of $x^2 - 3x + 2 = 0$ are $x^* = 1, 2$. If $f(x)$ can be rewritten as $(x - x^*)^m g(x)$ then x^* is called a *multiple root* with multiplicity m. When $m = 2$, x^* is said to be a *double root*.

1.7.3 Convergence Criteria

In both iterative and noniterative methods of solving a problem when a specified algorithm is applied successively a sequence $\{x_n\}$ is obtained. This sequence is assumed to converge to an actual root (or a solution) x^*. A sequence $\{x_n\}$ is said to *converge* to the root x^* if

$$\lim_{n\to\infty} |x^* - x_n| = 0. \tag{1.19}$$

To check the convergence of the sequence and to accept the solution one or more appropriate criteria must be used. In the case of finding a root of the equation $f(x) = 0$, the final x_n should have the property

$$f(x_n) \approx 0 \text{ or } |f(x_n)| \le \delta_f, \tag{1.20}$$

where δ_f is a positive small quantity, $\delta_f \ll 1$ and is called *error tolerance* in f. This alone does not guarantee the convergence of $\{x_n\}$. For some functions which have slow variations near the roots, $|f(x)|$ may be less than δ_f for a certain range of values of x which are far away from x^*. Thus, to ensure the convergence another criterion is required. One may check whether the consequent iterates x_{n-1} and x_n are sufficiently close. The closeness is checked with either the absolute error $|x_n - x_{n-1}|$ is $\le \delta_x$ or the relative error $|x_n - x_{n-1}|/|x_n|$ is $\le \delta_x$, $\delta_x \ll 1$.

To summarize, the two useful termination criteria for an iteration process are

1. $|f(x_n)| \le \delta_f$ and
2. $|x_n - x_{n-1}| \le \delta_x$ or $|x_n - x_{n-1}|/|x_n| \le \delta_x$.

The process can be stopped when both the above conditions are met. The values of δ_f and δ_x are crucial. If these values are too small then the iteration may continue forever.

1.7.4 Rate of Convergence

How rapidly the sequence $\{x_n\}$ converge to an actual root in an iterative method? How does one measure it? A quantity describing the speed or rate of convergence of $\{x_n\}$ is the *order of convergence*, R. It is expressed in terms of the errors $e_n = x^* - x_n$ and e_{n+1} in successive approximations. It is defined through

$$\lim_{n\to\infty} \frac{|e_{n+1}|}{|e_n|^R} = A, \tag{1.21}$$

where R and A are two positive constants. The number A is called an *asymptotic error constant*. The values of R equal to 1 and 2 are referred as *linear* and *quadratic* order of convergence, respectively. When R is large the sequence $\{x_n\}$ converge rapidly to x^*. If R is small ($\ll 1$) then $\{x_n\}$ converges slowly. R can be used to characterize the efficiency of an iteration scheme.

1.7.5 Efficiency Index

Another quantity that can be used to characterize the efficiency of an iterative method is the efficiency index α. It is defined as

$$\alpha = R^{1/m}, \tag{1.22}$$

where R is the order of convergence of the method and m is the number of function and derivative evaluations in each step of the iteration. Like the order of convergence if the value of α is larger then the method is more efficient.

1.8 Concluding Remarks

In this chapter, some preliminaries related to the numerical methods and the error analysis are discussed. As the calculators and the computers perform computations with a finite precision arithmetic, an error occurs in the results. Errors also occur due to the approximations used in obtaining appropriate schemes or formulas to solve the problems numerically. To minimize the errors one can modify the schemes or introduce additional calculations in the numerical methods. When the calculations are done using a computer it is better to declare the variables and constants involved in double precision. For the numerical methods, it is important to study the accuracy of the results, stability of the schemes, advantages and disadvantages and the applicability of the methods concerned.

1.9 Bibliography

[1]   E. Maos, *e: The Story of a Number*. University Press, Hyderabad, 1994. pp.24.

[2]   J.R. Taylor, *An Introduction to Error Analysis*. University Science Books, California, 1982.

[3]   A. Gezerlis and M. Williams, *Am. J. Phys.* 89:51, 2021.

[4]   G. Strang, *Computational Science and Engineering*. Wellesley-Cambridge Press, Wellesley, 2007.

[5]   C. Roy, *Review of discretization error estimators in scientific computing*. In 48th AIAA Aerospace Sciences Meeting Including the New Horizons Forum and Aerospace Exposition, 2010. pp.126.

[6]   A. Greenbaum and T.P. Chartier, *Numerical Methods: Design, Analysis and Computer Implementation of Algorithms*. Princeton, University Press, 2012.

1.10 Problems

1.1   Write 382.382 in the form of $\sum d_{n-1} 10^{n-1}$.

1.2   What is the decimal equivalent of 11101.11101?

1.3   Try to express the number 1.9 in the binary representation. What do you observe?

1.4   Represent the binary number 11010.11 in decimal, octal and hexadecimal systems.

1.5   Compute $e^{0.1}$ and $e^{0.9}$ using first two terms in the Taylor series of e^x. Also, compute the percentage of the relative error in the approximation.

1.6   Calculate $f(x) = 1/(1 - x^3)$ for $x_1 = 0.8$ and $x_1' = 0.80005$ and then compute $g(f(x)) = f + f^2$. Observe the propagation of error in x.

1.7   Calculate the values of $f(x) = 1/x^2$ for $x = 0.1$, 0.10005, 10 and 10.00005. What do you notice?

1.8 Write a program to calculate $x_1, x_2, \ldots, x_{100}$ by iterating the equation $x_{n+1} = ax_n(1 - x_n)$ with $x_0 = 0.3$ and a given value of a. Calculate $x_1, x_2, \ldots, x_{100}$ for $a = 3, 3.3, 3.5$ and 3.9. What do you observe?

1.9 Verify the Rolle's and intermediate-value theorems for the function $x^3 - 2x^2 - x + 2$, $x \in [a, b]$, with an appropriate choice of a and b.

1.10 Verify the intermediate-value theorem for $\displaystyle\int_1^2 \operatorname{sech}(x)e^x dx$.

2

Solutions of Polynomial and Reciprocal Equations

2.1 Introduction

Finding the solutions of nonlinear equations is an important mathematical problem. The present chapter is restricted to the discussion on solutions of polynomial equations. The next chapter describes several iterative methods to find roots of general nonlinear equations. In everyday life quadratic equations occur in the calculations of kinetic energy of an object, surface area of a bounded space or objects, time to reach the maximum height of a moving object and the distance travelled by an object and so on. Cubic and higher-order polynomial equations are found in many places in mathematics, physics and engineering. Examples include potentials of anharmonic oscillators, van der Waals equation connecting pressure, volume and temperature of a substance, the compressibility of a gas, the chemical equilibria of water and carbon dioxide and the characteristic equation of a square matrix. The Hermite and the Legendre polynomials have notable applications in electromagnetic theory and quantum mechanics. For some details of applications of quadratic and cubic equations, one may refer to the refs. [1-3].

For polynomial equations, more information about the roots can be extracted without solving the equations. Exact formulas to determine all the roots of quadratic and cubic equations are available. For a quartic equation also formula exists for exact roots which will not be considered in this book. For polynomials of degree greater than four exact formulas for the roots are not available in the mathematical literature. However, there is a numerical procedure namely, Gräffe's root squaring method, which can be used to find all the roots of a given polynomial equation. Further, certain classes of polynomial equations can be solved exactly by rewriting them in terms of a new variable $y \to q/x$. The resulting equations are called *reciprocal equations*.

The present chapter first describes determining analytically

1. a region or an interval enclosing all the roots and
2. the number of real and complex roots.

Next, the determination of roots of quadratic and cubic equations is considered. Then, the Gräffe's root squaring method of finding real and complex roots of polynomial equations is presented. Finally, solving certain reciprocal equations is discussed. Though these methods are analytical methods, they are presented in this book since roots of cubic and lower-order polynomial equations are required in the numerical computation of certain quantities.

DOI: 10.1201/9781032649931-2

2.2 Determination of the Region Enclosing all the Roots

Let us consider a polynomial equation of degree n of the form

$$P(x) = x^n + a_1 x^{n-1} + a_2 x^{n-2} + \cdots + a_{n-1} x + a_n = 0, \tag{2.1}$$

where a_i's are all real and $a_n \neq 0$. If $a_n = 0$, $a_{n-1} = 0, \cdots, a_{n-p} = 0$ then Eq. (2.1) can be rewritten as

$$x^{p+1} \left(x^{n-(p+1)} + a_1 x^{n-(p+2)} + \cdots + a_{n-(p+1)} \right) = 0 \tag{2.2}$$

so that $x = 0$ is a root of order $p + 1$. Then, defining $n - (p+1) = m$, from the above equation one can write

$$x^m + a_1 x^{m-1} + \cdots + a_m = 0 \tag{2.3}$$

which is of the form of Eq. (2.1) and $a_m \neq 0$.

If $r_1, r_2, \cdots, r_n$ are the roots of Eq. (2.1) then it can be rewritten as

$$(x - r_1)(x - r_2) \ldots (x - r_n) = 0. \tag{2.4}$$

With the convenient notations

$$[r_i] = \sum_{i-1}^{n} r_i, \quad [r_i r_j] = \sum_{i=1}^{n} \sum_{j>i}^{n} r_i r_j, \quad [r_1 r_2 \cdots r_n] = \prod_{i=1}^{n} r_i \tag{2.5}$$

Eq. (2.4) takes the form

$$x^n - [r_i] x^{n-1} + [r_i r_j] x^{n-2} - \cdots + (-1)^n [r_1 r_2 \ldots r_n] = 0. \tag{2.6}$$

Now, comparison of Eqs. (2.1) and (2.6) gives $a_1 = -[r_i]$, $a_2 = [r_i r_j]$ and so on. If $|r_1| \gg |r_2| \gg \cdots \gg |r_n|$ then one may write

$$[r_i] = \sum r_i \approx r_1 \quad \text{and} \quad [r_i r_j] \approx r_1 r_2. \tag{2.7}$$

From Eq. (2.5), for any root r_k

$$r_k^2 \leq r_1^2 + r_2^2 + \cdots + r_n^2 = \left(\sum r_i \right)^2 - 2 \sum r_i r_j = a_1^2 - 2a_2$$

or

$$|r_k| \leq \sqrt{a_1^2 - 2a_2}. \tag{2.8}$$

Any root of Eq. (2.1) must lie in the interval $\left[-\sqrt{a_1^2 - 2a_2}, \sqrt{a_1^2 - 2a_2} \right]$. Therefore, one need not search for a root outside this interval. The complex roots occur in conjugate pairs since the coefficients a_i's in Eq. (2.1) are all assumed as real. If n in Eq. (2.1) is an odd integer then it always has one real root.

2.3 Descartes' Rule for Sign of Real Roots

The number of positive real roots of $P(x) = 0$ cannot exceed the number of sign changes in Eq. (2.1). The number of negative real roots cannot exceed the number of sign changes in $P(-x) = 0$.

Example:

The exact roots of

$$P(x) = x^3 + 2x^2 - x - 2 = 0 \tag{2.9}$$

are $x^* = -2, -1, 1$. According to the Descartes' rule, the number of sign changes in $P(x)$ is one and the number of sign changes in $P(-x) = -x^3 + 2x^2 + x - 2 = 0$ is 2. Therefore, the numbers of positive and negative real roots cannot exceed 1 and 2, respectively. In the above example, the prediction based on the Descartes' rule is in agreement with the exact result. However, the rule does not give the exact number of real roots but gives only their upper bounds. The exact number of real roots of Eq. (2.1) can be determined by the *Sturm's theorem* presented in the next section.

2.4   Determination of Exact Number of Real Roots – Sturm's Theorem

Let $P(x)$ is a polynomial of degree n and $P_1(x) = \mathrm{d}P(x)/\mathrm{d}x$ is its first-order derivative. Denote the remainder of $(-P)/P_1$ as $P_2(x)$, the remainder of $(-P_1)/P_2$ as $P_3(x)$ and so on until a constant is reached. The obtained sequence of functions

$$P(x), \ P_1(x), \ \ldots, \ P_n(x) \tag{2.10}$$

are called *Sturm functions* or *Sturm sequences*.

Theorem:

The number of real roots of the equation $P(x) = 0$ in the interval $[a, b]$ is equal to the difference between the number of changes of sign in the Sturm sequence at $x = a$ and b provided $P(a) \neq 0$ and $P(b) \neq 0$.

Example:

Consider the equation

$$P(x) = x^3 - 2x^2 - x + 2 = 0. \tag{2.11}$$

Its exact roots are $x = -1, 1, 2$. The number of real roots is 3. The Sturm functions for Eq. (2.11) are

$$\begin{aligned}
P(x) &= x^3 - 2x^2 - x + 2, \\
P_1(x) &= 3x^2 - 4x - 1, \\
P_2(x) &= \frac{2}{9}(7x - 8), \\
P_3(x) &= \frac{81}{49}.
\end{aligned}$$

For the interval $[-2, 3]$ the Sturm sequence at $x = -2$ and 3 are obtained as

$$\text{At } x = -2 \ : \ -12, \ 19, \ -44/9, \ 81/49.$$
$$\text{At } x = 3 \quad : \ 8, \ 14, \ 26/9, \ 81/49.$$

The number of sign changes at $x = -2$ is 3 while it is 0 at $x = 3$. The difference in the numbers 3 and 0 is 3. Therefore, the number of real roots lying in the interval $[-2, 3]$ is 3 which is in fact true. Next, consider the interval $[0, 3]$. The number of exact roots in this interval is two. The Sturm sequence at $x = 0$ and 3 are

$$\text{At } x = 0 \quad : \quad 2, -1, -16/9, 81/49.$$
$$\text{At } x = 3 \quad : \quad 8, 14, 26/9, 81/49.$$

The number of sign changes in the above two sequences are 2 and 0, respectively. Therefore, the number of real roots of Eq. (2.11) in the interval $[0, 3]$ must be 2 which is again true.

If $P(x) = 0$ has a multiple root then $P_n(x)$ in the Sturm sequence will be zero. In this case obtain a new sequence by dividing all the Sturm functions by $P_{n-1}(x)$. The new sequence can be used to obtain the number of real roots in an interval $[a, b]$ without taking multiplicity into account as shown in the following example.

Example:

Consider the equation

$$P(x) = x^3 - 4x^2 + 5x - 2 = 0. \tag{2.12}$$

Its exact roots are 1, 1, 2 and 1 is a double root. The Sturm functions are

$$\begin{aligned}
P(x) &= x^3 - 4x^2 + 5x - 2, \\
P_1(x) &= 3x^2 - 8x + 5, \\
P_2(x) &= \frac{2}{9}(x - 1), \\
P_3(x) &= 0.
\end{aligned}$$

The occurrence of $P_3(x) = 0$ implies that Eq. (2.12) has a multiple root. For the interval $[0, 3]$ the Sturm sequences at $x = 0$ and $x = 3$ are

$$\text{at } x = 0 \quad : \quad -2, 5, -2/9, 0,$$
$$\text{at } x = 3 \quad : \quad 4, 8, 4/9, 0,$$

respectively. The difference in the number of sign changes in the two sequences is 3. Thus, Eq. (2.12) has 3 real roots in the interval $[0, 3]$. Now, to check the multiplicity, divide P, P_1, P_2 by P_2 and obtain the new sequence

$$\frac{9}{2}\left(x^2 + 3x + 8\right), \ \frac{9}{2}(3x - 1), \ 1.$$

The sequence at $x = 0$ and 3 are:

$$\text{At } x = 0 \quad : \quad 36, -9/2, 1.$$
$$\text{At } x = 3 \quad : \quad 117, 36, 1.$$

The number of sign changes is 2 and hence the number of real roots lying in the interval $[0, 3]$ is 2. Further, one root must be a double root.

2.5 Roots of the Quadratic Equation

The formula for the roots of the quadratic equation

$$ax^2 + bx + c = 0 \tag{2.13}$$

is

$$x_{\pm} = \frac{-b \pm \sqrt{b^2 - 4ac}}{2a}. \qquad (2.14)$$

The roots are real if $b^2 - 4ac \geq 0$ and are complex conjugates if $b^2 - 4ac < 0$. Note that *the discriminant $b^2 - 4ac$ discriminates the roots of the quadratic equation.* The equivalent formulas for $x_{\pm}$ are

$$x_{\pm} = \frac{-2c}{b \pm \sqrt{b^2 - 4ac}}. \qquad (2.15)$$

When $|b| \approx \sqrt{b^2 - 4ac}$ care must be taken to avoid loss of precision. If $b > 0$ then the formulas for $x_{\pm}$ are

$$x_+ = \frac{-2c}{b + \sqrt{b^2 - 4ac}}, \quad x_- = \frac{-b - \sqrt{b^2 - 4ac}}{2a}. \qquad (2.16)$$

For $b < 0$ one should use the formulas

$$x_+ = \frac{-b + \sqrt{b^2 - 4ac}}{2a}, \quad x_- = \frac{-2c}{b - \sqrt{b^2 - 4ac}}. \qquad (2.17)$$

To illustrate this, consider Eq. (2.13) with $a = 1$, $b = 100.12$ and $c = 2.2345$. The exact roots with five decimal digit accuracy are $x_+ = -0.02232$ and $x_- = -100.09$. Now, compute the roots using Eq. (2.14) with five decimal digit floating point arithmetic. Here,

$$b^2 = 10024, \quad b^2 - 4ac = 10015, \quad \sqrt{b^2 - 4ac} = 100.07 \qquad (2.18)$$

and

$$x_+ = -0.025. \qquad (2.19)$$

The percentage of relative error is

$$\frac{|-0.02232 - (-0.025)|}{|0.02232|} \times 100 = 12\%. \qquad (2.20)$$

The obtained root differs from the exact root in the third decimal place. This loss of significant digits can be avoided by using the formula given by Eq. (2.16). This formula gives $x_+ = -0.02232$ which is accurate to five digits.

2.6   Solutions of the General Cubic Equation

All the roots of a cubic equation (a polynomial of degree 3) of the form

$$y^3 + ay^2 + by + c = 0 \qquad (2.21)$$

can be determined directly. Elimination of y^2 in Eq. (2.21) by the substitution

$$y = x - \frac{a}{3} \qquad (2.22)$$

gives the equation

$$x^3 - qx - r = 0, \qquad (2.23a)$$

where

$$q = \frac{1}{3}a^2 - b, \quad r = -\frac{2}{27}a^3 + \frac{1}{3}ab - c. \tag{2.23b}$$

The roots of Eq. (2.23) are given separately for the following five cases [4].

1. $27r^2 \leq 4q^3$, $q \neq 0$
2. $27r^2 > 4q^3$, $q \neq 0$
3. $q = 0, r \neq 0$
4. $q = 0, r = 0$
5. $q \neq 0, r = 0$

Case 1: $27r^2 \leq 4q^3$, $q \neq 0$

In this case, Eq. (2.23) has three real roots:

$$x_1 = 2\sqrt{q/3} \cos(\phi/3), \tag{2.24a}$$
$$x_2 = -2\sqrt{q/3} \cos[(\pi - \phi)/3], \tag{2.24b}$$
$$x_3 = -2\sqrt{q/3} \cos[(\pi + \phi)/3], \tag{2.24c}$$

where

$$\phi = \cos^{-1}\left[(3/q)^{3/2}r/2\right]. \tag{2.24d}$$

Case 2: $27r^2 > 4q^3$, $q \neq 0$

Equation (2.23) has one real root and two complex roots. Now, one has to consider four subclasses.

(a) $q, r > 0$

The real root is given by

$$x_1 = 2\sqrt{q/3} \cosh(\phi/3), \quad \phi = \cosh^{-1}\left[(3/q)^{3/2}r/2\right]. \tag{2.25}$$

To determine the pair of complex roots divide Eq. (2.23) by $(x-x_1)$ and obtain the quadratic equation

$$x^2 + x_1 x + (x_1^2 - q) = 0. \tag{2.26}$$

Then, the two complex roots are written as

$$x_{2,3} = \frac{1}{2}\left[-x_1 \pm \sqrt{x_1^2 - 4(x_1^2 - q)}\right]. \tag{2.27}$$

(b) $q < 0, r > 0$

The real root is given by

$$x_1 = 2(-q/3)^{1/2} \sinh(\phi/3), \quad \phi = \sinh^{-1}\left[(-3/q)^{3/2} r/2\right]. \tag{2.28}$$

Then, the two complex roots are obtained from Eq. (2.27).

(c) $q < 0$, $r < 0$

The change of variable $x \to -x$ in Eq. (2.23) gives the equation $x^3 - qx + r = 0$ which can be rewritten as $x^3 - qx - (-r) = 0$. This is simply the case (2b). After obtaining the roots change their signs.

(d) $q > 0$, $r < 0$

The substitution $x \to -x$ in Eq. (2.23) gives an equation with $q > 0$, $r > 0$ which is the case (2a).

Case 3: $q = 0$, $r \neq 0$

Equation (2.23) has one real root and a pair of complex roots. The real root is given by

$$x_1 = \text{sgn}(r)(|r|)^{1/3}, \tag{2.29}$$

where $\text{sgn}(r)$ is the sign of r. The complex roots are given by Eq. (2.27) with $q = 0$.

Case 4: $q = 0$, $r = 0$

In this case, Eq. (2.23) becomes $x^3 = 0$ and it has a real root $x = 0$ with multiplicity 3.

Case 5: $q \neq 0$, $r = 0$

When $r = 0$ and $q \neq 0$ Eq. (2.23) becomes $x(x^2 - q) = 0$. For $q > 0$, the equation has three real roots and are $x_1 = 0$, $x_{2,3} = \pm\sqrt{q}$. When $q < 0$ it has a real root $x_1 = 0$ and two pure imaginary roots $x_{2,3} = \pm i\sqrt{-q}$.

Finally, for all the cases the roots of Eq. (2.21) are obtained from Eq. (2.22).

Example:

The equilibrium points of an anharmonic oscillator are the roots of the equation $y^3 - y^2 - 4y + 4 = 0$. Find all the equilibrium points.

First, eliminate y^2 in the given equation and write it in the standard form, Eq. (2.23). For the given equation $a = -1$, $b = -4$, $c = 4$. Therefore, the change of variable $y = x + 1/3$ converts the given equation into

$$x^3 - qx - r = 0, \quad q = 13/3, \quad r = -70/27. \tag{2.30}$$

Since $27r^2(= 181.48\ldots) < 4q^3(= 325.48\ldots)$ Eq. (2.30) has three real roots. From Eq. (2.24d) ϕ is found as 2.414 radians. Then, the roots computed from Eqs. (2.24) are

$$x_1 = 1.66666\ldots, \quad x_2 = -2.33333\ldots, \quad x_3 = 0.66666\ldots.$$

Finally, the roots of the given equation are

$$y_1 = 2, \quad y_2 = -2, \quad y_3 = 1.$$

One can find all the roots of a cubic equation from a plot of y versus $f(y)$. The procedure is the following [5]. Draw a graph of y versus $f(y)$. The curve must pass through at least one point on the y-axis since at least one root should be real. Consider Fig. 2.1, where there is only one y intercept and hence only one real root. If a given equation has three real roots then $f(y)$ curve must intersect the y-axis at three points which are the exact roots of the equation. Complex roots can also be determined. Denote the location of the

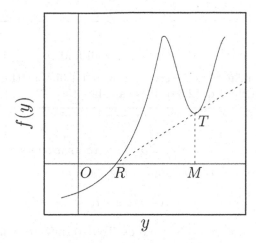

FIGURE 2.1
Determination of complex roots of a cubic equation.

real root as R and the origin as O. The distance OR is the real root. Draw a straight-line passing through the point R and also touching a point on the curve. This is the line passing through R and the local minimum of the curve. Mark the touching point on the curve as T and the corresponding y-coordinate as M as shown in Fig. 2.1. Define $OM = b$. Measure $TM(= f(OM))$, $RM = OM - OR$ and then compute

$$a = \sqrt{\frac{f(OM)}{OM - OR}}. \tag{2.31}$$

Now, the two complex roots are $a \pm \mathrm{i}b$.

2.7   Roots of Some Special Cubic Equations

The determination of roots of cubic equations by the formulas given in the previous section requires trigonometric, hyperbolic and their inverse functions evaluations of certain quantities. If the roots of a given equation has some special characteristic properties then they can be obtained in a simpler way. This is discussed in this section. As seen in Section 2.2 if $r_1, r_2, \ldots, r_n$ are the n roots of an nth-order polynomial equation then it can be rewritten as Eq. (2.6). For a cubic equation, (2.6) becomes

$$x^3 - (r_1 + r_2 + r_3)\, x^2 + (r_1 r_2 + r_2 r_3 + r_3 r_1)\, x - r_1 r_2 r_3 = 0\,. \tag{2.32}$$

Comparison of Eq. (2.32) with Eq. (2.21) with $y = x$ gives

$$a_1 = -\,(r_1 + r_2 + r_3)\,, \quad a_2 = r_1 r_2 + r_2 r_3 + r_3 r_1\,, \quad a_3 = -r_1 r_2 r_3. \tag{2.33}$$

These relations can be used to determine the roots of the cubic equation if some details of the roots are known.

For an early history of the solutions of certain special types of cubic equation one may refer to the refs. [5,6].

Example 1:

If one root of the equation $x^3 + 6x + 20 = 0$ is $1 - 3i$ find the other roots.

Since complex roots occur in pairs the second root is $1 + 3i$. The third root is then obtained from the first relation in the Eq. (2.33). It is calculated as -2.

Example 2:

Solve the equation $x^3 - 9x^2 + 23x - 15 = 0$ whose roots are in arithmetic progression.

Let the roots are $r - \alpha$, r, $r + \alpha$. The sum of the roots are

$$a_1 = -r - \alpha + r + r + \alpha = -9$$

which gives $r = 3$. Therefore, $x = 3$ is one root. To determine α consider the third relation in Eq. (2.33). One obtains $\alpha = \pm 2$. For $\alpha = -2$ the roots $r - \alpha$, r, $r + \alpha$ are $5, 3, 1$. For $\alpha = 2$ the roots are $1, 3, 5$. Thus, the roots are $1, 3, 5$.

2.8   Gräffe's Root Square Method

The Gräffe's root square method was introduced by Germinal Pierre Dandelin (1794-1847) in 1826 and independently by Karl Heinrich Gräffe (1799-1873) in 1837. It can be used to determine all real and complex roots of a given polynomial equation [4]. In this method, a new equation is formed whose roots are some higher power of the roots of the given equation. That is, if r_1, r_2, ..., r_n are the roots of the original equation then the roots of the new equation are r_1^s, r_2^s, ..., r_n^s. For large s the high powers of the roots are widely separated. In this case, the roots are obtained by a simple process.

First, separate the even and odd powers of x in Eq. (2.1) and square the both. This gives

$$x^{2n} - \left(a_1^2 - 2a_2\right) x^{2n-2} + \cdots + (-1)^n a_n^2 = 0. \tag{2.34}$$

Introducing the change of variable $z = -x^2$ Eq. (2.34) becomes

$$z^n + b_1 z^{n-1} + \cdots + b_{n-1} z + b_n = 0, \tag{2.35}$$

where

$$
\begin{aligned}
b_1 &= a_1^2 - 2a_2, \\
b_2 &= a_2^2 - 2a_1 a_3 + 2a_4, \\
b_3 &= a_3^2 - 2a_2 a_4 + 2a_1 a_5 - 2a_6, \\
&\vdots \\
b_{n-1} &= a_{n-1}^2 - 2a_{n-2} a_n, \\
b_n &= a_n^2.
\end{aligned}
\tag{2.36}
$$

Looking at Eqs. (2.36) one can easily identify a simple rule to compute the coefficients b_i's. This is left as an exercise to the readers.

2.8.1 Determination of Real Roots

Assume that $-r_1^2, -r_2^2, \ldots, -r_n^2$ are the roots of Eq. (2.35). Then, the roots of the Eq. (2.34) are $r_1, r_2, \ldots, r_n$. Repeating the above squaring process m times one has

$$z^n - B_1 z^{n-1} + B_2 z^{n-2} - \cdots + (-1)^n B_n = 0. \tag{2.37}$$

Its roots R_i are related to the roots of Eq. (2.34) as

$$R_i = -r_i^{2^m}, \quad i = 1, 2, \ldots, n. \tag{2.38}$$

If one assumes

$$|r_1| > |r_2| > \cdots > |r_n| \tag{2.39}$$

then

$$|R_1| \gg |R_2| \gg \cdots \gg |R_n|. \tag{2.40}$$

Referring to Eqs.(2.37) one can write

$$
\begin{aligned}
-B_1 &= \sum R_i \approx R_1, \\
B_2 &= \sum R_i R_j \approx R_1 R_2, \\
-B_3 &= \sum R_i R_j R_k \approx R_1 R_2 R_3, \\
&\vdots \\
(-1)^n B_n &= R_1 R_2 \cdots R_n.
\end{aligned}
\tag{2.41}
$$

From Eq. (2.41) the following simple formula for R_i's is obtained:

$$R_i = -\frac{B_i}{B_{i-1}}, \quad i = 1, 2, \ldots, n \tag{2.42}$$

where $B_0 = 1$. Use of Eq. (2.38) in Eq. (2.42) gives

$$\left| r_i^{2^m} \right| = \frac{|B_i|}{|B_{i-1}|}. \tag{2.43}$$

Taking logarithm on both sides of Eq. (2.43) leads to the equation

$$\ln|r_i| = 2^{-m} \left(\ln|B_i| - \ln|B_{i-1}| \right) = \alpha. \tag{2.44}$$

Then, $|r_i|$ is given by

$$|r_i| = e^\alpha. \tag{2.45}$$

In this way, the absolute value of each of the roots can be determined. The sign of a root is then determined by substituting it in the given equation. r_i is positive if $P(r_i) \approx 0$. r_i is negative if $P(-r_i) \approx 0$.

2.8.2 Determination of Complex Roots

Suppose, the given equation has a pair of complex roots. Let these roots are r_i and r_{i+1} and are given by $r_{i,i+1} = \alpha_i \exp(\pm i\phi_i)$. In this case the coefficients of x^{n-i} in the successive squaring fluctuate both in magnitude and sign. For sufficiently large m one has

$$\alpha_i^{2(2^m)} = \left| \frac{b_{i+1}}{b_{i-1}} \right| \tag{2.46}$$

or

$$\ln \alpha_i^2 = \frac{1}{2^m} \ln \left| \frac{b_{i+1}}{b_{i-1}} \right| = \beta. \tag{2.47}$$

The above equation gives

$$\alpha_i^2 = e^\beta. \tag{2.48}$$

First, determine all the real roots. Next, write the complex pair as $r_{i,i+1} = p \pm iq$. The sum of all the roots is then given by

$$r_1 + r_2 + \cdots + r_{i-1} + 2p + r_{i+2} + \cdots + r_n = -a_1. \tag{2.49}$$

That is,

$$p = \frac{1}{2} \left[-a_1 - \sum_R r_i \right], \tag{2.50}$$

where $\sum_R$ represents sum of all real roots. Since $p^2 + q^2 = \alpha_i^2$ the imaginary part q is determined from

$$q = \sqrt{\alpha_i^2 - p^2}. \tag{2.51}$$

2.8.3   Illustrative Examples

This subsection illustrates finding the solutions of three polynomial equations.

Example 1:

Find all the roots of the equation

$$x^3 + 2x^2 - x - 2 = 0 \tag{2.52}$$

by the Gräffe's root square method.

For the cubic equation of the form $x^3 + a_1 x^2 + a_2 x + a_3 = 0$ to determine the coefficients b_i's in Eq. (2.35) rewrite the given equation as

$$x^3 + a_2 x = - \left(a_1 x^2 + a_3 \right).$$

The left-side contains odd powers of x while the right-side has even powers of x. Squaring on both sides gives

$$x^6 + a_2^2 x^2 + 2a_2 x^4 = a_1^2 x^4 + 2a_1 a_3 x^2 + a_3^2.$$

Then, introducing the change of variable $z = -x^2$ the above equation becomes

$$z^3 + z^2(a_1^2 - 2a_2) + z(a_2^2 - 2a_1 a_3) + a_3^2 = 0.$$

Rewrite the above equation as

$$z^3 + b_1 z^2 + b_2 z + b_3 = 0.$$

The computation of b_i's is summarized in Table 2.1. This is with $m = 0$. Next, starting with the above equation and repeating the squaring process results in a cubic equation which corresponds to $m = 1$ and so on. Since the above obtained equation is similar to the given equation the b_i's for $m \geq 1$ can be obtained in a simple way (*how?*).

TABLE 2.1

Computation of the coefficients b_i's in Eq. (2.35) corresponding to Eq. (2.52). The first row gives the coefficients of x^3, x^2, x and x^0 in the given equation. The second and third rows give the coefficients of z^3, z^2, z and z^0. The addition of second and third rows gives the coefficients b_i's and is indicated.

1	a_1	a_2	a_3
1	a_1^2	a_2^2	a_3^2
	$-2a_2$	$-2a_1a_3$	
1	b_1	b_2	b_3

The coefficients a_i's and b_i's at first few squaring processes are given in Table 2.2. The roots can be calculated from this table. Let us calculate them for $m = 6$. $|r_1|$ is calculated as

$$\ln |r_1| = \frac{1}{2^6} \ln |b_1| = 0.6931$$

or

$$|r_1| = e^{0.6931}.$$

Next, $|r_2|$ is obtained from

$$\ln |r_2| = \frac{1}{2^6} \left(\ln |b_2| - \ln |b_1| \right) = 0.0108$$

as 1.011. Similarly, one obtains $|r_3| = 0.989$. Table 2.3 gives the computed absolute values of the roots as a function of m.

To determine the signs of the roots use the Descarte's rule and evaluate $P(x)$ for $x = \pm r_i$. Because $P(r_1) = 12$ and $P(-r_1) = 0$ the sign of the root is negative. $P(r_2) = 0.067$ and $P(-r_2) = 0.0218$. $P(r_2)$ and $P(-r_2)$ both are close to zero which implies that r_2 as well as $-r_2$ may be roots. Similar conclusion is drawn for the third root. The number of sign changes in $P(x)$ is one and that of $P(-x)$ is two. Therefore, Eq. (2.52) cannot have more than one positive real root and two negative real roots. Thus, $r_1 = -2$, $r_2 \approx -1.011$, $r_3 \approx 0.989$. The exact roots are -2, -1 and 1.

Example 2:

Find all the roots of the equation

$$x^3 - x^2 + 3x + 5 = 0. \tag{2.53}$$

The coefficients a_i's and b_i's obtained at first few squaring processes are given in table 2.4. The sign of the coefficient b_1 changes with m. That is, the coefficient b_1 oscillates which indicates that Eq. (2.53) has a pair of complex roots. At $m = 5$, the real root is obtained as $|r_3| = 1$. Since $P(r_3) = 8$ and $P(-r_3) = 0$ the real root is $r_3 = -1$. Next, let the complex roots are $r_{1,2} = p \pm iq$ with $\alpha^2 = p^2 + q^2$. Now,

$$\ln \alpha^2 = \frac{1}{2^5} \left(\ln |b_2| - \ln |b_0| \right) = 1.60944$$

TABLE 2.2
Calculated values of b_i's at first few squaring processes for Eq. (2.52). For each value of m the table is formed.

m	2^m		1	a_1	a_2	a_3
0	1		1	2	-1	-2
			1	4	1	4
				$+2$	$+8$	
1	2	b_i's $\rightarrow$	1	6	9	4
			1	36	81	16
				-18	-48	
2	4	b_i's $\rightarrow$	1	18	33	16
			1	324	1089	256
				-66	-576	
3	8	b_i's $\rightarrow$	1	258	513	256
			1	66564	263169	65536
				-1026	-132096	
4	16	b_i's $\rightarrow$	1	65538	131073	65536
			1	4.29522×10^9	1.71801×10^{10}	4.29496×10^9
				-2.62146×10^5	-8.59019×10^9	
5	32	b_i's $\rightarrow$	1	4.29495×10^9	8.58991×10^9	4.29496×10^9
			1	1.84465×10^{19}	7.37865×10^{19}	1.84466×10^{19}
				-1.71798×10^{10}	-3.68932×10^{19}	
6	64	b_i's $\rightarrow$	1	1.84465×10^{19}	3.68933×10^{19}	1.84466×10^{19}
			b_0	b_1	b_2	b_3

TABLE 2.3
Computed absolute values of the roots as a function of m for Eq. (2.52).

m	$|r_1|$	$|r_2|$	$|r_3|$	m	$|r_1|$	$|r_2|$	$|r_3|$
1	2.450	1.220	0.667	4	2.000	1.044	0.958
2	2.060	1.164	0.834	5	2.000	1.022	0.979
3	2.002	1.090	0.917	6	2.000	1.011	0.989

and so $\alpha^2 = 5$. From Eq. (2.50) p is calculated as 1. Then, from Eq. (2.51) $q = \sqrt{\alpha^2 - p^2} = 2$. Thus, the complex roots are $r_{2,3} \approx 1 \pm i2$. The exact roots are $x = 1$ and $1 \pm i2$.

TABLE 2.4

Calculated values of b_i's at first few squaring processes for Eq. (2.53).

m	2^m		1	a_1	a_2	a_3
0	1		1	-1	3	5
			1	1	9	25
				-6	$+10$	
1	2	b_i's $\rightarrow$	1	-5	19	25
			1	25	361	625
				-38	$+250$	
2	4	b_i's $\rightarrow$	1	-13	611	625
			1	169	373321	390625
				-1222	$+16250$	
3	8	b_i's $\rightarrow$	1	-1053	389571	390625
			1	1108809	1.51765×10^{11}	1.52587×10^{11}
				-779142	$+8.22656 \times 10^8$	
4	16	b_i's $\rightarrow$	1	329667	1.52587×10^{11}	1.52587×10^{11}
			1	1.0868×10^{11}	2.32827×10^{22}	2.32827×10^{22}
				-3.05174×10^{11}	-1.00605×10^{17}	
5	32	b_i's $\rightarrow$	1	-1.9649×10^{11}	2.3285×10^{22}	2.32827×10^{22}
			b_0	b_1	b_2	b_3

Example 3:

If all the powers of x in the given polynomial equation are multiples of some integer then re-scale the variable x and minimize the maximum power of x, otherwise the method may lead to incorrect result. As an example, consider the equation

$$x^4 + 5x^2 + 4 = 0 . \tag{2.54}$$

Its exact roots are $\pm i$ and $\pm 2i$. Consequently, one expects oscillation of the coefficients b_i's as m increases. Instead of it, no change in the signs of b_i's was found to occur for $m > 2$ (verify). At $m = 5$

$$b_1 = 8.589935 \times 10^9, \qquad b_2 = 1.844674 \times 10^{19},$$
$$b_3 = 3.689349 \times 10^{19}, \qquad b_4 = 1.844674 \times 10^9 .$$

From these b_i's the $|r_i|$'s are calculated to be

$$|r_1| = 2.044, \quad |r_2| = 1.957, \quad |r_3| = 1.022, \quad |r_4| = 0.979.$$

In all these cases $P(\pm |r_i|)$ are far from 0.

Let us proceed by minimizing the maximum power of x. Substitution of $x^2 = y$ in Eq. (2.54) gives

$$y^2 + 5y + 4 = 0.$$

Now, applying the method, the coefficients b_1 and b_2 at $m = 5$ are

$$b_1 = b_2 = 1.844674 \times 10^{19}.$$

Thus, $|r_1| = 4$ and $|r_2| = 1$. Since $P(-4) = 0$ and $P(-1) = 0$ the roots of $y^2 + 5y + 4 = 0$ are -4 and -1. Then, their square roots $\pm i2$ and $\pm i$ are the roots of Eq. (2.54).

The procedure described in this section can be applied to polynomial equations of arbitrary order whose roots are all real or at the most one pair of roots are complex. The procedure can be extended to equations possessing more than one pair of complex roots. This is left as an exercise to the reader.

2.9 Laguere's Method

Assume that the given polynomial of order n is of the form

$$P_n(x) = \prod_{i=1}^{n} (x - x_i). \tag{2.55}$$

By calculating the derivative of $\ln |P_n|$ one can show that

$$A = \frac{P_n'}{P_n} = \sum_{i=1}^{n} \frac{1}{(x - x_i)}. \tag{2.56}$$

The second derivative of $\ln |P_n|$ gives

$$B = \frac{-a^2 \ln |P_n|}{ax^2} = \left(\frac{P_n'}{P_n}\right)^2 - \frac{P_n''}{P_n} = \sum_{i=1}^{n} \frac{1}{(x - x_i)^2}. \tag{2.57}$$

Suppose x_1 is the root to be determined and assume that the other roots $x_2, \cdots, x_n$ are distant from x_1 and are close to a point say X. Define $a = x - x_1$ and $b = x - X$. Then,

$$A = \frac{1}{a} + \frac{(n-1)}{b}, \tag{2.58a}$$

$$B = \frac{1}{a^2} + \frac{(n-1)}{b^2}. \tag{2.58b}$$

Solving these two equations for a gives

$$a = \frac{n}{A \pm \sqrt{(n-1)(nB - A^2)}}. \tag{2.59}$$

In Eq. (2.59) '+' sign is used if A is positive otherwise use '−' sign. Start with a value x_0 and calculate A, B and a. Then, the next iteration is $x_1 = x_0 - a$ and repeats this until a is sufficiently small.

2.10 Reciprocal Equations

Certain types of polynomial equations of degree higher than three can also solved exactly. Introducing the change of variable $x \to 1/x'$ and then dropping the prime Eq. (2.1) becomes

$$a_n x^n + a_{n-1} x^{n-1} + \cdots + a_1 x + 1 = 0. \tag{2.60}$$

Equation (2.60) is a polynomial of degree n and is called *reciprocal equation* of Eq. (2.1). If $a_n = 1$ then Eq. (2.60) becomes

$$x^n + a_{n-1} x^{n-1} + \cdots + a_1 x + 1 = 0. \tag{2.61}$$

Comparison of Eqs. (2.1) and (2.61) gives $a_{n-1} = a_1$, $a_{n-2} = a_2$, ..., or simply

$$
\begin{aligned}
a_{n-r} = a_r, \qquad & r = 1, 2, \ldots, (n-1)/2 \quad \text{if } n \text{ is odd} \\
& r = 1, 2, \ldots, (n-2)/2 \quad \text{if } n \text{ is even.}
\end{aligned} \tag{2.62}
$$

That is, when $a_n = 1$ the coefficients of the terms equidistant from the beginning and end are equal in both sign and magnitude. If $a_n = -1$ then $a_{n-r} = -a_r$. Now, the coefficients of the terms equidistant from the beginning and end are equal in magnitude but opposite in sign. Reciprocal polynomials are noticed or used for example, in the deflection surface of the rectangular plates and cyclic error correcting codes, etc.

2.10.1 Some Properties of Reciprocal Equations

Some of the properties of the roots of reciprocal equations are summarized as follows:

(1) The roots of a reciprocal equation always occur in pairs like r_1, $1/r_1$, r_2, $1/r_2$, ... That is, if r is a root then $1/r$ is also a root.

(2) **Type-1: Reciprocal equation of degree four**

Equation (2.60) with $n = 4$ can be reduced to a quadratic equation in u by dividing it by x^2 and then by the change of variable $u = x + 1/x$. Then, the roots of the original equation are obtained by solving the equation $x^2 - ux + 1 = 0$ for each of the two values of u calculated from its quadratic equation.

(3) **Type-2: Reciprocal equation of odd degree n with $a_{n-r} = a_r$, $r = 1, 2, \ldots, (n+1)/2$**

For such equations, $x = -1$ is a root.

(4) **Type-3: Reciprocal equation of odd degree n with $a_{n-r} = -a_r$, $r = 1, 2, \ldots, (n+1)/2$**

In this case $x = 1$ is always a root.

(5) **Type-4: Reciprocal equation of even degree n with $a_{n-r} = -a_r$, $a_{n/2} = 0$**

For such equations $x = -1$ and 1 are always two roots.

Example:

Find the roots of the equation

$$x^4 + 2x^3 - x^2 + 2x + 1 = 0. \tag{2.63}$$

For Eq. (2.63), $n = 4$, $a_n = a_4 = 1$ and $a_3 = a_1$. That is, $a_{n-r} = a_r$, $r = 1, 2, \ldots, (n-2)/2$. The given equation is thus a type-1 reciprocal equation. One root of this equation is $x + 1/x$. Dividing Eq. (2.63) by x^2 one has

$$\left(x^2 + \frac{1}{x^2}\right) + 2\left(x + \frac{1}{x}\right) - 1 = 0.$$

Substitution of $u = x + 1/x$ in the above equation gives $u^2 + 2u - 3 = 0$. Its roots are $u = 1, -3$. When $u = 1$ the equation $x + 1/x = u$ becomes $x^2 - x + 1 = 0$ whose roots are

$$x = \frac{1 \pm i\sqrt{3}}{2}.$$

When $u = -3$ the equation $x + 1/x = u$ becomes $x^2 + 3x + 1 = 0$. Its roots are

$$x = \frac{-3 \pm \sqrt{5}}{2}.$$

Thus, the roots of Eq. (2.63) are given by $x = (1 \pm i\sqrt{3})/2$. One can easily verify the property (1) of the reciprocal equations for Eq. (2.63).

2.11   Concluding Remarks

As shown in Section 2.5, the well-known standard formula for the roots of a quadratic equation gives large error in the solution in certain cases. Therefore, it is necessary to know the applicability of even the theoretical formula involved in a method of solving a problem. The readymade analytical expressions available for all the roots of a cubic equation are presented in this chapter because computation of roots of cubic equations occurs in many scientific and engineering problems.

2.12   Bibliography

[1]   Pavithra Rajendran, Quadratic Equations in Real Life. https://study.com/learn/lesson/quadratic-equation-in-real-life-overview-examples.html (accessed on July 27, 2023).

[2]   Laurence Lavelle, Linear, Quadratic, and Cubic Equations With Applications in Chemistry. https://lavelle.chem.ucla.edu/wp-content/supporting-files/Chem14B/Linear_Quadratic_Cubic_Equations.pdf (accessed on July 27, 2023).

[3]   Zahedi, A.A. Kamil, Irvan, Jelita, H. Amin, A. Marwan, S. Suparni, *Math. Model. Eng. Prob.* 9:129, 2022.

[4]   L.A. Pipes and L.R. Harvill, *Applied Mathematics for Engineers and Physicists*. McGraw-Hill, Singapore, 1984. 3rd edition.

[5]   P.J. Nahin, *An Imaginary Tale: The Story of $\sqrt{-1}$*. University Press, Hyderabad, 2002.

[6]   V.M. Tikhomirov, *Stories About Maxima and Minima*. University Press, Hyderabad, 1998. English translation by A. Shenitzer.

2.13   Problems

2.1   Verify the Sturm's theorem for the following equations in the specified intervals.

(a) $x^2 - 1 = 0$, $[0, 2]$, $[-2, 2]$.

(b) $x^3 - 1 = 0$, $[0, 2]$, $[2, 3]$.

(c) $x^3 + 1 = 0$, $[0, 2]$, $[-2, 0]$.

(d) $(x - 1)(x - 2)(x - 3) = 0$, $[0, 4]$, $[0, 2]$.

2.2   Find the roots of the following equations using the formulas given in Section 2.6.

(a) $x^3 - 5x^2 + 6x = 0$.

(b) $x^3 - 2x^2 + x - 2 = 0$.

(c) $x^3 + x^2 + x - 5 = 0$.

(d) $x^3 - x^2 - 3x + 5 = 0$.

2.3   Develop a Python program to calculate the three roots of a given cubic equation. Then, find the roots of the equations given in the previous problem.

2.4   Find the value of α for which the roots of the equation $2x^3 + 6x^2 + 5x + \alpha = 0$ are in arithmetic progression.

2.5   If the roots of the equation $27x^3 + 42x^2 - 28x - 8 = 0$ are r/α, r and $r\alpha$ then find them.

2.6   Find all the roots of the following equations by Gräffe's root square method.

(a) $x^2 + 5x + 4 = 0$.

(b) $x^2 + 2x + 2 = 0$.

(c) $x^3 + 3x^2 + 3x + 1 = 0$.

(d) $x^3 + x - 2 = 0$.

2.7   Develop a C++ or a Python program for determining all the roots of a cubic equation by Gräffe's root square method.

2.8   Obtain all the roots of the following equations by solving their reciprocal equations.

(a) $x^3 + 2x^2 - 2x - 1 = 0$.

(b) $x^4 + \frac{9}{2}x^3 + 7x^2 + \frac{9}{2} + 1 = 0$.

(c) $x^5 + 3x^4 + x^3 + x^2 + 3x + 1 = 0$.

(d) $x^5 + 2x^4 + x^3 + x^2 + 2x + 1 = 0$.

2.9   The equilibrium points of a dynamical system are the roots of the equation $x^5 + 3x^4 - 5x^3 - 15x^2 + 4x + 12 = 0$. Find all its equilibrium points.

2.10   Write the characteristic equations of the following matrices and then find the eigenvalues.

a) $\begin{pmatrix} 3 & 2 & 1 \\ 1 & 0 & 3 \\ 5 & 2 & 1 \end{pmatrix}$.   b) $\begin{pmatrix} 3 & 2 & 1 & 0 \\ 1 & 8 & 3 & 5 \\ 0 & 2 & 1 & 4 \\ 1 & 5 & 3 & 1 \end{pmatrix}$.   c) $\begin{pmatrix} 2 & 3 & 1 & 4 \\ 5 & 0 & 2 & 3 \\ 4 & 8 & 1 & 1 \\ 2 & 2 & 0 & 1 \end{pmatrix}$.

3

Solution of General Nonlinear Equations

3.1 Introduction

An equation $f(x) = 0$ with $f(x)$ directly proportional to x is called a *linear equation*, otherwise it is *nonlinear*. An example of linear equation is $ax + b = 0$. Some of the nonlinear equations are $\sin x = 0$, $x^2 + x + 2 = 0$, $e^x - 1 = 0$. In many physical, mathematical and engineering problems it is often required to determine the roots (that is, zeros) of equations of the form $f(x) = 0$. For some simple equations, roots can be determined analytically. For example, ready-made formulas are available for linear equations and polynomial equations of degree two and three. However, general analytical methods are not available for nonlinear equations. There are numerous nonlinear equations occurring in physical problems. As an example, consider an elastic beam of length l whose one end is fixed and the other end is pinned. The natural frequency ω of the beam satisfies the equation

$$\tan\left(l\sqrt{\omega/c}\right) - \tanh\left(l\sqrt{\omega/c}\right) = 0, \tag{3.1}$$

where c is a constant. Analytical determination of ω from the above equation is difficult. In such a case one may look for a suitable numerical scheme. Nonlinear relations between two or more variables are very common in science. For some examples see Problems 13-19 at the end of the present chapter.

Most of the numerical methods are iterative and are based on the idea of successive approximation. Essentially, starting with one or more initial approximations to the root, using an iterative formula a sequence of approximations or iterates x_n is obtained. This sequence converges to an exact root of the given equation. Some of the basic questions on any iterative method are the following:

1. How does one proceed from the starting approximations?
2. When does one stop the iteration procedure?
3. What is the rate of convergence of the procedure?
4. What is the error involved in the approximation?
5. What are the advantages and disadvantages of the method?

Answers to these questions must be investigated in the iterative schemes.

The present chapter is devoted to general nonlinear equations. Several interesting numerical methods are available to find roots of general nonlinear equations. Some of them are:

1. Bisection method
2. False position method
3. Secant method
4. Newton–Raphson method

DOI: 10.1201/9781032649931-3

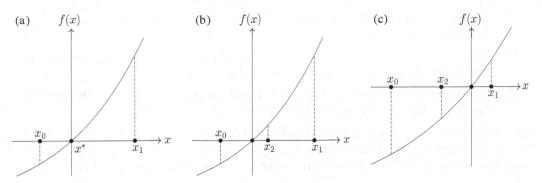

FIGURE 3.1
(a) Illustration of the choice of x_0 and x_1 enclosing a root x^* of a function $f(x)$. (b) $f(x_0)$ and $f(x_2)$ have opposite signs. (c) $f(x_1)$ and $f(x_2)$ have opposite signs.

5. Müller method

6. Chebyshev method

The above methods started with a good enough initial guess(es) to improve the solution of the given equation until some predetermined convergence criteria are satisfied. These methods have their own advantages and disadvantages. The salient features of the various methods are discussed in this chapter. Particularly, (i) basic idea, (ii) convergence rate and (iii) merits and demerits of them are presented. Finally, a comparative study of the methods considered is also made. The prime emphasis is on one-dimensional equations of the form

$$f(x) = 0, \tag{3.2}$$

where $f(x)$ is assumed to be continuous. The methods discussed in this chapter can in principle be extended to higher-dimensional equations also. A case study is considered for the Newton–Raphson method.

3.2 Bisection Method

The bisection method for Eq. (3.2) is based on the *intermediate-value theorem* (see Section 1.6). The theorem states that if $f(x)$ is a continuous function on some interval $[a, b]$ and $f(a)f(b) < 0$ then Eq. (3.2) has at least one real root or an odd number of real roots in the interval $[a, b]$.

3.2.1 Procedure

In the bisection method, two trial points, say, x_0 and x_1 are chosen such a way that signs of $f(x_0)$ and $f(x_1)$ are opposite with $f(x_0) < 0$ and $f(x_1) > 0$. An exact root x^* of $f(x)$ lies between x_0 and x_1. This is shown in Fig. 3.1a. The mid point x_2 of the interval $[x_0, x_1]$ is given by

$$x_2 = (x_0 + x_1)/2. \tag{3.3}$$

Then, the following three possibilities are analyzed:

(1) $|f(x_2)| < \delta_f$, where δ_f is the tolerance in $f(x)$: x_2 is a root of the given Eq. (3.2). Stop the calculation.

(2) $f(x_2) > 0$: x_2 lies right to x^* (Fig. 3.1b). The signs of $f(x_0)$ and $f(x_2)$ are opposite and hence the root lies in the interval $[x_0, x_2]$.

(3) $f(x_2) < 0$: x_2 lies left to x^* (Fig. 3.1c). The signs of $f(x_1)$ and $f(x_2)$ are opposite and hence the root lies in the interval $[x_2, x_1]$.

For the cases (2) and (3), a new interval is obtained. Its width is half of the previous interval. The above process is repeated for the new interval by relabelling it as $[x_0, x_1]$. The iteration is stopped when the width of the redefined interval is smaller than a preassumed small value δ_x.

The general formula of the method is

$$x_{n+1} = (x_n + x_{n-1})/2\,, \tag{3.4}$$

where x_n and x_{n-1} enclose the root x^*.

3.2.2   Termination Condition

How many iterations, N, are necessary to reach the Nth interval with width less than a tolerance δ_x? To compute this number, let us denote $[x_0, x_1]$ and $[x_{N-1}, x_N]$ as the initial and Nth intervals, respectively. The width of the Nth interval denoted as W_N is $|x_N - x_{N-1}|$ which is half of the width of the interval at $(N-1)$th iteration; one-fourth of the width of the interval at $(N-2)$th iteration and so on. That is,

$$W_N = |x_N - x_{N-1}| = |x_1 - x_0|/2^N\,. \tag{3.5}$$

Suppose, at Nth iteration W_N is less than or equal to δ_x, that is,

$$|x_1 - x_0|/2^N \le \delta_x\,. \tag{3.6}$$

Taking logarithm on both sides of Eq. (3.6) with equality sign gives

$$N = \text{Integer part of } \left[\frac{\ln|x_1 - x_0| - \ln\delta_x}{\ln 2} \right]. \tag{3.7}$$

In the above equation, integer part of right-side is taken since N is an integer. An important result is that the value of N depends on x_0, x_1 and δ_x and not on the functional form of $f(x)$. Therefore, if x_0 and x_1 enclose a root of two different functions $f_1(x)$ and $f_2(x)$ then these functions will have the same value for N. Equation (3.7) implies that convergence to an interval of an arbitrary tolerance is always guaranteed. The process can be stopped after $N + 1$ iterations.

If a given function varies very slowly near the actual root then even if the possibility $|f(x_2)| \le \delta_f$ is realized the approximate root may be far away from the actual root. On the other hand, if the given function varies very rapidly near the actual root then even if $|x_1 - x_0| \le \delta_x$, one may have $|f(x_2)| \gg \delta_f$. In such cases or for a higher accuracy, the iteration may be stopped when both the conditions $|f(x_2)| \le \delta_f$ and $|x_1 - x_0| \le \delta_x$ are satisfied. Write a program that will give a root when the above two conditions are realized.

3.2.3   Examples

Now, compute the roots of two equations by the bisection method.

TABLE 3.1

The bisection method: This table gives values of end points of the interval $[x_0, x_1]$, its mid point x_2 and the function values at x_0, x_1 and x_2 at each iteration for the equation $e^x - 1 = 0$. The trial points are $(x_0, x_1) = (-0.5, 0.4)$.

Iteration	x_0	x_1	x_2	$f(x_0)$	$f(x_1)$	$f(x_2)$
0	-0.50000	0.40000	$-\:-\:-\:-$	-0.39347	0.49182	$-\:-\:-\:-$
1	-0.50000	0.40000	-0.05000	-0.39347	0.49182	-0.04877
2	-0.05000	0.40000	0.17500	-0.04877	0.49182	0.19125
3	-0.05000	0.17500	0.06250	-0.04877	0.19125	0.06449
4	-0.05000	0.06250	0.00625	-0.04877	0.06449	0.00627
5	0.05000	0.00625	-0.02187	-0.04877	0.00627	-0.02163
6	-0.02187	0.00625	-0.00781	-0.02163	0.00627	-0.00778
7	-0.00781	0.00625	-0.00078	-0.00778	0.00627	-0.00078

Example 1:

Consider the equation $e^x - 1 = 0$. Its exact root is $x^* = 0$. The signs of $f(-0.5) = -0.39347$ and $f(0.4) = 0.49182$ are opposite and hence a root of the given function lies in the interval [-0.5,0.4]. Choose $x_0 = -0.5$, $x_1 = 0.4$, $\delta_x = 10^{-2}$ and $\delta_f = 10^{-5}$. It is desired to obtain a root which is a mid point of an interval whose width is less than δ_x. Then, $(N+1)$ iterations have to be performed, where N is given by Eq. (3.7), with the given starting points. For example, if $\delta_x = 10^{-2}$ and the starting points are $(-0.5, 0.4)$ then N value from Eq. (3.7) is computed as 6. Therefore, the process can be stopped after 7 iterations. The mid point at the 7th iteration is the desired root.

With $x_0 = -0.5$ and $x_1 = 0.4$ the mid point of this interval is $x_2 = -0.05$ and $f(x_2) = -0.04877$. Since $f(x_2) < 0$, x_0 is replaced by x_2 and the new interval is $[x_0, x_1] \to [-0.05, 0.4]$. Table 3.1 gives the result of successive iterations. At the 7th iteration the mid point (x_2) is -0.00078. Then, the new interval is $[-0.00078, 0.00625]$. Its width is less than the tolerance $\delta_x = 10^{-2}$. Therefore, $x = -0.00078$ is the approximation to the exact root with $\delta_x = 10^{-2}$. A small value of δ_x requires a greater number of iterations.

Example 2:

Consider the equation $x^3 - 1 = 0$. Its exact real root is 1. (x_0, x_1) are chosen as $(0.9, 1.1)$. Then, the number of iterations $N + 1$ is calculated as 5. At the first iteration $x_2 = (0.9 + 1.1)/2 = 1$. Now, checking the three possibilities (1)–(3) one realizes the possibility (1). Therefore, $x_2 = 1$ is the root. In this case, one need not do the remaining iterations. This kind of possibility arises whenever the points x_0 and x_1 are at equal distance from an exact root.

3.2.4 Limitations, Order of Convergence and Efficiency Index

The bisection method cannot be applied to equations with $f'(x^*) = 0$. In this case $f(x)$ does not change sign on either side of x^*. An example is $x^4 + x^2 = 0$ which has one real root $x^* = 0$ but $f(x)$ is positive on the entire real x-axis. When a given equation has multiple roots, bisection method may not be applicable because the function may not change sign at points on either side of the roots.

The error in the bisection method at nth iteration is the difference between the mid point x_n of the nth interval and the exact root x^*. That is,

$$e_n = x^* - x_n. \tag{3.8}$$

Since x^* and x_n are always contained in the initial interval one can write

$$|e_n| = |x^* - x_n| \leq \frac{|x_1 - x_0|}{2^{n+1}}. \tag{3.9}$$

Replacing n by $n+1$ in Eq. (3.9) gives

$$|e_{n+1}| \leq \frac{|x_1 - x_0|}{2^{n+2}} \quad \text{or} \quad e_{n+1} \propto e_n. \tag{3.10}$$

The *order of convergence* R of a sequence $\{x_n\}$ is defined through the equation

$$|e_{n+1}| = A|e_n|^R, \tag{3.11}$$

where A is a constant. From Eqs. (3.9)-(3.11) R is found to be 1. Thus, the rate of convergence of the sequence $\{x_n\}$ is linear in the bisection method. The efficiency index α in the bisection method is 1 since $R = 1$ and $m = 1$ (refer Eq. (1.22)).

3.3   Method of False Position

Like the bisection method, the false position method also begins with two appropriate starting points. However, instead of locating the mid point of the interval $[x_0, x_1]$ the points $X_0 = (x_0, f(x_0))$ and $X_1 = (x_1, f(x_1))$ in $x - f(x)$ plane are connected by a straight-line. The next approximate root is the point at which the straight-line intersects the x-axis.

3.3.1   Procedure

Consider two starting points x_0 and x_1 with $f(x_0) < 0$ and $f(x_1) > 0$. These two points $X_0 = (x_0, f(x_0))$ and $X_1 = (x_1, f(x_1))$ in $x - f(x)$ are joined by a straight-line. Then, the point x_2 at which the straight-line cuts the x-axis is determined. This is schematically represented in Fig. 3.2. The value of x_2 is obtained by constructing the equation for straight-line passing through X_0 and X_1. The straight-line equations at the points X_0, X_1 and X_2 are

$$f(x_0) = ax_0 + b, \quad f(x_1) = ax_1 + b, \quad 0 = ax_2 + b, \tag{3.12}$$

respectively. From Eqs. (3.12) x_2 is obtained as

$$x_2 = \frac{x_0 f(x_1) - x_1 f(x_0)}{f(x_1) - f(x_0)}. \tag{3.13}$$

Next, the following three possibilities (the same possibilities considered in the bisection method) are analyzed:

(1) $|f(x_2)| \leq \delta_f$, where δ_f is the tolerance in $f(x)$: x_2 is a root of the given Eq. (3.2). Stop the calculation.

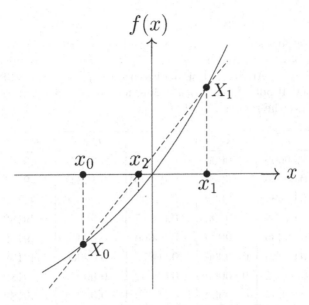

FIGURE 3.2
Illustration of the false position method. For description see the text.

(2) $f(x_2) > 0$: x_2 lies right to x^*. The signs of $f(x_0)$ and $f(x_2)$ are opposite and hence the root lies in the interval $[x_0, x_2]$.

(3) $f(x_2) < 0$: x_2 lies left to x^*. The signs of $f(x_1)$ and $f(x_2)$ are opposite and hence the root lies in the interval $[x_2, x_1]$.

When possibility (2) occurs the value of x_1 is replaced by x_2; when possibility (3) occurs then the value of x_0 is replaced by x_2. The formula given by Eq. (3.13) is repeated to get successive approximations to the root.

The general formula for $(n + 1)$th approximation is

$$x_{n+1} = \frac{x_{n-1} f(x_n) - x_n f(x_{n-1})}{f(x_n) - f(x_{n-1})}, \tag{3.14}$$

where x_n and x_{n-1} enclose the root x^*.

In the bisection method, the number of iterations required to reach the approximate root with a specified accuracy δ_x in x is predetermined. This number N is given by Eq. (3.7). But the number N is unknown in the false position method. Therefore, the following criteria may be checked to stop the iteration:

1. $|x_0 - x_2| \leq \delta_x$ if $f(x_2) > 0$ or $|x_2 - x_1| \leq \delta_x$ if $f(x_2) < 0$.
2. $|f(x_2)| \leq \delta_f$.

The process can be stopped when any one of these two conditions is met.

Example:

Let us find the root of the equation $e^x - 1 = 0$, the same equation considered for the bisection method, with $x_0 = -0.5$, $x_1 = 0.4$, $\delta_x = 0.01$ and $\delta_f = 10^{-5}$. Table 3.2 displays the summary of the calculation. At the end of 4th iteration the value of $|x_0 - x_2|$ is 0.00289 and is $< \delta_x$. Hence, the root is $x^* = -0.00066$. (In the bisection method, the root after 4

TABLE 3.2

The method of false position: Summary of iterations for the equation $e^x - 1 = 0$. The starting values are $x_0 = -0.5$ and $x_1 = 0.4$. The tolerances in x and $f(x)$ are chosen as $\delta_x = 10^{-2}$ and $\delta_f = 10^{-5}$. At the end of 4th iteration $|x_0 - x_2| = 0.00289 < \delta_x$. Hence, the root is $x^* = -0.00066$. If both δ_x and δ_f are chosen as 10^{-5} then at the end of 7th iteration $|f(x_2)| < \delta_f$ and hence the root is $x^* = 0.0$.

Iteration	x_0	x_1	x_2	$f(x_0)$	$f(x_1)$	$f(x_2)$
0	-0.50000	0.40000	$----$	-0.39347	0.49182	$----$
1	-0.50000	0.40000	-0.09999	-0.39347	0.49182	-0.09516
2	-0.09999	0.40000	-0.01894	-0.09516	0.49182	-0.01878
3	-0.01894	0.40000	-0.00355	-0.01876	0.49182	-0.00354
4	-0.00355	0.40000	-0.00066	-0.00354	0.49182	-0.00066
5	-0.00066	0.40000	-0.00012	-0.00066	0.49182	-0.00012
6	-0.00012	0.40000	-0.00002	-0.00012	0.49182	-0.00002
7	-0.00002	0.40000	0.00000	-0.00002	0.49182	0.00000

iterations is 0.00625.) For $\delta_x = \delta_f = 10^{-5}$ then at the end of 7th iteration $|f(x_2)| < \delta_f$ and the root is $x^* = 0.0$. In the bisection method, the root after 7 iterations is $x^* = -0.00078$. This shows that the convergence to an exact root in the false position method is much faster than the bisection method.

3.3.2   Limitation, Order of Convergence and Efficiency Index

Like the bisection method, the present method is also not applicable to the functions which do not cross the x-axis but touch it $x = x^*$.

The error e_{n+1} is given by

$$|e_{n+1}| = |x^* - x_{n+1}| \le |x_n - x_{n-1}| \,. \tag{3.15}$$

Replacing n by $n - 1$ in Eq. (3.15) one obtains

$$|e_n| = |x^* - x_n| \le |x_{n-1} - x_{n-2}| \,. \tag{3.16}$$

Then,

$$\frac{|e_{n+1}|}{|e_n|} \approx \frac{|x_n - x_{n-1}|}{|x_{n-1} - x_{n-2}|} = \text{constant} \,. \tag{3.17}$$

Thus, the order of convergence R is 1 for the false position method. Like the bisection method, the value of α for the false position method is 1.

3.4   Secant Method

Similar to the bisection and the false position methods the secant method also starts with two initial points x_0 and x_1. However, the interval $[x_0, x_1]$ need not enclose a root.

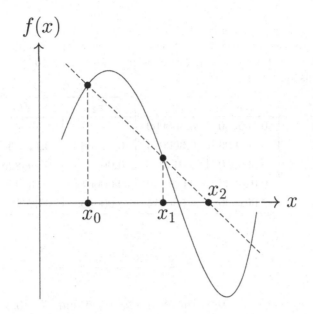

FIGURE 3.3
Illustration of the secant method. The interval $[x_0, x_1]$ not enclose a root.

3.4.1 Procedure

Consider two arbitrary points x_0 and x_1. In the first iteration, a straight-line passing through the two points $X_0 = (x_0, f(x_0))$ and $X_1 = (x_1, f(x_1))$ in the $x - f(x)$ plane is constructed. The point that cuts the x-axis is x_2. This x_2 is computed from Eq. (3.13). In the second iteration, x_0 and x_1 are replaced by x_1 and x_2, respectively. Then, x_2 is computed. The general formula is

$$x_{n+1} = \frac{x_{n-1}f(x_n) - x_n f(x_{n-1})}{f(x_n) - f(x_{n-1})}. \qquad (3.18)$$

In Eq. (3.18) the interval $[x_n, x_{n-1}]$ need not enclose the root x^*. The successive approximation to the root by the secant method is schematically illustrated in Fig. 3.3. The iteration is stopped when one of the following criteria is met:

(1) $|f(x_n) - f(x_{n-1})| \le \delta_s$, where δ_s is a preassumed small value.

(2) $|f(x_{n+1})| \le \delta_f$.

(3) Case (1) or (2) is not realized within N iterations.

When the first criterion is satisfied the process is terminated because division by a small quantity in Eq. (3.18) is encountered. In this case if both $|f(x_n)|$ and $|f(x_{n-1})|$ are less than δ_f then x_n and x_{n-1} are roots. x_n and x_{n-1} may or may not be equal. For example, if x_0 and x_1 are chosen as an exact root of the given equation then the case (1) occurs. Alternatively, if x_n is close to a root and x_{n-1} is close to another exact root then again case (1) occurs. If both $|f(x_n)|$ and $|f(x_{n-1})|$ are greater than δ_f then start with another set of starting values. When the case (2) occurs the convergent root is x_{n+1}. For the case (3) convergence is not obtained with the tolerance δ_f. This may occur if the given equation has no real root or δ_f is too small that a greater number of iterations is necessary or the starting values x_0 and x_1 are far away from the exact root.

TABLE 3.3

The secant method: Successive approximation to the root of the equation $f(x) = e^x - 1 = 0$ with $x_0 = 0.1$ and $x_1 = 0.3$. δ_f and δ_s are chosen as 10^{-5} and 10^{-9}, respectively. After 4 iterations the root is $x^* = 0.0$.

Iteration	x_0	x_1	x_2	$f(x_2)$
0	0.100000	0.300000	$----$	$------$
1	0.100000	0.300000	0.014037	0.14136e$-$01
2	0.300000	0.014037	0.001996	0.19982e$-$02
3	0.014037	0.001996	0.000014	0.13947e$-$04
4	0.001996	0.000014	0.000000	0.00000e+00

Example:

For the equation $f(x) = e^x - 1 = 0$, let us choose $x_0 = 0.1$ and $x_1 = 0.3$. For these starting values $f(x_0) > 0$ and $f(x_1) > 0$. The interval $[x_0, x_1]$ not enclose the root of the equation. Table 3.3 presents the numerical result, where $\delta_x = 10^{-2}$, $\delta_f = 10^{-5}$ and $\delta_s = 10^{-9}$. After 4 iterations $|f(x_2)|$ is $< \delta_f$ and hence the root is 0.0. For $x_0 = -0.5$ and $x_1 = 0.4$, the root is obtained after 4 iterations and is -0.9×10^{-5} (verify). Compare this value with the approximate root obtained by the bisection and the false position methods.

3.4.2   Limitation, Order of Convergence and Efficiency Index

Since the approximations x_n and x_{n+1} need not enclose the exact root, convergence of $\{x_n\}$ to x^* is not always guaranteed.

Now, determine the order of convergence of the secant method. Substitution of $x_{n+1} = x^* - e_{n+1}$, $x_n = x^* - e_n$ and $x_{n-1} = x^* - e_{n-1}$ in Eq. (3.18) gives

$$x^* - e_{n+1} = \frac{(x^* - e_{n-1})f(x^* - e_n) - (x^* - e_n)f(x^* - e_{n-1})}{f(x^* - e_n) - f(x^* - e_{n-1})}. \tag{3.19}$$

Replacement of e by $-e'$ and then dropping the prime in the above equation gives

$$x^* + e_{n+1} = \frac{(x^* + e_{n-1})f(x^* + e_n) - (x^* + e_n)f(x^* + e_{n-1})}{f(x^* + e_n) - f(x^* + e_{n-1})}. \tag{3.20}$$

Writing the Taylor series expansion of f about x^*, neglecting the higher powers of e and substituting $f(x^*) = 0$ one obtains

$$
\begin{aligned}
x^* + e_{n+1} &= \frac{(x^* + e_{n-1})\left(e_n f' + \frac{e_n^2}{2}f''\right)}{(e_n - e_{n-1})f'} \\
&\quad - \frac{(x^* + e_n)\left(e_{n-1}f' + \frac{e_{n-1}^2}{2}f''\right)}{(e_n - e_{n-1})f'} \\
&= x^* + \frac{f''\left(e_n^2 x^* + e_n^2 e_{n-1} - e_{n-1}^2 x^* - e_{n-1}^2 e_n\right)}{2f'(e_n - e_{n-1})}. \tag{3.21}
\end{aligned}
$$

In the limit $n \to \infty$, $e_n^2 x^* - e_{n-1}^2 x^* = 0$. Then,

$$x^* + e_{n+1} = x^* + \frac{f''}{2f'} e_n e_{n-1}$$

or

$$e_{n+1} = \frac{f''}{2f'} e_n e_{n-1}. \tag{3.22}$$

To estimate the order of convergence R, let us express the above equation in the standard form of Eq. (3.11). From Eq. (3.11) one can write

$$e_n = A e_{n-1}^R. \tag{3.23}$$

Inverting this equation, e_{n-1} is obtained as

$$e_{n-1} = \left(\frac{e_n}{A}\right)^{1/R}. \tag{3.24}$$

Replacing e_{n-1} in Eq. (3.22) with Eq. (3.24) gives

$$e_{n+1} = \frac{f'' e_n^{(1+R)/R}}{2f' A^{1/R}}. \tag{3.25}$$

Comparison of Eqs. (3.11) and (3.25) leads to

$$A = \left(\frac{f''}{2f'}\right)^{R/(R+1)}, \quad R = \frac{(1+R)}{R} \text{ or } R^2 - R - 1 = 0. \tag{3.26}$$

The roots of R are

$$R = \frac{1 \pm \sqrt{5}}{2}. \tag{3.27}$$

Since R must be greater than zero, one has $R = (1 + \sqrt{5})/2 \approx 1.62$. Thus, the order of convergence of the secant method is ≈ 1.62. Because of $m = 1$ and $R \approx 1.62$, the value of the efficiency index α is 1.62.

3.5   Newton–Raphson Method

If $f(x)$ and its first two derivatives $f'(x)$ and $f''(x)$ are continuous near a root x^* then this additional information about the nature of the function $f(x)$ can be used to develop algorithms that will produce sequences converging faster to x^* than the methods discussed earlier in this chapter. One such method is the Newton–Raphson method. This is one of the most widely used methods of solving nonlinear equations. In this section, first the features of the method are presented for one-dimensional equations. Then, the iterative rule for n-coupled equations is given. The case of $n = 2$ is illustrated. For a historical development of the Newton–Raphson method one may refer to the ref. [1].

3.5.1   One-Dimensional Equations

Unlike the earlier methods, the Newton–Raphson method starts with a single guess x_0. Figure 3.4 gives the graphical description of the method. A tangent line to $f(x)$ at x_0 is

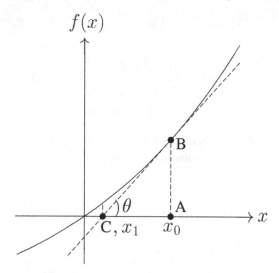

FIGURE 3.4
Graphical illustration of x_1 and x_2 for the Newton–Raphson method.

drawn. The point, say x_1, on the x-axis at which the tangent line intersects is the next approximation for the root. This process is repeated until a root with a desired accuracy is reached. From Fig. 3.4, $\tan \theta$ is written as

$$\tan \theta = \frac{AB}{AC} = \frac{f(x_0)}{x_0 - x_1}. \tag{3.28}$$

$\tan \theta$ represents $f'(x)$ at $x = x_0$. Replacing $\tan \theta$ by $f'(x_0)$ in Eq. (3.28) gives

$$x_1 = x_0 - \frac{f(x_0)}{f'(x_0)}. \tag{3.29}$$

Then, the general formula for nth to $(n+1)$th approximation is

$$x_{n+1} = x_n - \frac{f(x_n)}{f'(x_n)}. \tag{3.30}$$

This is the Newton–Raphson iterative rule for a root of a one-dimensional nonlinear equation.

3.5.2   Termination Criteria

In order to stop the iteration process certain possibilities are to be analyzed at each iteration. For this purpose, let us denote the tolerance in $|x_{n+1} - x_n|$ as δ_x; in $|f(x_n)|$ as δ_f and $|f'(x_n)|$ as δ_s with $\delta_s \ll \delta_f$. Then, the following possibilities are analyzed:

(1) $|f'(x_n)| \leq \delta_s$ and $|f(x_n)| \leq \delta_f$. The root is x_n.

(2) $|f'(x_n)| \leq \delta_s$ and $|f(x_n)| > \delta_f$. Division by a small number is encountered and the process is terminated.

(3) $|x_{n+1} - x_n| \leq \delta_x$ and $|f(x_{n+1})| \leq \delta_f$. The root is x_{n+1}.

(4)   No convergence in N iterations.

TABLE 3.4
The Newton–Raphson method: Successive approximation of the real root of $f(x) = e^x - 1 = 0$ with $x_0 = 0.5$, $\delta_x = 10^{-5}$, $\delta_f = 10^{-5}$ and $\delta_s = 10^{-9}$.

Iteration	x value	$f(x)$ value	$\|x_{n+1} - x_n\|$
0	0.500000	0.648721e+00	– – – –
1	0.106531	0.112412e+00	0.393469
2	0.005478	0.549316e−02	0.101053
3	0.000015	0.149012e−04	0.005463
4	0.000000	0.000000e−07	0.000015
5	0.000000	0.000000e+00	0.000000

Instead of checking the absolute error in case (3), one may check the relative error given by $|(x_{n+1} - x_n)/x_{n+1}|$. If the possibility (2) or (4) occurs for a large number of real values of x_0 then the real root may be $\pm\infty$ or the equation has no real root.

3.5.3   Systematic Procedure

The following is a systematic procedure of the Newton–Raphson method.

(1)   Choose the tolerance values δ_x, δ_f and δ_s and the maximum number of iterations N.

(2)   Choose a starting value of x, x_0.

(3)   Calculate $f(x_0)$ and $f'(x_0)$. If $|f'(x_0)| \leq \delta_s$ and $|f(x_0)| \leq \delta_f$ the root is x_0 and stops the process. If $|f'(x_0)| \leq \delta_s$ but $|f(x_0)| > \delta_f$ stop the process because division by a small quantity is encountered.

(4)   Compute x_1 using Eq. (3.30) and $f(x_1)$. If $|x_1 - x_0| \leq \delta_x$ and $|f(x_1)| \leq \delta_f$ the root is x_1 and stop the process. Otherwise, repeat the steps (3) and (4) with x_1 and so on.

(5)   Terminate the process if convergence is not realized in N iterations.

   In writing a computer language program to implement the Newton–Raphson method care must be taken to provide all the possible decision makings.

3.5.4   Examples

The following two examples illustrate the Newton–Raphson method.

Example 1:

Consider the equation $e^x - 1 = 0$. Let us choose $\delta_x = 10^{-5}$, $\delta_s = 10^{-9}$ and $\delta_f = 10^{-5}$. The derivative of the function is $f'(x) = e^x$. Table 3.4 gives the summary of the computation with $x_0 = 0.5$. After 5 iterations $|x_5 - x_4| = 0 < \delta_x$ and $|f(x_5)| = 0 < \delta_f$. Thus, the root is $x^* = 0$. For the starting value $x_0 = 2$ the same result is obtained after 6 iterations (verify).

Example 2:

Find a root of the equation $x^3 - 8 = 0$ with five decimal point accuracy with $x_0 = 2.5$.

The iteration can be stopped when the absolute difference between x_{n+1} and x_n, $|x_{n+1} - x_n|$, is $< 10^{-5}$. By the Newton–Raphson formula one has

$$x_1 = x_0 - \frac{f(x_0)}{f'(x_0)} = x_0 - \frac{x_0^3 - 8}{3x_0^2} = 2.0933333, \quad x_2 = 2.0040996.$$

The variation is in second decimal point and hence one has to continue the iteration. x_3 is 2.0000084. The difference between x_2 and x_3 is in third decimal. Next, x_4 is calculated as 2.0000000. Now, the difference between x_3 and x_4 is in sixth decimal place only. So, the iteration can be stopped. $x_4 = 2.0$ is the root with five decimal accuracy. The exact root is also 2.0.

3.5.5   Newton–Raphson Method for a Multiple Root

The Newton–Raphson method can be extended to compute a multiple root of the equation $f(x) = 0$. Assume that x^* is a root with multiplicity p. Then, x^* is also a root of $f'(x) = 0$ with multiplicity $(p - 1)$, a root of $f''(x) = 0$ with multiplicity $(p - 2)$, ..., a root of $f^{p-1}(x) = 0$ with multiplicity 1. In this case the formula

$$x_{n+1} = x_n - p\frac{f(x_n)}{f'(x_n)} \tag{3.31}$$

is useful. In fact, for $x_n \approx x^*$, the expressions

$$x_n - p\frac{f(x_n)}{f'(x_n)}, \quad x_n - (p-1)\frac{f'(x_n)}{f''(x_n)}, \quad \text{etc}, \tag{3.32}$$

all will have the same value.

3.5.6   Limitations

Let us point out the limitations of the Newton–Raphson method.

1. Division by Zero

An obvious limitation of the method is the occurrence of division by zero in the second term of right-side of Eq. (3.30). If $|f'(x_n)|$ is zero or too small then this term becomes infinity or diverge. In this case if $|f(x_n)| \le \delta_f$ then x_n is an approximation to the actual root x^*. Otherwise, use another starting value x_0.

Examples:

Consider the function $f(x) = x - \sin x = 0$. Its derivative is $f'(x) = 1 - \cos x$. The exact root of this equation is $x^* = 0$. If $x_0 = 0$ or close to zero then $f'(x_0) \approx 0$. Then, the second term in the right-side of Eq. (3.30) becomes infinity or very large. However, $x_0 = 0$ is the exact root. Next, consider the equation $f(x) = \sin x = 0$. Its exact roots are $n\pi$, $n = 0, \pm 1, \pm 2, \dots$. $f'(x)$ is $\cos x$. When $x_0 \approx (2n + 1)\pi/2$, $f'(x_0) \approx 0$. The iteration cannot be continued. Further, $x_0 \approx (2n + 1)\pi/2$ is not a root.

2. Sensitive Dependence on x_0

The method is sensitive to the starting values x_0. If the equation has more than one real root then the iterated values x converge to one of the roots. On the other hand, if x_0 is far from all real roots then x_n may diverge.

Example:

The roots of the equation $x^2 - 4x + 3 = 0$ are $x^* = 1, 3$. If x_0 is close to 1 then x_n's converge to $x^* = 1$. If x_0 is close to 3 then x_n converges to $x^* = 3$.

3. Cycling

In the Newton–Raphson method it is not always the case that after N iterations a root is reached. For some equations and for certain values of x_0 the iterations tend to repeat or almost repeat with a definite periodicity without approaching a root of the equation.

Example:

For the equation $x^3 - x - 3 = 0$ the initial guess $x_0 = 0$ generates $x_1 = -3$, $x_2 = -1.9615384$, $x_3 = -1.1471759$, $x_4 = -0.0065792$, $x_5 = -3.0003891$, $x_6 = -1.9618182$, $x_7 = -1.1474303$, $x_8 = -0.0072564$, $x_9 = -3.0004733$, ... The iterated values exhibit almost a period-4 cycle.

3.5.7 Order of Convergence and Efficiency Index

Defining $e_n = x^* - x_n$ and $e_{n+1} = x^* - x_{n+1}$ the rule (3.30) becomes

$$e_{n+1} = e_n + \frac{f(x^* - e_n)}{f'(x^* - e_n)}. \tag{3.33}$$

Writing the Taylor series expansions for $f(x^* - e_n)$ and $f'(x^* - e_n)$ and neglecting higher powers of e_n one has

$$\begin{aligned} e_{n+1} &= e_n - e_n \left(1 - \frac{e_n f''}{2f'}\right) \left(1 - e_n \frac{f''}{f'}\right)^{-1} \\ &= e_n - e_n \left(1 - \frac{e_n f''}{2f'}\right) \left(1 + e_n \frac{f''}{f'}\right) \\ &= -\frac{e_n^2 f''}{2f'}, \end{aligned} \tag{3.34}$$

where, in obtaining the last expression, the term e_n^3 is neglected. Then,

$$|e_{n+1}| = |e_n|^2 \left|\frac{f''}{2f'}\right|. \tag{3.35}$$

The order of convergence R is thus 2 (quadratic) for the Newton–Raphson method. That is, near a root, *the number of significant digits roughly doubles in each iteration*. Because of this the Newton–Raphson method is called a *powerful* method.

Next, find the value of the efficiency index α. At each iteration, the values of both $f(x)$ and $f'(x)$ are used so $m = 2$. Substitution of $R = 2$ and $m = 2$ in Eq. (1.22) gives $\alpha = \sqrt{2} = 1.41$.

3.5.8 Complex Roots of One-Dimensional Equations

The Newton–Raphson method can be used to compute complex roots of nonlinear equations. They can be obtained by choosing the starting value x_0 as a complex number and assuming

TABLE 3.5

The Newton–Raphson method: Successive approximation of the pair of complex roots of the equation $f(x) = x^2 + x + 1 = 0$. The initial guess is $x_0 = -0.4 + i0.6$. Here, $\delta_x = 10^{-5}$, $\delta_f = 10^{-5}$ and $\delta_s = 10^{-9}$. x_r and x_i denote real and imaginary parts of x, respectively. Similarly, f_r and f_i are the real and imaginary parts of $f(x)$, respectively. The exact roots are $x^* = -0.5 \pm i0.86603$.

Iteration	x_r	x_i	f_r	f_i
0	-0.400000	0.600000	$-\,-\,-\,-$	$-\,-\,-\,-$
1	-0.551351	0.908108	-0.072023	-0.093265
2	-0.502399	0.865684	0.000597	-0.004153
3	-0.499999	0.866022	0.000006	0.000002
4	-0.500000	0.866025	0.000000	0.000000

$f(x)$ and $f'(x)$ as complex functions. Imaginary part of x_0 should be sufficiently large. If it is zero or too small then the successive approximations $\{x_n\}$ are always real, if there is a real root, and convergence to a complex root never occurs. Since the complex roots always occur in pairs, if $x^* = a + ib$ is a root obtained by the Newton–Raphson method then the other root is its conjugate $a - ib$.

Example:

Let us compute the complex roots of the quadratic equation $f(x) = x^2 + x + 1 = 0$. Its exact roots are $x_{1,2}^* = (-1 \pm i\sqrt{3}\,)/2$. Table 3.5 summarizes the result for $x_0 = 0.2 + i0.2$ with $\delta_x = 10^{-5}$, $\delta_f = 10^{-5}$ and $\delta_s = 10^{-9}$. At the end of four iterations $|x_4 - x_3| < \delta_x$ and $|f(x_4)| < \delta_f$. The roots are $-0.5 \pm i0.86603$.

3.5.9   Roots of n-Dimensional Equations

Having studied the Newton–Raphson method for one-dimensional equations now deduce the formula for n-dimensional equations, that is, n-coupled first-order nonlinear equations with n variables.

Let

$$F(X) = 0 \tag{3.36}$$

be a set of n equations, where $X = (x^{(1)}, x^{(2)}, ..., x^{(n)})$ and $F = (f_1, f_2, ..., f_n)$. The Taylor series expansion of $F(X^*)$ is

$$F(X^*) \;=\; 0 = F(X_i) + (X^* - X_i)\frac{\partial F}{\partial X_i} + \cdots , \tag{3.37a}$$

where

$$\frac{\partial F}{\partial X_i} \;=\; F' = \sum_{j=1}^{n} \frac{\partial F}{\partial x_i^{(j)}} . \tag{3.37b}$$

Here, X_i represents the root at ith approximation or iteration. The elements of F' form a matrix J called *Jacobian matrix* of F at X_i. Keeping the first two terms in the right-side of Eq. (3.37a) one has

$$0 = F(X_i) + \delta X_i \cdot \frac{\partial F}{\partial X_i} , \tag{3.38}$$

where $\delta X_i = X^* - X_i$. From Eq. (3.38) one can write

$$\delta X_i = - \left(\frac{\partial F}{\partial X_i} \right)^{-1} F(X_i). \tag{3.39}$$

Then, the value of X at the $(i+1)$th iteration is

$$X_{i+1} = X_i + \delta X_i. \tag{3.40}$$

More precisely,

$$x_{i+1}^{(j)} = x_i^{(j)} + \delta x_i^{(j)}, \quad j = 1, 2, ..., n. \tag{3.41}$$

3.5.10   Roots of Two-Dimensional Equations

For the two-dimensional equations of the form

$$f(x, y) = 0, \quad g(x, y) = 0 \tag{3.42}$$

the Jacobian matrix J is given by

$$J = \begin{pmatrix} f_x & f_y \\ g_x & g_y \end{pmatrix}. \tag{3.43}$$

Then, Eq. (3.39) becomes

$$\begin{pmatrix} \delta x_i \\ \delta y_i \end{pmatrix} = -\frac{1}{\det J} \begin{pmatrix} g_y & -f_y \\ -g_x & f_x \end{pmatrix} \begin{pmatrix} f \\ g \end{pmatrix}, \tag{3.44a}$$

where

$$\det J = f_x g_y - g_x f_y. \tag{3.44b}$$

Referring the matrix multiplication in Eq. (3.44a) gives

$$\delta x_i = \frac{1}{\det J} (g f_y - f g_y), \quad \delta y_i = \frac{1}{\det J} (f g_x - g f_x). \tag{3.45}$$

Then, the Newton–Raphson iteration rule is

$$x_{i+1} = x_i + \delta x_i, \tag{3.46a}$$
$$y_{i+1} = y_i + \delta y_i. \tag{3.46b}$$

Example:

Consider the set of equations

$$x^2 + x - y = 0, \tag{3.47a}$$
$$xy - 2 = 0. \tag{3.47b}$$

An exact root of this system is $(x^*, y^*) = (1, 2)$. For the given system $f_x = 2x + 1$, $f_y = -1$, $g_x = y$, $g_y = x$. Table 3.6 summarizes the result, where $x_0 = 0.5$ and $y_0 = 1.5$. A root with the specified accuracy is arrived at 5th iteration. The obtained root is $(x, y) = (1, 2)$. For the initial guess $(x_0, y_0) = (1, 1)$ the same root is obtained after two iterations.

Newton–Raphson method is found applications in the studies of hydraulic problems [2,3], water distribution network [4], power flow problems [5,6], nonlinear regression models [7], thermal EHD lubrication model of line contacts [8], optimization of transmission control protocol [9] and data sciences and education [10], a few to mention. Methods to improve the order of convergence to a root have been reported [11-14]. Improvement and modifications of the Newton–Raphson method are proposed [15,16].

TABLE 3.6

The Newton–Raphson method: Successive approximation of a real root of the two-dimensional Eqs. (3.47). The initial values of x and y are $x_0 = 0.5$ and $y_0 = 1.5$. Here, $\delta_x = \delta_y = \delta_f = \delta_g = 10^{-5}$ and $\delta_s = 10^{-9}$. The root at the end of 5th iteration is $(x, y) = (1, 2)$.

Iteration	x value	y value	f value	g value
0	0.500000	1.500000	$----$	$----$
1	1.150000	2.446500	0.026000	0.813475
2	1.014876	2.054657	-0.009807	0.085222
3	1.000202	2.000793	-0.000186	0.001198
4	1.000000	2.000000	0.000000	0.000000
5	1.000000	2.000000	0.000000	0.000000

3.6   Müller Method

The methods considered so far approximated the function $f(x)$ in the neighbourhood of a root by a straight-line. Contrast to this the Müller method is based on approximating $f(x)$ by a quadratic polynomial.

The method of Müller uses three starting points x_0, x_1 and x_2 and assumes that x_2 is the best approximation to a root x^*. A parabola passing through these three points is constructed as shown in Fig. 3.5. Then, introduce a change of variable

$$t = x_1 - x_2 \tag{3.48}$$

and define t_0 and t_1 as

$$t_0 = x_0 - x_2 \quad \text{and} \quad t_1 = x_1 - x_2. \tag{3.49}$$

In terms of the variable t, the quadratic polynomial equation for parabola is

$$f = at^2 + bt + c. \tag{3.50}$$

To determine the unknowns a, b and c three equations are formulated from Eq. (3.50). These are Eq. (3.50) at $t = t_0$, t_1 and 0:

$$at_0^2 + bt_0 + c = f_0, \quad at_1^2 + bt_1 + c = f_1, \quad c = f_2, \tag{3.51}$$

where $f_0 = f(x_0)$, $f_1 = f(x_1)$ and $f_2 = f(x_2)$. Solving these equations one has

$$c = f_2, \quad a = \frac{e_1 t_0 - e_0 t_1}{t_0 t_1 (t_1 - t_0)}, \quad b = \frac{e_0 t_1^2 - e_1 t_0^2}{t_0 t_1 (t_1 - t_0)}, \tag{3.52a}$$

where

$$e_0 = f_0 - f_2, \quad e_1 = f_1 - f_2. \tag{3.52b}$$

Now, the roots of Eq. (3.50) can be easily determined. As mentioned in Section 2.5 if the roots z_1 and z_2 of a quadratic equation are small then roots calculated from the formula (2.14) are accurate to fewer decimal points only. Since $t = x - x_2$, at successive approximation

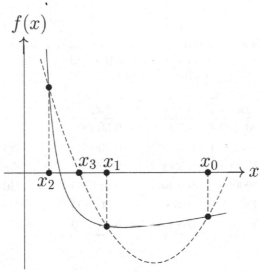

FIGURE 3.5
The Müller method: Construction of a parabola (dashed curve) passing through a set of three starting points x_0, x_1 and x_2.

the root t of Eq. (3.50) is required to decay to zero. Therefore, to avoid the loss of significant digits in the roots of Eq. (3.50) one can choose the formula

$$t = \frac{-2c}{b \pm \sqrt{b^2 - 4ac}}. \qquad (3.53)$$

Of these two roots the one which has the smallest absolute value is chosen. The smallest one can be identified by inspecting the sign of b. For $b > 0$, the positive sign gives the smallest root. For $b < 0$, use the negative sign. Then, x_3, the next approximation to x^*, is

$$x_3 = x_2 + t. \qquad (3.54)$$

Next, out of x_0, x_1 and x_2 find the two points which are closer to x_3. Relabel them as x_0 and x_1. Relabel x_3 as x_2. Repeat the above process until a root (x_2) with a desired accuracy is reached. The iteration can be stopped when $|x_3 - x_2| \leq \delta_x$ and $|f(x_3)| \leq \delta_f$.

The iteration rule is given by

$$x_{n+1} = x_n + (x_n - x_{n-1}) \frac{2c}{b \pm \sqrt{b^2 - 4ac}}. \qquad (3.55)$$

In Eq. (3.53), if $b^2 - 4ac < 0$ then t's are complex conjugates. That is, complex roots can also be determined by employing the Müller method. This is a great advantage of this method.

Assume that the given equation has both real and complex roots. For certain real values of the starting points x_0, x_1 and x_2 one may encounter complex values of t. Consequently, successive approximation may lead to a complex root. If one's aim is to find a real root then the imaginary part can be made zero and proceed with real numbers.

Example:

Find the root of the equation $f(x) = e^x - 1 = 0$. Choose $\delta_x = \delta_f = 10^{-5}$, $x_0 = 0.9$, $x_1 = 0.7$ and $x_2 = 0.5$.

TABLE 3.7
The Müller method: Successive approximation of the real root of equation $f(x) = e^x - 1 = 0$. The starting points are $x_0 = 0.9$, $x_1 = 0.7$, $x_2 = 0.5$ and $\delta_x = \delta_f = 10^{-5}$. After 5 iterations $|x_3 - x_2| < \delta_x$ and $|f(x_3)| < \delta_f$ so root is $x^* = 0$.

Iteration	x_0	x_1	x_2	x_3	$f(x_3)$
0	0.900000	0.700000	0.500000	$----$	$----$
1	0.900000	0.700000	0.500000	-0.246809	-0.218710
2	0.700000	0.500000	-0.246809	0.018589	0.018762
3	0.500000	-0.246809	0.018589	0.000404	0.000404
4	-0.246809	0.018589	0.000404	0.000000	0.000000
5	0.018589	0.000404	0.000000	0.000000	0.000000

Table 3.7 displays the numerical computation. The conditions $|x_3 - x_2| \le \delta_x$ and $|f(x_3)| \le \delta_f$ are met at the end of 5th iteration. Therefore, the root is $x^* = 0$.

Compared with other methods described earlier, the Müller method involves a lot of calculations at each stage of approximation. The order of convergence of the Müller method is 1.84. The efficiency index α is also 1.84.

3.7   Chebyshev Method

In the Müller method a parabola passing through a set of three starting points is constructed. In contrast to this in Chebyshev method a parabola passing through a single starting value is constructed. Assume a polynomial of degree 2 of the form

$$f(x) = ax^2 + bx + c = 0, \tag{3.56}$$

where a, b and c are to be determined. For the initial guess x_0, Eq. (3.56) is

$$f(x_0) = ax_0^2 + bx_0 + c. \tag{3.57}$$

Then, $f'(x_0)$ and $f''(x_0)$ are obtained as

$$f'(x_0) = 2ax_0 + b, \quad f''(x_0) = 2a. \tag{3.58}$$

Solving Eqs. (3.57) and (3.58) for a, b and c gives

$$a = \frac{1}{2}f''(x_0), \quad b = f'(x_0) - x_0 f''(x_0), \tag{3.59a}$$

$$c = f(x_0) - x_0 f'(x_0) + \frac{1}{2}x_0^2 f''(x_0). \tag{3.59b}$$

Then, substitution for a, b and c in Eq. (3.56) gives

$$f(x_0) + (x - x_0)f'(x_0) + \frac{1}{2}(x - x_0)^2 f''(x_0) = 0. \tag{3.60}$$

TABLE 3.8

The Chebyshev method: Successive approximation of the real root of the equation $f(x) = e^x - 1 = 0$. The initial guess is $x_0 = 0.5$. Here, $\delta_x = \delta_f = 10^{-5}$ and $\delta_s = 10^{-9}$. The root is $x^* = 0$.

Iteration	x value	$f(x)$ value	$|x_{n+1} - x_n|$
0	0.50000	$-----$	$----$
1	0.02912	0.29550E−01	0.47088
2	0.00001	0.80550E−05	0.02911
3	0.00000	0.17423E−15	0.00001

From Eq. (3.60), the nth approximation is written as

$$f(x_n) + (x - x_n) f'(x_n) + \frac{1}{2}(x - x_n)^2 f''(x_n) = 0. \tag{3.61}$$

Replacement of x by x_{n+1} gives

$$x_{n+1} = x_n - \frac{f(x_n)}{f'(x_n)} - \frac{1}{2}(x_{n+1} - x_n)^2 \frac{f''(x_n)}{f'(x_n)}. \tag{3.62}$$

Using the Newton–Raphson formula, Eq. (3.30), for $(x_{n+1} - x_n)$ in the right-side of Eq. (3.62) one obtains

$$x_{n+1} = x_n - \frac{f(x_n)}{f'(x_n)} - \frac{1}{2}\frac{f^2(x_n) f''(x_n)}{(f'(x_n))^3}, \quad f''(x_n) \neq 0. \tag{3.63}$$

When f'' is neglected assuming that $|f''| \ll 1$ then Eq. (3.63) becomes the Newton–Raphson rule (3.30). Thus, the iteration rule (3.63) can be considered as an improved version of the Newton–Raphson method.

The order of convergence of the Chebyshev method is 3. (The proof of this is left as an exercise to the reader.) At each iteration, the values of f, f' and f'' are required in the Chebyshev method. Thus, m in Eq. (1.22) is 3 and the efficiency index $\alpha = 3^{1/3} \approx 1.44$.

Example:

Compute the approximate real root of the equation $e^x - 1 = 0$ with $|x_{n+1} - x_n| \leq \delta_x = 10^{-5}$, $|f(x_{n+1})| \leq \delta_f = 10^{-5}$ and $\delta_s = 10^{-9}$.

Table 3.8 presents the result, where the starting value x_0 is 0.5. The significance of the third term in the right-side of Eq. (3.63) can be seen by comparing the tables 3.4 and 3.8. In the Newton–Raphson method for $x_0 = 0.5$ the root with $\delta_x = \delta_f = 10^{-5}$ is obtained at 5th iteration. In the Chebyshev method convergence is reached at 3rd iteration. That is, convergence is faster in the Chebyshev method.

3.8 Comparison of Iterative Methods

The applicability and efficacy of various iterative methods described in this chapter can be compared on the basis of

TABLE 3.9

A comparative study of various iterative methods of finding roots of nonlinear equations.

Method	Formula	No. of starting values	Order of conv.	Efficiency index	Reliability of convergence	Advantages and disadvantages
Bisection	$x_{n+1} = (x_n + x_{n-1})/2$	2	1	1	Convergence is guaranteed.	Applicable to equations in which $f(x)$ crosses the exact roots. Starting values must enclose a root.
False position	$x_{n+1} = \dfrac{x_{n-1}f(x_n) - x_n f(x_{n-1})}{f(x_n) - f(x_{n-1})}$	2	1	1	Convergence is guaranteed.	Applicable to equations in which $f(x)$ crosses the exact roots. Starting values must enclose a root.
Secant	$x_{n+1} = \dfrac{x_{n-1}f(x_n) - x_n f(x_{n-1})}{f(x_n) - f(x_{n-1})}$	2	1.62	1.62	Convergence is not guaranteed.	The starting values need not enclose a root. It can be applied to compute complex roots.
Newton–Raphson	$x_{n+1} = x_n - \dfrac{f(x_n)}{f'(x_n)}$	1	2	1.41	Fast convergence if x_0 is close to a root. Sensitive to x_0.	Both real and complex roots can be calculated. Easily extendable to coupled equations.
Chebyshev	$x_{n+1} = x_n - \dfrac{f(x_n)}{f'(x_n)} - \dfrac{1}{2}\dfrac{f^2(x_n)f''(x_n)}{[f'(x_n)]^3}$	1	3	1.44	Faster than the Newton–Raphson rule.	Complex roots can also be calculated. Easily extendable to coupled equations.
Müller	$x_{n+1} = x_n + (x_n - x_{n-1}) \times \dfrac{2c}{b \pm \sqrt{b^2 - 4ac}}$	3	1.84	1.84	Sensitive to the starting values.	Complex roots can be computed. Three starting points are required.

1. order of convergence,

2. efficiency index and

3. sensitive dependence on starting value(s).

The bisection and the false position methods are slow but convergence to an actual root is guaranteed. These methods are applicable only to equations $f(x) = 0$, where $f(x)$ crosses the exact roots. The secant, Newton–Raphson, Chebyshev and Müller methods can be applied to determine both real and complex roots. But in the first three methods, to obtain a complex root the starting value(s) must be complex. In the Müller method a complex root can be obtained for real starting values. Of the various methods studied in this chapter the Newton–Raphson method is very simple and is widely used. This method and its improvement, Chebyshev method, alone require one initial guess. Table 3.9 gives a brief comparison of various iterative methods.

3.9 Concluding Remarks

Among the various methods discussed in this chapter for numerically computing the roots of a nonlinear equation the Newton–Raphson method is often preferred because of the simplicity of the formula, easy to develop a computer program and the order of convergence is quadratic. The iterative methods, particularly the Newton–Raphson method is of great use in the field of fractals. For example, consider the equation $P(z) = z^2 + 1$. Its roots are not real but complex $z^* = \pm i1$. The set of z_0 (initial guess) converging to a particular root (for example, through the Newton–Raphson method) is called its *basin of attraction*. The boundary of the basin of attraction is not a simple curve but a complicated curve exhibiting a *self-similar* structure characterized by noninteger dimension. Such a pattern is called a *fractal*. Many complex functions have fractal basins of attraction. Iterative methods have applications in estimating regression models, for example, in finding the values of parameters in the Poisson, logistic and binomial regression models. Newton–Raphson type methods are used in stochastic programming and portfolio management.

3.10 Bibliography

[1] T.J. Ypma, *SIAM Rev.* 37:531, 1995.

[2] P.W. France, *Solution of various hydraulic problems using the Newton-Raphson method*. In the Proc. of 2nd International Conference on Computer Methods in Water Resources, Published by Computational Mechanics, Southampton, England, 1991. pp.3−70.

[3] D.M. Bonilla-Correa, O.E. Coronado-Hernandez, V.S. Fuertes-Miquel, M. Besharat and H.M. Ramos, *Water* 15:1304, 2023.

[4] E.W. Martin and G. Peters, *J. Inst. Water Engrs.* 17:115, 1963.

[5] A.K. Sameni, *Application of Newton-Raphson method in three-phase unbalanced power flow*. Toronto Metropolitan University, Thesis, 2010. https://doi.org/10.329 20/ryerson.14644656.v1 (accessed on June 10, 2023).

[6]   S. Nirupama, *Newton-Raphson Method to Solve Power Flow Problem: Electrical Engineering*. Engineering Notes, 22 August 2017. https://www.engineeringenotes. com/electrical-engineering/power-flow/newton-raphson-method-to-solve-power-flow-problem-electrical-engineering/25354 (accessed on June 10, 2023).

[7]   H.R. Bakari, T.M. Adegoke and A.M. Yahya, *Int. J. Math. Stat. Stu.* 4(4):21, 2016.

[8]   R. Wolff and A. Kubo, *J. Tribol.* 116:733, 1994.

[9]   J. Viji Priya and S. Suppiah, *J. Comp. Sci.* 9:566, 2013.

[10] B.C. Truong, N.V. Thuan, N.H. Hau and M. McAleer, *Adv. Dec. Sci.* 23(4), 2019.

[11] J. Kou, *Appl. Math. Comput.* 189:602, 2007.

[12] C. Chun, *Appl. Math. Comput.* 189:1384, 2007.

[13] M.A. Noor, *Appl. Math. Comput.* 191:128, 2007.

[14] C. Chun, *Appl. Math. Comput.* 191:193, 2007.

[15] S. Abbasbandy, *Appl. Math. Comput.* 145:887, 2003.

[16] W. Nazeer, A. Naseem, S.M. Kang and Y.C. Kwun, *J. Nonl. Sci. Appl.* 9:2923, 2016.

3.11   Problems

A. Bisection Method

3.1   For the following nonlinear equations calculate the number of iterations N necessary to obtain an interval with width less than the tolerance $\delta_x = 10^{-2}$ for the specified starting points. Find the root after $N + 1$ iterations.

(a) $e^x - 1 = 0$, $(x_0, x_1) = (-0.2, 0.1)$.
(b) $x\,e^x - 1 = 0$, $(x_0, x_1) = (0.5, 0.6)$.
(c) $0.5 \sin x + 0.75 \cos x = 0$, $(x_0, x_1) = (-1, -0.9)$
(d) $x^3 - 1 = 0$. $(x_0, x_1) = (0.9, 1.05)$.
(e) $x^3 - 8 = 0$, $(x_0, x_1) = (1.9, 2.2)$.

3.2   Write a few equations which cannot be solved by the bisection method.

3.3   Can you apply the bisection method to the equation $1 - \sin x = 0$? Why?

B. False Position Method

3.4   Compute a real root of the following nonlinear equations with the specified starting values. Stop the iteration when the width of the interval is less than the tolerance $\delta_x = 10^{-2}$ or $|f(x)| \leq \delta_f = 10^{-5}$.

(a) $e^x - 1 = 0$, $(x_0, x_1) = (-0.2, 0.1)$.
(b) $x\,e^x - 1 = 0$, $(x_0, x_1) = (0.5, 0.6)$.
(c) $0.5 \sin x + 0.75 \cos x = 0$, $(x_0, x_1) = (-1, -0.9)$.
(d) $x^3 - 1 = 0$, $(x_0, x_1) = (0.9, 1.05)$.
(e) $x^3 - 8 = 0$, $(x_0, x_1) = (1.9, 2.2)$.

3.5   List out the differences between the bisection and the false position methods.

C. Secant Method

3.6 Compute a real root of the following equations with the specified starting values. Stop the iteration when $|x_{n+1} - x_n| \leq \delta_x = 10^{-2}$ or $|f(x)| \leq \delta_f = 10^{-5}$.

(a) $e^x - 1 = 0$, $(x_0, x_1) = (-0.2, 0.1)$.

(b) $x\, e^x - 1 = 0$, $(x_0, x_1) = (0.5, 0.6)$.

(c) $0.5 \sin x + 0.75 \cos x = 0$, $(x_0, x_1) = (-1, -0.9)$.

(d) $x^3 - 1 = 0$, $(x_0, x_1) = (0.9, 1.1)$.

(e) $x^3 - 8 = 0$, $(x_0, x_1) = (1.0, 1.8)$.

D. Newton–Raphson Method

(i) Real roots of one-dimensional equations

Some of the following problems can also be solved by other methods. In all the following root finding problems choose $\delta_x = \delta_f = 10^{-5}$ and $\delta_s = 10^{-9}$.

3.7 Find a real root of the following equations with the given starting point.

(a) $e^x - 1 = 0$, $x_0 = 0.09$.

(b) $x\, e^x - 1 = 0$, $x_0 = 0.4$.

(c) $0.5 \sin x + 0.75 \cos x = 0$, $x_0 = -1$.

(d) $x^3 - 1 = 0$, $x_0 = 1.1$.

(e) $x^3 - 8 = 0$, $x_0 = 2.5$.

3.8 Find the square root of 11. [Hint: Assume that the square root of 11 as x. Then, one has the equation $x^2 - 11 = 0$. A root of this equation is the square root of 11. Find the root with $x_0 = 3$].

3.9 Determine the value of $(50)^{1/3}$. Choose the initial guess as 4.

3.10 Given the value of π as 3.14159 compute $1/\pi$.

3.11 Consider a unit line segment whose ends are marked as A and B and C is a point on the line segment such that $AB/AC = AC/CB$. Find the length of AC.

3.12 The motion of a particle along a one-dimensional line is described by the equation

$$x(t) = c\, v \left(1 - e^{-t/c}\right),$$

where x is the distance of the particle at time t from the origin, v is the initial velocity of the particle and c is a constant whose value is 5. If the initial velocity of the particle is $2\,\text{m/sec}$ find the time required to travel through a distance 5 m. Choose the initial guess of t as 2 sec. Compare the numerical result with the exact result.

3.13 In celestial mechanics, Kepler's equation is $y = x - \alpha \sin x$, where y is the planet's mean anomaly, x is its eccentric anomaly and α is the eccentricity of its orbit. If $y = 1$ and $\alpha = 0.75$ for a planet then find its eccentric anomaly x with the initial guess $x_0 = 1.5$.

3.14 The wave function of a quantum mechanical particle is given by

$$\psi = A \sin\left(\frac{2\pi x}{A} + \alpha\right).$$

If $\alpha = 0.5$, $A = 3$ and $\pi = 3.14159$ find the value of x at which a) $\psi = 0$ and b) $\psi = 0.5$. Choose the initial guess of x as 0.

3.15 The energy of the lowest molecular orbital occupied two electrons in the ground state of lithium hydride molecule is the negative root of the equation

$$0.75E^2 + 0.35E - 0.023 = 0.$$

Determine the lowest energy after three iterations with $E_0 = -0.5$ units.

3.16 The ripple factor of a full wave rectifier is given by

$$r = \sqrt{(I_{\text{rms}}/I_{\text{dc}})^2 - 1}.$$

If one wishes to have a ripple factor 0.482 determine the value of $(I_{\text{rms}}/I_{\text{dc}})$ with the starting value of it as 1.

3.17 The relation between p-n junction diode current, voltage and temperature is

$$I = I_0 \left(e^{V/V_T} - 1 \right).$$

If $I/I_0 = 1$ at room temperature and $V = 1\,\text{V}$ compute V_T with $V_{T,0} = 1.5\,\text{V}$ and $\delta_x = \delta_f = 10^{-2}$.

3.18 The frequency of oscillation of a LCR circuit is given by

$$f = \frac{1}{2\pi} \left(\frac{1}{LC} - \frac{R^2}{4L^2} \right)^{1/2}.$$

What value of R has to be chosen in order to have $f = \sqrt{0.75}/(2\pi)$, $L = 1\,\text{H}$, $C = 1\,\text{F}$? Choose the initial guess of R as $0.8\,\Omega$.

3.19 The Planck's radiation formula is given by

$$u = \frac{8\pi k_{\text{B}} T}{\hbar^4 c^4} \frac{x^5}{e^x - 1},$$

where $x = \hbar c/(k_{\text{B}} T \lambda)$ and $\hbar$, c k_{B} are constants. T and λ are the temperature and the wavelength of the radiation, respectively. Find the value of λ or x at which u is a maximum.

3.20 For the equation $\cos x = 0$, where x is in radian, can one use $x_0 = 0$ to find its root? Why?

3.21 Show that for the problem of finding the square-root of $A > 0$, the Newton–Raphson formula reduces to $x_{n+1} = (x_n^2 + A)/(2x_n)$.

3.22 For the equation $x^2 e^{-x} + x = 0$ simplify the Newton–Raphson formula for x_{n+1}.

3.23 A man is running at a speed of $3\,\text{m/sec}$ to catch a bus that is stopped at a traffic light. When he is still a distance of $1\,\text{m}$ from the bus the light changes and the bus starts to move away from the running man with a constant acceleration of $2\,\text{m/sec}$. If T is the time at which the man catches the bus then it is the root of the equation $T^2 - 3T + 1 = 0$. By the Newton–Raphson method with $T_0 = 2\,\text{sec}$ find the value of T after four iterations. Also, compute the exact time.

(ii) Complex roots of one-dimensional equations

3.24 Find the complex roots of the following equations with the given starting value. Use $\delta_x = \delta_f = 10^{-5}$ and $\delta_s = 10^{-9}$.

(a) $e^x + e^{-x} = 0$, $(x_{R0} + i x_{I0}) = (0.5 + i1)$.

(b) $x^3 + 1 = 0$, $(x_{R0} + i x_{I0}) = (0.5 + i1)$.

(c) $x^3 - 1 = 0$, $(x_{R0} + i x_{I0}) = (-0.5 - i1)$.

(d) $e^x + 1 = 0$, $(x_{R0} + i x_{I0}) = (0 + i3)$.

(e) $x^3 - 3x^2 + 7x - 5 = 0$, $(x_{R0} + i x_{I0}) = (1.0 + i1)$.

3.25   Can the Newton–Raphson method be used to find $\sqrt{-11}$? How?

(iii) Real roots of two-coupled equations

3.26   Compute a real root of the following equations. Choose $\delta_x = \delta_y = \delta_f = \delta_g = 10^{-5}$ and $\delta_s = 10^{-9}$.

(a) $x^3 + 2y + 3 = 0$, $\quad y^2 + 3xy + 2 = 0$, $\quad (x_0, y_0) = (1, -1.5)$.

(b) $x^2 + y - 4 = 0$, $\quad xy - 3 = 0$, $\quad (x_0, y_0) = (1, 3)$.

(c) $x^4 + y^3 - 8 = 0$, $\quad y^2 + 2xy + x - 4 = 0$, $\quad (x_0, y_0) = (0, 1.9)$.

(d) $x^2 + x - y = 0$, $\quad xy - 2 = 0$, $\quad (x_0, y_0) = (1, 3)$.

D. Müller Method

(i) Real roots of one-dimensional equations

3.27   Compute a real root of the following equations with the given starting values. Stop the iteration when $|x_{n+1} - x_n| \le 10^{-5}$ and $|f(x_{n+1})| \le 10^{-5}$ or $|t_0 t_1 (t_1 - t_0)| \le \delta_s = 10^{-9}$.

(a) $e^x - 1 = 0$, $\quad (x_0, x_1, x_2) = (0.3, 0.2, 0.1)$.

(b) $x e^x - 1 = 0$, $\quad (x_0, x_1, x_2) = (0.4, 0.5, 0.6)$.

(c) $0.5 \sin x + 0.75 \cos x - 0$, $\quad (x_0, x_1, x_2) - (-0.7, -0.8, -0.9)$.

(d) $x^3 - 1 = 0$, $\quad (x_0, x_1, x_2) = (0.8, 1.0, 1.2)$.

(ii) Complex roots of one-dimensional equations

3.28   Obtain a pair of complex roots of the following equations. Use the specified starting values. Stop the iteration when $|x_{n+1} - x_n| \le 10^{-5}$ and $|f(x_{n+1})| \le 10^{-5}$.

(a) $e^x + e^{-x} = 0$, $\quad (x_0, x_1, x_2) = (0 + i1.4, 0 + i1.5, 0 + i1.6)$.

(b) $x^3 + 1 = 0$, $\quad (x_0, x_1, x_2) = (0.4 + i0.8, 0.5 + i0.8, 0.6 + i0.8)$.

(c) $x^3 - 1 = 0$, $\quad (x_0, x_1, x_2) = (-0.4, -0.5, -0.6)$.

(d) $e^x + 1 = 0$, $\quad (x_0, x_1, x_2) = (0.2 + i2.8, 0 + i3, 0 + i3.2)$.

E. Chebyshev Method

3.29   Obtain the roots of the equations given in Problem 3.7.

3.30   Show that the order of convergence of the Chebyshev method is 3.

4

Solution of Linear Systems $AX = B$

4.1 Introduction

In many branches of physics and engineering, it is often required to solve a set of simultaneous linear equations. A general form of a system of n linear equations with n variables is given by

$$
\begin{aligned}
a_{11}x_1 + a_{12}x_2 + \cdots + a_{1n}x_n &= b_1, \\
a_{21}x_1 + a_{22}x_2 + \cdots + a_{2n}x_n &= b_2, \\
&\vdots \\
a_{n1}x_1 + a_{n2}x_2 + \cdots + a_{nn}x_n &= b_n,
\end{aligned}
\tag{4.1}
$$

where a_{ij}'s and b_i's are known constants and x_i's are to be determined. The above system of linear equations can conveniently be expressed in matrix form as

$$
AX = B.
\tag{4.2}
$$

Denote the elements of A as a_{ij} and B as b_i. When all b_i's are zero then the system (4.2) is said to be *homogeneous*. If at least one of the b_i's is nonzero then the system (4.2) is called *inhomogeneous*. An inhomogeneous system has a nontrivial solution *if and only if* the determinant of A is nonzero.

System (4.2) actually represents a wide class of problems. Some of them are the following:

(1) **Straight-forward problem**

The number of variables is equal to the number of equations, that is, A is a square matrix, $B \neq 0$ and $\det A \neq 0$. If $\det A = 0$ then the problem is a singular value and it belongs to class (3).

(2) **The least-squares**

The number of variables is less than the number of equations.

(3) **The singular-value problem**

The number of variables is more than the number of equations. Such systems have infinitely many solutions. The solution closest to the origin is called a *singular-value solution*.

(4) **The eigenvalue problem**

The number of variables is equal to the number of equations and $B = 0$. The trivial solution is $X = 0$. The condition for a nontrivial solution is $\det A = 0$. For $\det A \neq 0$ the only solution is $X = 0$.

DOI: 10.1201/9781032649931-4

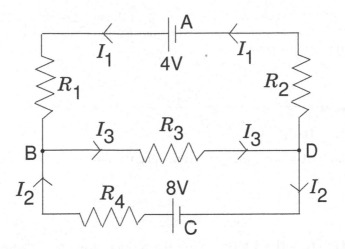

FIGURE 4.1
A simple dc circuit.

Systems of linear equations occur in optimal planning in industrial engineering, geometry, structural analysis in civil engineering and the representation of chemical reactions in equation forms. The equations of an equilibrium point of a system of first-order coupled autonomous linear differential equations essentially form a system of linear equations. In business management linear equations are used to compute cost, revenue, profit and loss in a business. In electrical engineering systems, coupled linear equations arise in writing the equations for the currents through various branches in circuits. Electrical networks contain number of resistances and circuits. Determination of currents flowing through various paths or branches of an electrical circuit and electrical networks involves solving a system of linear equations.

Consider the circuit shown in Fig. 4.1 [1]. Applying the Ohm's and Kirchhoff's law at the junctions B and D and the closed paths ABDA and CBDC gives

$$I_1 + I_2 - I_3 = 0, \tag{4.3a}$$
$$(R_1 + R_2)\,I_1 + R_3 I_3 = 4, \tag{4.3b}$$
$$R_4 I_2 + R_3 I_3 = 8. \tag{4.3c}$$

Solving these system of linear equations gives the values of the currents I_1, I_2 and I_3. An unbalanced chemical reaction can be written in a system of linear equations and solving it the reaction equation can be balanced (see Section 4.13).

In this chapter, the prime emphasis is on solving the class (1) type of problem. Solutions of such systems can be determined either by direct methods or iterative methods. Direct methods, in the absence of round-off and other errors, produce an exact solution after a finite number of arithmetic operations. On the other hand, the iterative methods start with an initial approximation to the solution, give a sequence of approximate solutions, which converge to the exact solution. Some advantages of iterative methods are:

1. They can be easily implemented on computers.

2. They are relatively insensitive to the growth of round-off errors.

Systems of classes (3) and (4) can also be solved for a nontrivial solution which will be presented at the end of this chapter. To start with, in the next section, the famous Cramer's rule for solving a system of simultaneous equations is discussed. This rule is more suitable for two- and three-coupled equations.

4.2   Cramer's Rule

Let us first obtain the Cramer's rule for determining the unknowns in the two-coupled linear equations of the form

$$a_{11}x_1 + a_{12}x_2 = b_1, \tag{4.4a}$$
$$a_{21}x_1 + a_{22}x_2 = b_2. \tag{4.4b}$$

For a nontrivial solution, the determinant of the coefficient matrix is nonzero. That is,

$$|A| = a_{11}a_{22} - a_{21}a_{12} \neq 0. \tag{4.5}$$

Solving Eqs. (4.4) for x_1 and x_2 gives

$$x_1 = \frac{a_{22}b_1 - a_{12}b_2}{a_{11}a_{22} - a_{21}a_{12}}, \quad x_2 = \frac{a_{11}b_2 - a_{21}b_1}{a_{11}a_{22} - a_{21}a_{12}}. \tag{4.6}$$

Defining

$$B_1 = \begin{pmatrix} b_1 & a_{12} \\ b_2 & a_{22} \end{pmatrix}, \quad B_2 = \begin{pmatrix} a_{11} & b_1 \\ a_{21} & b_2 \end{pmatrix} \tag{4.7}$$

the solution (4.6) can be expressed as

$$x_1 = \frac{|B_1|}{|A|}, \quad x_2 = \frac{|B_2|}{|A|}. \tag{4.8}$$

For a system of n linear equations

$$x_i = \frac{|B_i|}{|A|}, \tag{4.9}$$

The above process is known as *Cramer's rule*. Since the solution given by Eq. (4.9) involves the calculation of $(n + 1)$ determinants the method is inconvenient when $n > 3$.

4.3   Upper- and Lower-Triangular Systems

A system of linear Eqs. (4.2) with A being an upper-triangular coefficient matrix with all diagonal elements nonzero can be solved by the back-substitution algorithm. An upper-triangular system is of the form

$$
\begin{aligned}
a_{11}x_1 + a_{12}x_2 + \cdots + a_{1,n-1}x_{n-1} + a_{1n}x_n &= b_1, \\
a_{22}x_2 + \cdots + a_{2,n-1}x_{n-1} + a_{2n}x_n &= b_2, \\
&\vdots \\
a_{n-1,n-1}x_{n-1} + a_{n-1,n}x_n &= b_{n-1}, \\
a_{nn}x_n &= b_n,
\end{aligned} \tag{4.10}
$$

where $a_{ii} \neq 0$, $i = 1, 2, \ldots, n$. The last equation of the system (4.10) is solved first which gives

$$x_n = \frac{b_n}{a_{nn}}. \tag{4.11}$$

Then, solving the second last equation of the system (4.10) for x_{n-1} gives

$$x_{n-1} = \frac{b_{n-1} - a_{n-1,n}x_n}{a_{n-1,n-1}}, \tag{4.12}$$

where x_n is given by Eq. (4.11). Knowing x_n and x_{n-1} the third last equation of the system (4.10) can be used to solve x_{n-2} and so on. The general formula for x_k is

$$x_k = \frac{1}{a_{kk}} \left(b_k - \sum_{i=k+1}^{n} a_{ki}x_i \right), \quad k = n, n-1, \ldots, 2, 1. \tag{4.13}$$

In the above formula when $k = n$, the summation becomes $\sum_{i=n+1}^{n}$ which is interpreted as the sum over no terms and gives the value 0. Since a_{kk} occurs in the denominator in Eq. (4.13) the requirement is that all $a_{kk} \neq 0$. The above process of determining the solution of the system (4.10) is called *back-substitution*.

System (4.2) with A being a lower-triangular matrix can be solved by forward-substitution algorithm, where the formula for x_k is given by

$$x_k = \frac{1}{a_{kk}} \left(b_k - \sum_{i=1}^{k} a_{ki}x_i \right), \quad k = 1, 2, \ldots, n. \tag{4.14}$$

Example:

Consider the system

$$x_1 + 2x_2 + x_3 = 4, \tag{4.15a}$$
$$x_2 - x_3 = 3, \tag{4.15b}$$
$$x_3 = -1. \tag{4.15c}$$

Here, $a_{11} = 1$, $a_{12} = 2$, $a_{13} = 1$, $a_{22} = 1$, $a_{23} = -1$, $a_{33} = 1$, $b_1 = 4$, $b_2 = 3$ and $b_3 = -1$. The last subequation straight-forwardly gives $x_3 = -1$. Then, for $k = 2$ Eq. (4.13) gives

$$x_2 = \frac{1}{a_{22}} (b_2 - a_{23}x_3) = 1 \times (3 - 1) = 2. \tag{4.16}$$

When $k = 1$

$$x_1 = \frac{1}{a_{11}} (b_1 - a_{12}x_2 - a_{13}x_3) = 1 \times (4 - 2 \times 2 - 1 \times -1) = 1. \tag{4.17}$$

The solution of the system (4.15) is $(x_1, x_2, x_3) = (1, 2, -1)$.

Triangular linear systems occur in the least-squares problems [2] and splitting-based iterative schemes and in the Cholesky factorizations [3]. It is convenient to reduce a general system of linear equations into an upper- or lower-triangular system and obtain the solution by the back-substitution.

4.4   Gauss Elimination Method

The back-substitution algorithm described in the previous section can be applied without difficulty to solve triangular system of equations. *How does one proceed if A is not an upper- or lower-triangular form?* In this case, the system (4.2) can be transformed into an equivalent system with an upper-triangular form and then it can be solved by the back-substitution. This process of solving the system (4.2) is called *Gauss elimination*. The name is due to the fact that equations are combined to eliminate the unknowns. The method is straight forward, easily understandable and pedagogical.

Before going to describe the Gauss elimination method let us first recall the basic elementary transformations and the notion of augmented matrix, associated with the linear system of equations, which are essential to follow the underlying method.

4.4.1   Elementary Operations and Augmented Matrix

The solution of a given set of equations is unaffected by the following operations:

1.   multiplication of an equation by a nonzero constant,

2.   interchange of two equations and

3.   replacement of an equation by the sum of that equation and a multiple of any other equation.

Practically, it is convenient to work with the so-called *augmented matrix* of the system (4.2). The augmented matrix denoted by $[A, B]$ is given by

$$[A, B] = \begin{pmatrix} a_{11} & a_{12} & \cdots & a_{1n} & b_1 \\ a_{21} & a_{22} & \cdots & a_{2n} & b_2 \\ & & \vdots & & \\ a_{n1} & a_{n2} & \cdots & a_{nn} & b_n \end{pmatrix} = \begin{pmatrix} a_{11} & a_{12} & \cdots & a_{1n} & a_{1,n+1} \\ a_{21} & a_{22} & \cdots & a_{2n} & a_{2,n+1} \\ & & \vdots & & \\ a_{n1} & a_{n2} & \cdots & a_{nn} & a_{n,n+1} \end{pmatrix}. \quad (4.18)$$

It is a $n \times (n + 1)$ matrix with the elements of B forming the $(n + 1)$th column and the elements of A forming the rest. Since Eqs. (4.2) and (4.18) are equivalent, the elementary transformations listed above are also applicable to Eq. (4.18).

4.4.2   Systematic Procedure

Let us proceed to describe the various steps involved in the Gauss elimination method. For simplicity and to avoid writing many ellipses (...) the method is described for a system of three equations. Extension to a large number of equations is straight forward.

(i) Pivot Row and Pivot Element

Consider the system

$$a_{11}x_1 + a_{12}x_2 + a_{13}x_3 = b_1, \quad (4.19a)$$
$$a_{21}x_1 + a_{22}x_2 + a_{23}x_3 = b_2, \quad (4.19b)$$
$$a_{31}x_1 + a_{32}x_2 + a_{33}x_3 = b_3. \quad (4.19c)$$

Eliminating x_1 in Eqs. (4.19b) and (4.19c) and then eliminating x_2 in the modified Eq. (4.19c) the system can be brought into an upper-triangular form. For this purpose redefine the elements a_{ij} in Eqs. (4.19) as $a_{ij}^{(1)}$. The number 1 in the superscript of $a_{ij}^{(1)}$ refers to the value of a_{ij} at the first stage of elimination. To eliminate x_1 in Eq. (4.19b) multiply Eq. (4.19a) by $-a_{21}^{(1)}/a_{11}^{(1)}$ and add it to Eq. (4.19b). Similarly, to eliminate x_1 in Eq. (4.19c) multiply Eq. (4.19a) by $-a_{31}^{(1)}/a_{11}^{(1)}$ and add it to Eq. (4.19c). These give

$$a_{22}^{(2)} x_2 + a_{23}^{(2)} x_3 = b_2^{(2)}, \tag{4.20a}$$

$$a_{32}^{(2)} x_2 + a_{33}^{(2)} x_3 = b_3^{(2)}, \tag{4.20b}$$

where

$$a_{22}^{(2)} = a_{22}^{(1)} - \frac{a_{21}^{(1)}}{a_{11}^{(1)}} a_{12}^{(1)}, \quad a_{23}^{(2)} = a_{23}^{(1)} - \frac{a_{21}^{(1)}}{a_{11}^{(1)}} a_{13}^{(1)}, \tag{4.21a}$$

$$a_{32}^{(2)} = a_{32}^{(1)} - \frac{a_{31}^{(1)}}{a_{11}^{(1)}} a_{13}^{(1)}, \quad a_{33}^{(2)} = a_{33}^{(1)} - \frac{a_{31}^{(1)}}{a_{11}^{(1)}} a_{13}^{(1)}, \tag{4.21b}$$

$$b_2^{(2)} = b_2^{(1)} - \frac{a_{21}^{(1)}}{a_{11}^{(1)}} b_1^{(1)}, \quad b_3^{(2)} = b_3^{(1)} - \frac{a_{31}^{(1)}}{a_{11}^{(1)}} b_1^{(1)}. \tag{4.21c}$$

Now, the given system of equations becomes

$$a_{11}^{(1)} x_1 + a_{12}^{(1)} x_2 + a_{13}^{(1)} x_3 = b_1^{(1)}, \tag{4.22a}$$

$$a_{22}^{(2)} x_2 + a_{23}^{(2)} x_3 = b_2^{(2)}, \tag{4.22b}$$

$$a_{32}^{(2)} x_2 + a_{33}^{(2)} x_3 = b_3^{(2)}, \tag{4.22c}$$

Next, to eliminate x_2 in Eq. (4.22c) multiply Eq. (4.22b) by $-a_{32}^{(2)}/a_{22}^{(2)}$ and add it to Eq. (4.22c). This gives

$$a_{33}^{(3)} x_3 = b_3^{(3)}, \tag{4.23}$$

where

$$a_{33}^{(3)} = a_{33}^{(2)} - \frac{a_{32}^{(2)}}{a_{22}^{(2)}} a_{23}^{(2)}, \quad b_3^{(3)} = b_3^{(2)} - \frac{a_{32}^{(2)}}{a_{22}^{(2)}} b_2^{(2)}. \tag{4.24}$$

Then, the equivalent upper-triangular form of the system Eq. (4.19) is given by

$$a_{11}^{(1)} x_1 + a_{12}^{(1)} x_2 + a_{13}^{(1)} x_3 = b_1^{(1)}, \tag{4.25a}$$

$$a_{22}^{(2)} x_2 + a_{23}^{(2)} x_3 = b_2^{(2)}, \tag{4.25b}$$

$$a_{33}^{(3)} x_3 = b_3^{(3)}, \tag{4.25c}$$

The system (4.25) can be solved by back-substitution.

Let us generalize the above procedure for n number of equations. For simplicity focus the analysis on the augmented matrix. The formula for the elements a_{ij} after $(k+1)$th stage of elimination is given by

$$a_{ij}^{(k+1)} = a_{ij}^{(k)} - \frac{a_{ik}^{(k)}}{a_{kk}^{(k)}} a_{kj}^{(k)}, \tag{4.26a}$$

where

$$k = 1, 2, \ldots, n-1; \quad i = k+1, k+2, \ldots, n; \tag{4.26b}$$

$$j = k+1, k+2, \ldots, n+1. \tag{4.26c}$$

At kth stage of elimination, the unknown x_k is eliminated in the equations appearing below the kth equation using the kth equation. Therefore, at kth stage, the kth equation in the system is crucial or essential. Because of this at kth stage of elimination, the kth equation is called *pivot equation* (or *pivot row* if one works with augmented matrix). The coefficient $a_{kk}^{(k)}$ is called *pivot element*.

In order to use the method in most circumstances two modifications are necessary in the Gauss elimination method. These are described below.

(ii) Pivoting

If $a_{11}^{(1)} = 0$ then the first row in the augmented matrix does not help us to eliminate the elements in the first column appearing below $a_{11}^{(1)}$. Similarly, if $a_{22}^{(2)} = 0$ then elimination of the elements appearing below the second row in the second column using the second row is not possible. In general, if $a_{kk}^{(k)} = 0$ then the elements in the kth column appearing below the kth row cannot be eliminated using the kth row. *How to overcome this difficulty?* Whenever $a_{kk}^{(k)} = 0$ identify the row r with $a_{rk}^{(k)} \neq 0$, $r > k$ and interchange the rows r and k. After this interchange the new pivot element $a_{kk}^{(k)}$ is nonzero. This process is known as *pivoting*.

(iii) Partial Pivoting

To reduce the propagation of errors due to fixed-finite precision arithmetic, at kth stage of elimination process, irrespective of whether $a_{kk}^{(k)}$ is zero or not, the row r with $r \geq k$ for which the element $a_{rk}^{(k)}$ has the largest absolute value, that is,

$$\left| a_{rk}^{(k)} \right| \;=\; \text{max.} \left\{ \left| a_{kk}^{(k)} \right|, \left| a_{k+1,k}^{(k)} \right|, \ldots, \left| a_{nk}^{(k)} \right| \right\} \tag{4.27}$$

and the row k is interchanged. This process is known as *partial pivoting*. If all $a_{rk}^{(k)}$ are zero then the matrix A is singular.

4.4.3   Summary of Gauss Elimination Method

The various steps involved in the method are summarized below.

(1)   Write the augmented matrix for the given system of n equations.

(2)   *Pivoting*: At kth stage of elimination, find the row r with max.$|a_{rk}^{(k)}| \neq 0$, $r \geq k$. If $r \neq k$ interchange the rows r and k. If all $a_{rk}^{(k)}$ are zero then the matrix is singular and no solution exists.

(3)   Eliminate the elements $a_{rk}^{(k)}$ in all the rows $r > k$.

(4)   Repeat the steps (2) and (3) until the elimination is completed.

Example:

Find the point of intersection of the three straight-lines governed by the equations

$$x_1 + x_2 + x_3 \;=\; 2, \tag{4.28a}$$
$$2x_1 + x_2 + 2x_3 \;=\; 2, \tag{4.28b}$$
$$3x_1 - x_2 + 2x_3 \;=\; -1. \tag{4.28c}$$

in the $x_1 - x_2 - x_3$ space.

At the point of intersection, all the given equations must be simultaneously satisfied. Therefore, treat the above three equations as a system of coupled equations. Its solution is the point of intersection.

Let us solve the given set of equations applying the Gauss elimination method. The augmented matrix of the system (4.28) is

$$
\begin{pmatrix}
a_{11}^{(1)} & a_{12}^{(1)} & a_{13}^{(1)} & b_1^{(1)} \\
a_{21}^{(1)} & a_{22}^{(1)} & a_{23}^{(1)} & b_2^{(1)} \\
a_{31}^{(1)} & a_{32}^{(1)} & a_{33}^{(1)} & b_3^{(1)}
\end{pmatrix}
=
\begin{pmatrix}
1 & 1 & 1 & 2 \\
2 & 1 & 2 & 2 \\
3 & -1 & 2 & -1
\end{pmatrix}
\tag{4.29}
$$

Because $\left|a_{31}^{(1)}\right| = 3$ is the maximum of $\{\left|a_{11}^{(1)}\right| = 1,\ \left|a_{21}^{(1)}\right| = 2,\ \left|a_{31}^{(1)}\right| = 3\}$ interchange the rows 1 and 3. After this interchange the augmented matrix is (the pivot element is underlined)

$$
\begin{pmatrix}
\underline{3} & -1 & 2 & -1 \\
2 & 1 & 2 & 2 \\
1 & 1 & 1 & 2
\end{pmatrix} .
\tag{4.30}
$$

To eliminate the element $a_{21}^{(1)} = 2$ in the above matrix multiply the row 1 by $-a_{21}^{(1)}/a_{11}^{(1)} = -2/3$ and add it with the row 2. Similarly, multiplication of the row 1 by $-a_{31}^{(1)}/a_{11}^{(1)} = -1/3$ and adding it with the row 3 eliminates $a_{31}^{(1)}$. Then, the augmented matrix is

$$
\begin{pmatrix}
3 & -1 & 2 & -1 \\
2 + 3 \times -\dfrac{2}{3} & 1 - 1 \times -\dfrac{2}{3} & 2 + 2 \times -\dfrac{2}{3} & 2 - 1 \times -\dfrac{2}{3} \\
1 + 3 \times -\dfrac{1}{3} & 1 - 1 \times -\dfrac{1}{3} & 2 \times 1 - \dfrac{1}{3} & 2 - 1 \times -\dfrac{1}{3}
\end{pmatrix}
$$

$$
=
\begin{pmatrix}
3 & -1 & 2 & -1 \\
0 & 5/3 & 2/3 & 8/3 \\
0 & 4/3 & 1/3 & 7/3
\end{pmatrix} .
\tag{4.31}
$$

The next step is to eliminate $a_{32}^{(2)}$ (that is, x_2) in the last row of the above matrix. Since $\left|a_{22}^{(2)}\right| = 5/3 > \left|a_{32}^{(2)}\right| = 4/3$ no need for interchanging the rows. Elimination of $a_{32}^{(2)}$ leads to the matrix

$$
\begin{pmatrix}
3 & -1 & 2 & -1 \\
0 & 5/3 & 2/3 & 8/3 \\
0 & \dfrac{4}{3} - \dfrac{4}{5} \times \dfrac{5}{3} & \dfrac{1}{3} - \dfrac{4}{5} \times \dfrac{2}{3} & \dfrac{7}{3} - \dfrac{4}{5} \times \dfrac{8}{3}
\end{pmatrix}
$$

$$
=
\begin{pmatrix}
3 & -1 & 2 & -1 \\
0 & 5/3 & 2/3 & 8/3 \\
0 & 0 & -1/5 & 1/5
\end{pmatrix} .
\tag{4.32}
$$

In equational form, the above matrix is the upper-triangular system

$$
3x_1 - x_2 + 2x_3 = -1 ,
\tag{4.33a}
$$

$$
\frac{5}{3}x_2 + \frac{2}{3}x_3 = \frac{8}{3} ,
\tag{4.33b}
$$

$$
-\frac{1}{5}x_3 = \frac{1}{5} .
\tag{4.33c}
$$

The last equation gives $x_3 = -1$. Back-substitution of the value of x_3 in the second equation of (4.33) gives $x_2 = 2$. Then, substitution of $x_2 = 2$ and $x_3 = -1$ in the first equation of Eqs. (4.33) gives $x_1 = 1$. The solution is $(x_1, x_2, x_3) = (1, 2, -1)$.

Back-substitution is possible in the augmented matrix given by Eq. (4.32). The first back-substitution gives

$$
\begin{pmatrix}
a_{11}^{(1)} & a_{12}^{(1)} & a_{13}^{(1)} & b_1^{(1)} \\
 & a_{22}^{(2)} & 0 & b_2^{\prime(2)} = b_2^{(2)} - a_{23}^{(2)} b_3^{(3)} / a_{33}^{(3)} \\
 & & a_{33}^{(3)} & b_3^{(3)}
\end{pmatrix}
$$

$$
\implies
\begin{pmatrix}
3 & -1 & 1 & -1 \\
 & 5/3 & 0 & 10/3 \\
 & & -1/5 & 1/5
\end{pmatrix}. \tag{4.34}
$$

The second back-substitution gives

$$
\begin{pmatrix}
a_{11}^{(1)} & 0 & 0 & b_1^{(1)} - a_{12}^{(1)} b_2^{\prime(2)} / a_{22}^{(2)} - a_{13}^{(1)} b_3^{(3)} / a_{33}^{(3)} \\
 & a_{22}^{(2)} & 0 & b_2^{\prime(2)} \\
 & & a_{33}^{(3)} & b_3^{(3)}
\end{pmatrix}
$$

$$
=
\begin{pmatrix}
3 & 0 & 0 & 3 \\
 & 5/3 & 0 & 10/3 \\
 & & -1/5 & 1/5
\end{pmatrix}. \tag{4.35}
$$

Dividing each row by its diagonal element one obtains

$$
\begin{pmatrix}
1 & 0 & 0 & 1 \\
 & 1 & 0 & 2 \\
 & & 1 & -1
\end{pmatrix}. \tag{4.36}
$$

The last column gives the solution as $(x_1, x_2, x_3) = (1, 2, -1)$ and is the point of intersection. In the case of hand calculation, to avoid confusion, after obtaining the upper-triangular matrix the back-substitution can be done on the equational form of the matrix.

4.4.4   Significance of Pivoting

What is the significance of pivoting strategy? Is this process necessary? Pivoting is not necessary if the arithmetic operations in the Gauss elimination method are performed with infinite-precision accuracy. If the calculations are done without pivoting but with a finite-precision then the method will lead to an incorrect result in many circumstances. For example, the exact solution of the system

$$
0.0002x_1 + 1.572x_2 = 1.575, \tag{4.37a}
$$
$$
0.3210x_1 + 1.231x_2 = 6.046 \tag{4.37b}
$$

is $(x_1, x_2) = (15, 1)$. Now, apply the Gauss elimination method without and with pivoting using four-decimal floating arithmetic and observe the difference.

Solution without pivoting

Elimination of x_2 in Eq. (4.37a) by multiplying Eq. (4.37b) by $-0.321/0.0002$ and adding it with Eq. (4.37b) gives $x_2 = 0.9996$. Then, Eq. (4.37a) gives $x_1 = 20$ while the exact solution is $x_1 = 15$, $x_2 = 1$. The method incorrectly predicts the value of x_1. *What will happen with pivoting?*

Solution with pivoting

With pivoting the system (4.37) is rewritten as

$$0.3210x_1 + 1.231x_2 \;=\; 6.046\,, \tag{4.38a}$$
$$0.0002x_1 + 1.572x_2 \;=\; 1.575\,. \tag{4.38b}$$

Equation (4.38b) after elimination of x_1 becomes $1.571x_2 = 1.571$ giving $x_2 = 1$. Then, from Eq. (4.38a) $x_1 = 15$. The obtained solution is identical to the exact solution. This example clearly illustrates the necessity of pivoting for a general circumstance.

4.5 Gauss–Jordan Elimination Method

Wilhelm Jordan proposed a modification in the Gauss elimination method. He introduced additional calculations that avoid back-substitution by reducing the augmented matrix into a diagonal rather than an upper-triangular form. The additional steps are:

1. After interchanging the rows as per the relation (4.27) the pivot equation (that is, the pivot row in the augmented matrix) is divided by the pivot element thereby making the pivot element 1.

2. In the Gauss elimination scheme at the kth stage of elimination, the variable x_k (the element a_{rk}, $r > k$ in the augmented matrix) is eliminated in the equations appearing below the kth equation. The other equations, that is $r = 1, 2, \ldots, k-1$ are undisturbed. In the Gauss–Jordan method the variable x_k is eliminated in the equations $1, 2, \ldots, k-1$ also.

The final augmented matrix will be of the form

$$
\begin{pmatrix}
1 & 0 & \cdots & 0 & \alpha_1 \\
0 & 1 & \cdots & 0 & \alpha_2 \\
& & \vdots & & \\
0 & 0 & \cdots & 1 & \alpha_n
\end{pmatrix}. \tag{4.39}
$$

The last column elements of the final augmented matrix are the unknowns x_i's.

In the Gauss–Jordan method the formula (4.26) takes the form

$$a_{ij}^{(k+1)} \;=\; a_{ij}^{(k)} - a_{ik}^{(k)} a_{kj}^{(k)}\,, \tag{4.40a}$$

where

$$k \;=\; 1, 2, \ldots, n-1; \quad i = 1, 2, \ldots, n; \tag{4.40b}$$
$$j \;=\; k+1, k+2, \ldots, n+1. \tag{4.40c}$$

Remember that $a_{kk}^{(k)}$ is set into 1 by dividing the kth row by $a_{kk}^{(k)}$. Note that when the unknown x_k is eliminated in the Gauss–Jordan method, it is eliminated in all equations except in the kth equation. In Gauss elimination, x_k is eliminated only in the equations appearing below the kth equation. Further, in the Gauss–Jordan method all rows are normalized by dividing them by their pivot element resulting in identity matrix contrast to a triangular matrix in Gauss elimination.

Example:

Solve the system (4.28) by the Gauss–Jordan method.

The additional calculations involved in the Gauss–Jordan method over Gauss elimination are indicated by italics. The augmented matrix given by Eq. (4.29) after interchanging the first and third rows (since $|a_{31}| > |a_{11}|$ and $|a_{21}|$) is

$$\begin{pmatrix} 3 & -1 & 2 & -1 \\ 2 & 1 & 2 & 2 \\ 1 & 1 & 1 & 2 \end{pmatrix}. \tag{4.41}$$

Dividing the pivot row (row 1) by its pivot element 3, the above matrix becomes

$$\begin{pmatrix} 1 & -1/3 & 2/3 & -1/3 \\ 2 & 1 & 2 & 2 \\ 1 & 1 & 1 & 2 \end{pmatrix}. \tag{4.42}$$

Eliminating a_{21} and a_{31} as done earlier in Gauss elimination method, the augmented matrix becomes

$$\begin{pmatrix} 1 & -1/3 & 2/3 & -1/3 \\ & 5/3 & 2/3 & 8/3 \\ & 4/3 & 1/3 & 7/3 \end{pmatrix}. \tag{4.43}$$

Since $|a_{22}|$ is greater than $|a_{32}|$ interchange of rows is not necessary. *Dividing the second row by its pivot element 5/3, the above matrix becomes*

$$\begin{pmatrix} 1 & -1/3 & 2/3 & -1/3 \\ & 1 & 2/5 & 8/5 \\ & 4/3 & 1/3 & 7/3 \end{pmatrix}. \tag{4.44}$$

The next step is *elimination of a_{12} and a_{32} (that is, x_2 in the first and third equations of the system) in the first and third rows, respectively.* After this step, the augmented matrix is obtained as

$$\begin{pmatrix} 1 & 0 & 4/5 & 1/5 \\ & 1 & 2/5 & 8/5 \\ & & -1/5 & 1/5 \end{pmatrix}. \tag{4.45}$$

Next, divide the third row by its pivot element $-1/5$. Then, eliminate a_{13} and a_{23}. The augmented matrix now becomes

$$\begin{pmatrix} 1 & 0 & 0 & 1 \\ & 1 & 0 & 2 \\ & & 1 & -1 \end{pmatrix}. \tag{4.46}$$

Thus, the solution is

$$\begin{pmatrix} x_1 \\ x_2 \\ x_3 \end{pmatrix} = \begin{pmatrix} 1 \\ 2 \\ -1 \end{pmatrix}. \tag{4.47}$$

The solution is obtained without back-substitution. This is the advantage of Gauss–Jordan method over Gauss elimination.

4.6 Inverse of a Matrix by the Gauss–Jordan Method

A slight modification of the Gauss–Jordan method will provide an algorithm for finding the inverse of a matrix. To find the inverse of a matrix A the augmented matrix is written as $[A, I]$, where I is the unit matrix of order same as A. Performing Gauss–Jordan elimination on this augmented matrix finally gives the matrix $[I, A^{-1}]$. In this way, A^{-1} can be determined.

Example:

Find the inverse of the matrix

$$A = \begin{pmatrix} 1 & 0 & 2 \\ -1 & 1 & 2 \\ 1 & -2 & 0 \end{pmatrix} \tag{4.48}$$

applying the Gauss–Jordan method.

The augmented matrix is

$$[A, I] = \begin{pmatrix} 1 & 0 & 2 & 1 & 0 & 0 \\ -1 & 1 & 2 & 0 & 1 & 0 \\ 1 & -2 & 0 & 0 & 0 & 1 \end{pmatrix}. \tag{4.49}$$

The augmented matrix after eliminating a_{21} and a_{31} is

$$\begin{pmatrix} 1 & 0 & 2 & 1 & 0 & 0 \\ 0 & 1 & 4 & 1 & 1 & 0 \\ 0 & -2 & -2 & -1 & 0 & 1 \end{pmatrix}. \tag{4.50}$$

Since $|a_{32}| > |a_{22}|$ interchange the rows 2 and 3 and divide the second row by its pivot element $a_{22} = -2$. Then, eliminating a_{12} (which is already zero) and a_{32} leads to the augmented matrix

$$\begin{pmatrix} 1 & 0 & 2 & 1 & 0 & 0 \\ 0 & 1 & 1 & 1/2 & 0 & -1/2 \\ 0 & 0 & 3 & 1/2 & 1 & 1/2 \end{pmatrix}. \tag{4.51}$$

Finally, eliminating a_{13} and a_{23} one has

$$\begin{pmatrix} 1 & 0 & 0 & 2/3 & -2/3 & -1/3 \\ 0 & 1 & 0 & 1/3 & -1/3 & -2/3 \\ 0 & 0 & 1 & 1/6 & 1/3 & 1/6 \end{pmatrix}. \tag{4.52}$$

Thus, the inverse of A is

$$\begin{pmatrix} 2/3 & -2/3 & -1/3 \\ 1/3 & -1/3 & -2/3 \\ 1/6 & 1/3 & 1/6 \end{pmatrix}. \tag{4.53}$$

4.7   Triangular Factorization or Decomposition Method

Another method of solving coupled linear equations is the *triangular factorization* or *decomposition method*. In this method, the matrix A of the system $AX = B$, with $\det A \neq 0$ is expressed as a product of an upper-triangular matrix U with nonzero diagonal elements and the lower-triangular matrix L with all its diagonal elements being 1. That is,

$$A = LU. \tag{4.54}$$

Explicitly,

$$
\begin{pmatrix}
a_{11} & a_{12} & \cdots & a_{1n} \\
a_{21} & a_{22} & \cdots & a_{2n} \\
 & & \vdots & \\
a_{n1} & a_{n2} & \cdots & a_{nn}
\end{pmatrix}
=
\begin{pmatrix}
1 & 0 & \cdots & 0 \\
l_{21} & 1 & \cdots & 0 \\
 & & \vdots & \\
l_{n1} & l_{n2} & \cdots & 1
\end{pmatrix}
$$
$$
\times
\begin{pmatrix}
u_{11} & u_{12} & \cdots & u_{1n} \\
0 & u_{22} & \cdots & u_{2n} \\
 & & \vdots & \\
0 & 0 & \cdots & u_{nn}
\end{pmatrix}. \tag{4.55}
$$

How does A decompose into LU? Let us illustrate this for $n = 3$. Now, Eq. (4.55) becomes

$$
\begin{pmatrix}
a_{11} & a_{12} & a_{13} \\
a_{21} & a_{22} & a_{23} \\
a_{31} & a_{32} & a_{33}
\end{pmatrix}
=
\begin{pmatrix}
1 & 0 & 0 \\
l_{21} & 1 & 0 \\
l_{31} & l_{32} & 1
\end{pmatrix}
\begin{pmatrix}
u_{11} & u_{12} & u_{13} \\
0 & u_{22} & u_{23} \\
0 & 0 & u_{33}
\end{pmatrix}. \tag{4.56}
$$

Carrying out the matrix multiplication on the right of Eq. (4.56) and then equating the corresponding elements on both sides of this equation the unknown u's and l's are determined as

$$u_{11} = a_{11}, \quad u_{12} = a_{12}, \quad u_{13} = a_{13}, \quad l_{21} = a_{21}/u_{11}, \tag{4.57a}$$

$$u_{22} = a_{22} - u_{12}l_{21}, \quad u_{23} = a_{23} - l_{21}u_{13}, \tag{4.57b}$$

$$l_{31} = a_{31}/u_{11}, \quad l_{32} = (a_{32} - l_{31}u_{12})/u_{22}, \tag{4.57c}$$

$$u_{33} = a_{33} - l_{31}u_{13} - l_{32}u_{23}. \tag{4.57d}$$

The matrix B is not used in the decomposition of A into LU. Therefore, the obtained LU can be used to solve any number of systems $AX = B$ having same A but different B. This is an advantage of LU over the Gauss elimination and the Gauss–Jordan methods.

In terms of L and U the system $AX = B$ is $LUX = B$. Defining $UX = Y$ the equation $LUX = B$ is written as $LY = B$. Y can be determined by solving the system $LY = B$ and then X by solving the system $UX = Y$. The system $LY = B$ can be solved by forward-substitution while the system $UX = Y$ by back-substitution. It is easy to obtain the determinant of a matrix whose LU form is known. The determinant of A is simply the product of the diagonal elements of U.

Example:

Consider the system (4.28). From Eqs. (4.57) and (4.28) the u's and l's are obtained as

$$u_{11} = u_{12} = u_{13} = 1, \quad u_{22} = -1, \quad u_{23} = 0, \quad u_{33} = -1, \tag{4.58a}$$

$$l_{21} = 2, \quad l_{31} = 3, \quad l_{32} = 4. \tag{4.58b}$$

Thus, the matrices L and U are given by

$$L = \begin{pmatrix} 1 & 0 & 0 \\ 2 & 1 & 0 \\ 3 & 4 & 1 \end{pmatrix}, \quad U = \begin{pmatrix} 1 & 1 & 1 \\ 0 & -1 & 0 \\ 0 & 0 & -1 \end{pmatrix}. \tag{4.59}$$

The system $LY = B$ is

$$\begin{pmatrix} 1 & 0 & 0 \\ 2 & 1 & 0 \\ 3 & 4 & 1 \end{pmatrix} \begin{pmatrix} y_1 \\ y_2 \\ y_3 \end{pmatrix} = \begin{pmatrix} 2 \\ 2 \\ -1 \end{pmatrix}. \tag{4.60}$$

Solving the above system by forward-substitution gives

$$\begin{pmatrix} y_1 \\ y_2 \\ y_3 \end{pmatrix} = \begin{pmatrix} 2 \\ -2 \\ 1 \end{pmatrix}. \tag{4.61}$$

Then, the system $UX = Y$ becomes

$$\begin{pmatrix} 1 & 1 & 1 \\ 0 & -1 & 0 \\ 0 & 0 & -1 \end{pmatrix} \begin{pmatrix} x_1 \\ x_2 \\ x_3 \end{pmatrix} = \begin{pmatrix} 2 \\ -2 \\ 1 \end{pmatrix}. \tag{4.62}$$

The back-substitution in the above system of equations gives the solution as $(x_1, x_2, x_3) = (1, 2, -1)$.

4.8 Tridiagonal Systems

In certain problems, many coefficients in a given system of equations become zero. One such system of practical importance is the system of tridiagonal form

$$\begin{aligned} d_1 x_1 + c_1 x_2 &&&= b_1, \\ a_1 x_1 + d_2 x_2 + c_2 x_3 &&&= b_2, \\ a_2 x_2 + d_3 x_3 + c_3 x_4 &&&= b_3, \\ &\cdots\cdots && \vdots \\ a_{n-2} x_{n-2} + d_{n-1} x_{n-1} + c_{n-1} x_n &&&= b_{n-1}, \\ a_{n-1} x_{n-1} + d_n x_n &&&= b_n. \end{aligned} \tag{4.63}$$

From the appearance of Eq. (4.63) notice that interchanging of equations is not needed. Then, elimination of x_k, $k = 1, 2, \ldots, n-1$ in $(k+1)$th equation using kth equation gives the following system of equation

$$\begin{aligned} d_1' x_1 + c_1' x_2 &&&= b_1', \\ d_2' x_2 + c_2' x_3 &&&= b_2', \\ d_3' x_3 + c_3' x_4 &&&= b_3', \\ &\cdot\quad\cdot && \vdots \\ d_{n-1}' x_{n-1} + c_{n-1}' x_n &&&= b_{n-1}', \\ d_n' x_n &&&= b_n', \end{aligned} \tag{4.64}$$

where

$$d_1' = d_1, \quad b_1' = b_1, \quad c_k' = c_k, \quad k = 1, 2, \ldots, n-1 \tag{4.65a}$$

$$d_k' = d_k - \frac{a_{k-1} c_{k-1}'}{d_{k-1}'}, \quad k = 2, 3, \ldots, n \tag{4.65b}$$

$$b_k' = b_k - \frac{a_{k-1} b_{k-1}'}{d_{k-1}'}, \quad k = 2, 3, \ldots, n. \tag{4.65c}$$

System (4.64) is now in a very simple upper-triangular form and can be solved by back-substitution. The result is

$$x_n = \frac{b_n'}{d_n'}, \quad x_k = \frac{b_k' - c_k' x_{k+1}}{d_k'}, \quad k = n-1, n-2, \ldots, 1. \tag{4.66}$$

Example:

Consider the following system of equations

$$
\begin{aligned}
x_1 &+ 2x_2 & & & & & & & & & & = -1, \\
x_1 &- x_2 &+ x_3 & & & & & & & & & = 4, \\
& 2x_2 &+ 2x_3 &+ 4x_4 & & & & & & & = 10, \\
& & x_3 &- x_4 &+ 2x_5 & & & & & = 0, \\
& & & 4x_4 &- x_5 &+ x_6 & & & = 5, \\
& & & & x_5 &- x_6 & & = 3.
\end{aligned} \tag{4.67}
$$

Elimination of x_k in $(k+1)$th equation for $k = 1, 2, \ldots, 5$ gives

$$
\begin{aligned}
x_1 &+ 2x_2 & & & & & = -1, \\
&- 3x_2 &+ x_3 & & & & = 5, \\
& &\frac{8}{3}x_3 &+ 4x_4 & & & = \frac{40}{3}, \\
& & &- \frac{5}{2}x_4 &+ 2x_5 & & = -5, \\
& & & &\frac{11}{5}x_5 &+ x_6 & = -3, \\
& & & & &- \frac{16}{11}x_6 &= \frac{48}{11}.
\end{aligned} \tag{4.68}
$$

The last equation gives $x_6 = -3$. Substitution of $x_6 = -3$ in the 5th equation gives $x_5 = 0$. In this way, successive back-substitution leads to $x_4 = 2$, $x_3 = 2$, $x_2 = -1$ and $x_1 = 1$. Therefore, the solution is

$$(x_1, \ x_2, \ x_3, \ x_4, \ x_5, \ x_6) = (1, \ -1, \ 2, \ 2, \ 0, \ -3)$$

which is the exact solution.

Tridiagonal systems are realized in solving boundary-value problems of linear partial differential equations (see Chapter 13), interpolation problems, fluid simulation and scattering problems.

4.9 Counting Arithmetic Operations

A quantity useful to compare various direct methods of solving a system of linear equations is the total number of arithmetic operations involved. The method which requires relatively lesser number of arithmetic operations is preferable. Here, arithmetic operations refer only the operations such as addition, subtraction, multiplication and division. It is easy to compute the number of arithmetic operations involved in the direct methods considered so far. In this section, the number of arithmetic operations required in the Gauss and the Gauss–Jordan methods are computed. The same for the triangular factorization is left as an exercise (see Problem 4.15 at the end of this chapter).

4.9.1 Gauss Elimination Method

Let us calculate the number of arithmetic operations required for

1. eliminating the coefficients a_{k1}, $k = 2, 3, \ldots, n$,

2. complete elimination of the coefficients appearing below the diagonal elements and

3. back-substitution.

(i) Number of arithmetic operations for eliminating a_{k1}

The first equation of the system is used to eliminate x_1 in the remaining $(n-1)$ equations. When working with augmented matrix this is equivalent to eliminating a_{k1} in the rows $k = 2, 3, \ldots, n$ using the first row. To eliminate x_1 in the kth equation, the first equation is multiplied by $-a_{k1}/a_{11}$. That is, to eliminate a_{k1} in the kth row of the augmented matrix the first row is multiplied by $-a_{k1}/a_{11}$. Call this quantity as a multiplier, the calculation of this quantity requires one division operation. Therefore, for $(n-1)$ rows totally $(n-1)$ division operations are required for the calculation of $(n-1)$ multipliers. That is,

$$\text{number of division operations} = n - 1. \tag{4.69}$$

After determining the multiplier $-a_{k1}/a_{11}$ the first row of the system (4.2) is multiplied by it. Specifically, the elements a_{1k}, $k = 2, 3, \ldots, n$ and b_1 are multiplied by the multiplier. Then, the resultant row is added to the second row of the augmented matrix. So, for the second row, the number of multiplications is n and addition is also n. This procedure has to be repeated for $(n-1)$ rows appearing below the first row. Therefore,

$$\text{number of multiplication} = n(n-1), \tag{4.70a}$$
$$\text{number of division} = n(n-1). \tag{4.70b}$$

(ii) Number of operations for complete elimination

The numbers specified in Eqs. (4.70) are only for the elimination of a_{k1}, $k = 2, 3, \ldots, n$. The total number of operations for complete elimination of elements appearing below the diagonal elements are

$$\text{total division} = \sum (n-1) = \frac{1}{2}(n-1)n, \tag{4.71a}$$

$$\text{total multiplication} = \sum n(n-1) = \frac{1}{3}(n-1)n(n+1), \tag{4.71b}$$

$$\text{total addition} = \sum n(n-1) = \frac{1}{3}(n-1)n(n+1). \tag{4.71c}$$

(iii) Number of operations in back-substitution

In the back-substitution algorithm

$$\text{multiplication operation} \;=\; 1 + 2 + \cdots + (n-1)$$
$$=\; \frac{1}{2}(n-1)n\,, \tag{4.72a}$$
$$\text{subtraction (from } b_i\text{'s)} \;=\; 1 + 2 + \cdots + (n-1)$$
$$=\; \frac{1}{2}(n-1)n\,, \tag{4.72b}$$
$$\text{division (by the element } a_{kk}) \;=\; n\,. \tag{4.72c}$$

The total number of arithmetic operations, N, is then obtained, by combining Eqs. (4.71) and (4.72), as

$$\begin{aligned}
N \;=\;& \frac{1}{2}(n-1)n + \frac{1}{3}(n-1)n(n+1) + \frac{1}{3}(n-1)n(n+1) \\
& + \frac{1}{2}(n-1)n + \frac{1}{2}(n-1)n + n \\
\;=\;& \frac{1}{6}\left(4n^3 + 9n^2 - 7n\right)\,.
\end{aligned} \tag{4.73}$$

For $n = 2$, 3 and 4 the total number of arithmetic operations are 9, 28 and 62, respectively.

4.9.2   Gauss–Jordan Elimination Method

Next, count the number of arithmetic operations involved in the Gauss–Jordan method and compare it with that of the Gauss elimination method.

(i) Number of arithmetic operations to eliminate a_{k1}

Dividing the first row of the augmented matrix by its pivot element involves n division operation. Note that the element a_{11} need not be divided by a_{11} instead it can be set into 1. The multiplier is simply $-a_{k1}$ and so no operation is involved in calculating it. Then,

$$\text{division operation} \;=\; n\,, \tag{4.74a}$$
$$\text{multiplication operation} \;=\; (n-1)n\,, \tag{4.74b}$$
$$\text{addition operation} \;=\; (n-1)n\,. \tag{4.74c}$$

(ii) Number of operations for complete elimination

The total number of various operations involved in the elimination of all the elements below the diagonal element in the augmented matrix are

$$\text{division operation} \;=\; \sum n = \frac{1}{2}n(n+1)\,, \tag{4.75a}$$
$$\text{multiplication operation} \;=\; \sum (n-1)n = \frac{1}{3}(n-1)n(n+1)\,, \tag{4.75b}$$
$$\text{addition operation} \;=\; \sum (n-1)n = \frac{1}{3}(n-1)n(n+1)\,. \tag{4.75c}$$

In the Gauss–Jordan method back-substitution is avoided by additional calculations. Therefore, next count the number of operations in the additional calculations.

(iii) Number of operations in the additional calculations

Elimination of a_{12} in the first row of the augmented matrix involves

$$\text{multiplication operation} \quad = \quad (n-1) \cdot 1, \qquad (4.76a)$$
$$\text{addition operation} \quad = \quad (n-1) \cdot 1. \qquad (4.76b)$$

Eliminations of a_{13} and a_{23} in the first and second rows, respectively, of the augmented matrix involve

$$\text{multiplication operation} \quad = \quad (n-2) \cdot 2, \qquad (4.77a)$$
$$\text{addition operation} \quad = \quad (n-2) \cdot 2. \qquad (4.77b)$$

Therefore, the total number of additional calculations are

$$\begin{aligned}
N_a \quad &= \quad 2\sum_{k=1}^{n}(n+1-k)(k-1) \\
&= \quad 2\left[(n+1)\sum k - \sum(k^2 - k) - (n+1)\sum 1\right] \\
&= \quad n(n+1)^2 - \frac{1}{6}n(n+1)(2n+1) - n(n+1) \\
&= \quad \frac{1}{3}n\left(n^2 - 1\right).
\end{aligned} \qquad (4.78)$$

Combining the numbers in Eqs. (4.75) and (4.78) gives the total arithmetic operations as

$$N \quad = \quad \frac{1}{2}n(n+1)(2n-1). \qquad (4.79)$$

For $n = 2, 3$ and 4 the total number of arithmetic operations involved in the Gauss–Jordan method are 9, 30 and 70, respectively. For large n,

$$N(\text{Gauss}) \quad \approx \quad \frac{2}{3}n^3, \qquad (4.80a)$$
$$N(\text{Gauss–Jordan}) \quad \approx \quad n^3. \qquad (4.80b)$$

The Gauss–Jordan method, for large n, has additionally $n^3/3$ operations over the Gauss elimination method.

4.10 Iterative Methods

Having studied some direct methods of solving linear system of equations, in this section, the focus is on two indirect or iterative methods, namely, the *Jacobi* and *Gauss–Seidel* methods. For hand computation, the iterative methods have the advantage that they are *self-correcting if an error is made* and they can be used to *reduce the round-off error in the solutions* obtained by the direct methods.

4.10.1 Jacobi Method of Iteration

For the linear system (4.2) the Jacobi iteration scheme is

$$X^{(m+1)} \quad = \quad X^{(m)} + D^{-1}\left(B - AX^{(m)}\right), \quad m = 0, 1, 2, \ldots \qquad (4.81)$$

where $X^{(m)}$ is the mth approximation of X and D be the diagonal of A. When $X^{(m+1)} \approx X^{(m)}$ Eq. (4.81) gives

$$B = AX^{(m)} \tag{4.82}$$

and therefore $X^{(m+1)}$ is the solution of the given system. Since D is the diagonal of A, Eq. (4.81) is written as

$$
\begin{aligned}
x_i^{(m+1)} &= x_i^{(m)} + \frac{1}{a_{ii}} \left(b_i - \sum_{j=1}^{n} a_{ij} x_j^{(m)} \right), \\
&= \frac{1}{a_{ii}} \left(b_i - \sum_{\substack{j=1 \\ j \neq i}}^{n} a_{ij} x_j^{(m)} \right), \quad i = 1, 2, \ldots, n
\end{aligned}
\tag{4.83}
$$

where $x_i^{(0)}$ are the starting values of x_i and they may be chosen as zero. Successive approximations $X^{(m)}$ are assumed to converge towards the exact solution. Note that ith equation is used to solve ith unknown. Since each of the equation in the iterative formula is simultaneously changed by using the most recent set of x values the Jacobi method is also called the method of *simultaneous displacements*.

4.10.2   Gauss–Seidel Method of Iteration

In the Jacobi iteration formula (4.83) $X^{(m+1)}$ is evaluated from the known values of $X^{(m)}$, that is, the old coordinates are used to obtain the new coordinates. In the Gauss–Siedel method the new coordinates are used immediately as soon as they are available. The iterative formula here is

$$x_i^{(m+1)} = \frac{1}{a_{ii}} \left(b_i - \sum_{j<i} a_{ij} x_j^{(m+1)} - \sum_{j>i} a_{ij} x_j^{(m)} \right), \quad i = 1, 2, \ldots, n. \tag{4.84}$$

4.10.3   Convergence Criteria

The convergence of the Gauss–Seidel method is more rapid than the Jacobi method because the unknown $x_i^{(m+1)}$ is used in the calculation of all succeeding unknowns x_{i+r}, $r = 1, 2, \ldots, n - i$. The rate of the Gauss–Seidel method is roughly twice that of the Jacobi method.

Now, find out the condition for convergence of the Jacobi and the Gauss–Seidel methods. Consider the Jacobi iteration rule (4.83). Denote $\bar{X}$ as the exact solution. Then, the iteration rule with $\bar{X}$ is

$$\bar{x}_i = \frac{1}{a_{ii}} \left(b_i - \sum a_{ij} \bar{x}_j \right). \tag{4.85}$$

Subtraction of (4.83) from (4.85) gives

$$
\begin{aligned}
\xi_i^{(m+1)} &= \bar{x}_i - x_i^{(m+1)} \\
&= -\frac{1}{a_{ii}} \sum a_{ij} \xi_j^{(m)}, \quad \xi_j^{(m)} = \bar{x}_j - x_j^{(m)}.
\end{aligned}
\tag{4.86}
$$

For convergence the requirement is

$$\left\|\xi_i^{(m+1)}\right\| \leq \frac{1}{|a_{ii}|} \sum |a_{ij}| \left\|\xi_j^{(m)}\right\|$$

$$\leq \frac{1}{|a_{ii}|} \left(\sum |a_{ij}|\right) \left\|\xi_j^{(m)}\right\|$$

$$\leq A \left\|\xi_j^{(m)}\right\|, \tag{4.87}$$

where

$$A = \max_i \frac{1}{|a_{ii}|} \sum |a_{ij}|. \tag{4.88}$$

If $A < 1$ the successive $\|\xi_i\|$'s decrease. The condition $A < 1$ gives

$$\sum_{j \neq i} |a_{ij}| < |a_{ii}|. \tag{4.89}$$

A matrix with the above property is said to be *diagonally dominant*. The above condition is also the condition for the convergence of the Gauss–Seidel method. Both the methods give convergent solution if the coefficient matrix A is *strictly diagonally dominant*. That is,

$$|a_{ii}| > \sum_{j \neq i} |a_{ij}|, \quad i = 1, 2, \ldots, n. \tag{4.90}$$

In other words, in each row of A, the absolute value of the diagonal element must be greater than the sum of the other elements. The above criterion can also be expressed as

$$2|a_{ii}| > \sum_{j=1}^{n} |a_{ij}|, \quad i = 1, 2, \ldots, n. \tag{4.91}$$

If this condition is violated then successive iterations diverge from the actual solution. Note that if the above condition is not satisfied in the given order of appearance of the equations, one must check whether the above condition is satisfied if the order of occurrence of the equations is altered. An example for this is given below. The iteration can be stopped if the Euclidean distance between $X^{(m)}$ and $X^{(m+1)}$ given by

$$d = \left[\sum_{i=1}^{n} \left(x_i^{(m+1)} - x_i^{(m)} \right)^2 \right]^{1/2} \tag{4.92}$$

is less than a preassumed tolerance δ.

There are some examples of system of the form $AX = B$, where the matrix A does not have diagonal dominance but still the Jacobi and Gauss–Seidel methods do converge. For example, the Gauss–Seidel method will converge from any starting initial guess if A is symmetric and positive definite, that is, $A = A^T$ and $X^T AX > 0$ for all nonzero X. On the other hand, when A has diagonal elements that are all positive and off-diagonal elements that are all negative then both the methods will either converge or diverge.

Example 1:

Find the solution of the system

$$6x_1 - x_2 + 2x_3 = 2, \tag{4.93a}$$
$$x_1 + 5x_2 + x_3 = 10, \tag{4.93b}$$
$$2x_1 + x_2 + 7x_3 = -3 \tag{4.93c}$$

by (i) the Jacobi and (ii) the Gauss–Seidel methods.

In order to apply these two methods first verify whether the system is diagonally dominant. A system of three equations is said to be *diagonally stable* if the following conditions are satisfied:

$$2|a_{11}| > |a_{11}| + |a_{12}| + |a_{13}|, \tag{4.94a}$$

$$2|a_{22}| > |a_{21}| + |a_{22}| + |a_{23}|, \tag{4.94b}$$

$$2|a_{33}| > |a_{31}| + |a_{32}| + |a_{33}|. \tag{4.94c}$$

For the given system Eqs. (4.94) are $12 > 9$, $10 > 7$, $14 > 10$. Thus, the system is diagonally stable. Therefore, one can solve the system by the two methods.

Solution by the Jacobi method

The Jacobi iteration rule for a system of three equations is

$$x_1^{(m+1)} = \frac{1}{a_{11}} \left(b_1 - a_{12}x_2^{(m)} - a_{13}x_3^{(m)} \right), \tag{4.95a}$$

$$x_2^{(m+1)} = \frac{1}{a_{22}} \left(b_2 - a_{21}x_1^{(m)} - a_{23}x_3^{(m)} \right), \tag{4.95b}$$

$$x_3^{(m+1)} = \frac{1}{a_{33}} \left(b_3 - a_{31}x_1^{(m)} - a_{32}x_2^{(m)} \right). \tag{4.95c}$$

For the given system the above rule becomes

$$x_1^{(m+1)} = \frac{1}{6} \left(2 + x_2^{(m)} - 2x_3^{(m)} \right), \tag{4.96a}$$

$$x_2^{(m+1)} = \frac{1}{5} \left(10 - x_1^{(m)} - x_3^{(m)} \right), \tag{4.96b}$$

$$x_3^{(m+1)} = \frac{1}{7} \left(-3 - 2x_1^{(m)} - x_2^{(m)} \right). \tag{4.96c}$$

Table 4.1 displays the iterated values of x_1, x_2 and x_3. The initial values of x_1, x_2 and x_3 are all chosen as 0. After 7th iteration the solution with three decimal points accuracy is $(x_1, x_2, x_3) = (1, 2, -1)$ which is the exact solution.

Solution by the Gauss–Seidel method

The Gauss–Seidel iteration formula for a system of three equations is

$$x_1^{(m+1)} = \frac{1}{a_{11}} \left(b_1 - a_{12}x_2^{(m)} - a_{13}x_3^{(m)} \right), \tag{4.97a}$$

$$x_2^{(m+1)} = \frac{1}{a_{22}} \left(b_2 - a_{21}x_1^{(m+1)} - a_{23}x_3^{(m)} \right), \tag{4.97b}$$

$$x_3^{(m+1)} = \frac{1}{a_{33}} \left(b_3 - a_{31}x_1^{(m+1)} - a_{32}x_2^{(m+1)} \right). \tag{4.97c}$$

For the given system the algorithm is

$$x_1^{(m+1)} = \frac{1}{6} \left(2 + x_2^{(m)} - 2x_3^{(m)} \right), \tag{4.98a}$$

$$x_2^{(m+1)} = \frac{1}{5} \left(10 - x_1^{(m+1)} - x_3^{(m)} \right), \tag{4.98b}$$

$$x_3^{(m+1)} = \frac{1}{7} \left(-3 - 2x_1^{(m+1)} - x_2^{(m+1)} \right). \tag{4.98c}$$

TABLE 4.1
Successive iterated values of x_1, x_2 and x_3 of the system (4.93) by the Jacobi method. Exact solution is $(x_1, x_2, x_3) = (1, 2, -1)$.

Iteration number m	x_1	x_2	x_3
0	0.0000000	0.0000000	0.0000000
1	0.3333333	2.0000000	−0.4285714
2	0.8095238	2.0190476	−0.8095238
3	0.9396825	2.0000000	−0.9482993
4	0.9827664	2.0017234	−0.9827664
5	0.9945427	2.0000000	−0.9953223
6	0.9984408	2.0001559	−0.9984408
7	0.9995062	2.0000000	−0.9995768
8	0.9998589	2.0000141	−0.9998589
9	0.9999553	2.0000000	−0.9999617
10	0.9999872	2.0000013	−0.9999872
11	0.9999960	2.0000000	−0.9999965

TABLE 4.2
Successive iterated values of x_1, x_2 and x_3 of the system (4.93) by the Gauss–Seidel method. Exact solution is $(x_1, x_2, x_3) = (1, 2, -1)$.

Iteration number m	x_1	x_2	x_3
0	0.0000000	0.0000000	0.0000000
1	0.3333333	1.9333334	−0.8000001
2	0.9222223	1.9755557	−0.9742858
3	0.9873545	1.9973863	−0.9960136
4	0.9982356	1.9995556	−0.9994324
5	0.9997368	1.9999392	−0.9999161
6	0.9999619	1.9999908	−0.9999878
7	0.9999945	1.9999987	−0.9999982
8	0.9999992	1.9999998	−0.9999997

Table 4.2 gives the iterated values of x_1, x_2 and x_3 with the initial guess $x_1^{(0)}$, $x_2^{(0)}$ and $x_3^{(0)}$ all are 0. After 5 iterations the solution with three decimal point accuracy is $(x_1, x_2, x_3) = (1, 2, -1)$ which is the exact solution. In the Jacobi method after fifth iteration the solution is $(0.995, 2.000, -0.995)$. This clearly shows that convergence in the Gauss–Seidel algorithm is much faster than in the Jacobi method.

TABLE 4.3

First few iterated values of x_1, x_2 and x_3 of the system (4.99) by the Jacobi method.

Iteration number m	x_1	x_2	x_3
0	0.00000	0.00000	0.00000
1	−1.50000	−2.00000	10.00000
2	−35.50000	9.00000	21.50000
3	−81.25000	−172.00000	0.50000
4	82.75000	−488.50000	951.25000

Example 2:

Verify that the iterated values of the system

$$2x_1 + x_2 + 7x_3 \;=\; -3\,, \tag{4.99a}$$

$$6x_1 - x_2 + 2x_3 \;=\; 2\,, \tag{4.99b}$$

$$x_1 + 5x_2 + x_3 \;=\; 10 \tag{4.99c}$$

obtained by the Jacobi and the Gauss–Seidel methods diverge.

The above system is same as the system (4.93) but the order of occurrence of the equations is changed. The first few iterated values of x_1, x_2 and x_3 by the Jacobi method are given in table 4.3.

The divergence is due to the fact that the system is not diagonally dominant (verify). (The Gauss–Seidel algorithm also gives diverging values for x_1, x_2 and x_3. This is left as an exercise to the reader.) However, if one reorders the system (4.99) as in (4.93) then it is diagonally dominant. Then, both the methods give convergent solution as seen in the Example 1.

4.11   System $AX = B$ with A being Vandermonde Matrix

A system of equations $AX = B$ with upper- and lower-triangular forms of A discussed earlier in Section 4.3 are two special cases of a linear system of equations. Another special system is the one in which A is the so-called *Vandermonde matrix*, named after Alexandre-Théophile Vandermonde.

A Vandermonde matrix is an $N \times N$ matrix whose elements are expressed in a simple manner in terms of N numbers say $a_1, a_2, \ldots, a_N$. The N^2 components of a Vandermonde matrix A are simply a_i^{j-1}, $i, j = 1, 2, \ldots, N$. The elements of a Vandermonde matrix follow a geometric progression. The system takes the form

$$\begin{pmatrix} 1 & a_1 & a_1^2 & \cdots & a_1^{N-1} \\ 1 & a_2 & a_2^2 & \cdots & a_2^{N-1} \\ \vdots & \vdots & \vdots & \vdots & \vdots \\ 1 & a_N & a_N^2 & \cdots & a_N^{N-1} \end{pmatrix} \begin{pmatrix} x_1 \\ x_2 \\ \vdots \\ x_N \end{pmatrix} = \begin{pmatrix} b_1 \\ b_2 \\ \vdots \\ b_N \end{pmatrix}. \tag{4.100}$$

Such types of equations arise in the problem of constructing a polynomial interpolation function. Vandermonde matrices are realized in the discrete Fourier transform analysis of digital signal processing, matrix models of quantum field theories, representation theory of the symmetric groups, theory of BCH code and time series analysis.

The system (4.100) can be solved as follows [4]. Let P_j be the Lagrange polynomial of order $N - 1$ given by

$$P_j(a) = \prod_{\substack{n=1 \\ n \neq j}}^{N} \frac{a - a_n}{a_j - a_n} = \sum_{k=1}^{N} A_{jk} a^{k-1}. \tag{4.101}$$

From Eq. (4.101), it is easy to find that $P = 0$ at all $a = a_i$ with $i \neq j$ and is 1 at $a = a_j$. That is,

$$P_j(a_i) = \sum_{k=1}^{N} A_{jk} a_i^{k-1}. \tag{4.102}$$

Then, x_j's are given by

$$x_j = \sum_{k=1}^{N} A_{kj} b_k. \tag{4.103}$$

To get A_{jk} define a master polynomial

$$P_a = \prod_{n=1}^{N} (a - a_n). \tag{4.104}$$

Then, proceed to calculate its coefficients and obtain the numerators and denominators of the specific P_j's via synthetic division by the one supernumerary term. Note that the total procedure is of the order N^2. The system (4.100) can be solved by the Gauss–Jordan method also which requires operations of the order of N^3. The method described above is specially developed for the system (4.100).

Example:

Solve the system

$$x_1 + x_2 + x_3 = 2, \tag{4.105a}$$
$$x_1 - x_2 + x_3 = 4, \tag{4.105b}$$
$$x_1 + 3x_2 + 9x_3 = 16. \tag{4.105c}$$

In the form of Eq. (4.100) the above system is written as

$$\begin{pmatrix} 1 & 1 & 1 \\ 1 & -1 & 1 \\ 1 & 3 & 9 \end{pmatrix} \begin{pmatrix} x_1 \\ x_2 \\ x_3 \end{pmatrix} = \begin{pmatrix} 2 \\ 4 \\ 16 \end{pmatrix}. \tag{4.106}$$

The coefficients a's are $a_1 = 1$, $a_2 = -1$, $a_3 = 3$. Develop a Python program that reads the number of equations N, the coefficients $a_1, a_2, \ldots, a_N$ and $b_1, b_2, \ldots, b_N$ and solve the system $AX = B$ with A being Vandermonde matrix employing the method described in the present section. The solution obtained is $(x_1, x_2, x_3) = (1, -1, 2)$.

4.12   Ill-Conditioned Systems

Consider the system

$$0.1x + y \;=\; 2.1, \tag{4.107a}$$

$$x + 9y \;=\; 19. \tag{4.107b}$$

Equation (4.107) can be expressed in matrix form as $AX = B$ with

$$A = \begin{pmatrix} 0.1 & 1 \\ 1 & 9 \end{pmatrix}, \quad B = \begin{pmatrix} 2.1 \\ 19 \end{pmatrix}, \quad X = \begin{pmatrix} x \\ y \end{pmatrix}. \tag{4.108}$$

The solution of Eqs. (4.107) is $(x, y) = (1, 2)$. Now, slightly change the coefficient of x in Eq. (4.107a) as 0.11. The solution becomes $(x, y) = (10, 1)$. When the coefficient 0.1 is replaced by 0.111 the result is $(x, y) = (100, -9)$. Replacement of the coefficient of y in Eq. (4.107a) by 1.1 gives $(x, y) = (10, 1)$. A similar result can be noticed when the coefficient of x or y is altered in Eq. (4.107b). A small change in the coefficient of x or y in Eqs. (4.107) produces a large change in the solution. Such systems of linear equations are called *ill-conditioned systems* [6-8]. In general, in an ill-conditioned system of the form $AX = B$ a small change in the elements of A (as well as in B) leads to a large change in the solution. The corresponding matrix A is termed as an *ill-conditioned matrix*. Note that in certain problems of the form $AX = B$ one may wish to find the solution for fixed A (B) for slightly different set of B (A). In such a case ill-conditioning has to be taken into account.

Ill-conditioned system of equations or ill-conditioned matrices occur in the analysis of multiple scattering problems, inversion problem of the heat capacity of crystals and the determination of the momentum distribution of cosmic-ray muons [9], a few to mention.

4.12.1   Condition of a System of Linear Equations

For ill-conditioning, one has to find what happens to $X + \Delta X$ when A is changed into $A + \Delta A$ (or B into $B + \Delta B$) with infinitesimally small ΔA [7]. Replacement of A by $A + \Delta A$ and X by $X + \Delta X$ in $AX = B$ and neglecting $\Delta A \Delta X$ gives

$$A\Delta X + (\Delta A)X = 0. \tag{4.109}$$

Thus,

$$||\Delta X|| \leq ||A^{-1}|| \, ||\Delta A|| \, ||X|| \tag{4.110}$$

with the norms of ΔX and A are given by

$$||\Delta X|| = \left[\sum_{i=1}^{n} |\Delta x_i|^2 \right]^{1/2}, \quad ||A|| = \left[\sum_{i=1}^{n} \sum_{j=1}^{n} |a_{ij}|^2 \right]^{1/2}, \tag{4.111}$$

where x_i's are the components of X and a_{ij}'s are the elements of the $n \times n$ matrix A. Equation (4.111) can be rewritten as

$$\frac{||\Delta X||}{||X||} \leq ||A|| \, ||A^{-1}|| \frac{||\Delta A||}{||A||} \tag{4.112}$$

or

$$\frac{||\Delta X||}{||X||} \leq \kappa(A) \frac{||\Delta A||}{||A||}, \quad \kappa(A) = ||A|| \, ||A^{-1}||. \tag{4.113}$$

Amplification of small perturbation of A is determined by $\kappa(A)$ called the *condition number* [7]. A perturbation is not amplified if $\kappa(A)$ is small and the system is a well-conditioned one. For an ill-conditioned system $\kappa(A)$ will be large. The value of $\kappa(A)$ lies in the interval $[0, \infty]$. If the condition number is < 100 then the system can be treated as a well-conditioned system.

Another parameter one can consider is $\gamma(A)$ given by

$$\gamma(A) = \frac{|A|}{r_1 r_2 \cdots r_n}, \tag{4.114a}$$

where $|A|$ is the determinant of A and

$$r_i = \left[\sum_{i=1}^{n} |a_{ij}|^2 \right]^{1/2}, \quad i = 1, 2, \ldots, n. \tag{4.114b}$$

$\gamma(A)$ lies between -1 and 1. If $\gamma(A)$ is close to 0 then A is an ill-conditioned, otherwise, well-conditioned. $\gamma(A)$ describes how small the determinant of A.

Can one use $|A| \approx 0$ as a requirement for ill-conditioned behaviour of the system $AX = B$? Consider a system with $|A| \approx 0$. For any system of linear equations, multiplication of any equation by a number will not affect the solution but affect the value of $|A|$. For example, for Eqs. (4.107) the $|A|$ is -0.1, close to zero. Suppose, multiply Eqs. (4.107) by 10^5. Then, $|A| = -0.1 \times 10^{10}$ which is not close to zero. The point is that $|A|$ can be freely changed to any order without altering the solution. Thus, $|A| \approx 0$ is unsuitable to identify whether a system is ill-conditioned or well-conditioned [8]. *What is the condition to be used for identifying ill-conditioned system if A is fixed and B is varied?*

In the following subsection, let us consider a few systems, compute $\kappa(A)$ and $\gamma(A)$, identify the ill- or well-conditioned behaviour and verify the prediction.

4.12.2 Examples of Ill- and Well-Conditioned Systems

As a first example, consider the system (4.107) with A and B given by Eq. (4.108) and $(x, y) = (1, 2)$. For this A

$$|A| = -0.1, \quad A^{-1} = 10 \begin{pmatrix} -9 & 1 \\ 1 & -0.1 \end{pmatrix}, \quad ||A|| = \sqrt{83.01}, \tag{4.115a}$$

$$||A^{-1}|| = 10\sqrt{83.01}, \quad r_1 = \sqrt{1.01}, \quad r_2 = \sqrt{82} \tag{4.115b}$$

and

$$\kappa(A) = 830.01, \quad \gamma(A) = -0.01099. \tag{4.116}$$

As $\kappa(A)$ is very large the system is ill-conditioned. Further, $\gamma(A)$ is close to zero. Change the coefficient of x in Eq. (4.107a) to 0.11. The solution becomes $(x, y) = (10, 1)$ which is a large change. The system is thus verified as an ill-conditioned system.

The second system is

$$0.3x + y = 2.3, \tag{4.117a}$$

$$x + 3y = 7 \tag{4.117b}$$

with $(x, y) = (1, 2)$. With $\begin{pmatrix} 0.3 & 1 \\ 1 & 3 \end{pmatrix}$ one has

$$|A| = -0.1, \quad A^{-1} = 10 \begin{pmatrix} -3 & 1 \\ 1 & -0.3 \end{pmatrix}, \quad ||A|| = \sqrt{11.09}, \tag{4.118a}$$

$$||A^{-1}|| = 10\sqrt{11.09}, \quad r_1 = \sqrt{1.09}, \quad r_2 = \sqrt{10} \tag{4.118b}$$

and

$$\kappa(A) = 110.9, \quad \gamma(A) = -0.03029. \tag{4.119}$$

As $\kappa(A)$ is large and $\gamma(A)$ is ≈ 0 the system (4.117) is an ill-conditioned system. Changing 0.3 in Eq. (4.117a) by 0.31 and 0.33 give the solutions as $(x, y) = (1.42857, 1.85714)$ and $(x, y) = (10, -1)$, respectively. The changes in the solutions are large.

For the third system

$$x + 2y \quad = \quad 5, \tag{4.120a}$$
$$2.2x + 4.5y \quad = \quad 11.2 \tag{4.120b}$$

with $(x, y) = (1, 2)$ and $A = \begin{pmatrix} 1 & 2 \\ 2.2 & 4.5 \end{pmatrix}$ the values of $\kappa(A)$ and $\gamma(A)$ are obtained as 300.9 and 0.00893, respectively, implying that the given system is an ill-conditioned one. Change of 2.2 in Eq. (4.120b) to 2.3 leads to the solution as $(x, y) = (-1, 3)$ confirming the ill-conditioned behaviour of the system.

Next, consider two systems that are well-conditioned. The first example system is

$$3x + 5y \quad = \quad 13, \tag{4.121a}$$
$$-x + y \quad = \quad 1, \tag{4.121b}$$

where $(x, y) = (1, 2)$. For this system

$$|A| = 8, \quad A^{-1} = \frac{1}{8} \begin{pmatrix} 1 & -5 \\ 1 & 3 \end{pmatrix}, \quad \kappa(A) = 4.25, \quad \gamma(A) = 0.97014. \tag{4.122}$$

As $\kappa(A)$ is small and $\gamma(A)$ is not close to 0, the above system is a well-conditioned system. Changing the coefficient of x in Eq. (4.121a) to (a) 3.03 and (b) 3.3 leads to the solutions (a) $(x, y) = (0.99626, 1.99626)$ and (b) $(x, y) = (0.96386, 1.96386)$, respectively. The changes are very small.

The second well-conditioned system with the solution $(x, y) = (1, 2)$ is

$$0.1x + 0.2y \quad = \quad 0.5, \tag{4.123a}$$
$$0.3x + 0.4y \quad = \quad 1.1. \tag{4.123b}$$

The determinant of the coefficient matrix A of this system is -0.02 and is ≈ 0. A^{-1}, $\kappa(A)$ and $\gamma(A)$ are obtained as

$$A^{-1} = \begin{pmatrix} -20 & 10 \\ 15 & -5 \end{pmatrix}, \quad \kappa(A) = 15, \quad \gamma(A) = -0.17889. \tag{4.124}$$

Since $\kappa(A)$ is small and $\gamma(A)$ is not close to 0 the system (4.123) is a well-conditioned system even though $|A| \approx 0$. The change of the coefficient of x in Eq. (4.123a) to 0.101 and 0.11 change the solution to $(x, y) = (1.02041, 1.98469)$ and $(x, y) = (1.25, 1.8125)$, respectively. The changes in the solution are not large.

4.13 Homogeneous Systems with Equal Number of Equations and Unknowns

For systems of the form (4.2) with $B = 0$, $\det A \neq 0$ the unique solution is the trivial solution $X = 0$. Nontrivial solution exists when $\det A = 0$. To balance a chemical reaction equation one can set up homogeneous system of equations for the number of atoms of the elements in the chemical reactions. Finding the equilibrium points of autonomous coupled linear differential equations essentially involves solving a system of homogeneous equations.

Example 1:

Consider the unbalanced chemical reaction $Na_2O + H_2O \rightarrow NaOH$. To balance this reaction introduce the coefficients and determine them. For this purpose introduce the vector notation in the matrix form $\begin{pmatrix} Na \\ O \\ H \end{pmatrix}$. Now, write the given reaction in equation form as

$$x_1 \begin{pmatrix} 2 \\ 1 \\ 0 \end{pmatrix} \begin{pmatrix} Na \\ O \\ H \end{pmatrix} + x_2 \begin{pmatrix} 0 \\ 1 \\ 2 \end{pmatrix} \begin{pmatrix} Na \\ O \\ H \end{pmatrix} = x_3 \begin{pmatrix} 1 \\ 1 \\ 1 \end{pmatrix} \begin{pmatrix} Na \\ O \\ H \end{pmatrix}. \tag{4.125}$$

This gives the equations for the coefficients x_1, x_2 and x_3 as

$$2x_1 - x_3 = 0, \quad x_1 + x_2 - x_3 = 0, \quad 2x_2 - x_3 = 0. \tag{4.126}$$

The determinant of the coefficient matrix is 0. Elimination of x_1 in the last two subequations gives

$$2x_1 - x_3 = 0, \quad 2x_2 - x_3 = 0, \quad 2x_2 - x_3 = 0. \tag{4.127}$$

The last two subequations are identical and hence there are only two equations for the three variables which are

$$2x_1 - x_3 = 0, \quad 2x_2 - x_3 = 0. \tag{4.128}$$

Thus, $x_2 = x_1$ and $x_3 = 2x_2$. The choice $x_1 = 1$ gives $x_2 = 1$ and $x_3 = 2$. Then, from Eq. (4.125) the balanced reaction equation is $Na_2O + H_2O \rightarrow 2NaOH$.

Example 2:

Consider the system

$$x_1 + x_2 + x_3 = 0, \tag{4.129a}$$
$$5x_1 - x_2 + x_3 = 0, \tag{4.129b}$$
$$7x_1 - 2x_2 + x_3 = 0. \tag{4.129c}$$

The determinant of its coefficient matrix A is zero. It can have a nontrivial solution. The system after eliminating x_1 in the last two equations by Gauss elimination is

$$x_1 + x_2 + x_3 = 0, \tag{4.130a}$$
$$-6x_2 - 4x_3 = 0, \tag{4.130b}$$
$$-9x_2 - 6x_3 = 0. \tag{4.130c}$$

The third equation in (4.130) is a multiple (3/2 times) of the second equation. Therefore, essentially the system (4.130) has only two equations, say, (4.130a) and (4.130c). Elimination of x_3 in the first equation using the last equation gives the system

$$2x_1 - x_2 = 0, \tag{4.131a}$$
$$3x_2 + 2x_3 = 0. \tag{4.131b}$$

Now, there are two equations for three unknowns. Therefore, one variable is arbitrary. Let it be x_3 and $x_3 = \alpha$. Then, Eq. (4.131) gives $x_1 = -\alpha/3$, $x_2 = -2\alpha/3$. Thus, the solution is

$$X = \begin{pmatrix} x_1 \\ x_2 \\ x_3 \end{pmatrix} = \alpha \begin{pmatrix} -1/3 \\ -2/3 \\ 1 \end{pmatrix}. \tag{4.132}$$

4.14   Least-Squares Problem

Let the number of variables say m is less than the number of equations n. The solution of such systems can be obtained as follows.

From Eq. (4.2) one can write $X = A^{-1}B$. For the present case, A is not a square matrix. However, one can define a generalized inverse, which can be used to solve the set of equations. One can note that $A^T A$ is a square matrix with m rows and columns. Now, multiplication of Eq. (4.2) by A^T gives

$$A^T A X = A^T B. \tag{4.133}$$

Since, $A^T A$ is a square matrix, from Eq. (4.133) one can obtain

$$X = \left[A^T A\right]^{-1} A^T B. \tag{4.134}$$

Let us solve the system

$$x_1 + x_2 = 3, \quad 2x_1 + x_2 = 4, \quad x_1 - x_2 = -1 \tag{4.135}$$

by the above procedure. The matrix A for the given system is

$$A = \begin{pmatrix} 1 & 1 \\ 2 & 1 \\ 1 & -1 \end{pmatrix}. \tag{4.136}$$

Then,

$$\begin{aligned}
X &= \left[\begin{pmatrix} 1 & 2 & 1 \\ 1 & 1 & -1 \end{pmatrix} \begin{pmatrix} 1 & 1 \\ 2 & 1 \\ 1 & -1 \end{pmatrix} \right]^{-1} \begin{pmatrix} 1 & 2 & 1 \\ 1 & 1 & -1 \end{pmatrix} \begin{pmatrix} 3 \\ 4 \\ -1 \end{pmatrix} \\[2mm]
&= \begin{pmatrix} 6 & 2 \\ 2 & 3 \end{pmatrix}^{-1} \begin{pmatrix} 10 \\ 8 \end{pmatrix} \\[2mm]
&= \frac{1}{14} \begin{pmatrix} 3 & -2 \\ -2 & 6 \end{pmatrix} \begin{pmatrix} 10 \\ 8 \end{pmatrix} \\[2mm]
&= \begin{pmatrix} 1 \\ 2 \end{pmatrix}.
\end{aligned} \tag{4.137}$$

4.15   Singular-Value Problems

Is it possible to find a nontrivial solution for a homogeneous/inhomogeneous linear system with fewer equations than unknowns? The answer is yes. In this section, a method of solving this type of systems is outlined. Note that the method described in the previous section for more equations than variables does not work for more variables with a lesser number of equations. This is because for the latter case the size of the square matrix $A^T A$ is larger than the number of equations and $A^T A$ is always singular.

4.15.1   Homogeneous Systems

First, consider the case of two unknowns but one equation. Let it be

$$a_{11}x_1 + a_{12}x_2 = 0. \tag{4.138}$$

The nontrivial solutions are

1. $x_1 = 0$, $x_2 = $ arbitrary if $a_{12} = 0$.
2. $x_1 = a_{12}$, $x_2 = -a_{11}$, if $a_{11}, a_{12} \neq 0$.

Next, consider the case of n equations and m unknowns, where $n < m$. Then, the matrix A in $AX = 0$ is $n \times m$ matrix. Choose a nonzero element in the mth column of A and let it be say a_{im}. Now, construct a $n \times (m-1)$ matrix B with its jth column given by

$$b_j = a_j - \frac{a_{ij}}{a_{im}}a_m, \quad j = 1, 2, \ldots, m-1. \tag{4.139}$$

For each j, the ith element of b_j is

$$a_{ij} - \frac{a_{ij}}{a_{im}}a_{im} = a_{ij} - a_{ij} = 0. \tag{4.140}$$

That is, ith equation of $BX = 0$ is

$$0.x_1 + 0.x_2 + \cdots + 0.x_{n-1} = 0. \tag{4.141}$$

Ignoring the ith equation one can obtain the system

$$\widehat{B}X = 0 \tag{4.142}$$

which is a linear homogeneous system with $n-1$ equations in $m-1$ unknowns. Since Eq. (4.142) has fewer equations than unknowns it has nontrivial solution. Consequently, $BX = 0$ also has nontrivial solution and so the equivalent system $AX = 0$ has a nontrivial solution.

Example 1:

Consider the system

$$x_1 + x_2 + x_3 = 0, \tag{4.143a}$$
$$x_1 - x_2 + 2x_3 = 0. \tag{4.143b}$$

For the system (4.143) $n = 2$ and $m = 3$. Since $a_{13} = 1 \neq 0$ one may choose $i = 1$ and obtain

$$b_1 = a_1 - \frac{a_{11}}{a_{13}}a_3 = \begin{pmatrix} 1 \\ 1 \end{pmatrix} - \frac{1}{1}\begin{pmatrix} 1 \\ 2 \end{pmatrix} = \begin{pmatrix} 0 \\ -1 \end{pmatrix}, \tag{4.144a}$$

$$b_2 = a_2 - \frac{a_{12}}{a_{13}}a_3 = \begin{pmatrix} 0 \\ -3 \end{pmatrix}. \tag{4.144b}$$

Then, the system $BX = 0$ is

$$0.x_1 + 0.x_2 = 0, \qquad (4.145a)$$
$$-x_1 - 3x_2 = 0. \qquad (4.145b)$$

Ignoring the first equation of the system (4.145), one has $\widehat{B}X = 0$ as

$$x_1 + 3x_2 = 0. \qquad (4.146)$$

For simplicity choose $x_2 = 1$ which gives $x_1 = -3$. Then, from Eq. (4.143) x_3 is found to be 2.

Example 2:

Balance the chemical reaction equation $C_5H_8 + O_2 \rightarrow CO_2 + H_2O$.

To set up a system of equations for the given chemical reaction write it as

$$x_1 \begin{pmatrix} 5 \\ 8 \\ 0 \end{pmatrix} \begin{pmatrix} C \\ H \\ O \end{pmatrix} + x_2 \begin{pmatrix} 0 \\ 0 \\ 2 \end{pmatrix} \begin{pmatrix} C \\ H \\ O \end{pmatrix}$$

$$= x_3 \begin{pmatrix} 1 \\ 0 \\ 2 \end{pmatrix} \begin{pmatrix} C \\ H \\ O \end{pmatrix} + x_4 \begin{pmatrix} 0 \\ 2 \\ 1 \end{pmatrix} \begin{pmatrix} C \\ H \\ O \end{pmatrix}. \qquad (4.147)$$

From this equation write

$$5x_1 - x_3 = 0, \quad 4x_1 - x_4 = 0, \quad 2x_2 - 2x_3 - x_4 = 0. \qquad (4.148)$$

Suppose solve the above system by the Gauss–Jordan method. This gives

$$5x_1 - x_3 = 0, \quad 2x_2 - 2x_3 - x_4 = 0, \quad 4x_3 - 5x_4 = 0. \qquad (4.149)$$

There are three equations but four unknowns. Therefore, choose $x_1 = 1$ which gives $x_2 = 7$, $x_3 = 5$ and $x_4 = 4$. Then, the balanced chemical reaction is $C_5H_8 + 7O_2 \rightarrow 5CO_2 + 4H_2O$.

4.15.2   Inhomogeneous Systems

Next, consider the system $AX = B$ with $B \neq 0$ [5]. For a system with n equations involving m variables and when $m > n$, one can eliminate $(n-1)$ variables to obtain a single equation in $(m - n + 1)$ variables. Such an equation represents a hyper-surface in an $(m - n + 1)$-dimensional space, for example, a line on a plane or a plane in a three-dimensional space. The singular-value solution is defined as that point on the hyper-surface which lies closest to the origin. Once the solution to the single equation is obtained then the values of the remaining variables can be found by back-substitution.

Let the single equation is with two variables. This equation can be of the form, say,

$$\frac{x_1}{a} + \frac{x_2}{b} = 1. \qquad (4.150a)$$

The point which is closest to the origin will lie on the line

$$ax_1 - bx_2 = 0. \qquad (4.150b)$$

Thus, the singular-value problem is now reduced to the solution of the system (4.150). In matrix form, the system (4.150) is written as

$$\begin{pmatrix} 1/a & 1/b \\ a & -b \end{pmatrix} \begin{pmatrix} x_1 \\ x_2 \end{pmatrix} = \begin{pmatrix} 1 \\ 0 \end{pmatrix}. \tag{4.151}$$

The determinant D of the square matrix in Eq. (4.151) is

$$D = -\frac{(a^2 + b^2)}{ab} \neq 0. \tag{4.152}$$

Thus, the system (4.151) will have a nontrivial solution.

When the reduced single equation with three variables is, say, of the form

$$\frac{x_1}{a} + \frac{x_2}{b} + \frac{x_3}{c} = 1 \tag{4.153}$$

which is the equation of a plane, then the point on this plane which is closest to the origin lies on the planes

$$ax_1 - cx_3 = 0, \quad bx_2 - cx_3 = 0. \tag{4.154}$$

That is, one has the system

$$\begin{pmatrix} 1/a & 1/b & 1/c \\ a & 0 & -c \\ 0 & b & -c \end{pmatrix} \begin{pmatrix} x_1 \\ x_2 \\ x_3 \end{pmatrix} = \begin{pmatrix} 1 \\ 0 \\ 0 \end{pmatrix}. \tag{4.155}$$

The determinant of the square matrix in the system (4.155) is

$$D = \frac{(a^2 b^2 + b^2 c^2 + c^2 a^2)}{abc} \neq 0. \tag{4.156}$$

Thus, the system (4.155) can have a nontrivial solution.

Example:

Let us find the solution of the system

$$x_1 + x_2 + x_3 = 4, \tag{4.157a}$$
$$x_1 + 2x_2 - x_3 = 1. \tag{4.157b}$$

The variable x_3 in Eq. (4.157a) can be eliminated using (4.157b) giving the single equation

$$2x_1 + 3x_2 = 5 \tag{4.158}$$

which can be rewritten in the form of Eq. (4.150a) as

$$\frac{x_1}{5/2} + \frac{x_2}{5/3} = 1. \tag{4.159}$$

The equation corresponding to (4.150b) is $(5/2)x_1 - (5/3)x_2 = 0$. The result is the system

$$\begin{pmatrix} 2/5 & 3/5 \\ 5/2 & -5/3 \end{pmatrix} \begin{pmatrix} x_1 \\ x_2 \end{pmatrix} = \begin{pmatrix} 1 \\ 0 \end{pmatrix}. \tag{4.160}$$

The solution of the above system is

$$\begin{pmatrix} x_1 \\ x_2 \end{pmatrix} = \begin{pmatrix} 2/5 & 3/5 \\ 5/2 & -5/3 \end{pmatrix}^{-1} \begin{pmatrix} 1 \\ 0 \end{pmatrix} = \begin{pmatrix} 10/13 \\ 15/13 \end{pmatrix}. \tag{4.161}$$

Use of these values in Eq. (4.157a) gives $x_3 = 27/13$. The singular-value solution of the system (4.157) is

$$(x_1, \ x_2, \ x_3) = (10/13, \ 15/13, \ 27/13). \tag{4.162}$$

4.16   Concluding Remarks

In this chapter, methods of solving certain kinds of system of linear equations are discussed. Depending upon the nature of the problem one can choose an appropriate method. In scientific problems, while solving a system of linear equations it is important to check whether the concerned system is an ill-conditioned or well-conditioned system.

Sometimes it is required to find the solutions of several systems of equations $AX = B_q$, $q = 1, 2, \ldots, p$ with the same matrix A. In this case introduce the notation $B_q = (a_{1,m+q}, \ldots, a_{m,m+q})^T$. That is, now B_q is a $m \times q$ matrix with its qth column elements are B_q. The augmented matrix is of the order $m \times (m+1+q)$. Application of the Gauss–Jordan method to this augmented matrix finally gives a matrix with $a_{ii} = 1$ for $i = 1, 2, \ldots, m$ and $a_{ij} = 0$ for $i = 1, 2, \ldots, m$, $j \neq i, \leq m$. The $(m+1)$th column elements are the solution of the system with B_1, $(m+2)$th column elements are the solution of the system with B_2 and so on. Develop a program to solve two systems of equations with same A by Gauss–Jordan method. Verify the program by solving the systems

$$2x + 3y - z = 4, \quad -x + y - z = -1, \quad x + y + z = 3, \tag{4.163}$$
$$2x + 3y - z = 3, \quad -x + y - z = 0, \quad x + y + z = 0. \tag{4.164}$$

The solution of the first three equations is $(x, y, z) = (1, 1, 1)$ while the solution of the second set of three equations is $(x, y, z) = (1, 0, -1)$.

4.17   Bibliography

[1]   R. Odisho, Applications of systems of Linear equations to electrical networks. https://home.csulb.edu/~jchang9/m247/m247_poster_R_Odisho_sp09.pdf.

[2]   C.F. van Loan and G.H. Golub, *Matrix Computations*. The John Hopkins, University Press, 1996.

[3]   Y. Saad, *Iterative Methods for Sparse Linear Systems*. SIAM, Philadelphia, 2003.

[4]   W.H. Press, S.A. Teukolsky, W.T. Vetterling, B.P. Flannery, *Numerical Recipes in Fortran*. Foundation Books, New Delhi, 1993. Indian edition.

[5]   P.H. Borcherds, *Eur. J. Phys.* 16:201, 1995.

[6]   J.H. Mathews, *Numerical Methods for Mathematics, Science and Engineering*. Prentice Hall of India, New Delhi, 1998.

[7] L.N. Trefethen and D. Bau III, *Numerical Linear Algebra*. Society for Industrial and Applied Mathematics, Pennsylvania, 2002.

[8] A. Gezerlis and M. Williams, *Am. J. Phys.* 89:51, 2021.

[9] J.J. Torsti and A.M.Auerela, *Comput. Phys. Commun.* 4(1):27, 1972.

4.18 Problems

A. Cramer's Rule

4.1 Solve the following systems of equations by the Cramer's rule.

a) $x_1 + x_2 = -1$,
$2x_1 + 5x_2 = -8$.

b) $3x_1 + 5x_2 = 21$,
$2x_1 + 7x_2 = 25$.

c) $2x_1 + 8x_2 = -1$,
$3x_1 - 2x_2 = 2$.

B. Upper- and Lower-Triangular Systems

4.2 Show that the equation

$$
\begin{aligned}
x_1 + 2x_2 + x_3 &= 5, \\
0.x_2 + 3x_3 &- 6, \\
2x_3 &= 4
\end{aligned}
$$

has infinitely many solutions.

4.3 Find the solution of the upper-triangular system

$$
\begin{aligned}
x_1 + x_2 - x_3 &= 6, \\
3x_2 + 2x_3 &= 7, \\
x_3 &= -1.
\end{aligned}
$$

4.4 Find the solution of the lower-triangular system

$$
\begin{aligned}
3x_1 &= 6, \\
2x_1 + 3x_2 &= 7, \\
x_1 + x_2 + 3x_3 &= -6.
\end{aligned}
$$

C. Gauss Elimination Method

4.5 For the equation

$$
\epsilon x + By = C, \quad Dx + Ey = F,
$$

where ϵ is a sufficiently small number show that without pivoting $x \approx 0$ is a solution for any values of C and F. Obtain the solution by the Gauss elimination for $F = D + E$ and $C = B + \epsilon$.

4.6   Obtain the solutions of the following systems.

a) $x_1 + 2x_2 + x_3 = 2$,
    $2x_1 - x_2 - x_3 = 1$,
    $x_1 - x_2 - x_3 = 0$.

b) $2x_1 + x_2 - x_3 = 3$,
    $4x_1 - x_2 + 2x_3 = -1$,
    $x_1 + x_2 - x_3 = 2$.

c) $3x_1 - x_2 + 2x_3 = 11$,
    $2x_1 - 2x_2 + 3x_3 = 10$,
    $x_1 + x_2 - x_3 = 1$.

d) $3x_1 + x_3 = 3$,
    $2x_1 + 3x_2 - x_3 = 2$,
    $-x_1 + 2x_2 + x_3 = -1$.

4.7   Observe what happens when you try to solve the following systems by the Gauss elimination method. State the reason.

a) $x_1 + 2x_2 + 4x_3 = 10$,
    $2x_1 - 3x_2 - x_3 = 5$,
    $1.5x_1 + 3x_2 + 6x_3 = 15$.

b) $2x_1 + x_2 + x_3 = 5$,
    $3x_1 + 2x_2 - 3x_3 = 2$,
    $2x_1 + x_2 + x_3 = 7$.

D. Gauss–Jordan Method

4.8   Solve the equations given in Problem 4.6 by the Gauss–Jordan method.

4.9   Find the inverse of the following matrices by the Gauss–Jordan method.

a) $\begin{pmatrix} 1 & 2 & 1 \\ 0 & 3 & 1 \\ 2 & 5 & -3 \end{pmatrix}$.   b) $\begin{pmatrix} 0 & 1 & 1 \\ 1 & -2 & 3 \\ 5 & 2 & 1 \end{pmatrix}$.

4.10   Observe what happens when you apply the Gauss–Jordan method to find the inverse of the following matrices.

a) $\begin{pmatrix} 1 & 1 & 1 \\ 2 & 1 & 3 \\ 1 & 1 & 1 \end{pmatrix}$.   b) $\begin{pmatrix} 2 & 1 & 3 \\ 4 & 2 & 6 \\ 1 & 2 & 1 \end{pmatrix}$.

4.11   The inverse of A of a system of the form $AX = B$ is given by

$$A^{-1} = \begin{pmatrix} 1 & 0 & 2 \\ -1 & 1 & 3 \\ 2 & -2 & 5 \end{pmatrix}.$$

Find the solution X if

$$B = \begin{pmatrix} 2 \\ -1 \\ 1 \end{pmatrix}.$$

E. Triangular Factorization Method

4.12 Solve the systems in Problem 4.6 by the triangular factorization.

4.13 Calculate the determinant of the following matrices by writing them in LU form.

a) $\begin{pmatrix} 3 & -1 & 2 \\ 2 & 1 & 2 \\ 1 & 1 & 1 \end{pmatrix}$. b) $\begin{pmatrix} 1 & 0 & 2 \\ -1 & 1 & 2 \\ 1 & -2 & 0 \end{pmatrix}$. c) $\begin{pmatrix} 1 & 2 & 1 \\ 0 & 3 & 1 \\ 2 & 5 & -3 \end{pmatrix}$.

F. Counting Arithmetic Operations in Direct Methods

4.14 Show that the number of arithmetic operations involved to find the inverse of a given matrix of order $n \times n$ by the Gauss–Jordan method is $(16n^3 - 9n^2 - n)/6$.

4.15 Show that the number of arithmetic operations involved in the triangular factorization is same as the number involved in the Gauss elimination method.

G. Iterative Methods

4.16 Obtain the solutions of the following systems by applying (i) the Jacobi method and (ii) the Gauss–Seidel method. When the calculation is done using computer use the initial guess as zero for all unknowns in these systems and stop the iteration if the distance given by (4.92) is 10^{-5}. For hand calculation find the solution with two-decimal point accuracy after 2 iterations starting with the specified guess.

a) $5x_1 + x_2 = 3$,
$x_1 - 7x_2 = 15$. $\left(x_1^{(0)}, x_2^{(0)}\right) = (0,0)$.

b) $5x_1 + x_2 - x_3 = 10$,
$x_1 - 4x_2 + x_3 = -10$,
$x_1 + x_2 - 4x_3 = 15$. $\left(x_1^{(0)}, x_2^{(0)}, x_3^{(0)}\right) = (2, 2.5, -3.75)$.

c) $6x_1 - x_2 - x_3 = -4$,
$x_1 + 6x_2 - x_3 = 2$,
$x_1 + x_2 + 4x_3 = 2$. $\left(x_1^{(0)}, x_2^{(0)}, x_3^{(0)}\right) = (-0.6, 0.6, 0.6)$.

d) $3x_1 + x_2 + x_3 = 0.4$,
$x_1 - 4x_2 - x_3 = 1.0$,
$x_1 + x_2 - 5x_3 = 0$. $\left(x_1^{(0)}, x_2^{(0)}, x_3^{(0)}\right) = (0, 0, 0)$.

e) $6x_1 - x_2 - x_3 - x_4 = 3$,
$2x_1 + 5x_2 + x_3 - x_4 = 7$,
$x_1 - 2x_2 + 4x_3 + x_4 = 4$,
$x_1 + x_2 + x_3 + 3x_4 = 6$.
$\left(x_1^{(0)}, x_2^{(0)}, x_3^{(0)}, x_4^{(0)}\right) = (0.9, 0.9, 0.9, 0.9)$.

H. Systems $AX = B$ with A being Vandermonde Matrix

4.17 Solve the following systems of equations, where the matrix A is Vandermonde.

a) $x_1 + x_2 + x_3 = 3$,
$x_1 + 4x_2 + 16x_3 = 36$,
$x_1 - 2x_2 + 4x_3 = 6$.

b) $x_1 + 2x_2 + 4x_3 = 0,$
   $x_1 + x_2 + x_3 = 2,$
   $x_1 - x_2 + 4x_3 = -3.$

c) $x_1 + 3x_2 + 9x_3 = 12,$
   $x_1 + 5x_2 + 25x_3 = 16,$
   $x_1 - 2x_2 + 4x_3 = -5.$

I. Homogeneous Systems with Equal Number of Equations and Unknowns

4.18   Obtain the nontrivial solution of the following systems.

a) $x_1 + x_2 + 2x_3 = 0,$
   $2x_1 + 3x_2 + 5x_3 = 0,$
   $-2x_1 + x_2 - x_3 = 0.$

b) $x_1 + x_2 + 2x_3 = 0,$
   $3x_1 + x_2 + 5x_3 = 0,$
   $-x_1 + x_2 - x_3 = 0.$

c) $2x_1 + 3x_2 + 7x_3 = 0,$
   $3x_1 + x_2 + 7x_3 = 0,$
   $5x_1 - 2x_2 + 8x_3 = 0.$

J. Inhomogeneous Systems with More Equations than unknowns

4.19   Find the solution of the following systems.

a) $2x_1 + x_2 = 0,$
   $x_1 + x_2 = 1,$
   $3x_1 + 2x_2 = 1.$

b) $x_1 - x_2 = -4,$
   $2x_1 + x_2 = -2,$
   $3x_1 - 2x_2 = -10.$

c) $3x_1 + 2x_2 = 9,$
   $4x_1 - x_2 = 1,$
   $5x_1 + 3x_2 = 14.$

K. Singular-Value Problems

4.20   Determine the nontrivial solution of the following equations.

a) $x_1 + 2x_2 - x_3 = 0,$
   $-2x_1 - x_2 + x_3 = 0.$

b) $x_1 + x_2 + x_3 = 0,$
   $3x_1 + x_2 + 2x_3 = 0.$

c) $2x_1 - x_2 + x_3 = 0,$
   $-5x_1 - 11x_2 + 2x_3 = 0.$

5

Curve Fitting

5.1 Introduction

In many scientific and engineering experiments number of values of two variables are often measured. For example, in an experiment, by varying the volume (V) of a cylinder one may measure the pressure (P) of steam in the cylinder. Assume that the law or rule connecting the two variables P and V is not known. To get a rough idea about how P varies with V a graph between the measured P and V may be drawn and then a conclusion on the dependence of P on V can be made. Sometimes, one may also wish to know the values of P for the values of V that are not considered in the experiment. *How does one determine the corresponding values of P without performing the experiment?* This is possible by constructing an appropriate mathematical relation between P and V from the given experimental data. Alternatively, from the given set of values of P and V an approximate value of the unknown P for a given value of V not in the set can be estimated by an extrapolation or interpolation which will be discussed later in Chapter 6.

The general problem of construction of an appropriate mathematical equation which fit a set of given data is called *curve fitting*. Let us assume that a set of data (x_k, y_k), $k = 1, 2, ..., n$ representing the values of two variables is given. The goal of the curve fitting is to determine a formula

$$y = f(x). \tag{5.1}$$

How does one choose an appropriate form of f(x)? Usually, by inspecting the given data and the physical situation a particular form of $f(x)$ is chosen. For example, in Fig. 5.1a y appears to be varying linearly with x. Therefore, the linear relation $y = ax + b$ can be tried. For the data in Fig. 5.1b an exponential function $y = be^{ax}$ is suitable. Methods are available to determine the constants a and b. The method of curve fitting can be extended to functions of several variables also. In this chapter, the least-squares curve fitting to linear and certain nonlinear functions of one variable which are often encountered in physical and engineering problems is studied. Many examples of physical quantities exhibiting linear and nonlinear relations are given in the problems section at the end of this chapter.

5.2 Method of Least-Squares

Let (x_k, y_k), $k = 1, 2, ..., n$ be n sets of observations and it is required to make a fit to the function

$$y = F(x, c_1, c_2, ..., c_m), \tag{5.2}$$

where F is some function of x which depends linearly on the parameters $\{c_i\}$. That is,

$$F = c_1\phi_1(x) + c_2\phi_2(x) + \cdots + c_m\phi_m(x), \tag{5.3}$$

DOI: 10.1201/9781032649931-5

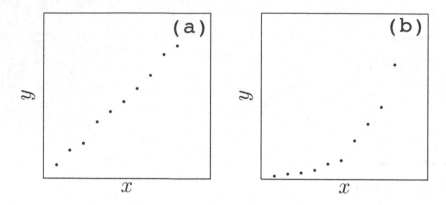

FIGURE 5.1
Data points y varying roughly (a) linearly and (b) nonlinearly with x.

where $\{\phi_i(x)\}$ are a priory selected set of functions and $\{c_i\}$ are to be determined. Normally, m is small compared with the number, n, of the data points. The functions $\{\phi_i(x)\}$ may be simply x or x^k or $\cos(k\pi x)$ and so on.

5.2.1 Basic Idea

The basic idea of curve fitting is to choose the parameters $\mathbf{c} = \{c_i\}$ in such a way that the *deviation errors* or the *residuals* e_k given by

$$e_k = y_k - F(x_k, \mathbf{c}), \quad k = 1, 2, ..., n \tag{5.4}$$

are minimum. In other words, the difference between the observed or given y and the value of y calculated from the relation (5.2) to be determined is made as small as possible. Here e_k's can take both positive and negative values. In order to give equal weightage to positive and negative e_k's one can consider the following two kinds of errors:

1. Average Error:

$$E_{av} = \frac{1}{n} \sum_{k=1}^{n} |e_k|. \tag{5.5}$$

2. Root-Mean-Square (rms) Error:

$$E_{rms} = \left[\frac{1}{n} \sum_{k=1}^{n} e_k^2 \right]^{1/2}. \tag{5.6}$$

The best fit is then obtained by minimizing one of these two quantities. The choice E_{av} will lead to a system of linear equations for the unknowns $\{c_i\}$ but with constraints and an analytical method of solving them not exists. For more details see Section 5.2.3.

Consideration of E_{rms} will lead to a system of linear equations for $\{c_i\}$ without any constraints as shown in Sections 5.2.2 and 5.3. Because of solving a set of coupled linear equations is simpler than the equations obtained from the minimization of E_{av}, the case of E_{rms} is generally preferred. Further, the system of linear equations obtained using E_{rms} can be solved either directly or iteratively as discussed in Chapter 4.

From Eq. (5.6) it is clear that the E_{rms} will be minimum *if and only if the quantity*

$$E(\mathbf{c}) = \sum_{k=1}^{n} e_k^2 \tag{5.7}$$

is a minimum. Hence, *the best curve fit to a set of data is that for which the sum of the squares of the residuals is a minimum.* This is known as *least-squares criterion* and the resulting approximation $F(x, \mathbf{c})$ is called a *least-squares approximation* to the given data.

5.2.2   Derivation of Equations for the Unknowns $\{c_i\}$

It is desired to find $\{c_i\}$ for which $E(\mathbf{c})$ given by Eq. (5.7) is a minimum. Recall that the slope $df(x)/dx = f'(x)$ of a single variable function $f(x)$ is zero at the minimum value of f. A function $f(\mathbf{X})$ of several variables is minimum when $\nabla f(\mathbf{X}) = 0$, where ∇ is the gradient operator. Thus, the necessary condition for the function E to be minimum is

$$\nabla E(\mathbf{c}) = 0, \tag{5.8}$$

where the gradient operator ∇ here takes the form

$$\nabla = \mathbf{i}_1 \frac{\partial}{\partial c_1} + \mathbf{i}_2 \frac{\partial}{\partial c_2} + \cdots + \mathbf{i}_m \frac{\partial}{\partial c_m} \tag{5.9}$$

with $\mathbf{i}_j$'s being unit vectors. Equation (5.8) is satisfied *if and only if* all the components of ∇E are identically zero: $\partial E / \partial c_i = 0$. That is,

$$\frac{\partial E}{\partial c_i} = \frac{\partial}{\partial c_i} \sum_{k=1}^{n} [y_k - F(x_k, \mathbf{c})]^2 = 0, \quad i = 1, 2, ..., m. \tag{5.10}$$

From Eq. (5.3) one has $\partial F / \partial c_i = \phi_i$. Therefore, Eqs. (5.10) become

$$-2 \sum_{k=1}^{n} [y_k - F(x_k, \mathbf{c})] \phi_i(x_k) = 0. \tag{5.11}$$

Then, using Eq. (5.4) in (5.11) one gets

$$\sum_{k=1}^{n} e_k \phi_i(x_k) = 0, \quad i = 1, 2, ..., m. \tag{5.12}$$

That is,

$$\mathbf{e} \cdot \mathbf{\Phi}_i = 0, \quad i = 1, 2, ..., m \tag{5.13}$$

where

$$\mathbf{e} = [e_1, e_2, ..., e_n]^{\text{T}}, \tag{5.14a}$$
$$\mathbf{\Phi}_i = [\phi_i(x_1), \phi_i(x_2), ..., \phi_i(x_m)]. \tag{5.14b}$$

Equation (5.13) implies that the error vector $\mathbf{e}$, for all n, should be normal or orthogonal to each of the n vectors $\mathbf{\Phi}_i$. Because of this, the m equations, namely, Eqs. (5.11) are called *normal equations*.

5.2.3   Least Absolute Residuals

In the case of E_{av} the quantity $E(\mathbf{c}) = \sum_{k=1}^{n} |e_k|$. For the best fit, it is desired to minimize

$$E(\mathbf{c}) = \sum_{k=1}^{n} |y_k - c_1\phi_1(x_k) - c_2\phi_2(x_k) - \cdots - c_m\phi_m(x_k)|. \tag{5.15}$$

For simplicity consider $c_3 = c_4 = \cdots = c_m = 0$, $\phi_1 = x$ and $\phi_2 = 1$. Then,

$$E(\mathbf{c}) = \sum_{k=1}^{n} |y_k - c_1 x_k - c_2| = \sum_{k=1}^{n} z_k. \tag{5.16}$$

Minimization of E_{av} is minimizing z_k with respect to c_1, c_2 and $z_1, z_2, \cdots, z_n$ subjected to

$$z_k \geq y_k - c_1 x_k - c_2 \text{ for } k = 1, 2, \cdots, n \tag{5.17a}$$

$$z_k \geq -(y_k - c_1 x_k - c_2) \text{ for } k = 1, 2, \cdots, n. \tag{5.17b}$$

The result will be constraint equations. After minimization, each z_k must be equal to $|y_k - c_1 x_k - c_2|$. This kind of problem can be solved by a linear programming package. For the least absolute deviations approach analytically solving the problem is not available in the mathematics literature. Further, another advantage of using E_{rms} over E_{av} is that in the case of former e_k is squared, that is, the deviation errors are enlarged. For more details, one may refer to the refs. [1,2].

5.3   Least-Squares Straight-Line Fit

In this section how to make a linear or straight-line fit to a given set of n data is described. Essentially, the problem is to determine the values of the constants a and b in the function

$$y = f(x, a, b) = ax + b. \tag{5.18}$$

Simple formulas for a and b can be obtained by solving the normal equations of a and b. From Eqs. (5.10) and (5.18) the normal equations are written as

$$\frac{\partial E}{\partial a} = \frac{\partial}{\partial a} \sum_{k=1}^{n} (y_k - ax_k - b)^2 = 0, \tag{5.19a}$$

$$\frac{\partial E}{\partial b} = \frac{\partial}{\partial b} \sum_{k=1}^{n} (y_k - ax_k - b)^2 = 0. \tag{5.19b}$$

Performing the partial derivatives in Eqs. (5.19) one has

$$-2 \sum (y_k - ax_k - b) x_k = 0, \tag{5.20a}$$

$$-2 \sum (y_k - ax_k - b) = 0, \tag{5.20b}$$

where the suffices in the summations are dropped for simplicity. Equations (5.20) can be rewritten as

$$a \sum x_k^2 + b \sum x_k = \sum x_k y_k, \tag{5.21a}$$

$$a \sum x_k + nb = \sum y_k. \tag{5.21b}$$

The above set of equations are linear in the unknowns a and b and can be easily solved. The equation for a is obtained by multiplying Eq. (5.21a) by n, Eq. (5.21b) by $\sum x_k$ and then subtracting one from another. Similarly, equation for b is obtained by multiplying Eq. (5.21a) by $\sum x_k$, Eq. (5.21b) by $\sum x_k^2$ and then subtracting one from another. The equations for a and b are obtained as

$$a = \frac{n \sum x_k y_k - \sum x_k \sum y_k}{n \sum x_k^2 - \left(\sum x_k\right)^2}, \tag{5.22a}$$

$$b = \frac{\sum x_k^2 \sum y_k - \sum x_k \sum x_k y_k}{n \sum x_k^2 - \left(\sum x_k\right)^2}. \tag{5.22b}$$

If y is expected to fall on a straight-line through the origin, $y = ax$, and if the measurement of y all have the same uncertainties, then show that the best estimate for the constant a is $\sum x_k y_k / \sum x_k^2$.

To assess the fitness of the obtained relation one may inspect a plot of the given data with the obtained curve and from the value of the *coefficient of determination* r^2 which is defined as

$$r^2 = \frac{S_t - S_r}{S_t}, \quad S_t = \sum_{k=1}^{n} (y_k - \bar{y})^2, \quad S_r = \sum_{k=1}^{n} [y_k - (ax_k + b)]^2 \tag{5.23}$$

with $\bar{y}$ being the average value of y_k. For a perfect fit $S_r = 0$ and hence $r^2 = 1$. For a poor fit r^2 will be far from the value 1.

One can also determine the uncertainty or standard deviation σ_y in the numbers y_1, y_2, ..., y_n. The given set of values y_k are normally distributed about the true value $ax_k + b$ with width parameter σ_y. Thus, the deviations $y_k - (ax_k + b)$ are normally distributed. Then, σ_y is given by

$$\sigma_y = \left[\frac{1}{n} \sum (y_k - (ax_k + b))^2 \right]^{1/2}. \tag{5.24}$$

In Eq. (5.24) a and b are unknown. They must be replaced by the best estimates, namely, (5.22) and this replacement would slightly reduce the value of σ_y given by Eq. (5.24). It can be shown that this reduction is compensated if n in the denominator of Eq. (5.24) is replaced by $(n-2)$. Now, σ_y is

$$\sigma_y = \left[\frac{1}{(n-2)} \sum (y_k - (ax_k + b))^2 \right]^{1/2}. \tag{5.25}$$

As long as n is sufficiently large the difference between n and $(n-2)$ is unimportant. On the other hand, $(n-2)$ is reasonable if only two data points, say, (x_1, y_1) and (x_2, y_2) are used. In this case, the fitted line passes exactly through both points. Because it is always possible to find a straight-line passing through any two given points, one cannot deduce anything about the reliability of the estimate. Since both points fall exactly on the best line, the summation terms in Eqs. (5.24) and (5.25) are zero. The formula (5.24) gives $\sigma_y = 0$ which is absurd whereas with $(n-2)$ Eq.(5.25) gives $\sigma_y = 0/0$, undetermined, and is correct. Essentially, there is only $(n-2)$ degrees of freedom. The uncertainties in a and b are given by

$$\sigma_a = \sigma_y \sqrt{n/\Delta}, \quad \sigma_b = \sigma_y \sqrt{\left(\sum x_k^2\right)/\Delta}, \quad \Delta = n \sum x_k^2 - \left(\sum x_k\right)^2. \tag{5.26}$$

What is the expression for σ_y in the case of a straight-line passing through the origin? In this case

$$a = \frac{\sum x_k y_k}{\sum x_k^2}. \tag{5.27}$$

To draw a straight-line passing through the origin only one additional point is required. Hence,

$$\sigma_y = \left[\frac{1}{n-1} \sum (y_k - ax_k)^2 \right]^{1/2}. \tag{5.28}$$

Example:

The following table gives the measured output voltage (y) of an electronic circuit as a function of the applied input voltage (x). Find the least-squares straight-line fit for the data given.

x in V	0.1	0.2	0.3	0.4	0.5	0.6	0.7
y in V	0.16	0.21	0.23	0.30	0.36	0.39	0.46

For the above data, the values of x_k^2, $x_k y_k$, $\sum x_k$, $\sum y_k$, $\sum x_k^2$ and $\sum x_k y_k$ are given in the following table:

	x_k	y_k	x_k^2	$x_k y_k$
	0.1	0.16	0.01	0.016
	0.2	0.21	0.04	0.042
	0.3	0.23	0.09	0.069
	0.4	0.30	0.16	0.120
	0.5	0.36	0.25	0.180
	0.6	0.39	0.36	0.234
	0.7	0.46	0.49	0.322
$\sum$	2.8	2.11	1.40	0.983

Now, from Eqs. (5.22) the coefficients a and b are calculated as

$$\begin{aligned}
a &= \frac{7 \times 0.983 - 2.8 \times 2.11}{7 \times 1.4 - 2.8 \times 2.8} \\
&= 0.49643, \\
b &= \frac{1.4 \times 2.11 - 2.8 \times 0.983}{7 \times 1.4 - 2.8 \times 2.8} \\
&= 0.10286 \, \text{V}.
\end{aligned}$$

Thus, the least-squares straight-line fit is $y = 0.49643x + 0.10286$. Figure 5.2 shows the graph of x versus y. Solid circles are the given data and the straight-line is the best fit obtained. The straight-line is drawn by generating 100 points of (x_k, y_k) from the obtained fit. (This is followed in other examples in the present chapter.) The value of the coefficient of determination r^2 is estimated as 0.987 which is close to the perfect fit value 1.

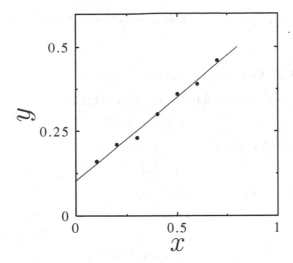

FIGURE 5.2

Graphical verification of fitness of the least-squares straight-line fit of the data given in the example. Solid circles are the given data and straight-line is the best fit given by $y = 0.49643x + 0.10286$. The units of x and y are in V.

5.4   Curve Fitting to Exponential and Power-Law Functions

In many physical problems nonlinear relations between two or more variables are encountered. For example, the number of particles travelled a distance x from a source obeys the relation $N = N_0 e^{-\lambda x}$, where N_0 and λ are constants. The absorption spectrum of crystals is of exponential form. Suppose, it is desired to fit a nonlinear curve for a set of given data. In the case of linear function $y = ax + b$ the normal Eqs. (5.21) are also linear in a and b. Consequently, the normal equations are easily solved and simplified formulas for a and b are obtained. *What is the situation for nonlinear functions?* One may consider two types of nonlinear functions in Eq. (5.2):

(1)   nonlinear in x but linear in $\{c_i\}$ and

(2)   nonlinear in both x and $\{c_i\}$.

The functions $y = c_1 x + c_2 x^2 + c_3 x^3$, $c_1 x^{3/2}$ and $c_1 \sin x + c_2$ are nonlinear in x but linear in the unknowns to be determined. In contrast, the functions $y = ax^b$, be^{ax} and $a/(x + b)$ are nonlinear in x as well as in the unknowns.

For case (1), the unknowns $\{c_i\}$ in the normal Eqs. (5.11) are still linear and hence they can be easily determined. This will be done for polynomial and trigonometric polynomials in Sections 5.6 and 5.8, respectively. For case (2), the normal Eqs. (5.11) are obviously nonlinear. Consequently, solving them is difficult and often one has to employ a suitable numerical scheme which requires time-consuming computations and also one has to make a good starting values of the unknowns. However, for certain nonlinear functions of the case (2), by suitable change of variables and parameters a linear relation can be obtained. For the redefined constants, the normal equations are linear.

In this section, curve fitting to the often realized exponential and power-law nonlinear functions by transforming them into linear function is demonstrated. These two functions occur in many physical problems, for example, see Problems 5.6–5.10, 5.13 and 5.14 at the end of this chapter.

5.4.1 Exponential Function $y = b\,e^{ax}$

Let the given set of data be fitted into the exponential relation

$$y = f(x) = b\,e^{ax}. \tag{5.29}$$

Taking logarithm on both sides of Eq. (5.29) results in

$$\ln y = \ln b + ax. \tag{5.30}$$

Introducing the change of variables and constants

$$Y = \ln y, \quad X = x, \quad A = a, \quad B = \ln b \tag{5.31}$$

Eq. (5.30) is rewritten as

$$Y = AX + B. \tag{5.32}$$

(Note: In Eq. (5.29) y must be greater than zero because logarithmic function is undefined for $y < 0$. If the values of y are not greater than 0 then y can be redefined. If all y_k's are negative then assume the relation $y = -b\,e^{ax}$. In this case Eqs. (5.30)-(5.31) read as

$$\ln(-y) = \ln b + ax \tag{5.33}$$

and

$$Y = \ln(-y), \quad X = x, \quad A = a, \quad B = \ln b, \tag{5.34}$$

respectively.) That is, under the change of variables given by Eq. (5.31) the nonlinear Eq. (5.29) is transformed into the linear Eq. (5.32). For the data set (X_k, Y_k) the straight-line fit, Eq. (5.32), can be obtained by the least-squares approximation. This gives the unknowns A and B in Eq. (5.32). Then, the unknown a is simply A and b is e^B. The following example illustrates the curve fitting to the exponential function (5.29).

Example:

A particle is ejected from a source and its measured velocities at a few values of time t are given below. Fit the given data to $v = b\,e^{at}$.

t in sec.	0	1	2	3	4	5
v in m/sec.	12.6	4.5	2.4	1.1	0.6	0.1

Fitting the above data to the given nonlinear function is equivalent to fitting the transformed data X, Y (Eq. (5.31)) to the linear relation (5.32) and determining the constants A and B. Then, the values of a and b in the given nonlinear function can be obtained from the transformation given by Eq. (5.31). The least-squares calculation is summarized below.

$X_k = t_k$	$Y_k = \ln v_k$	X_k^2	$X_k Y_k$
0	2.53370	0	0.00000
1	1.50408	1	1.50408
2	0.87547	4	1.75094
3	0.09531	9	0.28593
4	−0.51083	16	−2.04330
5	−2.30259	25	−11.51293
$\sum$ 15	2.19514	55	−10.01528

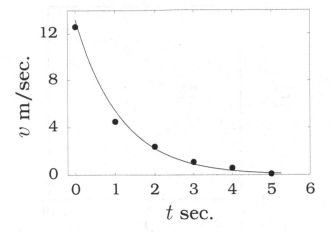

FIGURE 5.3
Graphical verification of fitness of exponential fit of the given data. Solid circles are the given data and the continuous curve is the obtained best fit.

From Eqs. (5.22) the constants A and B are obtained as (here $n = 6$)

$$A = \frac{6 \times -10.01528 - 15 \times 2.19514}{6 \times 55 - 15 \times 15} = -0.88589,$$

$$B = \frac{55 \times 2.19514 - 15 \times -10.01528}{6 \times 55 - 15 \times 15} = 2.58059.$$

The constants a and b are then calculated as

$$a = A = -0.88589\,\text{sec}^{-1},$$
$$b = e^B = 13.20494\,\text{m/sec}.$$

That is, the exponential fit to the given data is

$$v = 13.20494\,e^{-0.88589t}\,\text{m/sec}.$$

Figure 5.3 illustrates the correctness of the obtained best fit. Here the given data are represented by solid circles while the continuous curve is the computed best exponential fit. The given data falls closely to the continuous curve implying that the obtained fit is the best one.

5.4.2 Power-Law Function $y = b\,x^a$

The number of earthquakes of size x or greater, where x is the energy released in different geographic regions or even across the world as a whole, is found to be proportional to $x^{-0.8}$ to x^{-1}. Power-law distributions are increasingly being used by reinsurance companies and governments to assess the risks posed by natural hazards. The number of wildfires per unit area per year per eco-region versus the area of the wildfire and the number of rockfalls per unit volume versus their volume have exhibited power-law relation [3–5].

Like the exponential function (5.29) the power-law function $y = bx^a$ can be transformed into a linear relation by suitable change of variables and constants. The following example illustrates curve fitting to the power-law relation.

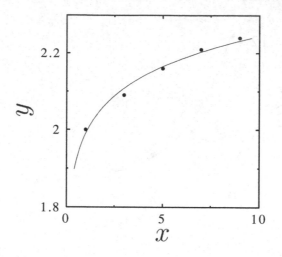

FIGURE 5.4
Graphical verification of fitness of power-law fit of the given set of data. Solid circles are
the given data and the continuous curve is the obtained best fit.

Example:

A nonlinear oscillator is subjected to an external periodic force of fixed amplitude and
frequency. The following table gives the ratio (G) of the amplitudes of the oscillation of the
oscillator and the applied force calculated as a function of a control parameter α by solving
the underlying equation of motion. Obtain the power-law fit $G = b\alpha^a$ to the given data
after linearizing them.

α	1	3	5	7	9
G	2.00	2.09	2.16	2.21	2.24

For illustrative purpose, define $x = \alpha$ and $y = G$. The nonlinear function $y = bx^a$ can be
converted into the linear form $Y = AX + B$ under the change of variables and constants
$Y = \ln y$, $X = \ln x$, $A = a$ and $B = \ln b$. From the given data one has

$$\sum X_k = 6.851185,$$

$$\sum Y_k = 3.799888,$$

$$\sum X_k^2 = 12.411600,$$

$$\sum X_k Y_k = 5.364400$$

and

$$a = A = 0.05214, \quad B = 0.68854, \quad b = e^B = 1.9908.$$

Thus, the power-law fit is $y = 1.9908\,x^{0.05214}$. Figure 5.4 depicts the plot of x versus given y
(marked by solid circles) and the obtained best power-law relation (continuous curve). The
deviation between the computed fit and the given data is negligible.

TABLE 5.1

Linearization of certain nonlinear functions. For all the cases the linearized relation is $Y = AX + B$.

Function $y = f(x)$	Example problems	New variables and constants			
		$X =$	$Y =$	$A =$	$B =$
$y = b\,e^{ax}$, $y > 0$	6–10	x	$\ln y$	a	$\ln b$
$y = -b\,e^{ax}$, $y < 0$	11	x	$\ln(-y)$	a	$\ln b$
$y = b\,e^{\pm ax^2}$	12	$\pm x^2$	$\ln y$	a	$\ln b$
$y = b\,x^a$, $y > 0$	13–15	$\ln x$	$\ln y$	a	$\ln b$
$y = -b\,x^a$, $y < 0$	16	$\ln x$	$\ln(-y)$	a	$\ln b$
$y = a\ln x + b$	17, 18	$\ln x$	y	a	b
$y = b\,x\,e^{ax}$	19	x	$\ln(y/x)$	a	$\ln b$
$y = 1/(ax + b)$	20	x	$1/y$	a	b
$y = x/(bx + a)$	21	$1/x$	$1/y$	a	b
$y = 1/(c + be^{ax})$ c must be given	22	x	$\ln(1/y - c)$	a	$\ln b$
$y = (a/x) + b$	23	x	xy	a	b
$y = b\,a^x$, $y > 0$	24	x	$\ln y$	$\ln a$	$\ln b$
$y = -b\,a^x$, $y < 0$	25	x	$\ln(-y)$	$\ln a$	$\ln b$
$y = a\,x^2 + b$	26	x^2	y	a	b
$y = ax^2 + bx$	27	x	y/x	a	b
$y = ag(x) + b$	28	$g(x)$	y	a	b

In addition to the exponential and power-law functions there exist many nonlinear functions which are convertible into a linear form. Some of the nonlinear functions and their corresponding linear relations and the change of variables and constants are summarized in Table 5.1. Draw graphs of x versus y for the nonlinear functions listed in Table 5.1 for fixed values of the constants a, b and c.

5.5 Curve Fitting to a Sigmoid Function

A sigmoid curve is an 'S' shape curve and is represented by mathematical functions called sigmoid functions. Some of the sigmoid functions are the logistic function $1/(1 + e^{-x})$, hyperbolic tangent function $(e^x - e^{-x})/(e^x + e^{-x})$ and the error function $\mathrm{erf}(x) = (2/\sqrt{\pi}) \int_0^x e^{-t^2}\, dt$. Figure 5.5 shows the plot of the logistic function. The sigmoid logistic function maps any real value to the range $[0, 1]$ as seen in Fig. 5.5.

Sigmoid functions are used as an activation function of artificial neurons and cumulative distribution functions. It is also used to find the probability of a binary variable in binary classification. Certain experimentally measurable variables in dynamical systems are found to display sigmoid-type variation with a control parameter, an example is the dependence of normalized transmittance of crystals on input intensity. Another example is the reflection

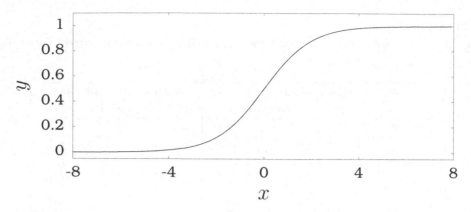

FIGURE 5.5

x versus $y(x) = 1/(1 + e^{-x})$.

probability of neutrons in the diffusion of neutrons through a moderating material slab (see Section 20.8 and Fig. 20.7).

For simplicity consider the sigmoid curve of the form appearing for $x > 0$ in Fig. 5.5. Let us describe the fitting a given data set to a sigmoid function. A choice of the sigmoid function is

$$y(x) \propto \frac{1}{1 + be^{ax}}, \quad x > 0. \tag{5.35}$$

If x is not > 0 then introduce the change of variable $x' = x - x_{\min}$ where $x_{\min}$ is the minimum value of x. Since the right-side of Eq. (5.35) has the exponential term, by appropriately redefining the variables x and y and the constants, an exponential fit can be constructed and then it can be converted to the sigmoid function.

Assume that $x > 0$ and convert the range of values of y into, for example, the range $[1, 2]$ (why?) by introducing the change of variable

$$y' = 1 + \frac{y - y_{\min}}{y_{\max} - y_{\min}}, \tag{5.36}$$

where $y_{\min}$ and $y_{\max}$ are the minimum and the maximum values of y, respectively. In this case, an appropriate sigmoid function is

$$y' = \frac{2}{1 + be^{ax}}. \tag{5.37}$$

This equation can be rewritten as

$$y'' = \frac{2}{y'} - 1 = be^{ax}. \tag{5.38}$$

Taking logarithm on Eq. (5.38) gives

$$\ln y'' = \ln b + ax. \tag{5.39}$$

Defining

$$X = x, \quad Y = \ln y'', \quad A = a, \quad B = \ln b \tag{5.40}$$

FIGURE 5.6
A system consisting of m oscillators coupled in one way. Larger solid circles represent oscillators. Arrows indicate the direction of coupling. jth oscillator is coupled to $(j + 1)$th oscillator only.

Eq. (5.39) takes the form

$$Y = AX + B. \tag{5.41}$$

For the given data set (x_k, y_k), $k = 1, 2, \ldots, n$ with x_k's are arranged in an ascending order $y_{\min} = y_1$ and $y_{\max} = y_n$ (generally). When $y = y_n (= y_{\max})$ then in Eq. (5.36) $y'_n = 2$ and y''_n (refer Eq. (5.38)) becomes 0 and $Y_n = \ln y''_n = \ln 0$ which is undefined. So, discard (X_n, Y_n) (as well as any other pair of data giving $Y_k = \ln 0$) and consider the remaining set of data. For this data set make a straight-line fit, Eq. (5.41), and find the values of A and B and then $a = A$ and $b = e^B$. From Eqs. (5.36) and (5.38) the relevant fit is

$$y = y_{\min} + (y_{\max} - y_{\min})\left(-1 + \frac{2}{1 + b\,e^{ax}}\right). \tag{5.42}$$

Example:

In a system of one-way coupled m oscillators (refer Fig. 5.6) the output gain measured in the first 10 oscillators is given below. A rough graph of the oscillator number (n) versus the gain (G) shows a sigmoid curve. Fit the given data to an appropriate sigmoid function.

n	1	2	3	4	5	6	7	8	9	10
G	0.7	1.58	1.9	2.45	2.74	2.8	2.9	2.93	2.96	3

Here, n is > 0 while $G \in [0.7, 3]$. First, convert G into the range $[1, 2]$ using Eq. (5.36) by taking G as y and G' as y'. The result is $(n, G') = (1, 1)$, $(2, 1.38261)$, $(3, 1.52174)$, $(4, 1.76087)$, $(5, 1.88696)$, $(6, 1.91304)$, $(7, 1.95652)$, $(8, 1.96957)$, $(9, 1.98261)$ and $(10, 2)$. For these data assume the sigmoid function as

$$G' = \frac{2}{1 + b\,e^{an}}. \tag{5.43}$$

With $G'' = (2/G') - 1$, $X = n$, $Y = \ln G''$, $A = a$ and $B = \ln b$ the relation between X and Y is $Y = AX + B$. For $(n, G') = (10, 2)$ the value of $Y = \ln G'' = \ln 0$ so discard this data for curve fit. A and B are computed as $A = -0.59048$ and $B = 0.44363$ and so $a = -0.59048$ and $b = e^B = 1.55835$. Then, referring to Eq. (5.42) one can write

$$\begin{aligned}
G &= G_{\min} + (G_{\max} - G_{\min})\left(-1 + \frac{2}{1 + b\,e^{an}}\right) \\
&= -1.6 + \frac{4.6}{1 + 1.55835\,e^{-0.59048n}}.
\end{aligned} \tag{5.44}$$

Figure 5.7 compares the obtained fit with the given data. The obtained fit is satisfactory.

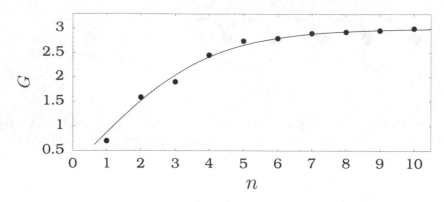

FIGURE 5.7
Plot of n versus G with solid circles and continuous line representing the given data and the obtained fit (Eq. (5.44)), respectively.

5.6   Polynomial Fit

There are certain physically measurable quantities obeying polynomial relations. For example, the square of magnetic rigidity of a particle as a function of temperature T is given by

$$H_r^2 = A\left(BT + T^2\right),\tag{5.45}$$

where A and B are constants. Thus, it is of interest to study the polynomial fit to the given data. In this sections the problem of polynomial fit using polynomial regression is described.

It is desired to fit the given data (x_k, y_k), $k = 1, 2, ..., n$ to the polynomial $P(x)$ of degree m that is given by

$$y = P(x) = c_1 + c_2 x + c_3 x^2 + \cdots + c_{m+1} x^m.\tag{5.46}$$

Referring to Section 5.2.2 the normal equations for the function of the form

$$y = c_1 \phi_1(x) + c_2 \phi_2(x) + \cdots + c_{m+1} \phi_{m+1}(x)\tag{5.47}$$

are obtained as

$$\sum_{k=1}^{n} \left[\phi_i\left(x_k\right) y_k - \phi_i\left(x_k\right) P\left(x_k\right)\right] = 0, \quad i = 1, 2, \ldots, m+1.\tag{5.48}$$

For the polynomial function, Eq. (5.46), the ϕ's are given by

$$\phi_i = x^{i-1}, \quad i = 1, 2, ..., m+1.\tag{5.49}$$

Now, the normal Eqs. (5.48) become

$$\sum_{j=1}^{m+1} c_j \sum_{k=1}^{n} x_k^{i+j-2} = \sum_{k=1}^{n} y_k x_k^{i-1}, \quad i = 1, 2, ..., m+1.\tag{5.50}$$

Defining

$$\sum_{k=1}^{n} y_k x_k^{i-1} = b_i\tag{5.51}$$

Eqs. (5.50) are rewritten as

$$\sum_{j=1}^{m+1} c_j \sum_{k=1}^{n} x_k^{i+j-2} = b_i, \quad i = 1, 2, ..., m+1. \tag{5.52}$$

This generates $m+1$ equations for $m+1$ unknown c's. The Gauss–Jordan method discussed in Section 4.5 can be used to solve the above simultaneous equations. For this purpose Eqs. (5.52) can be re-expressed in the form $Ac = B$ (refer Eq. (4.18), Section 4.4), where c is the unknown to be determined, that is,

$$\sum_{j=1}^{m+1} a_{ij} c_j = b_i, \quad i = 1, 2, ..., m+1, \tag{5.53a}$$

with

$$a_{ij} = \sum_{k=1}^{n} x_k^{i+j-2}. \tag{5.53b}$$

In vector form Eq. (5.53a) is given by $Ac = B$, where A is a matrix of order $(m+1) \times (m+1)$ and B is a column matrix with $m+1$ rows. Note that when $m = 1$ the normal equations for the straight-line fit are recovered (verify).

Example:

(a) Derive the normal equations for the quadratic polynomial

$$y(x) = P(x) = c_1 + c_2 x + c_3 x^2.$$

(b) The potential energy $V(x)$ of a system measured at five values of x are given below. The plot of $V(x)$ suggests a polynomial form of $V(x)$. Fit the data to the quadratic polynomial.

x in cm	0	1	2	3	4
V in $g\,cm^2/sec^2$	1	2	7	16	29

(a) The normal equations for the given quadratic polynomial function are obtained from Eq. (5.53a) by substituting $m = 2$. They are

$$a_{11}c_1 + a_{12}c_2 + a_{13}c_3 = b_1,$$
$$a_{21}c_1 + a_{22}c_2 + a_{23}c_3 = b_2,$$
$$a_{31}c_1 + a_{32}c_2 + a_{33}c_3 = b_3,$$

where the coefficients a_{ij}'s and b_i's are (from Eqs. (5.51) and (5.53b))

$$a_{11} = n, \quad a_{12} = \sum x_k, \quad a_{13} = \sum x_k^2, \quad b_1 = \sum y_k ,$$

$$a_{21} = \sum x_k, \quad a_{22} = \sum x_k^2, \quad a_{23} = \sum x_k^3, \quad b_2 = \sum x_k y_k ,$$

$$a_{31} = \sum x_k^2, \quad a_{32} = \sum x_k^3, \quad a_{33} = \sum x_k^4, \quad b_3 = \sum x_k^2 y_k.$$

(b) For convenience define $y = V$. For the given data the coefficients a_{ij}'s and b_i's are calculated as

$$a_{11} = 5, \quad a_{12} = 10, \quad a_{13} = 30, \quad b_1 = 55 ,$$
$$a_{21} = 10, \quad a_{22} = 30, \quad a_{23} = 100, \quad b_2 = 180 ,$$
$$a_{31} = 30, \quad a_{32} = 100, \quad a_{33} = 354, \quad b_3 = 638.$$

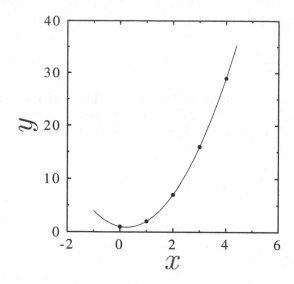

FIGURE 5.8

Graphical verification of fitness of the polynomial fit $y(x) = c_1 + c_2 x + c_3 x^2$ of the given data. Solid circles are the given data and the continuous curve is the obtained best fit.

Solving the normal equations gives

$$c_1 = 1, \quad c_2 = -1, \quad c_3 = 2.$$

Thus, the polynomial fit is $y(x) = 1 - x + 2x^2$. Figure 5.8 shows the graph of x versus y for the given data and the best fit obtained. *What are the units of c_1, c_2 and c_3?* Suppose, the problem is to find c_2 and c_3 with the value of c_1 given as 1. In this case the given polynomial can be transformed into the linear equation $Y = c_2 + c_3 x$ by the change of variable $Y = (y - c_1)/x$. Then, the least-squares straight-line fit can be performed to the given data.

5.7   Curve Fitting to Gaussian Functions

Gaussian functions are realized in many branches of physics, mathematics and engineering, for example, in quantum mechanics and optics. Fluctuations in the values of observables in most of the real experiments obey Gaussian statistics. The height distribution of a population, the distribution of the sum of a throw of a set of six-sided dice and voltage fluctuations in a power supply, the photoluminescence spectrum of certain crystals are of the Gaussian form [6-7].

The functional form of the Gaussian distribution is

$$G(x) = \frac{1}{\sigma\sqrt{2\pi}} e^{-1/(2\sigma^2)(x-x_0)^2}, \tag{5.54}$$

where x_0 is the average of the measurement and σ is the standard deviation of the measurement. Often $G(x)$ is related to the probabilities of measurements of the values of certain variables. Suppose, the probability of outcome of a measurement follows the Gaussian distribution. Denote the probability of getting a value of a variable, say, x in a measurement in the range of $x \to x + \Delta x$ is $P(x \to x + \Delta x) = \Delta x G(x)$, that is,

$$P(x \to x + \Delta x) = \frac{\Delta x}{\sigma \sqrt{2\pi}} e^{-1/(2\sigma^2)(x-x_0)^2}. \tag{5.55}$$

Out of M measurements the expected number of measured values within a range $[x, x+\Delta x]$ is $MP(x \to x + \Delta x)$. This is written as

$$m(x) = MP(x \to x + \Delta x) = \frac{M\Delta x}{\sigma \sqrt{2\pi}} e^{-1/(2\sigma^2)(x-x_0)^2}. \tag{5.56}$$

When x versus m follows Eq. (5.56), it is desired to find the values of x_0 and σ through the curve fit. Taking logarithm on both sides of Eq. (5.56) gives

$$\ln m = \ln(M\Delta x) - \ln\left(\sigma\sqrt{2\pi}\right) - \frac{x_0^2}{2\sigma^2} + \frac{x_0}{\sigma^2}x - \frac{1}{2\sigma^2}x^2. \tag{5.57}$$

With

$$y = \ln(m), \quad c_1 = \ln(M\Delta x) - \ln\left(\sigma\sqrt{2\pi}\right) - \frac{x_0^2}{2\sigma^2}, \tag{5.58a}$$

$$c_2 = \frac{x_0}{\sigma^2}, \quad c_3 = -\frac{1}{2\sigma^2} \tag{5.58b}$$

Eq. (5.57) is rewritten as

$$y = c_1 + c_2 x + c_3 x^2. \tag{5.59}$$

In Eq. (5.58) M and Δx are known values while x_0 and σ are the unknown to be determined through the curve fit. The values of c_1, c_2 and c_3 can be calculated from the polynomial fit discussed in Section 5.6. After this the values of x_0 and σ can be obtained from the relations

$$x_0 = -\frac{c_2}{2c_3}, \tag{5.60a}$$

$$\sigma = \frac{1}{\sqrt{2\pi}} \exp\left[\ln(M\Delta x) + \frac{c_2^2}{4c_3} - c_1\right]. \tag{5.60b}$$

Example 1:

A power supply was set to 5 V and 500 measurements of voltage at a sampling rate of, say, 1 MHz were made. The values of V varied in the interval [4.9588 V, 4.9656 V]. This interval was divided into 18 bins with equal size $\Delta V = 0.0004$ V and the number of values of V denoted as m in each bin noted are given in the table below.

V	4.9588	4.9592	4.9596	4.96	4.9604	4.9608	4.9612	4.9616	4.962
m	1	1	2	5	12	24	40	53	63
V	4.9624	4.9628	4.9632	4.9636	4.964	4.9644	4.9648	4.9652	4.9656
m	65	61	55	45	37	21	12	2	1

Fit the data to the Gaussian function

$$m = \frac{M\Delta V}{\sigma \sqrt{2\pi}} e^{-1/(2\sigma^2)(V-V_0)^2}, \quad M = 500.$$

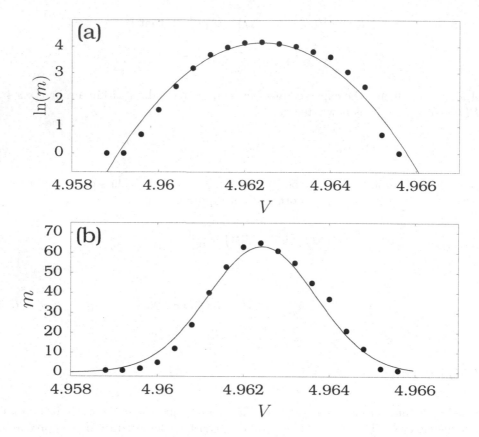

FIGURE 5.9

(a) Quadratic polynomial fit of V versus $\ln(m)$. (b) Plot of given V versus m (solid circles) and the obtained corresponding Gaussian fit (continuous line).

Fitting the given data to the quadratic Eq. (5.59), as in the example in Section 5.6, with $x = V$ and $y = \ln(m)$ gives

$$
\begin{aligned}
c_1 &= -9075343.58068, \\
c_2 &= 3657621.57493\,\mathrm{V}^{-1}, \\
c_3 &= -368531.21193\,\mathrm{V}^{-2}.
\end{aligned}
$$

Then, substitution of these values of c_1, c_2 and c_3, $M = 500$ and $\Delta x = \Delta V = 0.0004$ in Eq. (5.60) gives $V_0 = x_0 = 4.96243\,\mathrm{V}$ and $\sigma = 0.00126\,\mathrm{V}$. Figures 5.9a and b show the plots of V versus $\ln(m)$ and V versus m, respectively. The closeness of the given data with the fitted Gaussian function indicates that the fluctuations in the voltage follow the Gaussian statistics.

Example 2:

The height, H, of 1000 girls of a particular age in a city was measured. The values of H varied in the interval $[110.3\,\mathrm{cm},\ 111.8\,\mathrm{cm}]$. This interval was divided into 16 bins with equal size $\Delta H = 0.1\,\mathrm{cm}$ and the number of values of H denoted as m in each bin noted are given in the table below.

H in cm	110.3	110.4	110.5	110.6	110.7	110.8	110.9	111.0
m	1	5	10	47	64	123	157	165
H in cm	111.1	111.2	111.3	111.4	111.5	111.6	111.7	111.8
m	159	129	75	33	20	8	3	1

Fit the data to the Gaussian function

$$m = \frac{M\Delta H}{\sigma\sqrt{2\pi}}e^{-1/(2\sigma^2)(H-H_0)^2}, \quad M = 1000.$$

Fitting the given data to the quadratic Eq. (5.59) with $x = H$ and $y = \ln(m)$ gives

$$
\begin{aligned}
c_1 &= -111889.95601, \\
c_2 &= 2015.58007 \, \text{cm}^{-1}, \\
c_3 &= -9.07673 \, \text{cm}^{-2}.
\end{aligned}
$$

Then, substitution of these values of c_1, c_2 and c_3, $M = 1000$ and $\Delta x = \Delta V = 0.1$ in Eq. (5.60) gives $H_0 = x_0 = 111.03014 \, \text{cm}$ and $\sigma = 0.24126 \, \text{cm}$.

5.8   Trigonometric Polynomial Fit

In many phenomena, particularly, in oscillatory processes periodic variation of certain physically measurable quantities is encountered. For example, solutions of certain differential equations, intensity of light in an experiment and planetary motion appear periodic or almost periodic. Consider the so-called van der Pol oscillator equation

$$\ddot{x} - b\left(1 - x^2\right)\dot{x} + x = 0. \tag{5.61}$$

Exact analytical solution of this equation is not known. A numerical solution of it is shown in Fig. 5.10 for $b = 0.4$. Because $x(t)$ appears as a periodic function of t, one can try to fit the data to an approximate trigonometric polynomial. The least-squares procedure developed earlier for linear function can be readily extended to trigonometric polynomials. This section defines trigonometric polynomials and discusses curve fitting of a given set of data to it.

5.8.1   Trigonometric Polynomials

Consider the Fourier series

$$f(x) = \frac{a_0}{2} + \sum_{i=1}^{\infty}\left(a_i\cos ix + b_i\sin ix\right), \tag{5.62}$$

where a_0, a_i's and b_i's are constants and are computed from the Euler's formula

$$a_i = \frac{1}{2\pi}\int_0^{2\pi} f(x)\cos ix\,dx, \tag{5.63a}$$

$$b_i = \frac{1}{2\pi}\int_0^{2\pi} f(x)\sin ix\,dx, \quad i = 1, 2, ... \tag{5.63b}$$

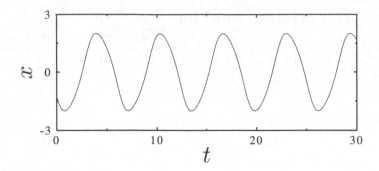

FIGURE 5.10
The numerical solution of the van der Pol oscillator Eq. (5.61).

The Fourier series $f(x)$ given by Eq. (5.62) is a convergent series in the interval $[0, 2\pi]$ if $f'(x)$ is piece-wise continuous on $[0, 2\pi]$. Therefore, in practical problems it is sufficient to consider first few terms in the summation of Eq. (5.62). When the running index i in Eq. (5.62) is chosen as $1, 2, \ldots$ up to M then the Fourier series becomes

$$f(x) = \frac{a_0}{2} + \sum_{i=1}^{M} (a_i \cos ix + b_i \sin ix) \tag{5.64}$$

then it is called a *trigonometric polynomial* of order M. When $f(x)$ is an even function, that is, $f(-x) = f(x)$ for all x then the Fourier series Eq. (5.62) becomes a cosine series

$$f_c(x) = \frac{a_0}{2} + \sum_{i=1}^{\infty} (a_i \cos ix). \tag{5.65}$$

On the other hand, the series reduces to a sine series

$$f_s(x) = \frac{a_0}{2} + \sum_{i=1}^{\infty} (b_i \sin ix) \tag{5.66}$$

if $f(x)$ is an odd function, that is, $f(-x) = -f(x)$ for all x.

5.8.2　Trigonometric Polynomial Approximation

Let (x_k, y_k), $k = 1, 2, \ldots, n$ be a given set of n data, where $y_k = f(x_k)$ and x_k are equally spaced. The values of x_k are given by

$$x_k = \frac{k}{n} 2\pi \tag{5.67}$$

so that $k = 1$ gives $x_1 = 2\pi/n$ and $k = n$ gives $x_n = 2\pi$. In other words, y is assumed as a periodic function of x with period 2π. It is desired to fit the data to the trigonometric function given by Eq. (5.64). The coefficients a_0, a_i's and b_i's can be determined by minimizing the quantity

$$\sum_{k=1}^{n} [y_k - f(x_k)]^2. \tag{5.68}$$

The coefficients are given by

$$a_i = \frac{2}{n} \sum_{k=1}^{n} y_k \cos i x_k, \quad i = 0, 1, ..., M \tag{5.69a}$$

$$b_i = \frac{2}{n} \sum_{k=1}^{n} y_k \sin i x_k, \quad i = 1, ..., M. \tag{5.69b}$$

The fitting is possible provided $2M + 1 \leq n$.

Example:

The numerically computed solution of $y''(x) + y(x) = 0$, $y(0) = 1$ and $y'(0) = 0$ at four values of x are given below. Fit the data to an appropriate trigonometric polynomial.

x	$\pi/2$	π	$3\pi/2$	2π
y	0	-1	0	1

Since the number of (x, y) pairs is $(n =)4$ the maximum value of M according to $2M + 1 \leq n$ is 1. That is, the data can be fitted to the function of the form

$$f(x) = \frac{a_0}{2} + a_1 \cos x + b_1 \sin x.$$

The values of the coefficients a_0, a_1 and b_1 are estimated from Eqs. (5.69) and are

$$a_0 = \frac{2}{4} \sum_{k=1}^{4} y_k = 0,$$

$$a_1 = \frac{2}{4} \sum_{k=1}^{4} y_k \cos x_k = 1,$$

$$b_1 = \frac{2}{4} \sum_{k=1}^{4} y_k \sin x_k = 0.$$

Thus, the obtained trigonometric polynomial fit for the given data is simply $f(x) = \cos x$ which is the exact solution of the given equation.

5.9   Least-Squares Curve Fitting to $y = af(x) + bg(x)$

Assume that the functional relation is not a linear or linearizable or polynomial or trigonometric polynomial. In such a case if the constants a and b are still linear then linear normal equations can be easily derived. This is illustrated in this section.

Let the function to which the given data is to be fitted is of the form

$$y = af(x) + bg(x). \tag{5.70}$$

The error quantity $E(a, b)$ (Eq. (5.7)) for the function given by Eq. (5.70) is

$$E(a, b) = \sum_{k=1}^{n} (y_k - af - bg)^2. \tag{5.71}$$

The necessary conditions for E to be minimum are

$$\frac{\partial E}{\partial a} = -2 \sum (y_k - af - bg) f = 0, \tag{5.72a}$$

$$\frac{\partial E}{\partial b} = -2 \sum (y_k - af - bg) g = 0. \tag{5.72b}$$

The above equations are rewritten as

$$a \sum f^2 + b \sum fg = \sum y_k f, \tag{5.73a}$$

$$a \sum fg + b \sum g^2 = \sum y_k g \tag{5.73b}$$

which are *linear in a and b*. Defining

$$c_{11} = \sum f^2, \quad c_{12} = \sum fg, \quad c_{22} = \sum g^2,$$

$$b_1 = \sum y_k f, \quad b_2 = \sum y_k g \tag{5.74}$$

Eqs. (5.73) become

$$ac_{11} + bc_{12} = b_1, \quad ac_{12} + bc_{22} = b_2. \tag{5.75}$$

Solving the above equations gives

$$a = \frac{b_1 c_{22} - b_2 c_{12}}{c_{11} c_{22} - c_{12}^2}, \tag{5.76a}$$

$$b = \frac{b_2 c_{11} - b_1 c_{12}}{c_{11} c_{22} - c_{12}^2}. \tag{5.76b}$$

Example:

Obtain the least-squares fit to the function $y = a\,e^{-2x} + b \sin x$ for the following data.

x	0	1	2	3	4
y	4.8	−1	−1.5	−0.3	1.4

For the given function $f(x) = e^{-2x}$ and $g(x) = \sin x$. For the given data Eqs. (5.74) become

$$c_{11} = \sum e^{-4x_k} = 1.018657,$$

$$c_{12} = \sum e^{-2x_k} \sin x_k = 0.130631,$$

$$c_{22} = \sum \sin^2 x_k = 2.12756,$$

$$b_1 = \sum y_k e^{-2x_k} = 4.636918,$$

$$b_2 = \sum y_k \sin x_k = -3.307277.$$

The constants a and b are then calculated as $a = 4.78904$ and $b = -1.84854$. Figure 5.11 depicts x versus y for the given data and the obtained fit.

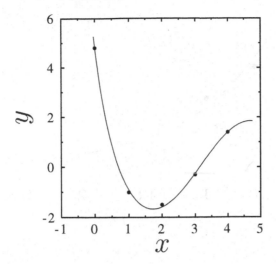

FIGURE 5.11

Graphical verification of fitness for the function $y(x) = ae^{-2x} + b\sin x$ of the given data. Solid circles are the given data and the continuous curve is the obtained best fit.

5.10 Least-Squares Curve Fit to $y = f(x)g(\mathbf{c}, x)$

Let us consider the functional relation between x and $y(x)$ as

$$y = f(x)g(\mathbf{c}, x), \tag{5.77}$$

where $f(x)$ is independent of the unknown parameters and $g(\mathbf{c}, x)$ can be linearizable or polynomial or trigonometric polynomial or for which the normal equations are linear in $\mathbf{c}$. In this case rewrite this equation with $y' = y/f(x)$ as $y' = g(\mathbf{c}, x)$. The curve fitting of the given data set to this new equation can be done.

Relations of the form of Eq. (5.77) are realized in many fields of science. For example, the famous *Planck's black body radiation formula* is

$$E(\nu, t) = \frac{8\pi\nu^2}{c^3} \frac{h\nu}{e^{h\nu/(K_B T)} - 1}. \tag{5.78}$$

For visible and ultraviolet regions (higher values of ν) $E(\nu, T)$ is approximated as

$$E(\nu, t) \propto a\nu^3 e^{-b\nu/T}, \tag{5.79}$$

where a and b are constants. $E(\nu, T)$ given by Eq. (5.79) is the *Wien's formula*. In certain nonlinear systems driven by a weak external periodic driving signal and noise the so-called *signal-to-noise ratio* (SNR) scales with the noise intensity D as

$$SNR = \frac{b}{D^2} e^{a/D}. \tag{5.80}$$

Example:

The numerically computed SNR values of a driven nonlinear oscillator for 8 values of noise intensity D are given below (units of D and SNR are dropped for simplicity). Assuming the relation (5.80) compute the values of a and b.

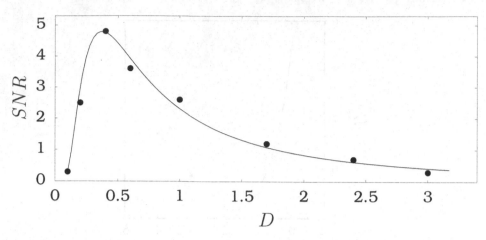

FIGURE 5.12
Plot of given SNR versus D (solid circles) and the obtained corresponding fit (Eq. (5.83))
(continuous line).

D	0.1	0.2	0.4	0.6	1	1.7	2.4	3
SNR	0.3	2.5	4.8	3.6	2.6	1.2	0.7	0.3

Rewrite the given relation as

$$(SNR)D^2 = b\,e^{a/D}. \tag{5.81}$$

Define

$$X = 1/D, \quad Y = \ln\left[(SNR)D^2\right], \quad A = a, \quad B = \ln b. \tag{5.82}$$

Then, Eq. (5.81) takes the form $\ln Y = AX + B$. For the given data set the straight-line fit
gives $a = A = -0.73978$, $B = 1.57735$ and $b = e^B = 4.84213$. So, the obtained fit is

$$SNR = \frac{4.84213}{D^2}e^{-0.73978/D}. \tag{5.83}$$

Figure 5.12 shows the plot of given SNR (solid circles) and the SNR obtained (continuous
line) from the constructed mathematical relation (5.83). The closeness of the given SNR
with the obtained SNR indicates that the Eq. (5.83) is an appropriate relation between D
and SNR.

5.11   Concluding Remarks

Curve fitting to a variety of relations between two variables is presented. For certain nonlinear relations under suitable change of variables and constants, the nonlinear relations can
be transformed to the linear relation. For a data set following, say, a straight-line relation,
it is possible to make a fit to any other functional relation also. But the deviation of the
data obtained from the constructed fit from the given data will be large. Therefore, first
draw a graph of the given data and identify an appropriate functional relation for the given
data. After obtaining the mathematical relation by the least-squares approximation draw

the graph of the computed data and the given data and check the closeness of these two. It is important to compute the coefficient of determination r^2 to assess the fitness of the arrived fit. Further, it is appropriate to specify the uncertainties in the constants involved in the fit.

5.12 Bibliography

[1] *Least absolute deviations.* https://en.wikipedia.org/wiki/Least_absolute_deviations (accessed on January 10, 2023).

[2] *Why we use 'least squares' regression instead of 'least absolute deviations' regression.* https://www.bradthiessen.com/html5/docs/ols.pdf (accessed on January 10, 2023).

[3] B.D. Malamud, *Phys. World* August 2004. pp.31–35.

[4] S. Hergarten, *Natural Hazards and Earth System Sciences* 4:309, 2004.

[5] M. Mitzenmacher, *Internet Maths.* 1:226, 2004.

[6] J. Taylor, *An Introduction to Error Analysis.* University Science Books, Sausalito, 1982. 2nd edition.

[7] M.W. Ray, *The Phys. Teach.* 61:224, 2023.

5.13 Problems

A. Linear Fit

5.1 A particle of charge q is accelerated to 5 different potentials of V volts and the corresponding masses m are measured. The ratio m/m_0, where m_0 is the rest mass, as a function of V is given below. Fit the data to the relation $m = m_0(aV + b)$ and find a and b.

V(in volts)	1	2	3	4	5
m/m_0	1.5	2.0	2.5	3	3.5

5.2 The volume V of a gas measured at various temperatures T is listed below. Assuming the linear relation $V = V_0(1 + aT)$, where V_0 is the volume of the gas at $0°$C and a is the expansion coefficient, determine V_0 and a.

$T°$ C	70	90	110	130	150
V m^3	2.084	2.108	2.132	2.156	2.180

5.3 The latent heat l_H of vaporization of steam is given below as a function of temperature T. Fit the data to the relation $l_H = aT + b$.

$T°$ C	40	50	60	70	80
l_H J/kg	920	880	870	835	780

5.4   The measured values of current (I) flowing through a resistor for various values of applied voltage (V) are given below. Assuming the linear relation $V = IR$ find the value of resistance R.

V (in volts)	0.5	1.0	1.5	2.0	2.5
I (in ampere)	0.005	0.01	0.015	0.02	0.025

5.5   The zero-pressure molar heat capacities C in KJ/mol.K as a function of temperature T in Kelvin for water are given below. Perform the straight-line fit to the data.

T	10	20	30	40	50
C	32.258	32.281	32.305	32.332	32.360

B. Curve Fit to Nonlinear Functions That Can Be Converted into a Linear Function

5.6   In an experiment, molecules are ejected from a source of particles and a beam is formed by means of a diaphragm. The number of particles N travelled a distance x without collisions is given below as a function of x. Assume that N follows the relation $N = N_0 e^{-\lambda x}$. Determine N_0 and λ.

x (in m)	0.5	1.0	1.5	2.0	2.5
N	687	472	325	223	153

5.7   The probability P of finding a particle as a function of energy E is given in the table below. If P is assumed to obey the relation $P = be^{-aE}$ determine a and b.

E in eV	0.5	0.6	0.7	0.8	0.9
P	0.1839	0.1506	0.1233	0.1009	0.0827

5.8   The amount of charge q discharged through a resistor R in an RC circuit at various time are given below. Assuming the relation $q = q_0 e^{-\tau t}$ determine the relaxation time τ and the initial amount of charge q_0 stored in the capacitor.

t sec	0.05	0.10	0.15	0.20	0.25
$q \times 10^{-5}$ coulomb	9.87	9.75	9.63	9.5	9.4

5.9   In an experiment, a metal foil of thickness t is placed between a β-particles source and a β-detector. The number of particles, N, reaching the detector is counted. The experiment is repeated with metal foils of different thickness. The data obtained are listed below. Assuming that N varies according to the relation $N = N_0 e^{-\mu t}$ determine the unknowns N_0 and μ.

t in minutes	0.001	0.010	0.020	0.030	0.040
N	780	623	485	377	294

5.10   The measured number of particles N emitted from a sample at 5 instants of time are given below. The decay law is $N = N_0 e^{-\lambda t}$, where N_0 and λ are the number of particles present in the sample at time $t = 0$ and decay rate, respectively, and the unit of time is hour. Curve fitting the data to the above relation determine the values of N_0 and λ.

t (hours)	20	40	60	80	100
$N(t)$	8200	6700	5500	4500	3700

5.11   Obtain the exponential fit $y = be^{ax}$ to the following data.

x	0.2	0.4	0.6	0.8	1.0
y	-1.5	-2.8	-4.5	-7.8	-11.6

5.12   Find the fit $y = be^{ax^2}$ to the given data.

x	0.3	0.5	0.7	0.9	1.1
y	2	2.25	2.5	2.9	3.7

5.13   The resistivity R of a conductor as a function of temperature T is given below. Show that R varies as bT^a and find the constants a and b.

$T° \,C$	20	40	60	80	100
R ohms	2.2	1.6	1.3	1.1	1.0

5.14   The pressure (P) of helium measured at 5 different volumes (V) is listed below. Fitting the data to the power-law relation $P = bV^a$ determine the constants a and b.

V m^3	0.02	0.06	0.1	0.14	0.3
P bar	23.0	4.4	2.0	1.2	0.4

5.15   Find the fit $y = bx^a$ to the following data by the least-squares approximation after linearizing the data.

x	0.4	0.8	1.2	1.6	2.0
y	1.9	0.7	0.4	0.2	0.1

5.16   Obtain the fit $y = -bx^a$ to the following data.

x	1	2	3	4	5
y	-0.700	-0.990	-1.212	-1.400	-1.565

5.17   The electric potential V as a function of perpendicular distance r from a long straight-wire of cross-sectional radius a is given in the following table. Fitting the data to the relation $V = -K \ln(r/a)$ determines the unknowns K and a.

r m	1	2	3	4	5
V volts	−0.55	−1.1	−1.40	−1.65	−1.85

5.18   Obtain the fit $y = a \ln x + b$ to the following data after linearizing the data.

x	0.25	0.40	0.55	0.70	0.85
y	−3.65	−2.25	−1.30	−0.55	0.01

5.19   Assuming that the data given below vary according to the relation $y = b x e^{ax}$ find the approximate values of a and b.

x	1	2	3	4	5
y	0.45	0.20	0.07	0.02	0.01

5.20   For the following data obtain a least-squares fit of the form $y = 1/(ax + b)$.

x	0.0	0.25	0.5	0.75	1.0
y	2.0	1.3	1.0	0.8	0.6

5.21   Apply the least-squares method fit to the following data to the nonlinear relation $y = x/(b + ax)$.

x	0.25	0.5	0.75	1.00	1.25
y	0.2	0.3	0.4	0.5	0.6

5.22   Find the fit $y = 1/(c + be^{ax})$, where $c = 2$ to the given data.

x	1	2	3	4	5
y	0.34	0.35	0.36	0.37	0.38

5.23   For the following data fit the nonlinear function $y = (a/x) + b$.

x	−4	−6	−8	−10	−12
y	−2.75	−2.50	−2.37	−2.30	−2.25

5.24   Obtain the fit $y = ba^x$ to the following data after linearizing them.

x	0	1	2	3	4
y	0.5	1.0	2.0	4.0	8.0

5.25   Find the fit $y = -ba^x$ to the given data.

x	0	1	2	3	4
y	−0.25	−0.75	−2.25	−6.75	−20.25

5.26   Obtain the fit $y = ax^2 + b$ to the given data after linearizing them.

x	0	−1	−2	−3	−4
y	3	2	−1	−6	−13

5.27 Find the fit $y = ax^2 + bx$ to the following data.

x	1	2	3	4	5
y	1.5	2.0	1.5	0.0	−2.5

5.28 The experimentally measured kinetic energy E of beta particles as a function of v/c, where v is the velocity of the particles and c is the velocity of light is given below. Assume that E varies according to $E = a \left[\frac{1}{\sqrt{1-v^2/c^2}} - b \right]$. By suitable change of variables transform the above nonlinear equation into a linear equation. Then, fit the data to the linearized equation and determine a and b.

v/c	0.0	0.2	0.4	0.6	0.8
E MeV	0.0	0.01054	0.04655	0.1278	0.3407

5.29 The ratio l/l_0 of a rod, where l_0 and l are the lengths at rest in a reference frame moving with velocity v and in another frame, respectively, as a function of v/c, where c is the velocity of light, is given below. Assuming that l varies as $l = l_0\sqrt{b + a(v/c)^2}$ determine the unknowns a and b.

v/c	0.0	0.2	0.4	0.6	0.9
l/l_0	1.0	0.9798	0.9165	0.80	0.4359

C. Polynomial Fit

5.30 During the calibration of a thermocouple, the readings of emf and temperature are as follows:

$T°$ C	50	100	150	200	250	300
emf mV	122.5	170.0	232.5	310.0	402.5	510.0

(a) Assuming the polynomial relation $emf = A + BT + CT^2$ determine the calibration constants A, B and C. (b) What are the units of A, B and C?

5.31 The lattice constants l_c of solid argon as a function of temperature T are given below. (a) Assuming that l_c varies according to the polynomial relation $l_c = C_1 + C_2T + C_3T^2$ find the approximate values of C_1, C_2 and C_3. (b) Identify the units of C_1, C_2 and C_3.

$T°$ K	0	20	40	60	80
l_c Å	5.30	5.31	5.33	5.38	5.44

5.32 The isothermal compressibility (K) of water as a function of temperature (T) is given below. Fit the data to the function $K = C_1 + C_2T + C_3T^2$ and state the units of C's.

T° C	10	30	50	70	90
K (GPa)$^{-1}$	0.480	0.450	0.440	0.450	0.467

5.33 The magnetic rigidity of a particle is found to obey the relation $H_r = A(T^2 + BT)^{1/2}$, where A and B are constants and T is the kinetic energy of the particle. For the given data using a polynomial fit determine the values of A and B.

T units	20	30	40	50
H_r units	68399.357	101764.320	135125.84	168485.97

5.34 The number of arithmetic operations (N) involved in solving a system of simultaneous linear equations of order n is given below as a function of n. Assuming the relation $N = \left(C_1 + C_2 n + C_3 n^2 + C_4 n^3\right)/6$ determine the values of C_1, C_2, C_3 and C_4.

n	2	3	4	5	6	7
N	9	28	62	115	191	294

D. Trigonometric Polynomials

5.35 In the following some sets of values of x and y are given. Develop a computer program to construct the best trigonometric polynomials for the given set of (x, y) values.

(a)

x	0	$\pi/2$	π	$3\pi/2$
y	0.0	0.5	0.0	-0.5

(b)

x	0	$\pi/2$	π	$3\pi/2$
y	0.7	1.5	1.3	0.5

6

Interpolation and Extrapolation

6.1 Introduction

Suppose, the values of a function $y = f(x)$ at a set of points $x_1, x_2, \ldots, x_{n+1}$ with $x_1 < x_2 < \cdots < x_{n+1}$ are given but the analytic expression for $f(x)$ is not known. The $f(x_i)$'s might be obtained in an experiment. For example, one has measured resistivity R of a resistor at the temperatures (T), say, $20°C$, $30°C$, $40°C$, $50°C$ and $60°C$ and wishes to know the functional form of the dependence of R on T or to compute R for an arbitrary temperature. To estimate $f(x)$ for an arbitrary x using the given data set $(x_i, y_i = f(x_i))$, $i = 1, 2, \ldots, n$ one has to construct a suitable functional form of $f(x)$. A most common functional form is a polynomial. Rational function approximation is also useful in certain cases. When the desired (target) x at which $f(x)$ is to be determined lies inside the interval $[x_1, x_{n+1}]$ then the problem is called an *interpolation*. When the target x falls outside the interval $[x_1, x_{n+1}]$ it is an *extrapolation*. The order of the interpolation/extrapolation is n if the number of points used is $n + 1$.

The interpolation and the extrapolation are utilized in the stock analysis to visualize the near future trend in the stock market. Nowadays, interpolation is utilized in computer graphics and data engineering to minimize the number of data sufficient to represent patterns in photos and videos.

How are the interpolation or the extrapolation different from curve fitting? Curve fitting is a smooth process in which the fitted curve need not pass through all the given n data points (x_i, y_i). In contrast to curve fitting the interpolation function perfectly passes through all the given data points. Further, in curve fitting the number of coefficients in the analytical expression of $y(x)$ is much less than the data points. For example, for a straight-line fit $y = ax + b$ or exponential fit $y = be^{-ax}$, the number of coefficients is always two and is independent of number of data points. On the other hand, in interpolation, the number of coefficients in the analytical expression of $y(x)$ is equal to the number of given data points.

An interpolation essentially consists of two stages. They are:

1. Construction of an interpolating function from the given set of data.

2. Calculation of value of the function at the desired point x from the constructed function.

In this chapter, the following interpolation approximations are presented.

1. Newton polynomial interpolation

2. Lagrange polynomial interpolation

3. Vandermonde approximation

4. Rational function interpolation

For other types of interpolation such as Chebyshev, Hermite and Legendre polynomials see the refs. [1–2]. Restrict ourselves to interpolation in one-dimension, that is, $y = f(x)$.

DOI: 10.1201/9781032649931-6

For interpolation in two or more dimensions refer the ref. [3]. The interpolation methods described in this chapter are also the methods for extrapolation.

6.2   Newton Interpolation Polynomial

Consider the following set of polynomials:

$$
\begin{aligned}
P_1(x) &= a_1 + a_2\,(x - x_1)\,, \\
P_2(x) &= a_1 + a_2\,(x - x_1) + a_3\,(x - x_1)\,(x - x_2)\,, \\
&\ \ \vdots \\
P_n(x) &= a_1 + a_2\,(x - x_1) + \cdots + a_{n+1}\,(x - x_1)\,(x - x_2)\cdots(x - x_n)\,.
\end{aligned}
\tag{6.1}
$$

The $P_n(x)$ can be rewritten as

$$
P_n(x) = \sum_{i=1}^{n+1} a_i \prod_{j=1}^{i-1} (x - x_j)\,.
\tag{6.2}
$$

The polynomial given by Eq. (6.2) is called a *Newton polynomial* of order n with n centres or nodes $x_1, x_2, \ldots, x_n$. The x_i's need not be equally spaced. For a given $n+1$ data $(x_i, y_i = f(x_i))$ the polynomial $P_n(x)$ can be constructed. The coefficients a_i's can be computed as follows.

6.2.1   Linear Interpolation

Linear interpolation means one wish to construct the polynomial

$$
P_1(x) = a_1 + a_2\,(x - x_1)\,,
\tag{6.3}
$$

where a_1 and a_2 are to be determined. To calculate a_1 and a_2 two sets of data $(x_1, f(x_1))$ and $(x_2, f(x_2))$ are sufficient. When $x = x_1$ Eq. (6.3) gives

$$
a_1 = f(x_1)\,.
\tag{6.4a}
$$

When $x = x_2$ one has

$$
f(x_2) = a_1 + a_2\,(x_2 - x_1) = f(x_1) + a_2\,(x_2 - x_1)
$$

or

$$
a_2 = \frac{f(x_2) - f(x_1)}{(x_2 - x_1)}\,.
\tag{6.4b}
$$

The truncation error $E_1(x)$ in the approximation is

$$
E_1(x) = f(x) - P_1(x)\,,
\tag{6.5}
$$

where $f(x)$ is the exact or true function and $P_1(x)$ is its polynomial approximation. To obtain an expression for the truncation error consider

$$
g(t) = f(t) - P_1(t) - [f(x) - P_1(x)]\,\frac{(t - a)(t - b)}{(x - a)(x - b)}\,.
\tag{6.6}
$$

Assume that $g(t)$ is zero at the two points $t = a$ and $t = b$. $g''(t)$ is given by

$$g''(t) = f''(t) - \frac{2\left[f(x) - P_1(x)\right]}{(x-a)(x-b)}. \tag{6.7}$$

According to Rolle's theorem (refer Section 1.6) if $g(t)$ is a continuous function on some interval $[a, b]$, differentiable on $[a, b]$ and $g(a) = g(b) = 0$ then there is at least one point c inside $[a, b]$ with $g''(c) = 0$. Since $g(t)$ given by Eq. (6.6) is 0 at $t = a = x_1$ and $t = b = x_2$ one can set $g''(c) = 0$. Then, Eq. (6.7) with the replacement $f(x) - P_1(x) = E_1(x)$ gives

$$E_1(x) = \frac{1}{2}(x - x_1)(x - x_2) f''(c), \tag{6.8}$$

where $c \in [x_1, x_2]$. E_1 is zero if $f(x)$ is a polynomial of order less than 2. E_1 can be calculated only if the analytical expression for $f(x)$ is known.

To obtain a bound for the truncation error denote the absolute maximum value of $f''(x)$ on $x \in [x_1, x_2]$ as $f''_m(x)$. Then,

$$|E_1| \le \frac{1}{2} f''_m(x) \max\left|(x - x_1)(x - x_2)\right|, \quad x \in [x_1, x_2]. \tag{6.9}$$

The maximum value of $u(x) = |(x - x_1)(x - x_2)|$ corresponds to the value of $x(= x_m)$ at which $u'(x) = 0$. This gives $x_m = (x_1 + x_2)/2$. Now,

$$\begin{aligned}
|E_1| &\le \frac{1}{2}\left|\left(\frac{x_1}{2} + \frac{x_2}{2} - x_1\right)\left(\frac{x_1}{2} + \frac{x_2}{2} - x_2\right)\right| f''_m(x) \\
&\le \frac{1}{8}(x_2 - x_1)^2 f''_m(x). \tag{6.10}
\end{aligned}$$

6.2.2 Examples

Now, present two examples to illustrate the linear interpolation.

Example 1:

The values of e^x at $x = 0, 1$ are 1 and 2.71828, respectively. Compute $e^{0.61}$. Obtain a bound on the truncation error.

Choosing $x_1 = 0$ and $x_2 = 1$ the values of a_1 and a_2 are calculated as

$$\begin{aligned}
a_1 &= f(x_1) = e^0 = 1, \\
a_2 &= \frac{f(x_2) - f(x_1)}{(x_2 - x_1)} = \frac{2.71828 - 1}{1} = 1.71828.
\end{aligned}$$

The linear polynomial approximation of e^x is

$$P_1(x) = a_1 + a_2 x = 1 + 1.71828x.$$

$e^{0.61}$ is then evaluated as 2.04815 while the exact value is 1.84043.

For the given function $f''(x)$ is e^x. Its maximum value in $x \in [0, 1]$ is e^1. Therefore,

$$|E| \le \frac{1}{8}(x_2 - x_1)^2 f''_m(x) = \frac{1}{8}e^1 = 0.3397852.$$

Given the data set $(x, f(x)) = (0, 1)$ and $(1, 2.71828)$ how do you compute the value of x at which $f(x)$ is, say, 2?

Example 2:

The value of $f(x) = e^x$ is given below at four values of x. Calculate $e^{0.61}$ using the linear interpolation.

x	0.0	0.5	0.7	1.0
e^x	1.0	1.64872	2.01375	2.71828

For the linear interpolation, only two data points are needed. One can use the nodes $(0, 1)$ or $(0, 0.7)$ or $(0, 0.5)$ or $(0.5, 0.7)$. The target point $x = 0.61$ is close to the nodes $(0.5, 0.7)$. Therefore, use of these nodes would produce relatively a higher accurate result than the other set of nodes. Using the nodes $(0.5, 0.7)$ the values of a_1 and a_2 are computed as

$$a_1 = 1.64872, \quad a_2 = 1.82520.$$

Then,

$$e^x = P_1(x) = 1.64872 + 1.82520(x - 0.5).$$

For $x = 0.61$

$$e^{0.61} = 1.849492$$

which is more accurate than the value obtained in the previous example using the nodes $(0, 1)$.

6.2.3   Calculation of Coefficients for Higher-Order Polynomials

Let us calculate the coefficients a_1, a_2 and a_3 for three-points set $(x_i, f(x_i))$, $i = 1, 2, 3$. Then, inspecting the sequence a_1, a_2, a_3 one can write the other coefficients. For $n + 1$ points the order of the polynomial is n. Therefore, for 3 points set the Newton polynomial is

$$P_2(x) = a_1 + a_2 (x - x_1) + a_3 (x - x_1)(x - x_2). \tag{6.11}$$

Substitution of $x = x_1$ in Eq. (6.11) gives

$$a_1 = f(x_1). \tag{6.12a}$$

Next, substituting $x = x_2$ one obtains

$$a_2 = \frac{f(x_2) - f(x_1)}{(x_2 - x_1)}. \tag{6.12b}$$

Then, $x = x_3$ in Eq. (6.11) gives

$$a_3 = \frac{1}{(x_3 - x_1)}\left[\frac{f(x_3) - f(x_2)}{(x_3 - x_2)} - \frac{f(x_2) - f(x_1)}{(x_2 - x_1)}\right]. \tag{6.12c}$$

Derivation of Eq. (6.12c) is left as an exercise to the reader.

Example:

Estimate $e^{0.61}$ using second-order Newton polynomial interpolation formula to the following table of data.

TABLE 6.1
Divided-difference table for the function $f(x)$.

i	x_i	$f[x_i]$	$f[\ ,\ ]$	$f[\ ,\ ,\ ]$	$\cdots$
1	x_1	$f[x_1]$			
2	x_2	$f[x_2]$	$f[x_1, x_2]$		
3	x_3	$f[x_3]$	$f[x_2, x_3]$	$f[x_1, x_2, x_3]$	
$\vdots$	$\vdots$	$\vdots$	$\vdots$	$\vdots$	$\vdots$

x	0.0	0.5	0.7
e^x	1.0	1.64872	2.01375

The coefficients a_1, a_2 and a_3 in Eq. (6.11) are obtained as

$$a_1 = f(x_1) = f(0) = 1,$$

$$a_2 = \frac{f(x_2) - f(x_1)}{(x_2 - x_1)} = \frac{f(0.5) - f(0)}{0.5 - 0} = 1.29744,$$

$$a_3 = \frac{1}{0.7 - 0}\left[\frac{f(0.7) - f(0.5)}{(0.7 - 0.5)} - \frac{f(0.5) - f(0)}{(0.5 - 0)}\right]$$

$$= [1.82515 - 1.29744]/0.7$$

$$= 0.7538714.$$

Then,

$$P_2(x) = 1 + 1.29744x + 0.7538714x(x - 0.5).$$

For $x = 0.61$ the result is $P_2(0.61) = 1.84202$.

In Eqs. (6.12) observe that a_1 is simply $f(x_1)$; a_2 and a_3 are functions of differences of $f(x_i)$'s. In view of this, the coefficients a_i's can be re-expressed in terms of the so-called *first-order divided-differences* as described below.

6.2.4 First-Order Divided-Differences

The divided-differences for a function $f(x)$ are defined as follows:

$$f[x_i] = f(x_i), \tag{6.13a}$$

$$f[x_{i-1}, x_i] = \frac{f[x_i] - f[x_{i-1}]}{(x_i - x_{i-1})}, \tag{6.13b}$$

$$f[x_{i-2}, x_{i-1}, x_i] = \frac{f[x_{i-1}, x_i] - f[x_{i-2}, x_{i-1}]}{(x_i - x_{i-2})} \tag{6.13c}$$

and so on. Let us call $f[x_{i-1}, x_i]$ as first divided-differences, $f[x_{i-2}, x_{i-1}, x_i]$ as second divided-differences and so on. Using $f[\ ]$'s a divided-difference Table 6.1 can be constructed.

In terms of the above-defined divided-differences the coefficients a_1, a_2 and a_3 given by Eqs. (6.12) are rewritten as

$$a_1 = f[x_1], \quad a_2 = f[x_1, x_2], \quad a_3 = f[x_1, x_2, x_3]. \tag{6.14}$$

TABLE 6.2
Divided-difference table for four nodes.

i	x_i	$f[x_i]$	$f[\ ,\ ]$	$f[\ ,\ ,\ ]$	$f[\ ,\ ,\ ,\ ]$
1	x_1	$f[x_1]$			
2	x_2	$f[x_2]$	$f[x_1, x_2]$		
3	x_3	$f[x_3]$	$f[x_2, x_3]$	$f[x_1, x_2, x_3]$	
4	x_4	$f[x_4]$	$f[x_3, x_4]$	$f[x_2, x_3, x_4]$	$f[x_1, x_2, x_3, x_4]$

In general

$$a_i = f[x_1, x_2, \ldots x_i] . \tag{6.15}$$

The a_i's are simply the diagonal elements in Table 6.1. The nodes x_i's need not be equally spaced. When the a_i's are equally spaced the polynomial $P_n(x)$ is called *Gregory–Newton interpolation polynomial*.

6.2.5 Error in the Approximation

If $f(x)$ is the true function and $P_n(x)$ is the Newton polynomial approximation then the error term $E_n(x)$ in $f(x) = P_n(x) + E_n(x)$ is given by

$$E_n(x) = \left(\prod_{i=1}^{n+1} (x - x_i) \right) \frac{f^{(n+1)}(c)}{(n+1)!} , \tag{6.16}$$

where $c \in [x_1, x_{n+1}]$. Calculation of $E_n(x)$ is possible if an analytical expression for $f(x)$ is known.

6.2.6 Examples

The following two examples illustrate the use of the divided-difference table.

Example 1:

The values of $f(x) = e^x$ at four nodes (which are not equally spaced) with five decimal accuracy are given below. Compute $e^{0.61}$ by constructing a third-order Newton interpolating polynomial by forming a divided-difference table.

x	0.0	0.5	0.7	1.0
e^x	1.0	1.64872	2.01375	2.71828

The third-order Newton interpolating polynomial is given by

$$\begin{aligned} P_3(x) &= a_1 + a_2 (x - x_1) + a_3 (x - x_1)(x - x_2) \\ &\quad + a_4 (x - x_1)(x - x_2)(x - x_3) . \end{aligned} \tag{6.17}$$

To avoid confusion or mistakes in calculating the entries in the difference table first write the table in terms of $f[\ ]$'s. For four nodes Table 6.2 is the corresponding divided-difference table.

TABLE 6.3
Divided-differences for function e^x calculated based on 4 nodes.

i	x_i	$f[x_i]$	$f[\,,\,]$	$f[\,,\,,\,]$	$f[\,,\,,\,,\,]$
1	x_1	1.00000			
2	x_2	1.64872	1.29744		
3	x_3	2.01375	1.82515	0.75387	
4	x_4	2.71828	2.34843	1.04657	0.29270

Now, replace each $f[\ ]$'s by their numerical number. When computing the value of $f[x_j, \ldots, x_k]$ remember that the denominator is $x_k - x_j$. When calculating the values of $f[\ ]$'s in 5th and 6th columns one may make a mistake by writing

$$f[x_1, x_2, x_3] = \frac{f[x_2, x_3] - f[x_1, x_2]}{(x_3 - x_2)}.$$

This is wrong. The denominator in $f[\ ]$'s is always the difference between the values of last x and first x in the arguments of $f[\ ]$'s. The correct $f[x_1, x_2, x_3]$ is

$$f[x_1, x_2, x_3] = \frac{f[x_2, x_3] - f[x_1, x_2]}{(x_3 - x_1)}.$$

Table 6.3 shows the divided-differences for the given function. From this table the coefficients a_1 to a_4 are obtained as

$$a_1 = 1, \quad a_2 = 1.29744, \quad a_3 = 0.75387, \quad a_4 = 0.29270.$$

The value of the polynomial at $x = 0.61$ is calculated as 1.84026 which is very close to the exact value 1.84043. $P(0.61)$ is calculated as a function of order n and the result is presented in Table 6.4. The accuracy increases with increase in the order n. The percentage of relative error is computed using

$$\text{Percentage of relative error} = 100 \left| (e^x - P_n(x))/e^x \right|.$$

TABLE 6.4
Calculated values of $P_n(x)$ at $x = 0.61$ and percentage of relative error for various order n of the polynomial (6.2) for e^x. The nodes are equally spaced in the interval $[0, 1]$. For $n = 2$ the nodes are $x_1 = 0$, $x_2 = 0.5$ and $x_3 = 1$; for $n = 3$ the nodes are $x_1 = 0$, $x_2 = 1/3$, $x_3 = 2/3$, $x_4 = 1$ and so on.

Order n	$P_n(x)$	% of error	Order n	$P_n(x)$	% of error
2	1.84792	0.417	5	1.84043	0.000
3	1.84017	0.014	6	1.84043	0.000
4	1.84041	0.001	7	1.84043	0.000

TABLE 6.5
Same as Table 6.4 but at the target value $x = 1.1$. The exact value of $e^{1.1}$ is 3.00417.

Order n	$P_n(x)$	% of error	Order n	$P_n(x)$	% of error
2	2.98269	0.715	5	3.00414	0.001
3	3.00130	0.096	6	3.00416	0.000
4	3.00386	0.010	7	3.00417	0.000

The constructed polynomial can be used to evaluate its value at a point falling outside the interval $[0, 1]$. The value of $P_3(1.1)$ (extrapolated value) computed from Eq. (6.17) is 3.00201 while the exact value is 3.00417. Table 6.5 gives the extrapolated value of e^x at $x = 1.1$ as a function of order n.

Example 2:

From the table given below, form the difference table and determine the maximum degree of the polynomial to which the data can be interpolated.

x	0	1	2	3	4	5	6
$f(x)$	2	3	10	29	66	127	218

Table 6.6 shows the divided-differences for the given data set. The divided-differences $f[\ ,\ ,\ ]$ are all identical, that is, the 4th and the other higher divided-differences become zero. Therefore, the coefficients a_5, a_6 and a_7 are zero. For a 7-node data set the maximum degree (or order) of the Newton polynomial is 6. However, for the given data set the maximum degree of the polynomial is 3. In general, if the nth divided-differences are all identical then the degree of the polynomial to which a data set can be interpolated is n.

6.2.7   Other Uses of Newton Polynomials

Derivatives and integral values of an unknown function at a desired value of x lying in the interval $[x_1, x_{n+1}]$ can also be obtained from Eq. (6.2). For example, consider $P_2(x)$. Its

TABLE 6.6
Divided-differences for the given data set in Example 2.

i	x_i	$f[x_i]$	$f[\ ,\ ]$	$f[\ ,\ ,\ ]$	$f[\ ,\ ,\ ,\ ]$
1	0	2			
2	1	3	1		
3	2	10	7	3	
4	3	29	19	6	1
5	4	66	37	9	1
6	5	127	61	12	1
7	6	218	91	15	1

TABLE 6.7

Table 6.1 in terms of a two-dimensional array $a_{i,j}$.

i	x_i	$f[x_i]$	$f[\,,\,]$	$f[\,,\,,\,]$	$\cdots$
1	x_1	$a_{1,1}$			
2	x_2	$a_{2,1}$	$a_{2,2}$		
3	x_3	$a_{3,1}$	$a_{3,2}$	$a_{3,3}$	
$\vdots$	$\vdots$	$\vdots$	$\vdots$	$\vdots$	$\vdots$

derivative is

$$P_2'(x) = a_2 + a_3\left[(x - x_1) + (x - x_2)\right].\qquad(6.18)$$

In this way, derivative of $P_n(x)$ can be constructed (see Problem 6.4 at the end of this chapter).

6.2.8 Program Aspect to Construct Newton Polynomials

For developing a program to construct Newton polynomials the Table 6.1 can be rewritten in terms of a two-dimensional array $a_{i,j}$ leading to Table 6.7.

Comparison of the Tables 6.1 and 6.7 gives

$$a_{i,1} = f[x_i],\quad a_{i,2} = f[x_{i-1}, x_i],\qquad(6.19a)$$
$$a_{i,3} = f[x_{i-2}, x_{i-1}, x_i]\qquad(6.19b)$$

and so on. The above set of equations can be arranged in a simpler form as

$$a_{i,1} = y_i = f[x_i],\quad i = 1, 2, \ldots, n+1\qquad(6.20a)$$
$$a_{i,j} = f[x_{i-j+1}, x_{i-j+2}, \ldots, x_i],\quad i = 2, 3, \ldots, n+1,\ j \le i.\qquad(6.20b)$$

Using Eqs. (6.13) a_{ij}'s are given by

$$a_{i,1} = y_i,\quad i = 1, 2, \ldots, n+1\qquad(6.21a)$$
$$a_{i,j} = \frac{(a_{i,j-1} - a_{i-1,j-1})}{x_i - x_{i-j+1}},\quad \begin{matrix} j = 2, 3, \ldots, n+1, \\ i = j, j+1, \ldots, n+1. \end{matrix}\qquad(6.21b)$$

Then, the Newton polynomial of order n is given by

$$\begin{aligned} P_n(x) = &\ a_{1,1} + a_{2,2}(x - x_1) + a_{3,3}(x - x_1)(x - x_2) \\ &\cdots + a_{n+1,n+1}(x - x_1)(x - x_2)\cdots(x - x_n). \end{aligned}\qquad(6.22)$$

6.3 Gregory–Newton Interpolation Polynomials

Suppose, the nodes are equally spaced: $x_{i+1} = x_1 + ih$, $i = 1, 2, \ldots, n$. In this case, it is convenient to introduce forward- and backward-difference operators which will avoid division in (6.13).

TABLE 6.8
Forward-difference table for a function $f(x)$.

i	x_i	$f(x_i)$	Δf	$\Delta^2 f$	$\cdots$
1	x_1	f_1			
2	x_2	f_2	Δf_1		
3	x_3	f_3	Δf_2	$\Delta^2 f_1$	
$\vdots$	$\vdots$	$\vdots$	$\vdots$	$\vdots$	$\vdots$

6.3.1   Gregory–Newton Forward-Difference Interpolation

Define the forward-difference operator $\Delta f(x_i)$ as

$$\Delta f(x_i) = \Delta f_i = f(x_i + h) - f(x_i) = f_{i+1} - f_i. \tag{6.23}$$

In general, an nth order forward-difference operator $\Delta^n f_i$ is given by

$$\Delta^n f_i = \sum_{k=0}^{n} (-1)^k \frac{n!}{k!(n-k)!} f_{i+n-k}. \tag{6.24}$$

Then, the forward-difference Table 6.8 similar to Table 6.1 can be formulated. The a_i's given by Eqs. (6.14) and (6.15) are written in terms of the forward-difference operators as

$$
\begin{aligned}
a_1 &= f_1, \\
a_2 &= \frac{(f_2 - f_1)}{(x_2 - x_1)} = \frac{1}{h} \Delta f_1, \\
a_3 &= \frac{f_3 - f_2}{2h^2} - \frac{f_2 - f_1}{2h^2} = \frac{1}{2h^2} (\Delta f_2 - \Delta f_1) = \frac{1}{2h^2} \Delta^2 f_1, \\
&\vdots \\
a_{n+1} &= \frac{1}{n! h^n} \Delta^n f_1.
\end{aligned}
\tag{6.25}
$$

Now, the Newton polynomial $P_n(x)$ given by Eq. (6.2) takes the form

$$
\begin{aligned}
P_n(x) &= a_1 + a_2 (x - x_1) + a_3 (x - x_1)(x - x_2) + \cdots \\
&\quad \cdots + a_n (x - x_1)(x - x_2) \cdots (x - x_{n-1}) \\
&= f_1 + \frac{(x - x_1)}{h} \Delta f_1 + \frac{(x - x_1)(x - x_2)}{2! h^2} \Delta^2 f_1 + \cdots \\
&\quad + \frac{(x - x_1) \cdots (x - x_n)}{n! h^n} \Delta^n f_1.
\end{aligned}
\tag{6.26}
$$

This polynomial is called *Gregory–Newton forward-difference interpolation polynomial*. The error term in the approximation is given by Eq. (6.16).

6.3.2   Gregory–Newton Backward-Difference Interpolation

Like the forward-difference operator define the backward-difference operator $\nabla f(x_i)$ as

$$\nabla f(x_i) = \nabla f_i = f(x_i) - f(x_i - h). \tag{6.27}$$

TABLE 6.9
Backward-difference table for the function $f(x)$ whose values are specified at 4 nodes. Similarly, a table for n nodes can be constructed.

i	x_i	$f(x_i)$	∇f	$\nabla^2 f$	$\nabla^3 f$
1	x_1	$f(x_1)$	∇f_2	$\nabla^2 f_3$	$\nabla^3 f_4$
2	x_2	$f(x_2)$	∇f_3	$\nabla^2 f_4$	
3	x_3	$f(x_3)$	∇f_4		
4	x_4	$f(x_4)$			

The nth order backward-difference operator $\nabla^n f_i$ is given by

$$\nabla^n f_i = \sum_{k=0}^{n} (-1)^n \frac{n!}{k!(n-k)!} f_{i-k}. \tag{6.28}$$

In this case, the backward-difference quantities form the Table 6.9. The Gregory–Newton backward-difference interpolation polynomial with $n+1$ data is then obtained as

$$\begin{aligned} P_n(x) &= f_{n+1} + \frac{1}{h}(x - x_{n+1})\nabla f_{n+1} \\ &\quad + \frac{1}{2!h^2}(x - x_{n+1})(x - x_n)\nabla^2 f_{n+1} \\ &\quad + \cdots \\ &\quad + \frac{1}{n!h^n}(x - x_{n+1})\cdots(x - x_2)\nabla^n f_{n+1}. \end{aligned} \tag{6.29}$$

Example:

The value of $f(x) = e^x$ is given below at 4 values of x. Calculate $e^{0.61}$ using the Gregory–Newton forward- and backward-difference polynomials.

x	0.0	0.25	0.50	0.75
$f(x)$	1.0	1.28403	1.64872	2.117

The forward-difference Table 6.10 is constructed. This table gives

$$h = 0.25, \quad n+1 = 4, \quad f_1 = 1, \quad \Delta f_1 = 0.28403,$$
$$\Delta^2 f_1 = 0.08066, \quad \Delta^3 f_1 = 0.02293.$$

Substitution of these values in the polynomial

$$\begin{aligned} P(x) &= f_1 + \frac{1}{h}(x - x_1)\Delta f_1 + \frac{1}{2!h^2}(x - x_1)(x - x_2)\Delta^2 f_1 \\ &\quad + \frac{1}{3!h^3}(x - x_1)(x - x_2)(x - x_3)\Delta^3 f_1 \end{aligned}$$

gives

$$\begin{aligned} P(x) &= 1 + 1.13612x + 0.64528x(x - 0.25) \\ &\quad + 0.24459x(x - 0.25)(x - 0.5), \\ P(0.61) &= 1 + 0.69303 + 0.14170 + 0.00591 = 1.84064. \end{aligned}$$

TABLE 6.10
The forward-difference table for the given four-points data set for $f(x) = e^x$.

i	x_i	f	Δf	$\Delta^2 f$	$\Delta^3 f$
1	0.0	1.0			
2	0.25	1.28403	0.28403		
3	0.50	1.64872	0.36469	0.08066	
4	0.75	2.11700	0.46828	0.10359	0.02293

Next, apply the Gregory–Newton backward-difference formula. In this case, Table 6.11 is obtained. For $n + 1 = 4$, that is, $n = 3$ the polynomial is

$$P(x) = f_4 + \frac{1}{h}(x - x_4)\nabla f_4 + \frac{1}{2!h^2}(x - x_4)(x - x_3)\nabla^2 f_4$$
$$+ \frac{1}{3!h^3}(x - x_4)(x - x_3)(x - x_2)\nabla^3 f_4. \tag{6.30}$$

Table 6.11 gives

$$h = 0.25, \quad f_4 = 2.11700, \quad \nabla f_4 = 0.46828,$$
$$\nabla^2 f_4 = 0.10359, \quad \nabla^3 f_4 = 0.2293.$$

Substitution of these values in Eq. (6.30) gives

$$P(x) = 2.117 + 1.87312(x - 0.75) + 0.82872(x - 0.75)(x - 0.5)$$
$$+ 0.24459(x - 0.75)(x - 0.5)(x - 0.25).$$

For $x = 0.61$

$$P(0.61) = 2.117 - 0.26224 - 0.01276 - 0.00135 = 1.84065.$$

6.4   Lagrange Interpolation Polynomials

Joseph Louis Lagrange introduced a method slightly different from the Newton polynomial method to construct a polynomial passing through a given set of points. Instead of the form

TABLE 6.11
The backward-difference table for the given four-points data set for the function e^x.

i	x_i	f	∇f	$\nabla^2 f$	$\nabla^3 f$
1	0.0	1.0	0.28403	0.08066	0.02293
2	0.25	1.28403	0.36469	0.10359	
3	0.50	1.64872	0.46828		
4	0.75	2.11700			

given by Eq. (6.2) Lagrange assumed the polynomial of the form

$$P(x) \;=\; a_1 + a_2 x + \cdots + a_{n+1} x^n = \sum_{i=1}^{n+1} a_i x^{i-1}. \tag{6.31}$$

First, consider the method for the simplest case of two points.

6.4.1 Lagrange Linear Polynomial

For a set of two points, the polynomial to be constructed is a linear function

$$P_1(x) = a_1 + a_2 x. \tag{6.32}$$

Note the difference between this $P_1(x)$ and the $P_1(x)$ given by Eq. (6.3). For $x = x_1$ and $x = x_2$ Eq. (6.32) becomes

$$f(x_1) \;=\; P_1(x_1) = a_1 + a_2 x_1, \tag{6.33a}$$
$$f(x_2) \;=\; P_1(x_2) = a_1 + a_2 x_2. \tag{6.33b}$$

Solving Eqs. (6.33) for a_1 and a_2 gives

$$a_1 \;=\; \frac{x_2 f(x_1) - x_1 f(x_2)}{x_2 - x_1}, \tag{6.34a}$$

$$a_2 \;=\; \frac{f(x_2) - f(x_1)}{x_2 - x_1}. \tag{6.34b}$$

Substitution of the above expressions for a_1 and a_2 in Eq. (6.32) leads to the following functional form for $P_1(x)$:

$$\begin{aligned} P_1(x) \;&=\; \frac{x_2 f(x_1) - x_1 f(x_2) + x\,[f(x_2) - f(x_1)]}{(x_2 - x_1)} \\[2mm] &=\; f(x_1)\frac{(x - x_2)}{(x_1 - x_2)} + f(x_2)\frac{(x - x_1)}{(x_2 - x_1)}. \end{aligned} \tag{6.35}$$

Defining

$$L_1 = \frac{(x - x_2)}{(x_1 - x_2)}, \quad L_2 = \frac{(x - x_1)}{(x_2 - x_1)} \tag{6.36}$$

Eq. (6.35) is rewritten as

$$P_1(x) = f(x_1) L_1(x) + f(x_2) L_2(x). \tag{6.37}$$

The error term $E_1(x)$ is same as the one given for the Newton polynomial interpolation.

Example:

Compute the value of $f(x) = e^x$ at $x = 0.61$ using the data $(x, f(x)) = (0, 1), (1, 2.71828)$.

The functions L_1 and L_2 are calculated as

$$L_1(x = 0.61) \;=\; \frac{(x - x_2)}{(x_1 - x_2)} = \frac{0.61 - 1.0}{0.0 - 1.0} = 0.39.$$

$$L_2(x = 0.61) \;=\; \frac{(x - x_1)}{(x_2 - x_1)} = \frac{0.61 - 0.0}{1.0 - 0.0} = 0.61.$$

Then,

$$
\begin{aligned}
P_1(x = 0.61) &= f(x_1) L_1(x) + f(x_2) L_2(x) \\
&= 0.39 \times 1 + 0.61 \times 2.71828 \\
&= 2.04815.
\end{aligned}
$$

6.4.2 Higher-Order Lagrange Polynomials

The Lagrange interpolation polynomials based on $n+1$ points can be obtained in a way described for two points. They are given by

$$
f(x) = P_n(x) = \sum_{i=1}^{n+1} f(x_i) L_i(x), \tag{6.38a}
$$

where

$$
L_i(x) = \prod_{\substack{j=1 \\ j \neq i}}^{n+1} \left(\frac{x - x_j}{x_i - x_j} \right), \quad i = 1, 2, \ldots, n+1. \tag{6.38b}
$$

In Eq. (6.38a) the term L_i is made of product of n terms with each being of the form $(x - x_j)/(x_i - x_j)$. The factor $(x - x_j)/(x_i - x_i)$ is excluded in the product in Eq. (6.38b) for obvious reason. The functions L_i are called the *Lagrange fundamental polynomials* or *Lagrange coefficients polynomials*. Each L_i is a polynomial of order n. $P_n(x)$ based on $n+1$ points is called *nth order Lagrange polynomial*. The error in the approximation is same as the one given by Eq. (6.16) for the Newton polynomial of order n.

Example:

The values of $f(x) = e^x$ at three nodes with five decimal accuracy are given below. Compute $e^{0.61}$ and $e^{1.1}$ by constructing the Lagrange polynomial $P_2(x)$.

x	0.0	0.50	1
$f(x)$	1.0	1.64872	2.71828

For a three-points data set, the Lagrange polynomial is

$$
P_2(x) = f(x_1) L_1(x) + f(x_2) L_2(x) + f(x_3) L_3(x).
$$

L_i's in the above equation are computed as

$$
\begin{aligned}
L_1(x = 0.61) &= \frac{(x - x_2)(x - x_3)}{(x_1 - x_2)(x_1 - x_3)} \\
&= \frac{0.11 \times -0.39}{-0.5 \times -1.0} \\
&= -0.08580. \\
L_2(x = 0.61) &= \frac{(x - x_1)(x - x_3)}{(x_2 - x_1)(x_2 - x_3)} = 0.95160. \\
L_3(x = 0.61) &= \frac{(x - x_1)(x - x_2)}{(x_3 - x_1)(x_3 - x_2)} = 0.1342.
\end{aligned}
$$

TABLE 6.12
The numerically computed value of $e^{0.61}$ by the Lagrange interpolating polynomial approximation.

Order n	$P_n(x)$ value	% of error	Order n	$P_n(x)$ value	% of error
2	1.84792	0.407	5	1.84043	0.000
3	1.84017	0.014	6	1.84043	0.000
4	1.84041	0.001	7	1.84043	0.000

For $x = 0.61$

$$
\begin{aligned}
P_2(0.61) &= 1 \times -0.08580 + 1.64872 \times 0.95160 + 2.71828 \times 0.1342 \\
&= 1.84792.
\end{aligned}
$$

The exact value is 1.84043. Accuracy can be improved by considering more points in the interval $[0, 1]$. Table 6.12 displays the computed $P_n(0.61)$ for e^x as a function of order n. The value of $e^{1.1}$ computed with the given 3 data points is 2.98269 whereas the exact value is 3.00417.

6.4.3 Interpolation with Large Number of Nodes

In the Newton and Lagrange polynomial interpolations the spacing between the successive nodes in the interval $[a, b]$ need not be the same. Suppose, the number of nodes is large, say, greater than 50 and the nodes are equally spaced. The fitted polynomial perfectly passes through all the nodes. However, for the x values between the nodes near the end points a and b the interpolated values of the constructed polynomial may highly deviate from the true values and the plot of x versus $P_n(x)$ will show wiggles. In such a case if the choice of the nodes is free then it is better to choose more points near the end points a and b. Now, the spacings between the successive nodes are not the same. The $P_n(x)$ obtained with a large value of n does not produce wiggles [4]. One choice for unequally spaced nodes with clustering near the end points a and b is the Chebyshev points. For the interval $[-1, 1]$ these points are given by

$$
x_i = -\cos\left(\frac{(i-1)\pi}{n}\right), \quad i = 1, 2, \ldots, n+1. \tag{6.39}
$$

Here, $x_1 = -1$ and $x_{n+1} = 1$. For the interval $[a, b]$ the Chebyshev points x_i given by Eq. (6.39) can be rescaled as

$$
x_i = \frac{1}{2}(x_i + 1)(b - a) + a. \tag{6.40}
$$

Let us illustrate the above with an example in the case of Lagrange interpolation.

Consider the function

$$
f(x) = \frac{1}{1 + \sqrt{1 - x^2}}, \quad x \in [0, 1]. \tag{6.41}
$$

In Fig. 6.1a continuous line represents the exact $f(x)$ given by Eq. (6.41). The filled circles are the nodes ($n+1 = 11$) chosen for the Lagrange interpolation. The order of the polynomial

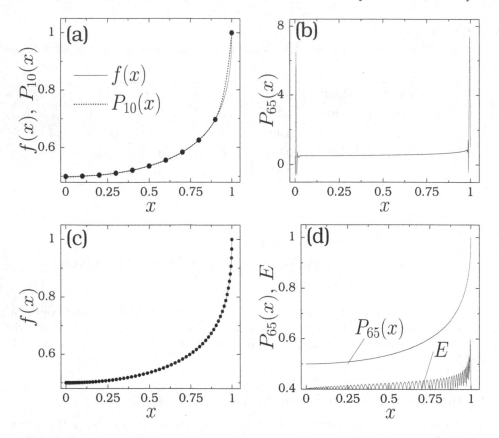

FIGURE 6.1

(a) x versus $f(x)$ (continuous line) and x versus $P_{10}(x)$. The 11 nodes used to construct the Lagrange polynomial $P_{10}(x)$ are the filled circles. (b) x versus $P_{65}(x)$ obtained from uniformly spaced 66 nodes. (c) x versus $f(x)$ (continuous line) and the nonuniformly spaced Chebyshev points. (d) Variation of $P_{65}(x)$ constructed with the Chebyshev nodes. E is the percentage of relative error in $P_{65}(x)$. For convenience E in this subplot is shifted by 0.4.

obtained from these 11 nodes is $n = 10$. The Lagrange polynomial $P_{10}(x)$ is constructed. The numerical values of $P_{10}(x)$ are computed for $m = 501$ target values of x with $x_i = (i-1)\Delta x$, $i = 1, 2, \ldots, m$ and $\Delta x = 1/(m-1)$. $P_{10}(x)$ is shown in Fig. 6.1a as the dotted line. A smooth variation of $P_{10}(x)$ is clearly seen near the end points.

Next, $P_{65}(x)$ is constructed with equally spaced 66 nodes. Then, $P_{65}(x)$ is calculated for $m = 501$ target values of x. The variation of the obtained $P_{65}(x)$ is plotted in Fig. 6.1b, where the wiggles are clearly seen near the end points $x = 0$ and $x = 1$. To eliminate the occurrence of wiggles in the constructed Lagrange polynomial, now, choose the 66 nodes as the Chebyshev points using Eqs. (6.39) and (6.40). For the function $f(x)$ given by Eq. (6.41) the generated nodes are shown in Fig. 6.1c as filled circles and $f(x)$ by the continuous line. The spacings between the successive nodes are not the same and moreover the nodes are clustered near the end points of the interval $[0, 1]$. Using these 66 nodes $P_{65}(x)$ is constructed. Figure 6.1d presents x versus $P_{65}(x)$ with 501 target values of x. The obtained polynomial is now a smooth curve between all the nodes without wiggles. The percentage of relative errors $E(x)$ shown in Fig. 6.1d is less than 0.2.

6.5 Calculation of Polynomial Coefficients

It is possible to construct an interpolating polynomial of the form

$$y(x) = c_1 + c_2 x + c_3 x^2 + \cdots + c_n x^{n-1} \tag{6.42}$$

from the given n data points (x_i, y_i). Substituting $x = x_1, x_2, \ldots, x_n$ and $y = y_1, y_2, \ldots, y_n$ in Eq. (6.42) generate a system of n linear equations for the coefficients c_i's which in matrix form is given by

$$\begin{pmatrix} 1 & x_1 & x_1^2 & \cdots & x_1^{n-1} \\ 1 & x_2 & x_2^2 & \cdots & x_2^{n-1} \\ \vdots & \vdots & \vdots & \vdots & \vdots \\ 1 & x_n & x_n^2 & \cdots & x_n^{n-1} \end{pmatrix} \begin{pmatrix} c_1 \\ c_2 \\ \vdots \\ c_n \end{pmatrix} = \begin{pmatrix} y_1 \\ y_2 \\ \vdots \\ y_n \end{pmatrix}. \tag{6.43}$$

The first $(n \times n)$ matrix in the left-side of Eq. (6.43) is known as *Vandermonde matrix* (see Section 4.11). The system (6.43) can be solved by the Gauss–Jordan method. However, for Vandermonde matrix system a special method is available. For details see Section 4.11 and also the ref. [3].

The method is tested for the function $y = e^x$ using the data $(x, y) = (0, 1)$, $(0.15, 1.16183)$, $(0.5, 1.64872)$, $(0.7, 2.013750)$, $(1, 2.71828)$. The polynomial constructed is

$$P(x) = 1 + 0.99929x + 0.50743x^2 + 0.14396x^3 + 0.06761x^4. \tag{6.44}$$

Don't compare the polynomial (6.44) with the Taylor series of e^x. The value of $P(0.61)$ is calculated as 1.84041 while the exact value is $e^{0.61} = 1.84043$.

6.6 Cubic Spline Interpolation

Another interpolation method of interest is the cubic spline interpolation. It uses polynomial of degree 3 to connect points. This method assumes that for a given data set $(x_0, f(x_0)), \ldots, (x_n, f(x_n))$ the second derivatives $f''(x_0)$ and $f''(x_n)$ are 0. The remaining second derivatives are obtained by solving the set of equations

$$(x_i - x_{i-1}) f''(x_{i-1}) + 2(x_{i+1} - x_{i-1}) f''(x_i) + (x_{i+1} - x_i) f''(x_{i+1})$$

$$= \frac{6}{(x_{i+1} - x_i)} [f(x_{i+1}) - f(x_i)]$$

$$+ \frac{6}{(x_i - x_{i-1})} [f(x_{i-1}) - f(x_i)], \quad i = 1, 2, \ldots, n-1. \tag{6.45}$$

Then, the cubic function for each interval is given by

$$\begin{aligned} f_i(x) &= \frac{f''(x_{i-1})}{6(x_i - x_{i-1})} (x_i - x)^3 + \frac{f''(x_i)}{6(x_i - x_{i-1})} (x - x_{i-1})^3 \\ &+ \left| \frac{f(x_{i-1})}{(x_i - x_{i-1})} - \frac{1}{6} f''(x_{i-1})(x_i - x_{i-1}) \right| (x_i - x) \\ &+ \left| \frac{f(x_i)}{(x_i - x_{i-1})} - \frac{1}{6} f''(x_i)(x_i - x_{i-1}) \right| (x - x_{i-1}). \end{aligned} \tag{6.46}$$

6.7　Rational Function Approximation

In the early sections, polynomial approximations of continuous functions are described. A polynomial approximation is not always the best approximation. For functions behaving as rational functions (that is, quotient of two polynomials), assuming finite values for very large values of x and changing rapidly in certain small regions like poles the polynomial approximation gives poor results. In these cases, rational function approximation is of useful. In this section, a rational function approximation, namely, Padé approximation is described.

6.7.1　Rational Function

A rational function is a quotient of two polynomials and is written as

$$R_{n,m}(x) = \frac{P_n(x)}{Q_m(x)} = \frac{\sum_{i=0}^{n} a_i x^i}{\sum_{i=0}^{m} b_i x^i}. \qquad (6.47)$$

The numerator and denominator are polynomials of order n and m, respectively. Functions such as e^x, $\log(1+x)$, $\sin x$, etc. can be approximated as rational functions. For example, e^x can be approximated as

$$e^x \approx \frac{1+\frac{x}{2}}{1-\frac{x}{2}}, \quad \frac{1+\frac{x}{2}+\frac{x^2}{12}}{1-\frac{x}{2}+\frac{x^2}{12}}, \quad \frac{1+\frac{x}{2}+\frac{x^2}{12}+\frac{x^3}{120}}{1-\frac{x}{2}+\frac{x^2}{12}-\frac{x^3}{120}}, \ldots \qquad (6.48)$$

Consider a kind of rational function called *Padé approximant*. It is a rational function which can also be expressed as the power series expansion

$$f(x) = \sum_{i=0}^{\infty} c_i x^i, \quad R(0) = f(0) \qquad (6.49)$$

and

$$\frac{\mathrm{d}^i}{\mathrm{d}x^i} R(x)\big|_{x=0} = \frac{\mathrm{d}^i}{\mathrm{d}x^i} f(x)\big|_{x=0}, \quad i = 0, 1, 2, \ldots, n+m. \qquad (6.50)$$

6.7.2　Analytical Construction of Rational Functions

The problem is to construct an analytical expression for $R_{n,m}(x)$ using the given analytic expression $f(x)$. For example, it is desired to find $R_{n,m}(x)$ for, say, e^x or $\sin x$. For this purpose, consider the equation

$$f(x) - R_{n,m}(x) = 0. \qquad (6.51)$$

Use of Eqs. (6.47) and (6.49) in Eq. (6.51) gives

$$\left(\sum_{i=0}^{m} b_i x^i \right) \left(\sum_{i=0}^{\infty} c_i x^i \right) - \sum_{i=0}^{n} a_i x^i = 0. \qquad (6.52)$$

c_i's are obtained from the power series approximation of $f(x)$ given by Eq. (6.49). Now, a_i's and b_i's are to be determined. Equating the coefficients of various powers of x^i in Eq. (6.52) one gets $m+n+1$ linear systems of equations for the unknowns a_i's and b_i's. These equations are given below, where $b_0 = 1$:

$$
\begin{aligned}
c_0 - a_0 &= 0, \\
b_1 c_0 + c_1 - a_1 &= 0, \\
b_2 c_0 + b_1 c_1 + c_2 - a_2 &= 0, \\
&\;\;\vdots \\
b_m c_{n-m} + b_{m-1} c_{n-m+1} + \cdots + c_n - a_n &= 0
\end{aligned}
\tag{6.53}
$$

and

$$
\begin{aligned}
b_m c_{n-m+1} + b_{m-1} c_{n-m+2} + \cdots + b_1 c_n + c_{n+1} &= 0, \\
&\;\;\vdots \\
b_m c_n + b_{m-1} c_{n+1} + \cdots + b_1 c_{n+m-1} + c_{n+m} &= 0.
\end{aligned}
\tag{6.54}
$$

Equations (6.54) involve b_i's and c_i's only and hence they can be solved first. Then, Eqs. (6.53) can be used successively to compute a_i's. For the special case of $m = n$ Eqs. (6.53) and (6.54) take the form [3]

$$
a_0 = c_0, \quad b_0 = 1, \quad \sum_{i=1}^{n} b_i c_{n-i+j} = -c_{n+j}, \quad j = 1, 2, \ldots, n
\tag{6.55a}
$$

$$
\sum_{i=0}^{j} b_i c_{j-i} = a_j, \quad j = 1, 2, \ldots, n.
\tag{6.55b}
$$

Example:

Construct the Padé approximations $R_{1,1}, R_{2,2}, R_{3,3}$ and $R_{4,4}$ for e^x.

The series expansion of e^x is

$$
e^x = \sum_{i=0}^{\infty} \frac{1}{i!} x^i = \sum_{i=0}^{\infty} c_i x^i .
\tag{6.56}
$$

Here $c_i = 1/i!$. First few c_i's are $c_0 = 1$, $c_1 = 1$, $c_2 = 1/2$, $c_3 = 1/6$ and $c_4 = 1/24$. For $R_{1,1}$ the values of m and n are 1. Now, Eqs. (6.55) become

$$
a_0 = c_0, \quad b_0 = 1, \quad b_1 c_1 = -c_2, \quad b_0 c_1 + b_1 c_0 = a_1 .
\tag{6.57}
$$

Substituting the values of c's, $b_0 = 1$ and $a_0 = c_0 = 1$ the above equations become $b_1 = -1/2$, $1 + b_1 = a_1$. That is,

$$
a_0 = 1, \quad b_0 = 1, \quad b_1 = -1/2, \quad a_1 = 1/2 .
\tag{6.58}
$$

The approximation $R_{1,1}$ is then

$$
R_{1,1} = \frac{a_0 + a_1 x}{b_0 + b_1 x} = \frac{1 + \frac{1}{2} x}{1 - \frac{1}{2} x} .
\tag{6.59}
$$

For $R_{2,2}$, $m = n = 2$ and Eqs. (6.55) are

$$
a_0 = c_0, \quad b_0 = 1, \quad b_1 c_2 + b_2 c_1 = -c_3, \quad b_1 c_3 + b_2 c_2 = -c_4,
\tag{6.60a}
$$

$$
b_0 c_1 + b_1 c_0 = a_1, \quad b_0 c_2 + b_1 c_1 + b_2 c_0 = a_2.
\tag{6.60b}
$$

TABLE 6.13
Comparison of the exact value of e^x with the rational function approximations $R_{1,1}$, $R_{2,2}$, $R_{3,3}$, $R_{4,4}$ at several values of x.

x	e^x	$R_{1,1}$	$R_{2,2}$	$R_{3,3}$	$R_{4,4}$
-0.5	0.60653	0.60000	0.60656	0.60653	0.60653
-0.4	0.67032	0.66667	0.67033	0.67032	0.67032
-0.3	0.74082	0.73193	0.74082	0.74082	0.74082
-0.2	0.81873	0.81818	0.81873	0.81873	0.81873
-0.1	0.90484	0.90476	0.90484	0.90484	0.90484
0.0	1.00000	1.00000	1.00000	1.00000	1.00000
0.1	1.10517	1.10526	1.10517	1.10517	1.10517
0.2	1.22140	1.22222	1.22140	1.22140	1.22140
0.3	1.34986	1.35294	1.34985	1.34986	1.34986
0.4	1.49182	1.50000	1.49180	1.49182	1.49182
0.5	1.64872	1.66667	1.64865	1.64872	1.64872

Substituting the values of c's the above set of equations become

$$a_0 = 1, \quad b_0 = 1, \quad \frac{b_1}{2} + b_2 = -\frac{1}{6}, \quad \frac{b_1}{6} + \frac{b_2}{2} = -\frac{1}{24}, \tag{6.61a}$$

$$1 + b_1 = a_1, \quad \frac{1}{2} + b_1 + b_2 = a_2. \tag{6.61b}$$

The solution of the above system of equations is

$$a_0 = 1, \quad b_0 = 1, \quad a_1 = \frac{1}{2}, \quad a_2 = \frac{1}{12}, \quad b_1 = -\frac{1}{2}, \quad b_2 = \frac{1}{12}. \tag{6.62}$$

Then,

$$R_{2,2} = \frac{a_0 + a_1 x + a_2 x^2}{b_0 + b_1 x + b_2 x^2} = \frac{1 + \frac{1}{2}x + \frac{1}{12}x^2}{1 - \frac{1}{2}x + \frac{1}{12}x^2}. \tag{6.63}$$

Proceeding further with $m = n = 3, 4$ one obtains

$$R_{3,3} = \frac{1 + \frac{1}{2}x + \frac{1}{10}x^2 + \frac{1}{120}x^3}{1 - \frac{1}{2}x + \frac{1}{10}x^2 - \frac{1}{120}x^3}, \tag{6.64}$$

$$R_{4,4} = \frac{1 + \frac{1}{2}x + \frac{3}{28}x^2 + \frac{1}{84}x^3 + \frac{1}{1680}x^4}{1 - \frac{1}{2}x + \frac{3}{28}x^2 - \frac{1}{84}x^3 + \frac{1}{1680}x^4}. \tag{6.65}$$

Table 6.13 displays the values of e^x, $R_{1,1}$, $R_{2,2}$, $R_{3,3}$ and $R_{4,4}$ for several values of x. Successive approximations converge to the exact value of e^x.

6.7.3   Rational Function Interpolation

Stoer and Bulirsch [5] described an algorithm to compute the value of a function using the given data set (x_i, y_i) by rational function approximation. The algorithm is the following.

TABLE 6.14
Illustration of construction of various rational functions R's.

x_i	y_i		R's		
x_1	$y_1 = R_1$				
x_2	$y_2 = R_2$	$R_{1,2}$			
x_3	$y_3 = R_3$	$R_{2,3}$	$R_{1,2,3}$		
x_4	$y_4 = R_4$	$R_{3,4}$	$R_{2,3,4}$	$R_{1,2,3,4}$	
x_5	$y_5 = R_5$	$R_{4,5}$	$R_{3,4,5}$	$R_{2,3,4,5}$	
$\vdots$	$\vdots$	$\vdots$	$\vdots$	$\vdots$	$\vdots$

Suppose $(x_i, y_i = f(x_i))$ are n given data points. Let us denote R_i be the value of the function $f(x)$ at x_i, that is, $R_i = f(x_i)$. Then, define $R_{1,2}$ as the value of f at x of a rational function approximation, passing through both (x_1, y_1) and (x_2, y_2). Similarly, $R_{2,3}$, $R_{3,4}$, ..., $R_{1,2,3}$, $R_{2,3,4}$, ... can be defined. In this way, one can construct $R_{i(i+1)...(i+m)}$. The R's form the Table 6.14.

The recurrence relation for R's is given by [3]

$$R_{i,(i+1),\cdots,(i+m)} = R_{(i+1),\cdots,(i+m)}$$
$$+ \frac{R_{(i+1),\cdots,(i+m)} - R_{i,\cdots,(i+m-1)}}{\left(\frac{x-x_i}{x-x_{i+m}}\right)\left(1 - \frac{R_{(i+1),\cdots,(i+m)} - R_{i,\cdots,(i+m-1)}}{R_{(i+1),\cdots,(i+m)} - R_{(i+1),\cdots,(i+m-1)}}\right) - 1}. \quad (6.66)$$

The recurrence relation starts with $R_i = y_i$. Now, define

$$C_{m,i} = R_{i,\cdots,(i+m)} - R_{i,\cdots,(i+m-1)}, \quad (6.67a)$$
$$D_{m,i} = R_{i,\cdots,(i+m)} - R_{(i+1),\cdots,(i+m)} \quad (6.67b)$$

which satisfies the relation

$$C_{m+1,i} - D_{m+1,i} = C_{m,i+1} - D_{m,i}. \quad (6.68)$$

Using Eqs. (6.67) and (6.68) one can obtain the recurrence relations

$$D_{m+1,i} = \frac{C_{m,i+1}\left(C_{m,i+1} - D_{m,i}\right)}{\left(\frac{x-x_i}{x-x_{i+m+1}}\right)D_{m,i} - C_{m,i+1}}, \quad (6.69a)$$

$$C_{m+1,i} = \frac{\left(\frac{x-x_i}{x-x_{i+m+1}}\right)D_{m,i}\left(C_{m,i+1} - D_{m,i}\right)}{\left(\frac{x-x_i}{x-x_{i+m+1}}\right)D_{m,i} - C_{m,i+1}}. \quad (6.69b)$$

At each level m, the C's and D's are the corrections which make the interpolation one order higher. For n data points $R_{1,\cdots,n}$ is the final approximation. It is the sum of any y_i and a set of C's and/or D's. The value of $f(0.61) = e^{0.61}$ is computed employing the rational function approximation. The five data points $(x, y) = (0, 1)$, $(0.15, 1.16183)$, $(0.5, 1.64872)$, $(0.7, 2.01375)$, $(1.0, 2.71828)$ are used. The result obtained is $f(0.61) = 1.84043$. For $x = 1.1$ the numerically computed value of $f(1.1)$ is 3.00407 while the exact value is 3.00417. Tabulate the numerical values of $e^{0.61}$ obtained by rational function approximation as a function of order n and also the percentage of relative error.

6.8 Concluding Remarks

If a given set of data displays a significant trend in the variation then the value of the data at points that are missing in the set can be computed either by an appropriate curve fitting or by an interpolation technique. Curve fitting needs a good choice of the mathematical function relating the variables involved. In contrast to this, in the case of interpolation, one may able to construct, for example, a polynomial passing through all the given data points.

In the case of constructing an interpolation polynomial if the nodes are equally spaced then one can construct Gregory–Newton forward or backward interpolating polynomials and then evaluate the value of the function at a specified node value. For unequally spaced nodes the Newton interpolation polynomial or the Lagrange interpolation polynomial can be used to compute the value of the function at a desired node value. The Newton and the Lagrange interpolation can also be applied to equally spaced nodes. The problem one may encounter when the target points are near the end nodes is pointed out in Section 6.4.3. How to overcome this problem is also illustrated with an example.

6.9 Bibliography

[1] J.H. Mathews, *Numerical Methods for Mathematics, Science and Engineering*. Prentice–Hall, New Delhi, 1998.

[2] M.K. Jain, R.K. Iyengar and R.K. Jain, *Numerical Methods for Scientific and Engineering Computation*. New Age International Pvt Ltd, New Delhi, 1993.

[3] W.H. Press, S.A. Teukolsky, W.T. Vetterling, B.P. Flannery, *Numerical Recipes in Fortran*. Foundation Books, New Delhi, 1993. Indian edition.

[4] A. Gezerlis and M. Williams, *Am. J. Phys.* 89:51, 2021.

[5] J. Stoer and R. Bulirsch, *Introduction to Numerical Analysis*, Springer, New York, 1980.

6.10 Problems

A. Newton and Gregory–Newton Polynomial Interpolations

6.1 The values of some functions $f(x)$ at two values of x are given below. For these functions

    i) calculate $f(x)$ at the specified target points using Newton linear interpolation polynomial,

    ii) compare the interpolated value with the exact value by calculating percentage of relative error and

    iii) calculate bound on truncation error.

    a) $f(x) = e^x$: $(x, f(x)) = (-1, 0.36788)$, $(-2, 0.13535)$, target $x = -1.5$.

    b) $f(x) = 1/(1 + x)$: $(x, f(x)) = (2.0.33333)$, $(3, 0.25)$, target $x = 2.5$.

c) $f(x) = \sin x :$ $(x, f(x)) = (1.57080, 1), (2.35619, 0.70711)$ target $x = 2$.

6.2 The values of some functions $f(x)$ at four values of x are given below. For these functions

i) construct the divided-difference table,

ii) construct the third-order Newton polynomial, compute the values of the function at the target points and

iii) compare the numerical approximation with the exact value by calculating percentage of relative error.

a) $f(x) = e^x$, target points $x = -1.5$ and -3.1.

x	0	−1	−1.75	−3.0
$f(x)$	1	0.36788	0.17377	0.04979

b) $f(x) = \ln(1 + x)$, target points $x = 0.3$ and -0.1.

x	0	0.25	0.5	0.75
$f(x)$	0	0.22314	0.40547	0.55962

c) $f(x) = \sin x$, target points $x = 2$ and -0.2.

x	0	0.78540	1.57080	2.35619
$f(x)$	0	0.70711	1.00000	0.70711

6.3 In Table 6.7 note that if all the elements in $(i + m)$th column are zero then the elements in the next successive columns are also zero. Therefore, if $|a_{i,m}| < \delta$, $i = m, m + 1, \cdots, n + 1$ where δ is a preassumed small positive value then one need not calculate the remaining elements and jump to calculate the polynomial value at the target x. In this case, the Newton polynomial is of order $m - 1$. Develop a Python program implementing this aspect in the Newton polynomial interpolation.

6.4 Write the derivative of $P_3(x)$. For the functions given in Problem 6.2 compute $P'_3(x)$ at the given target values. Compare the numerical value with the exact value.

6.5 Calculate the step size h to be used in the tabulation of $f(x) = e^x$ in the interval $[0, 1]$ so that the linear interpolation will be correct to 5 decimal places.

6.6 Show that the error in the linear interpolation for the case of equispaced tabular data does not exceed $1/8$ of the second difference.

For Problems 6.7–6.11

i) construct the divided-difference table,

ii) construct the third-order Newton polynomial, compute the value of the variable at the target point and

iii) construct the Gregory–Newton forward- and backward-difference tables and then compute the value of the variable at the target points.

6.7 The measured value of current I in an electronic circuit as a function of applied voltage V is given below. It is required to know the current for the applied voltage 1.65 V.

V in volts	1.0	1.5	2.0	2.5
I in amperes	0.1	0.15	0.2	0.25

6.8   The distance of a particle measured at various time is given below. Calculate the distance of the particle at $t = 0.9$ minute.

Time in minutes	0	0.5	1.0	1.5
Distance in meter	0	2.2	4.0	5.9

6.9   The volumes V of a gas measured at various temperatures T are listed below. Calculate the volume at $T = 100°C$.

$T°C$	90	110	130	150
V in m^3	2.108	2.13	2.157	2.180

6.10   The probabilities P of finding a particle with energy $E = 0.5$, 0.7, 0.9 and 1.1 units are 0.184, 0.123, 0.083, 0.052, respectively. Compute P for $E = 0.6$.

6.11   The mass of radioactive decaying sample ^{222}Rn measured at few times (in units of days) is given below. Compute the mass at the end of 12 days.

Time in days	0	5	10	15
Mass in mg	1	0.42	0.17	0.08

6.12   The values of some functions $f(x)$ at four node values are given below.

   i) Construct the Gregory–Newton forward-difference interpolation polynomial. Then, compute the values of the function at the specified node values.

   ii) Construct the Gregory–Newton backward-difference interpolation polynomial. Then, compute the values of the function at the target points.

   a) $f(x) = e^{-x}$, target points $x = 1.5$ and 3.1.

x	0	1	2	3
$f(x)$	1	0.36788	0.13535	0.04979

   b) $f(x) = 1/(1 + x)$, target points $x = 2.5$ and -0.1.

x	0	1	2	3
$f(x)$	1	0.5	0.33333	0.25

   c) $f(x) = (\sin x)/x$, target points $x = 0$, and 0.8.

x	-0.1	0.1	0.3	0.5
$f(x)$	0.99833	0.99833	0.98507	0.95885

B. Lagrange Interpolation Polynomials

6.13   Apply the Lagrange interpolation for Problems 6.7–6.11.

C. Analytical Construction of Rational Function $R_{n,n}(x)$

6.14   Verify that for the function $1/(1 - x)$ with $\frac{1}{1-x} = 1 + x + x^2 + \cdots$ the rational approximation $R_{1,1}$ is also $1/(1 - x)$.

6.15 The series expansion of $\sin x$ is given by $\sin x = x - \frac{x^3}{3!} + \frac{x^5}{5!} - \cdots$. Show that $R_{1,1} = x$, $R_{2,2} = x \left/ \left(1 - \frac{x^2}{6}\right)\right.$ and $R_{3,3} = \left(x - \frac{13}{60}x^3\right) \left/ \left(1 - \frac{x^2}{20}\right)\right.$. Tabulate the values of $\sin x$, $R_{1,1}$, $R_{2,2}$ and $R_{3,3}$ for $x = 0.1, 0.2, 0.3, 0.4, 0.5$.

6.16 The Maclaurin series for $(\tan^{-1}\sqrt{x})/\sqrt{x}$ is given by

$$\frac{\tan^{-1}\sqrt{x}}{\sqrt{x}} = 1 - \frac{x}{3} + \frac{x^2}{5} - \frac{x^3}{7} + \cdots .$$

Show that

$$R_{1,1} = \left(1 + \frac{4}{15}x\right) \left/ \left(1 + \frac{3}{5}x\right)\right.,$$

$$R_{1,1} = \left(1 + \frac{7}{9}x + \frac{64}{945}x^2\right) \left/ \left(1 + \frac{10}{9}x + \frac{5}{21}x^2\right)\right. .$$

6.17 Show that for $\cos x$

$$R_{2,2} = \left(1 - \frac{5}{12}x^2\right) \left/ \left(1 + \frac{1}{12}x^2\right)\right. .$$

D. Rational Function Interpolation/Extrapolation

6.18 For the data set given below compute the values of the function at $x = -1.5$ and -3.1 by rational function approximation.

x	0	-1	-1.75	-3.0
$f(x)$	1	0.36788	0.17377	0.04979

6.19 For the data set given below compute the values of the function at $x = 0.3$ and -0.1 by rational function approximation.

x	0	0.25	0.5	0.75
$f(x)$	0	0.22314	0.40547	0.55962

6.20 For the data set given below compute the values of the function at $x = 2$ and -0.2 by rational function approximation.

x	0	0.78540	1.57080	2.35619
$f(x)$	0	0.70711	1.00000	0.70711

7

Eigenvalues and Eigenvectors

7.1 Introduction

Consider a system of the form

$$A\mathbf{X} = \lambda\mathbf{X}, \tag{7.1}$$

where A is a $n \times n$ square matrix, λ is a scalar and $\mathbf{X} = (x_1, x_2, \ldots, x_n)^{\mathrm{T}}$. The solution $\mathbf{X}$ of the system (7.1) is called an *eigenvector* of the matrix A and λ is the corresponding *eigenvalue*. Any multiple of an eigenvector $\mathbf{X}$ will also be an eigenvector. λ and $\mathbf{X}$ are called *eigenpair* of A. Equation (7.1) can also be written as

$$(A - \lambda I)\mathbf{X} = 0 \tag{7.2}$$

which implies that the product of the matrix $(A - \lambda I)$ and the nonzero vector $\mathbf{X}$ is the zero vector. System (7.2) will have a nontrivial solution (that is, at least one x_i is nonzero) *if and only if*

$$\det(A - \lambda I) = \begin{vmatrix} a_{11} - \lambda & a_{12} & \cdots & a_{1n} \\ a_{21} & a_{22} - \lambda & \cdots & a_{2n} \\ & & \cdots & \\ a_{n1} & a_{n2} & \cdots & a_{nn} - \lambda \end{vmatrix} = 0. \tag{7.3}$$

When the above determinant is expanded one gets the following polynomial of degree n:

$$P(\lambda) = (-1)^n \left(\lambda^n + b_1\lambda^{n-1} + \cdots + b_{n-1}\lambda + b_n\right). \tag{7.4}$$

This equation is called *characteristic equation* of the matrix A and it has n roots. These roots when substituted in Eq. (7.2) give the corresponding nontrivial solution $\mathbf{X}$, that is, the eigenvector $\mathbf{X}$. The system (7.1) has nontrivial solutions $\mathbf{X}$ for certain values of λ only and are the roots of Eq. (7.4). These *certain values* of λ are called *eigenvalues*. *Eigen* is a German word which means *certain*.

Equation (7.4) suggests that roots of a polynomial of degree n can be viewed as eigenvalues of a matrix. For example, the *companion matrix* C of the equation

$$f(\lambda) = \lambda^3 + a_1\lambda^2 + a_2\lambda + a_3 = 0 \tag{7.5}$$

is

$$C = \begin{pmatrix} 0 & 1 & 0 \\ 0 & 0 & 1 \\ -a_3 & -a_2 & -a_1 \end{pmatrix}. \tag{7.6}$$

The eigenvalues of C are the roots of Eq. (7.5). Therefore, the roots of the equation

$$\lambda^n + a_1\lambda^{n-1} + \cdots + a_{n-1}\lambda + a_n = 0 \tag{7.7}$$

DOI: 10.1201/9781032649931-7

are the eigenvalues of the matrix

$$
C = \begin{pmatrix}
0 & 1 & 0 & \cdots & 0 \\
0 & 0 & 1 & \cdots & 0 \\
0 & 0 & 0 & \cdots & 0 \\
 & & & \cdots & \\
0 & 0 & 0 & \cdots & 1 \\
-a_n & -a_{n-1} & -a_{n-2} & \cdots & -a_1
\end{pmatrix}. \tag{7.8}
$$

Simple algorithms are not available to find all eigenpairs of arbitrary $n \times n$ matrices. However, efficient and easily understandable methods are developed for some special cases. Depending upon the nature of the given matrix and the requirement one can choose a suitable method. For example, power method is useful to compute the absolute largest eigenvalue and the corresponding eigenvector of an arbitrary matrix. For a symmetric matrix, Jacobi method enables us to calculate all the eigenpairs. For a symmetric tridiagonal matrix QL algorithm is efficient to determine all the eigenvalues. The method of Rutishauser is of useful for arbitrary matrices of small order. In this chapter, the power, Jacobi, QL and Rutishauser methods are described. To start with, in the next section some of the basic properties of eigenvalues and eigenvectors are summarized which are essential for our discussion.

7.2 Some Basic Properties of Eigenvalues and Eigenvectors

In this section, some basic properties of eigenvalues and eigenvectors of square matrices are given without proof. For details see the refs. [1-2].

1. For a real square matrix A the eigenvalues are the roots of the characteristic polynomial $P(\lambda) = \det(A - \lambda I)$. The eigenvalues can be real or complex.

2. If λ is an eigenvalue of a matrix A then the associated eigenvector $\mathbf{X}$ is defined through the relation $A\mathbf{X} = \lambda\mathbf{X}$.

3. For each distinct eigenvalue λ there exists at least one eigenvector.

4. The largest eigenvalue in modulus of a matrix cannot exceed the largest sum of the moduli of the elements along any row or column.

5. An eigenvalue whose absolute value is greater than any other eigenvalue is called the *dominant eigenvalue*. The associated eigenvector is called the *dominant eigenvector*.

6. If the largest magnitude of components of an eigenvector is unity then the eigenvector is said to be *normalized*. An eigenvector $\mathbf{X} = (x_1, x_2, \ldots, x_n)^{\mathrm{T}}$ can be normalized by constructing a new vector $V = (x_1, x_2, \ldots, x_n)/\beta$, where $\beta = x_j$ with $x_j =$ maximum of $\{ |x_1|, |x_2|, \ldots, |x_n| \}$.

7. If $(\lambda, \mathbf{X})$ is an eigenpair of a matrix A then $(\lambda - \alpha, \mathbf{X})$ is an eigenpair of the matrix $A - \alpha I$, where α is a constant.

8. If $(\lambda \neq 0, \mathbf{X})$ is an eigenpair of a matrix A then an eigenpair of A^{-1} is $(1/\lambda, \mathbf{X})$.

9. If $(\lambda \neq 0, \mathbf{X})$ is an eigenpair of a matrix A then $(1/(\lambda - \alpha), \mathbf{X})$ with $\alpha \neq \lambda$ is an eigenpair of the matrix $(A - \alpha I)^{-1}$. The eigenvalues of diagonal, upper-triangular and lower-triangular matrices are simply their diagonal elements.

10.   A matrix A is said to be *symmetric* if $A^T = A$. The eigenvalues of a real symmetric matrix are all real numbers. Eigenvectors corresponding to distinct eigenvectors of a symmetric matrix are orthogonal, that is, $x_i \cdot x_j = 0$, for $i \neq j$.

7.3   Applications of Eigenvalues and Eigenvectors

Eigenvalues and eigenvectors have applications in the field of vibration analysis, electric circuits, control theory, linear stability analysis and quantum mechanics. Let us enumerate some of the notable applications of them [3-8].

1.   Eigenvalues arise in the construction of the solution of linear ordinary differential equations with constant coefficients.

2.   The stability of an equilibrium point of a dynamical system depends on the nature of the eigenvalues of a square matrix involved in the linear stability analysis.

3.   In the vibrational analysis eigenvalues (that is, the eigenfrequencies) are the allowed frequencies of vibrations and the corresponding eigenmodes of vibrations are the allowed possible meaningful bounded solutions.

4.   In mechanics, the study of rotation of a rigid body about its centre of mass involves the tensor of moment of inertia. The principal axes are defined by the eigenvalues of the moment of inertia tensor. The stress tensor can be decomposed into a diagonal tensor by considering the eigenvectors as a basis and the eigenvalues on the diagonal.

5.   In quantum mechanics, eigenvalues are the allowed values of certain experimentally measurable quantities such as energy, orbital angular momentum and spin and the eigenvectors (eigenfunctions) represent the allowed states.

6.   Eigenvalues and the eigenvalue analysis are employed in oil firm to identify the location of an oil reserve.

7.   A theoretical upper limit of information that can be transmitted in a communication channel can be obtained using the eigenvalues involved in the concerned problem.

8.   In linear transformations, the building blocks are the eigenvalues and the eigenvectors. They denote the scaling factor and the direction of the transformation applied.

9.   Eigenvalues and the eigenvectors have applications in image compression, dimensionality reduction in machine learning and in the Google page rank algorithm.

10.   The facial recognition schemes like the eigenfaces make use of the eigenvalues and the eigenvectors.

7.4   The Power Method

To determine the dominant eigenvalue and the associated normalised eigenvector of a square matrix of order n there is a simple method called *power method*. It is an iterative method. The iteration rule is obtained as follows.

Let $\mathbf{X}_0$ is the initial guess of the exact dominant eigenvector $\mathbf{V}_1$ corresponding to the dominant eigenvalue λ_1, where λ's are ordered as $|\lambda_1| > |\lambda_2| > \ldots > |\lambda_n|$. $\mathbf{X}_0$ is written as

$$\mathbf{X}_0 = \sum_{i=1}^{n} b_i \mathbf{V}_i. \tag{7.9}$$

Assume that $\mathbf{X}_1$ is the next approximation of $\mathbf{V}_1$ and α_1 is the corresponding approximation to the associated eigenvalue. From the eigenvalue equation one may write

$$A\mathbf{X}_0 = \alpha_1 \mathbf{X}_1 + \delta \mathbf{X}_1. \tag{7.10}$$

Neglecting $\delta \mathbf{X}_1$, from Eq. (7.10) $\mathbf{X}_1$ is written as

$$\mathbf{X}_1 - \frac{1}{\alpha_1} A\mathbf{X}_0 = \frac{1}{\alpha_1} (b_1 A\mathbf{V}_1 + b_2 A\mathbf{V}_2 \mid \cdots \mid b_n A\mathbf{V}_n). \tag{7.11}$$

Because $A\mathbf{V}_i = \lambda_i \mathbf{V}_i$, $i = 1, 2, \ldots, n$

$$\mathbf{X}_1 = \frac{\lambda_1}{\alpha_1} \left(b_1 \mathbf{V}_1 + b_2 \frac{\lambda_2}{\lambda_1} \mathbf{V}_2 + \cdots + b_n \frac{\lambda_n}{\lambda_1} \mathbf{V}_n \right). \tag{7.12}$$

$\mathbf{X}_1$ is normalised because of the term α_1 in the denominator in Eq. (7.12). Repeating the above process the kth approximation is obtained as

$$\mathbf{X}_k = \frac{\lambda_1^k}{\alpha_1 \alpha_2 \cdots \alpha_k} \left[b_1 \mathbf{V}_1 + b_2 \left(\frac{\lambda_2}{\lambda_1} \right)^k \mathbf{V}_2 + \cdots + b_n \left(\frac{\lambda_n}{\lambda_1} \right)^k \mathbf{V}_n \right]. \tag{7.13}$$

$\mathbf{X}_k$ is again normalized. Because λ_1 is the dominant eigenvalue all the terms except the first term in the right-side of Eq. (7.13) vanish so that

$$\lim_{k \to \infty} \mathbf{X}_k = \lim_{k \to \infty} \frac{b_1 \lambda_1^k}{\alpha_1 \cdots \alpha_k} \mathbf{V}_1. \tag{7.14}$$

For normalized $\mathbf{X}_k$ and $\mathbf{V}_1$ their largest component will have the value 1. Therefore,

$$\lim_{k \to \infty} \frac{b_1 \lambda_1^k}{\alpha_1 \cdots \alpha_k} = 1. \tag{7.15}$$

Then, Eq. (7.14) becomes

$$\lim_{k \to \infty} \mathbf{X}_k = \mathbf{V}_1. \tag{7.16}$$

Next, from Eq. (7.15)

$$\lim_{k \to \infty} \frac{b_1 \lambda_1^{k-1}}{\alpha_1 \cdots \alpha_{k-1}} = 1. \tag{7.17}$$

Dividing Eq. (7.15) by Eq. (7.17) gives

$$\lim_{k \to \infty} \alpha_k = \lambda_1. \tag{7.18}$$

From the above analysis, the iterative formula for the dominant eigenvector is written as

$$\mathbf{X}_{k+1} = \frac{1}{\alpha_{k+1}} A\mathbf{X}_k, \tag{7.19}$$

where α_{k+1} is the absolute largest coordinate of $A\mathbf{X}_k$. The iteration can be started with

$\mathbf{X}_0 = (1, 1, \ldots, 1)^{\mathrm{T}}$. The sequences $\{\alpha_k\}$ and $\{\mathbf{X}_k\}$ converge to the exact dominant eigenvalue and the dominant eigenvector. The iteration is stopped if

$$|\alpha_{k+1}| - |\alpha_k| < \delta \quad \text{and} \quad \| \mathbf{X}_{k+1} - \mathbf{X}_k \| < \delta, \tag{7.20}$$

where δ is a preassumed small positive number and

$$\| \mathbf{X}_{k+1} - \mathbf{X}_k \| = \left[\sum_{i=1}^{n} (x_{i,k+1} - x_{i,k})^2 \right]^{1/2}. \tag{7.21}$$

The rate of convergence of $\{\mathbf{X}_k\}$ to $\mathbf{V}_1$ is linear and is governed by the term $(\lambda_2/\lambda_1)^k$. Similarly, the rate of convergence of the sequence $\{\alpha_k\}$ to λ_1 is also linear.

Example:

Consider the matrix

$$\begin{pmatrix} 1 & 0 & 0 \\ 2 & 1 & 3 \\ 5 & 2 & 2 \end{pmatrix}. \tag{7.22}$$

With the choice $\mathbf{X}_0 = (1, 1, 1)^{\mathrm{T}}$ Eq. (7.19) becomes

$$\mathbf{X}_1 = \frac{1}{\alpha_1} A \mathbf{X}_0 = \frac{1}{\alpha_1} \begin{pmatrix} 1 & 0 & 0 \\ 2 & 1 & 3 \\ 5 & 2 & 2 \end{pmatrix} \begin{pmatrix} 1 \\ 1 \\ 1 \end{pmatrix} = \frac{1}{\alpha_1} \begin{pmatrix} 1 \\ 6 \\ 9 \end{pmatrix}.$$

The absolute largest component of the above vector $(1, 6, 9)^{\mathrm{T}}$ is 9. So, $\alpha_1 = 9$. Then,

$$\mathbf{X}_1 = \begin{pmatrix} 1/9 \\ 2/3 \\ 1 \end{pmatrix}.$$

The second iteration gives

$$\mathbf{X}_2 = \frac{1}{\alpha_2} A \mathbf{X}_1 = \frac{1}{\alpha_2} \begin{pmatrix} 1/9 \\ 35/9 \\ 35/9 \end{pmatrix},$$

where $\alpha_2 = 35/9 = 3.88889$. That is,

$$\mathbf{X}_2 = \begin{pmatrix} 1/35 \\ 1 \\ 1 \end{pmatrix}.$$

The third iteration gives

$$\mathbf{X}_3 = \frac{1}{\alpha_3} A \mathbf{X}_2 = \frac{1}{\alpha_3} \begin{pmatrix} 1/35 \\ 142/35 \\ 145/35 \end{pmatrix}$$

TABLE 7.1
The numerically computed dominant eigenpair of the matrix A given by Eq. (7.22) at each iteration (k) of the power method. δ in Eq. (7.20), stopping criterion for the iterations, is chosen as 10^{-5}. The starting eigenvector is $\mathbf{X}_0 = (1,1,1)^{\mathrm{T}}$.

k	Eigen-value	Components of eigenvector			$\|\alpha_{k+1} -\alpha_k\|$	$\|\mathbf{X}_{k+1} -\mathbf{X}_k\|$
		x_1	x_2	x_3		
0	0.00000	1.00000	1.00000	1	$----$	$----$
1	9.00000	0.11111	0.66667	1	9.00000	0.94933
2	3.88889	0.02857	1.00000	1	5.11111	0.34340
3	4.14286	0.00690	0.97931	1	0.25397	0.02996
4	3.99310	0.00173	1.00000	1	0.14975	0.02133
5	4.00864	0.00043	0.99871	1	0.01553	0.00183
6	3.99957	0.00011	1.00000	1	0.00907	0.00133
7	4.00054	0.00003	0.99992	1	0.00097	0.00011
8	3.99997	0.00001	1.00000	1	0.00057	0.00008
9	4.00003	0.00000	0.99999	1	0.00006	0.00001
10	4.00000	0.00000	1.00000	1	0.00004	0.00001
11	4.00000	0.00000	1.00000	1	0.00000	0.00000

with $\alpha_3 = 145/35 = 4.14286$. The result is

$$\mathbf{X}_3 = \begin{pmatrix} 1/145 \\ 142/145 \\ 1 \end{pmatrix}.$$

Table 7.1 displays the successive approximation of the dominant eigenvalue and the corresponding eigenvector. If δ is chosen as 10^{-1} then the process is to be stopped after five iterations. At the end of five iterations $\lambda_1 = \alpha_5 = 4.00864$, where the exact dominant eigenvalue is 4. The numerically computed eigenvector is $\mathbf{X}_5 = (0.00043, 0.99871, 1.0)^{\mathrm{T}}$ whereas the exact eigenvector is $\mathbf{V}_1 = (0, 1, 1)^{\mathrm{T}}$. 11 iterations are required to obtain λ_1 and $\mathbf{X}$ with $\delta = 10^{-5}$. At the end of 11 iterations $\lambda_1 = 4$ and $\mathbf{X} = (0, 1, 1)^{\mathrm{T}}$.

7.5 Eigenpairs of Symmetric Matrices – Jacobi Method

For symmetric matrices, a few methods are available to find all the eigenpairs. In this section, the Jacobi method is described. This method is straight forward and simpler than the other more efficient methods. The method is reliable to produce accurate eigenpairs for all real symmetric matrices. The method consists of a sequence of orthogonal similarity transformations. Let us first define plane rotations, similarity transformation and orthogonal transformation which are essential to follow the Jacobi method.

7.5.1   Similarity and Orthogonal Transformations

Let A and B are two matrices and S is another non-singular matrix, that is, $\det A \neq 0$ such that

$$B = S^{-1}AS. \tag{7.23}$$

The matrices A and B satisfying Eq. (7.23) are said to be *similar* and this relation is called a *similarity transformation*. The matrix S is called a *similarity matrix*. A matrix C is said to be *orthogonal* if $C^{-1} = C^{T}$. If S in Eq. (7.23) is orthogonal then the transformation is an *orthogonal similarity transformation*. Further, when $S^{-1} = S^{T}$ then $S^{T}S = S^{-1}S = I$, unit matrix.

For a real symmetric matrix A, it is possible to find a real orthogonal matrix S such that $S^{-1}AS$ is a diagonal matrix:

$$D = S^{-1}AS = S^{T}AS. \tag{7.24}$$

What can one do with the transformation (7.24)? Interestingly, as the eigenvalues of a diagonal matrix D are its diagonal elements, the eigenvalues of A are the eigenvalues of D and there exists a relation between the eigenvectors of A and D. This is shown below.

Multiplication of Eq. (7.24) by $S^{T}X$ gives

$$
\begin{aligned}
DS^{T}\mathbf{X} &= S^{T}ASS^{T}\mathbf{X} \\
&= S^{T}ASS^{-1}\mathbf{X} \\
&= S^{T}A\mathbf{X} \\
&= S^{T}\lambda\mathbf{X} \\
&= \lambda S^{T}\mathbf{X},
\end{aligned} \tag{7.25}
$$

where λ is the eigenvalue of A corresponding to the eigenvector $\mathbf{X}$. Defining

$$\mathbf{Y} = S^{T}\mathbf{X}\left(= S^{-1}\mathbf{X}\right) \tag{7.26}$$

Eq. (7.25) is rewritten as

$$D\mathbf{Y} = \lambda\mathbf{Y} \tag{7.27}$$

which implies that eigenvalues of D are simply the eigenvalues of A. The eigenvectors $\mathbf{Y}$ of D can be easily determined since D is a diagonal. Once $\mathbf{Y}$ is known then the eigenvectors $\mathbf{X}$ of A are determined from the relation (obtained from Eq. (7.26))

$$\mathbf{X} = S\mathbf{Y} \tag{7.28}$$

Thus, if a given real symmetric matrix A is diagonalized into D by an orthogonal similarity matrix S through the orthogonal similarity transformations (7.24) then its eigenvalues are the diagonal elements of D and the eigenvectors are given by Eq. (7.28). The next subsection describes how to diagonalize A.

7.5.2   Jacobi Rotation

Matrix A can be diagonalized by applying a series of orthogonal similarity transformation as described below. The idea is that by applying a similarity transformation $S^{T}AS$ to A one of its off-diagonal elements is made zero. Successive transformations make the values of the off-diagonal elements smaller and smaller until the matrix is diagonal. *How does one determine the matrix S which does the above?*

Let us choose the form of S as

$$
S = \begin{pmatrix}
1 & & & & & & \\
& \ddots & & & & & \\
& & \cos\theta & \cdots & \sin\theta & & \\
& & \vdots & 1 & \vdots & & \\
& & -\sin\theta & \cdots & \cos\theta & & \\
& & & & & \ddots & \\
& & & & & & 1
\end{pmatrix}. \tag{7.29}
$$

S is called *Jacobi rotation*. In S

1.   all the diagonal elements are unity except for two rows say i and k,
2.   all off-diagonal elements are zero except for two elements and
3. θ is the angle of rotation to be determined.

Using S the matrix A is transformed into A' by

$$
A' = S^{\mathrm{T}} A S. \tag{7.30}
$$

$S^{\mathrm{T}} A$ alters only the elements in the rows i and k and AS changes only the elements in the columns i and k.

7.5.3   Setting of Two Off-Diagonal Elements to Zero

θ is chosen to make two off-diagonal elements in A' to zero. Out of the various off-diagonal elements, first the absolute largest off-diagonal elements are made zero. Suppose, $|a_{ik}|$ (and $|a_{ki}|$ since A is symmetric) be the absolute largest among the off-diagonal elements. The 2×2 submatrix of A with the elements a_{ii}, a_{ik}, a_{ki} and a_{kk} is transformed into a diagonal form by the Jacobi rotation. The elements a_{ik} and a_{ki} are made to zero. Now, write

$$
S_1 = \begin{pmatrix}
\cos\theta & \sin\theta \\
-\sin\theta & \cos\theta
\end{pmatrix}. \tag{7.31}
$$

To find the value of θ which can make a_{ik} and a_{ki} to zero consider the transformation $A_1 = S_1^{\mathrm{T}} A S_1$. This is worked out as

$$
\begin{aligned}
A_1 &= S_1^{\mathrm{T}} A S_1 \\
&= \begin{pmatrix}
\cos\theta & -\sin\theta \\
\sin\theta & \cos\theta
\end{pmatrix}
\begin{pmatrix}
a_{ii} & a_{ik} \\
a_{ki} & a_{kk}
\end{pmatrix}
\begin{pmatrix}
\cos\theta & \sin\theta \\
-\sin\theta & \cos\theta
\end{pmatrix} \\
&= \begin{pmatrix}
A_{11} & A_{12} \\
A_{21} & A_{22}
\end{pmatrix},
\end{aligned} \tag{7.32a}
$$

where

$$
A_{11} = a_{ii}\cos^2\theta + a_{kk}\sin^2\theta - a_{ik}\sin 2\theta, \tag{7.32b}
$$

$$
A_{12} = \frac{1}{2}(a_{ii} - a_{kk})\sin 2\theta + a_{ik}\cos 2\theta, \tag{7.32c}
$$

$$
A_{21} = \frac{1}{2}(a_{ii} - a_{kk})\sin 2\theta + a_{ik}\cos 2\theta, \tag{7.32d}
$$

$$
A_{22} = a_{ii}\sin^2\theta + a_{kk}\cos^2\theta + a_{ik}\sin 2\theta. \tag{7.32e}
$$

In obtaining Eqs. (7.32) the relations $a_{ik} = a_{ki}$, $\cos 2\theta = \cos^2 \theta - \sin^2 \theta$ and $\sin 2\theta = 2 \sin \theta \cos \theta$ are used. Setting $A_{12} = A_{21} = 0$ in Eqs. (7.32) gives

$$\tan 2\theta = 2a_{ik}/(a_{kk} - a_{ii}). \tag{7.33}$$

The most suitable stable reduction is given by the smallest rotation. Therefore, choose the root in the interval $-\pi/4 \le \theta \le \pi/4$. The desired root is given by

$$\theta = \begin{cases} \frac{1}{2} \tan^{-1} \left[2a_{ik}/(a_{kk} - a_{ii}) \right], & \text{if } a_{ii} \ne a_{kk} \\[2mm] \pi/4, & \text{if } a_{ii} = a_{kk}, \ a_{ik} > 0 \\[2mm] -\pi/4, & \text{if } a_{ii} = a_{kk}, \ a_{ik} < 0. \end{cases} \tag{7.34}$$

The round-off error in the calculation of θ using the formula (7.34) for $a_{ii} \ne a_{kk}$ seriously affect the further computation which then leads to an inaccurate result. To minimize the round-off error let us obtain an alternate formula for $\cos \theta$ and $\sin \theta$ using trigonometric identities. This is described in the next subsection.

7.5.4   More Accurate Calculation of $\cos \theta$ and $\sin \theta$

Define

$$\phi = (a_{kk} - a_{ii}) / (2a_{ik}) \tag{7.35}$$

and consider the identity

$$\phi = \frac{1}{\tan 2\theta} = \frac{\cos 2\theta}{\sin 2\theta} = \frac{\cos^2 \theta - \sin^2 \theta}{2 \sin \theta \cos \theta}. \tag{7.36}$$

From Eq. (7.36) one has

$$\phi = \frac{1 - \tan^2 \theta}{2 \tan \theta}$$

or

$$\tan^2 \theta + 2\phi \tan \theta - 1 = 0. \tag{7.37}$$

The smaller root of this equation using the formula for quadratic equation given in Section 2.5 is

$$\begin{aligned} \tan \theta &= -\phi \pm \left(\phi^2 + 1 \right)^{1/2} \\ &= \frac{\text{sign}(\phi)}{|\phi| + (\phi^2 + 1)^{1/2}}, \end{aligned} \tag{7.38a}$$

where $\text{sign}(\phi) = 1$ for $\phi \ge 0$ and -1 for $\phi < 0$. Then,

$$\begin{aligned} \cos \theta &= 1/ \left(1 + \tan^2 \theta \right)^{1/2}, \tag{7.38b} \\ \sin \theta &= \cos \theta \tan \theta. \tag{7.38c} \end{aligned}$$

Instead of using the θ given by Eq. (7.34), to compute $\cos \theta$ and $\sin \theta$ Eqs. (7.38) can be used to minimize the round-off error. When $a_{ii} = a_{kk}$ from Eq. (7.35) and (7.38) $\phi = 0$, $\tan \theta = 1$, $\cos \theta = \sin \theta = 1/\sqrt{2}$ and $\theta = \pi/4$.

7.5.5 Effect of Successive Jacobi Rotation

The transformation $S_1^T A S_1$ gives a new matrix A_1 in which two off-diagonal elements are set to zero. The next transformation is applied to the matrix A_1. Another matrix $A_2 = S_2^T A_1 S_2$ is obtained in which two off-diagonal elements of A_1 are made zero. At rth rotation

$$
\begin{aligned}
A_r &= S_r^T S_{r-1}^T \cdots S_1^T A \, S_1 S_2 \cdots S_{r-1} S_r \\
&= (S_1 S_2 \cdots S_r)^T A \, (S_1 S_2 \cdots S_r) \\
&= S^{-1} A S .
\end{aligned}
\tag{7.39}
$$

As $r \to \infty$, $A_r \to D$, a diagonal matrix. Its diagonal elements are the eigenvalues of A. The matrix of eigenvalues of A is given by

$$
X = S_1 S_2 \cdots S_r .
\tag{7.40}
$$

In the Jacobi method, each rotation is designed to make the absolute largest off-diagonal elements to zero. If the calculations are done using computer then one can proceed to set the elements (a_{12}, a_{21}) to zero then set (a_{13}, a_{31}) to zero and so on. In this way searching of largest off-diagonal element can be avoided.

7.5.6 Examples

Let us determine the eigenpairs of two matrices employing the Jacobi method.

Example 1:

Compute all the eigenpairs of the symmetric matrix

$$
A = \begin{pmatrix} 1 & 2 & 1 \\ 2 & 1 & 1 \\ 1 & 1 & 3 \end{pmatrix}.
\tag{7.41}
$$

The off-diagonal elements are $a_{12} = a_{21} = 2$, $a_{13} = a_{31} = 1$, $a_{23} = a_{32} = 1$. Among them, the absolute largest element is $a_{12} = a_{21} = 2$. Therefore, first the off-diagonal elements a_{12} and a_{21} in A can be made zero. The Jacobi rotation matrix which do this is given by

$$
S_1 = \begin{pmatrix} \cos\theta & \sin\theta & 0 \\ -\sin\theta & \cos\theta & 0 \\ 0 & 0 & 1 \end{pmatrix}.
$$

$\sin\theta$ and $\cos\theta$ are determined using Eqs. (7.35) and (7.38). For the given matrix A these equations become

$$
\begin{aligned}
\phi &= (a_{22} - a_{11}) / (2a_{12}) = 0 , \\
\tan\theta &= 1, \quad \cos\theta = 1/\left(1 + \tan^2\theta\right)^{1/2} = 1/\sqrt{2}, \quad \sin\theta = 1/\sqrt{2} .
\end{aligned}
$$

Then,

$$
S_1 = \begin{pmatrix} 1/\sqrt{2} & 1/\sqrt{2} & 0 \\ -1/\sqrt{2} & 1/\sqrt{2} & 0 \\ 0 & 0 & 1 \end{pmatrix}
$$

and

$$A_1 = S_1^T A S_1 = \begin{pmatrix} -1 & 0 & 0 \\ 0 & 3 & \sqrt{2} \\ 0 & \sqrt{2} & 3 \end{pmatrix}.$$

In A_1 the elements $a_{12} = a_{21} = 0$. The rotation has made a_{13} and a_{31} to zero which is unexpected, however, a nice effect of S_1.

Next, perform the above analysis for the matrix A_1. Among its off-diagonal elements, the absolute largest is $a_{23}(= a_{32}) = \sqrt{2}$. Therefore, the off-diagonal elements of the submatrix

$$\begin{pmatrix} a_{22} & a_{23} \\ a_{32} & a_{33} \end{pmatrix}$$

in A_1 are chosen to set zero by a Jacobi rotation. The matrix S_2 is

$$S_2 = \begin{pmatrix} 1 & 0 & 0 \\ 0 & \cos\theta & \sin\theta \\ 0 & -\sin\theta & \cos\theta \end{pmatrix}.$$

$\cos\theta$ and $\sin\theta$ are calculated as $1/\sqrt{2}$. Thus,

$$S_2 = \begin{pmatrix} 1 & 0 & 0 \\ 0 & 1/\sqrt{2} & 1/\sqrt{2} \\ 0 & -1/\sqrt{2} & 1/\sqrt{2} \end{pmatrix}.$$

The second similarity transformation $A_2 = S_2^T A_1 S_2$ gives

$$A_2 = \begin{pmatrix} -1 & 0 & 0 \\ 0 & 3 - \sqrt{2} & 0 \\ 0 & 0 & 3 + \sqrt{2} \end{pmatrix}.$$

The matrix A_2 is diagonal and its diagonal elements are the eigenvalues. In order to verify the above result consider the characteristic equation of the given matrix A. It is

$$\lambda^3 - 5\lambda^2 + \lambda + 7 = 0.$$

The three eigenvalues

$$\lambda_1 = -1, \quad \lambda_{2,3} = 3 \pm \sqrt{2}$$

satisfy the above characteristic equation. The eigenvectors are the column vectors of the product matrix $S = S_1 S_2$. S is obtained as

$$S = \begin{pmatrix} 1/\sqrt{2} & 1/2 & 1/2 \\ -1/\sqrt{2} & 1/2 & 1/2 \\ 0 & -1/\sqrt{2} & 1/\sqrt{2} \end{pmatrix}.$$

The eigenpairs are thus

$$\left[-1, \left(1/\sqrt{2}, -1/\sqrt{2}, 0 \right)^T \right], \left[3 + \sqrt{2}, \left(1/2, 1/2, -1/\sqrt{2} \right)^T \right],$$
$$\left[3 - \sqrt{2}, \left(1/2, 1/2, 1/\sqrt{2} \right)^T \right].$$

Example 2:

In the Jacobi method, a similarity transformation is used to set off-diagonal elements to zero in the given or its equivalent matrix. But the transformation may produce nonzero value for off-diagonal elements which are set to zero earlier. Consequently, more iterations than the number of pairs of nonzero off-diagonal elements may be required to set all the off-diagonal elements to zero. This happens in the matrix

$$A = \begin{pmatrix} 1 & 0 & 1 \\ 0 & 1 & 2 \\ 1 & 2 & 1 \end{pmatrix}. \tag{7.42}$$

When the elements a_{23} and a_{32} are tried to set zero the resultant matrix is

$$A_1 = \begin{pmatrix} 1 & -1/\sqrt{2} & 1/\sqrt{2} \\ -1/\sqrt{2} & -1 & 0 \\ 1/\sqrt{2} & 0 & 3 \end{pmatrix}.$$

Though the elements a_{23} and a_{32} are zero in A_1, the elements a_{12} and a_{21} which are zero in A now become nonzero. This problem is overcome in the Householder's and Given's method, where the zero off-diagonal elements remain zero in subsequent iterations. For details of this method see for example refs. [9–11]. Proceeding the Jacobi method with the matrix A_1 after 7 rotations the diagonal matrix obtained is

$$A_7 = \begin{pmatrix} 1 & 0 & 0 \\ 0 & -1.23607 & 0 \\ 0 & 0 & 3.23607 \end{pmatrix}.$$

Thus, the eigenvalues are $1, -1.23607, 3.23607$. The eigenvector matrix is computed as

$$\begin{pmatrix} 0.89443 & 0.31623 & 0.31623 \\ -0.44721 & 0.63246 & 0.63246 \\ 0.00000 & -0.70711 & 0.70711 \end{pmatrix}.$$

7.6 Eigenvalues of Symmetric Tridiagonal Matrices – QL Method

A *symmetric tridiagonal matrix* is of the form

$$A = \begin{pmatrix} a_1 & d_1 & 0 & \cdots & & & 0 \\ d_1 & a_2 & d_2 & \cdots & & & 0 \\ 0 & d_2 & a_3 & \cdots & & & 0 \\ & & & \cdots & & & \cdot \\ & & & & a_{n-2} & d_{n-2} & 0 \\ & & & & d_{n-2} & a_{n-1} & d_{n-1} \\ & & & & 0 & d_{n-1} & a_n \end{pmatrix}. \tag{7.43}$$

All the eigenvalues of A can be computed by the QL method. In quantum mechanics, the problem of finding numerical solution of time-independent Schrödinger equation can be converted into the eigenvalue problem $(A - \lambda I)X = 0$, with A being symmetric tridiagonal. The QL method is of great use in solving such problems.

In the QL method, the given matrix $A_1 = A$ is factorized into the form

$$A_1 = Q_1 L_1, \tag{7.44}$$

where $Q_1 = Q$ and $L_1 = L$ are orthogonal and lower-triangular matrices, respectively. To write A_1 into the form of Eq. (7.44), a matrix P_{n-1} is constructed in such a way that the element in the location $(n-1, n)$ in the matrix $P_{n-1}A$ is zero. That is,

$$P_{n-1}A = \begin{pmatrix} a_1 & d_1 & 0 & & & & \\ d_1 & a_2 & d_2 & & & & \\ 0 & d_2 & a_3 & & & & \\ & & & \cdots & & & \\ & & & & a_{n-2} & d_{n-2} & 0 \\ & & & & q_{n-2} & p_{n-1} & 0 \\ & & & & r_{n-2} & q_{n-1} & p_n \end{pmatrix}. \tag{7.45}$$

Similarly, by a suitable P_{n-2} the element in the position $(n-2, n-1)$ of $P_{n-1}A$ can be made zero. Repetition of the above process $(n-1)$ times gives

$$P_1 P_2 \cdots P_{n-1} A = \begin{pmatrix} p_1 & 0 & 0 & & & & & \\ q_1 & p_2 & 0 & & & & & \\ r_1 & q_2 & p_3 & & & & & \\ & & & \cdots & & & & \\ & & & & q_{n-3} & p_{n-2} & 0 & 0 \\ & & & & r_{n-3} & q_{n-3} & p_{n-1} & 0 \\ & & & & 0 & r_{n-2} & q_{n-1} & p_n \end{pmatrix}. \tag{7.46}$$

Now, define

$$Q_1 = P_{n-1}^{\mathrm{T}} P_{n-2}^{\mathrm{T}} \cdots P_1^{\mathrm{T}} \tag{7.47}$$

and form the matrix

$$A_2 = L_1 Q_1. \tag{7.48}$$

The matrix Q_1 is orthogonal, therefore, $Q_1^{\mathrm{T}} A_1 = Q_1^{\mathrm{T}} Q_1 L_1 = L_1$. Thus, $A_2 = Q_1^{\mathrm{T}} A_1 Q_1 = Q_1^{-1} A_1 Q_1$ which implies that A_2 is similar to A_1 and has the same eigenvalues. In general $A_{k+1} = Q_k^{\mathrm{T}} A_k Q_k$ so that A_{k+1} is similar to A_k.

Let us consider the case of making the element A_{pq} and A_{qp} to zero. In this case, P_1 has the form

$$P_1 = \begin{pmatrix} 1 & \cdots & & & & & \\ 0 & \cdots & & & & & \\ & & c & \cdots & s & & \\ & & & \cdots & & & \\ & & -s & \cdots & c & & \\ & & & & & \cdots & \\ & & & & & & 1 \end{pmatrix}. \tag{7.49}$$

The tridiagonal form of A_2 implies that it also has zeros below the lower diagonal. The detailed calculation shows that the terms r_j are used only to compute these zero elements. In writing a computer program the r_j's need not be stored. For each P_j it is enough to store the coefficients s_j and c_j. Further, it is not necessary to compute and store Q explicitly. Instead s_j's and c_j's can be used to find the product

$$A_2 = LQ = LP_{n-1}^{\mathrm{T}} P_{n-2}^{\mathrm{T}} \cdots P_1^{\mathrm{T}}. \tag{7.50}$$

To speed up the process a *shifting technique* can be employed. The idea is that if λ_j is an eigenvalue of A then $\lambda_j - s_i$ is an eigenvalue of $A - s_i I$. This is implemented in the step

$$A_i - s_i I = Q_i L_i \tag{7.51}$$

then form

$$A_{i+1} = L_i Q_i, \quad \text{for } i - 1, 2, \ldots k_j \tag{7.52a}$$

where

$$\lambda_j = s_1 + s_2 + \cdots + s_{k_j}. \tag{7.52b}$$

The correct shift required at each step can be found by using the four elements in the upper-left corner of the matrix. For example, for the matrix $\begin{pmatrix} a_1 & d_1 \\ d_1 & a_2 \end{pmatrix}$ the eigenvalues x_1 and x_2 are the roots of equation

$$x^2 - (a_1 + a_2)\, x + a_1 a_2 - d_1^2 = 0. \tag{7.53}$$

The root which is closer to a_1 is the value of s_i in Eq. (7.51).

The QL iteration with shifting is repeated until $d_1 \approx 0$ which produces $\lambda_1 = s_1 + s_2 + \cdots + s_{k_1}$. Repeating the process with the lower $(n-1)$ rows gives $d_2 \approx 0$ and the next eigenvalue λ_2. Next iteration gives $d_{n-2} \approx 0$ and the eigenvalue λ_{n-2}. The quadratic formula can be used to find the last two eigenvalues.

Example:

Let us apply the QL method to the matrix

$$A = \begin{pmatrix} 1 & -1 & 0 \\ -1 & 2 & -1 \\ 0 & -1 & 1 \end{pmatrix}. \tag{7.54}$$

Its exact eigenvalues are $\lambda = 0, 1, 3$.

For the given matrix $a_1 = 1$, $a_2 = 2$, $d_1 = -1$ and the quadratic Eq. (7.53) is $x^2 - 3x + 1 = 0$. Its roots are $x_1 = 2.61803$ and $x_2 = 0.38197$. The root close to $a_1 = 1$ is x_2. Therefore, the shift is $s_1 = 0.38197$. The first shifted matrix is

$$A_1 - s_1 I = \begin{pmatrix} 0.61803 & -1 & 0 \\ -1 & 1.61803 & -1 \\ 0 & -1 & 0.61803 \end{pmatrix}.$$

The LQ matrix is computed as

$$A_2 = L_1 Q_1 = \begin{pmatrix} 0 & 0.52573 & 0 \\ 0.52573 & 0.61803 & -0.85065 \\ 0 & -0.85065 & 2.23607 \end{pmatrix}.$$

$|d_1| = 0.52573$ in A is not less than the preassumed small value $\delta = 10^{-5}$. Therefore, perform second iteration with A_2. Now, $a_1 = 0$, $a_2 = 0.61803$, $d_1 = 0.52573$ and the roots of Eq. (7.53) are $x_1 = 0.91884$, $x_2 = -0.30081$ and the second shift is $s_2 = -0.30081$. The second shifted matrix is

$$A_2 - s_2 I = \begin{pmatrix} 0.30081 & 0.52573 & 0 \\ 0.52573 & 0.91884 & -0.85065 \\ 0 & -0.85065 & 2.53687 \end{pmatrix}.$$

The LQ matrix is obtained as

$$A_3 = L_2 Q_2 = \begin{pmatrix} -0.07668 & -0.06711 & 0 \\ -0.06711 & 0.94706 & -0.25379 \\ 0 & -0.25379 & 2.88615 \end{pmatrix}.$$

In the third iteration the shift is $s_3 = -0.08107$. Next,

$$A_4 = L_3 Q_3 = \begin{pmatrix} -0.00009 & 0.00001 & 0 \\ 0.00001 & 1.00358 & -0.08565 \\ 0 & -0.08565 & 2.99623 \end{pmatrix}.$$

The shift s_4 is -0.00009. The LQ matrix is obtained as

$$A_5 = L_4 Q_4 = \begin{pmatrix} 0 & 0 & 0 \\ 0 & 1.00041 & -0.02860 \\ 0 & -0.02860 & 2.99959 \end{pmatrix}.$$

In the next iteration $s_5 = 0$ hence an eigenvalue is obtained as

$$\lambda_1 = s_1 + s_2 + s_3 + s_4 + s_5 = 0.$$

Replace the first diagonal element in A_5 by λ_1 and write A_5 as

$$A_5 = \begin{pmatrix} 0 & 0 & 0 \\ 0 & 1.00041 & -0.02860 \\ 0 & -0.02860 & 2.99959 \end{pmatrix}.$$

The LQ matrix is

$$A_6 = L_5 Q_5 = \begin{pmatrix} 0 & 0 & 0 \\ 0 & 1.00005 & -0.00953 \\ 0 & -0.00953 & 2.99995 \end{pmatrix}.$$

The last two eigenvalues are determined using the 2×2 right-side corner matrix. The roots of the quadratic equation are $x_1 = 3$ and $x_2 = 1$. Then, $\lambda_2 = \lambda_1 + x_1 = 3$ and $\lambda_3 = \lambda_1 + x_2 = 1$. Therefore, the eigenvalues are 0, 1 and 3.

7.7   Eigenpairs of General Real Matrices

For general real matrices, the method proposed by Rutishauser is of use to find all eigenpairs. In his method a given matrix A_1 is decomposed into lower and upper triangular matrices as described in Section 4.7. Let us call these matrices as L_1 and U_1, respectively. Then,

$$A_1 = L_1 U_1, \qquad (7.55)$$

where $l_{ii} = 1$. A matrix A_2 is next constructed through equation

$$A_2 = U_1 L_1. \qquad (7.56)$$

Multiplication of right-side of Eq. (7.56) by $I = U_1 U_1^{-1}$ gives

$$A_2 = U_1 L_1 U_1 U_1^{-1} = U_1 A_1 U_1^{-1}. \qquad (7.57)$$

From Eq. (7.57) A_1 is expressed as

$$A_1 = U_1^{-1} A_2 U_1 \qquad (7.58)$$

which is the similarity transformation. Equation (7.58) implies that A_1 and A_2 are equivalent and hence have same eigenvalues. Decomposing A_2 into $L_2 U_2$ a new matrix $A_3 = U_2 L_2$ is constructed. This process is continued until an upper-triangular matrix is arrived. The diagonal elements of the final matrix are the eigenvalues of the given matric A_1. Then the associated eigenvectors are obtained by solving the linear system $A_1 \mathbf{X} = \lambda \mathbf{X}$.

Example:

Consider the matrix

$$A_1 = \begin{pmatrix} 4 & 1 \\ -1 & 1 \end{pmatrix}. \qquad (7.59)$$

The equation $A_1 = L_1 U_1$ is

$$\begin{pmatrix} 4 & 1 \\ -1 & 1 \end{pmatrix} = \begin{pmatrix} 1 & 0 \\ l_{21} & 1 \end{pmatrix} \begin{pmatrix} u_{11} & u_{12} \\ 0 & u_{22} \end{pmatrix}$$

$$= \begin{pmatrix} u_{11} & u_{12} \\ l_{21}u_{11} & l_{22}u_{12} + u_{22} \end{pmatrix}.$$

Comparison of the elements on both sides of the matrices in the above equation gives

$$L_1 = \begin{pmatrix} 1 & 0 \\ -1/4 & 1 \end{pmatrix}, \quad U_1 = \begin{pmatrix} 4 & 1 \\ 0 & 5/4 \end{pmatrix}.$$

Repeating the process 4 more times gives

$$A_2 = \begin{pmatrix} 15/4 & 1 \\ -5/16 & 5/4 \end{pmatrix}, \quad A_3 = \begin{pmatrix} 11/3 & 1 \\ -1/9 & 4/3 \end{pmatrix},$$

$$A_4 = \begin{pmatrix} 40/11 & 1 \\ -13/297 & 13/9 \end{pmatrix}, \quad A_5 = \begin{pmatrix} 3.62432 & 1 \\ -0.01753 & 1.45648 \end{pmatrix}.$$

The magnitude of the element a_{21} decreases and in the limit $r \to \infty$ one expects a_{21} in A_r to be ≈ 0. After 5 iterations the eigenvalues are 3.62432 and 1.45648, where the exact values are 3.61803 and 1.38196.

7.8   Concluding Remarks

In Chapter 4, the linear system of equations of the form $AX = B$ is considered. When $B = 0$ the system $AX = 0$ is a system of homogeneous equations. The eigenvalue equation can be rewritten in the homogeneous form $(A - \lambda I)X = 0$. In the present chapter, numerical computation of eigenvalues and eigenvectors of certain types of matrices are presented. Using the power method dominant eigenpair can be obtained. The Jacobi method is for symmetric matrices. For symmetric tridiagonal matrices, the QL method is useful. A method for general real matrices is also considered.

The eigenvalues of a matrix in a problem can be zero, real positive, real negative, complex conjugate with negative real part, complex conjugate with positive real part, pure imaginary, magnitude being small and magnitude being large. Some of the eigenvalues may be identical and in some cases all can be different and so on. *What are the significances of these cases?* The significances of these cases depend on the nature of the problem concerned. In certain problems real eigenvalues are meaningful.

Consider the anharmonic oscillator equation $\ddot{x} + d\dot{x} + ax + bx^3 = 0$. It can be rewritten as a system of first-order equations as

$$\dot{x} = y = P(x,y), \quad \dot{y} = -dy - ax - bx^3 = Q(x,y). \tag{7.60}$$

Its equilibrium points (x^*, y^*) are the roots of the equation $P(x,y) = 0$ and $Q(x,y) = 0$. Depending upon the values of the parameters d, a and b the given system can admit one or three real equilibrium points. The stability of an equilibrium point is determined by the nature of the eigenvalues of the Jacobian matrix

$$J = \begin{pmatrix} \partial P/\partial x & \partial P/\partial y \\ \partial Q/\partial x & \partial Q/\partial y \end{pmatrix} \tag{7.61}$$

evaluated at (x^*, y^*). For lower-dimensional systems, the eigenvalues can be determined analytically. For higher-dimensional systems often they need to be computed numerically.

An equilibrium point is stable, that is nearby trajectories approach it in the long-time limit, only if all the eigenvalues have negative real part. It at least one eigenvalue has a positive real part then the equilibrium point becomes unstable and the nearby trajectories diverge from its neighbourhood. An equilibrium point can be further classified into a stable star, an unstable star, a stable node, an unstable node, a saddle, a stable focus, an unstable focus and a centre in terms of the nature of the eigenvalues. For details refer the ref. [8].

7.9   Bibliography

[1]   E. Kreyszig, *Advanced Engineering Mathematics*. John Wiley, New York, 1999. 8th edition.

[2]   A. Jeffrey, *Advanced Engineering Mathematics*. Academic Press, San Diego, 2003. Indian reprint.

[3]   Yandasaketh, *Applications of Eigenvalues and Eigenvector*. https://www.geeksforgeeks.org/applications-of-eigenvalues-and-eigenvectors/ (accessed on June 4, 2023).

[4]   *Some Applications of the Eigenvalues and Eigenvectors of a square matrix.* https://sthcphy.files.wordpress.com/2020/05/eigenvalue-applications.pdf (accessed on June 4, 2023).

[5]   *Applications of Eigenvalues and Eigenvectors.* https://www.sheffield.ac.uk/media/32039/download?attachment (accessed on June 4, 2023).

[6]   *Eigenvalues and Eigenvectors.* https://en.wikipedia.org/wiki/Eigenvalues_and_eigenvectors.

[7]   Ajitesh Kumar, *Why & When to use Eigenvalues & Eigenvectors?* https://vitalflx.com/why-when-use-eigenvalue-eigenvector/ (accessed on June 4, 2023).

[8]   M. Lakshmanan and S. Rajasekar, *Nonlinear Dynamics: Integrability, Chaos and Patterns.* Springer, Berlin, 2002.

[9]   J.H. Mathews, *Numerical Methods for Mathematics Science and Engineering.* Prentice-Hall of India, New Delhi, 2005.

[10]   M.K. Jain, S.R.K. Iyengar, R.K. Jain, *Numerical Methods for Scientific and Engineering Computation.* New Age International, New Delhi, 1993.

[11]   W.H. Press, S.A. Teukolsky, W.T. Vetterling, B.P. Flannery, *Numerical Recipes in Fortran.* Foundation Books, New Delhi, 1993. Indian edition.

7.10   Problems

A. Power Method

In Problems 7.1–7.3 for hand calculation obtain the eigenpair after 3 iterations. If the calculation is done using a computer choose $\delta = 10^{-5}$.

7.1   For the following cubic equations

    i) obtain their dominant root by finding the dominant eigenvalue of their companion matrix and

    ii) obtain their smallest real root if they exist (hint: substitute $z = 1/y$ in the cubic equations).

    a) $z^3 - 16z^2 + 68z - 80 = 0$.   b) $z^3 - z^2 + 10z - 10 = 0$.

    c) $z^3 - 5z^2 + 3z - 9 = 0$.

7.2   Compute the dominant eigenpair of the following matrices.

a) $\begin{pmatrix} 2 & 1 & 1 \\ 1 & 1 & 0 \\ 1 & -1 & 0 \end{pmatrix}$ b) $\begin{pmatrix} 1 & 1 & 1 \\ 2 & 0 & 3 \\ 2 & 1 & 1 \end{pmatrix}$ c) $\begin{pmatrix} 2 & 0 & 0 & 2 \\ 1 & 1 & 0 & 1 \\ 1 & 1 & 1 & 0 \\ 2 & 3 & 4 & 1 \end{pmatrix}$

7.3   Obtain the dominant eigenpair of the matrix

$$A = \begin{pmatrix} 2 & 0 & -1 \\ 0 & 2 & 0 \\ 0 & 0 & 1 \end{pmatrix}.$$

By inspecting $\mathbf{X}_0$, $\mathbf{X}_1$, $\mathbf{X}_2$ and $\mathbf{X}_3$ write the eigenvector after large number of iterations.

B. Jacobi Method

7.4 Obtain all the eigenpairs of the following symmetric matrices.

a) $\begin{pmatrix} 1 & 2 \\ 2 & 1 \end{pmatrix}$ b) $\begin{pmatrix} \sqrt{2} & -1 \\ -1 & \sqrt{2} \end{pmatrix}$ c) $\begin{pmatrix} 3 & 1 \\ 1 & 3 \end{pmatrix}$

7.5 Compute all the eigenpairs of the following symmetric matrices.

a) $\begin{pmatrix} 1 & \sqrt{2} & 2 \\ \sqrt{2} & 3 & \sqrt{2} \\ 2 & \sqrt{2} & 1 \end{pmatrix}$ b) $\begin{pmatrix} 1 & 0 & 2 \\ 0 & 1 & 0 \\ 2 & 0 & 1 \end{pmatrix}$

c) $\begin{pmatrix} -2 & 3 & 0 \\ 3 & -2 & 1 \\ 0 & 1 & 0 \end{pmatrix}$

7.6 Develop a Python program to compute the eigenvalues and the corresponding eigenvectors by the Jacobi method. Then, modify the program so that the eigenvalues are arranged in ascending order in the final diagonalized matrix and the corresponding eigenvectors are normalized.

7.7 Find the eigenpairs of the matrix

$$A = \begin{pmatrix} -2 & 3 & 0 \\ 3 & -2 & 1 \\ 0 & 1 & 0 \end{pmatrix}$$

(a) by searching for the largest off-diagonal element and
(b) without searching for the largest off-diagonal element.

Compare the two approaches.

C. QL Method

7.8 Compute the eigenvalues of the following symmetric tridiagonal matrices. Also, compute the associated eigenvectors.

a) $\begin{pmatrix} 2 & 1 \\ 1 & 2 \end{pmatrix}$ b) $\begin{pmatrix} 2 & 1 & 0 \\ 1 & 3 & 2 \\ 0 & 2 & -2 \end{pmatrix}$ c) $\begin{pmatrix} 1 & -2 & 0 \\ -2 & -2 & 1 \\ 0 & 1 & 1 \end{pmatrix}$

7.9 Find the eigenpairs of the following symmetric tridiagonal matrices.

a) $\begin{pmatrix} 3 & 1 & 0 \\ 1 & 3 & 1 \\ 0 & 1 & 3 \end{pmatrix}$ b) $\begin{pmatrix} 3 & -1 & 0 \\ -1 & 3 & -1 \\ 0 & -1 & 3 \end{pmatrix}$

c) $\begin{pmatrix} 1 & -1 & 0 & 0 \\ -1 & 1 & -1 & 0 \\ 0 & -1 & 1 & -1 \\ 0 & 0 & -1 & 1 \end{pmatrix}$

7.10 Compute the eigenpairs of the following symmetric tridiagonal matrices.

a) $\begin{pmatrix} 2 & -1 & 0 & 0 \\ -1 & 3 & -2 & 0 \\ 0 & -2 & 4 & -3 \\ 0 & 0 & -3 & 5 \end{pmatrix}$ b) $\begin{pmatrix} 2 & -2 & 0 & 0 \\ -2 & 2 & -2 & 0 \\ 0 & -2 & 2 & -2 \\ 0 & 0 & -2 & 2 \end{pmatrix}$

8

Numerical Differentiation

8.1 Introduction

What is meant by numerical differentiation? It is defined as a mathematical process of determining the numerical value of an nth-order derivative $f^{(n)}(x)$ at a target point x_0 using the values of $f(x)$ given at several nodal points of x near x_0. The goal of the numerical differentiation is to find approximate formulas for the computation of derivatives of $f(x)$, using only the given data set, say, $(x_i, f(x_i))$, $i = 0, 1, 2, \ldots, N$.

In many phenomena in science and engineering derivatives of a function describing variation of a physical variable with respect to one or more control parameters are helpful to understand them [1-2]. It is well known that differentiation of position $x(t)$ gives velocity $v(t)$, velocity to acceleration $a(t)$, momentum ma (mass × acceleration) to force F, charge q to current I and energy E to power P. In mathematics, differentiation makes a cumulative distribution into a probability density, volumes into areas and areas into lengths.

Consider the motion of a particle in a potential well. Let its displacement from the equilibrium point is given by $x(t) = \sin t$. Then, the velocity $v(t)$ of the particle is $v(t) = \mathrm{d}x/\mathrm{d}t = \cos t$ and it changes continuously with time t. From the above expressions of $x(t)$ and $v(t)$ the position and velocity of the particle at time $t = 1$ and 1.01 are

$$x(1) = 0.8414709, \quad v(1) = 0.5403023, \quad x(1.01) = 0.8468318, \quad v(1.01) = 0.5318607. \quad (8.1)$$

The explicit forms of $x(t)$ and $v(t)$ are not known but only the numerical values of x at discrete values of t are given. For example, the values of $x(0)$, $x(0.01), \ldots, x(2.0)$ are given.

How does one calculate $v(t)$ from the time series of $x(t)$? Since the velocity is the rate of change of position, one may approximate

$$v(t) = \frac{\mathrm{d}x}{\mathrm{d}t} = \frac{x(t+h) - x(t)}{h} + O(h). \quad (8.2)$$

Equation (8.2) is obtained from the Taylor series expansion of $x(t+h)$ about $x(t)$ by assuming that $|h| \ll 1$. Using the values of $x(1)$ and $x(1.01)$ in Eq. (8.2), $v(1)$ is computed as 0.53609 while the exact value is 0.5403023. The approximation of $\mathrm{d}x/\mathrm{d}t$ given by Eq. (8.2) is inaccurate. Thus, more accurate formulas are required to determine $v(t) = \mathrm{d}x/\mathrm{d}t$. Therefore, it is important to develop and study good approximate formulas to evaluate first, second and higher-order derivatives of a function $f(x)$ whose exact form is not known but only its numerical values are given at discrete values of x. This is the goal of numerical differentiation. The present chapter first develops some simple numerical formulas to compute integer order derivatives of a function $f(x)$ and illustrate their efficiency. Next, computation of fractional order derivatives is considered.

DOI: 10.1201/9781032649931-8

8.2 Formulas for First-Order Derivative

Methods involving difference quotient approximations, called *finite-difference* approximations, were used for derivatives first by Euler in 1768. Such simple formulas for evaluating numerically the first-, second- and higher-order derivatives can be derived using Taylor series expansion of a given function $f(x)$. Runge, Richardson and Liebmann were the first to explore the finite-differences for derivatives. To start with, first presents the formulas for first-order derivative.

8.2.1 Forward- and Backward-Difference Formulas

It is desired to evaluate the derivative f' of the function $f(x)$ at x_0, where x, $x_0 \in [a, b]$. For this purpose, consider the Taylor series expansion of $f(x_0 + h)$ about the point x_0:

$$f(x_0 + h) = f(x_0) + hf'(x_0) + \frac{h^2}{2!}f''(x_0) + \frac{h^3}{3!}f^{(3)}(x_0)$$
$$+ \frac{h^4}{4!}f^{(4)}(x_0) + \frac{h^5}{5!}f^{(5)}(x_0) + \frac{h^6}{6!}f^{(6)}(x_0) + \cdots . \tag{8.3}$$

Equation (8.3) after neglecting the terms containing h^3 and higher powers of h becomes

$$f(x_0 + h) = f(x_0) + hf'(x_0) + \frac{h^2}{2!}f''(x_0). \tag{8.4}$$

Solving Eq. (8.4) for $f'(x_0)$ gives

$$f'(x_0) = \frac{f(x_0 + h) - f(x_0)}{h} - \frac{h}{2!}f''(x'), \tag{8.5}$$

where $x' \in [a, b]$ with $x_0, x_0 + h \in [a, b]$. The first term in the right-side of the above equation is called *forward-difference formula* and the second term is the *truncation error*. In Eq. (8.5) the error is of the order of h, that is, the approximation is only first-order accurate.

On the other hand, from the Taylor series expansion

$$f(x_0 - h) = f(x_0) - hf'(x_0) + \frac{h^2}{2!}f''(x_0) - \frac{h^3}{3!}f^{(3)}(x_0)$$
$$+ \frac{h^4}{4!}f^{(4)}(x_0) - \frac{h^5}{5!}f^{(5)}(x_0) + \frac{h^6}{6!}f^{(6)}(x_0) - \cdots . \tag{8.6}$$

$f'(x_0)$ is obtained as

$$f'(x_0) = \frac{f(x_0) - f(x_0 - h)}{h} + \frac{h}{2!}f''(x') . \tag{8.7}$$

This is known as *backward-difference formula* and is also first-order accurate in h.

Another formula in terms of forward-difference quantities can be obtained. Replacing h by $2h$ in Eq. (8.3) gives

$$f(x_0 + 2h) = f(x_0) + 2hf'(x_0) + 2h^2 f''(x_0) + \frac{4h^3}{3}f^{(3)}(x_0) + \cdots . \tag{8.8}$$

Equation (8.8) $- 4 \times$ Eq. (8.3) results in

$$f'(x_0) = \frac{-3f(x_0) + 4f(x_0 + h) - f(x_0 + 2h)}{2h}, \tag{8.9}$$

where the terms containing higher powers of h are neglected. The error term is $O(h^2)$.

8.2.2   Central-Difference Formulas

Subtraction of Eq. (8.6) from Eq. (8.3) gives

$$f(x_0 + h) - f(x_0 - h) = 2hf'(x_0) + \frac{1}{3}h^3 f^{(3)}(x_0) + \frac{2}{5!}h^5 f^{(5)}(x_0) + \cdots . \qquad (8.10)$$

From this equation, $f'(x_0)$ is obtained as

$$f'(x_0) = \frac{f(x_0 + h) - f(x_0 - h)}{2h} - \frac{h^2}{6}f^{(3)}(x_0) - \frac{1}{5!}h^4 f^{(5)}(x_0) - \cdots . \qquad (8.11)$$

The second and the other successive terms in the right-side of the above equation contain even powers of h. Neglecting the terms containing h^2 and other higher-order terms result in

$$f'(x_0) = \frac{f(x_0 + h) - f(x_0 - h)}{2h} . \qquad (8.12)$$

Because the two nodes used to compute $f'(x_0)$ are symmetrically situated from x_0, the formula, Eq. (8.12), is called a *two-point central-difference formula*. The value of the first-order derivative of a function $f(x)$ at a point x_0 is thus given by the difference between the values of the function evaluated at the points $x_0 - h$ and $x_0 + h$ lying to the left and right of x_0, respectively, divided by $2h$. In this formula the truncation error term is $E_{\text{trun}} = -(h^2/6)f^{(3)}(x')$, where $x' \in [a, b]$ with $x_0 + h$, $x_0 - h \in [a, b]$. The error is of the order of h^2 and is exact for polynomials of degree ≤ 2. Like the forward and backward formulas, the central-difference formula also needs only two values of f. The central-difference approximation has a better accuracy than the forward- and backward-difference formulas and hence it is most commonly used.

Equation (8.10) with terms up to h^5 is

$$f(x_0 + h) - f(x_0 - h) = 2hf'(x_0) + \frac{1}{3}h^3 f^{(3)}(x_0) + \frac{1}{60}h^5 f^{(5)}(x_0) . \qquad (8.13)$$

Replace h by $2h$ in Eq. (8.13) and obtain

$$f(x_0 + 2h) - f(x_0 - 2h) = 4hf'(x_0) + \frac{8}{3}h^3 f^{(3)}(x_0) + \frac{8}{15}h^5 f^{(5)}(x_0) . \qquad (8.14)$$

Multiplication of Eq. (8.13) by 8 and then subtracting it from (8.14) lead to

$$f'(x_0) = \frac{1}{12h}\{8[f(x_0 + h) - f(x_0 - h)] - [f(x_0 + 2h) - f(x_0 - 2h)]\}$$

$$+ \frac{1}{30}h^4 f^{(5)}(x') . \qquad (8.15)$$

The first term in the right-side of Eq. (8.15) is a *four-point central-difference formula* while the second term is the error. Note that the value of the derivative of $f(x)$ at x_0 is determined using the four values of f at equally spaced nodes with two on left-side of x_0 and another two on right-side of x_0. The error term is proportional to fifth derivative of f. That is, the formula (8.15) is exact for polynomial of degree ≤ 4.

Assume that the values of $f(x_i)$, $i = 0, 1, 2, \ldots, N$ are given. Forward-difference formula is not useful to find $f'(x_N)$ while $f'(x_0)$ cannot be computed using backward-difference formula. On the other hand, central-difference formula cannot be applicable at both x_0 and x_N. For $f'(x_0)$ the formula (8.9) can be used. To obtain a formula for $f'(x_N)$, replace x_0 by x_N and h by $-h$ in Eq. (8.9) and obtain

$$f'(x_N) = \frac{1}{2h}[-4f(x_N - h) + f(x_N - 2h) + 3f(x_N)] . \qquad (8.16)$$

8.3 Formulas for Second-Order Derivative

Formulas for second-order and higher-order derivatives of $f(x)$ can also be derived from Taylor series. Let us derive the formulas for second-order derivative.

Adding of Eqs. (8.3) and (8.6) results in

$$f(x_0 + h) + f(x_0 - h) = 2f(x_0) + h^2 f''(x_0) + \frac{2}{4!} h^4 f^{(4)}(x_0)$$
$$+ \frac{2}{6!} h^6 f^{(6)}(x_0) + \cdots. \qquad (8.17)$$

From the above equation $f''(x_0)$ is obtained as

$$f''(x_0) = \frac{1}{h^2}[f(x_0 + h) - 2f(x_0) + f(x_0 - h)] - \frac{h^2}{12} f^{(4)}(x'), \qquad (8.18)$$

where the series is truncated at fourth derivative. The first term in the right-side of Eq. (8.18) is an approximate formula for second derivative and the second term is the truncation error and is of the order of h^2. Equation (8.8)$-2\times$Eq. (8.3) gives the formula

$$f''(x_0) = \frac{1}{h^2}[f(x_0) - 2f(x_0 + h) + f(x_0 + 2h)]. \qquad (8.19)$$

A formula with the truncation error of the order of h^4 is obtained as follows. Equation (8.17) with terms up to h^6 is written as

$$f(x_0 + h) + f(x_0 - h) = 2f(x_0) + h^2 f''(x_0)$$
$$+ \frac{2}{4!} h^4 f^{(4)}(x_0) + \frac{2}{6!} h^6 f^{(6)}(x_0). \qquad (8.20)$$

Replacing h by $2h$ in Eq. (8.20) gives

$$f(x_0 + 2h) + f(x_0 - 2h) = 2f(x_0) + 4h^2 f''(x_0)$$
$$+ \frac{32}{4!} h^4 f^{(4)}(x_0) + \frac{128}{6!} h^6 f^{(6)}(x_0). \qquad (8.21)$$

Multiplication of Eq. (8.20) by 16 and then subtracting it from Eq. (8.21) give

$$f''(x_0) = \frac{1}{12h^2}[-f(x_0 + 2h) + 16f(x_0 + h) - 30f(x_0)$$
$$+ 16f(x_0 - h) - f(x_0 - 2h)] + \frac{23}{2160} h^4 f^{(6)}(x'). \qquad (8.22)$$

8.4 Formulas for Third-Order Derivative

Formulas for third-order derivative with the error term of the order of h and h^2 can be obtained as follows. Replacing h by $3h$ in Eq. (8.3) gives

$$f(x_0 + 3h) = f(x_0) + 3hf'(x_0) + \frac{9}{2} h^2 f''(x_0) + \frac{9}{2} h^3 f^{(3)}(x_0)$$
$$+ \frac{27}{8} h^4 f^{(4)}(x_0) + \cdots. \qquad (8.23)$$

Equation (8.23)$-3\times$Eq. (8.8) gives

$$f(x_0 + 3h) - 3f(x_0 + 2h) = -2f(x_0) - 3hf'(x_0) - \frac{3}{2}h^2 f''(x_0)$$

$$+ \frac{1}{2}h^3 f^{(3)}(x_0) + \cdots. \qquad (8.24)$$

Then, Eq. (8.24)$+3\times$Eq. (8.3) leads to the formula

$$f^{(3)}(x_0) = \frac{1}{h^3}\left[-f(x_0) + 3f(x_0 + h) - 3f(x_0 + 2h) + f(x_0 + 3h)\right] + O(h). \qquad (8.25)$$

Next, consideration of Eq. (8.14)$-2\times$Eq. (8.13) gives the central-difference formula

$$f^{(3)}(x_0) = \frac{1}{2h^3}\left[f(x_0 + 2h) - f(x_0 - 2h)\right.$$

$$\left. -2f(x_0 + h) + 2f(x_0 - h)\right] + O(h^2). \qquad (8.26)$$

8.4.1   Computation of a Highly Accurate Result

Generally, the numerical value of $f'(x)$ from an approximation is expected to approach the true value of $f'(x)$ in the limit $h \to 0$. Consequently, accuracy in the numerical computation of derivatives can be highly improved by combining an approximation, for example the central-difference formula, and an extrapolation method like Richardson extrapolation technique. Denote $y(h)$ and $y(\alpha h)$ are the two approximate values of y obtained by a method of order p with step size h and αh, respectively. Then,

$$y(h) = y + ch^p + O\left(h^{p+q}\right), \qquad (8.27a)$$

$$y(\alpha h) = y + c\alpha^p h^p + O\left(h^{p+q}\right). \qquad (8.27b)$$

Multiplication of Eq. (8.27a) by α^p and then subtracting from Eq. (8.27b) give

$$y = \frac{y(h)\alpha^p - y(\alpha h)}{\alpha^p - 1} + O\left(h^{p+q}\right). \qquad (8.28)$$

The above formula is known as *Richardson extrapolation*. The first term in the right-side of Eq. (8.28) is a more accurate value of y with error term of the order of h^{p+q}. For a two-point formula $p = q = 2$ in Eq. (8.28). If the above extrapolation scheme is applied to a two-point formula for the computation of f' then the local truncation error is $O(h^4)$.

For convenience define $h = h_1$, $\alpha h = h_2$, $\alpha = h_2/h_1$. Then,

$$y(\text{accurate}) = \frac{(h_2/h_1)^p y(h_1) - y(h_2)}{(h_2/h_1)^p - 1}. \qquad (8.29)$$

Suppose, f_1' and f_2' are the values of f' obtained at a point x with step size h_1 and h_2 using a two-point formula. Then, a more accurate value of f' is

$$f'(\text{accurate}) = \frac{(h_2/h_1)^2 f_1' - f_2'}{(h_2/h_1)^2 - 1}. \qquad (8.30)$$

Example:

Compute the first and second derivatives of $f(x) = \sin x$ at $x = 1$ using the two-point central-difference formula for two-step sizes $h = 0.05$ and 0.1. Then, obtain the accurate value of $f'(1)$ by employing Richardson extrapolation to the two computed values of f'. Also, compute the bounds of truncation error in the two-point central-difference formula.

1. Calculation of first derivative

The two-point formula for $f'(x_0)$ is

$$f'(x_0) = \frac{f(x_0 + h) - f(x_0 - h)}{2h}.$$

For $f(x) = \sin x$ with $h = 0.05$, the above formula at $x_0 = 1$ becomes

$$f'(1) = \frac{\sin(1.0 + 0.05) - \sin(1.0 - 0.05)}{2 \times 0.05} = 0.5400772.$$

The exact value of $f'(1)$ is $\cos 1 = 0.5403023$. The percentage of relative error is 0.042. With $h = 0.1$

$$f'(1) = \frac{1}{2 \times 0.1}(\sin(1.0 + 0.1) - \sin(1.0 - 0.1)) = 0.5394022.$$

The percentage of relative error is 0.167.

A more accurate result is then obtained using Richardson's extrapolation (Eq. (8.30)). With $h_1 = 0.1$, $h_2 = 0.05$, $f'_1 = 0.5394022$ and $f'_2 = 0.5400772$

$$f'(\text{accurate}) = \frac{0.5394022 \times (0.05/0.1)^2 - 0.5400772}{(0.05/0.1)^2 - 1} = 0.5403022$$

which is very close to the exact value and the percentage of relative error is 1.47×10^{-5}.

Next, proceed to compute the bounds of truncation error. The truncation error is

$$E_{\text{trun}} = -\frac{1}{6}h^2 f^{(3)}(x'),$$

where $x' \in [a, b]$ with $x_0 + h$, $x_0 - h \in [a, b]$. For $h = 0.1$, $[a, b] \to [0.9, 1.1]$. Since, $|\sin(1.1)| > |\sin(0.9)|$ one can choose $x' = 1.1$. Now, $|f^{(3)}(x')| \le |-\sin(1.1)| = 0.8912073$. Then,

$$|E_{\text{trun}}| \le \left|-\frac{1}{6}h^2 f^{(3)}(1.1)\right| = 0.00148534.$$

For $h = 0.05$, $[a, b] \to [0.95, 1.05]$ and

$$|E_{\text{trun}}| \le \frac{1}{6}h^2\left|f^{(3)}(1.05)\right| = 0.000361426.$$

2. Calculation of second derivative $f''(1)$

The two-point formula for $f''(x_0)$ is

$$f''(x_0) = \frac{1}{h^2}[f(x_0 + h) - 2f(x_0) + f(x_0 - h)].$$

With $h = 0.05$

$$f''(1) = \frac{1}{0.05^2}[\sin(1.05) - 2\sin(1) + \sin(0.95)] = -0.841296.$$

The exact value of $f''(1) = -\sin(1)$ is -0.841471. The percentage of relative error is 0.021.

Next, with $h = 0.1$

$$f''(1) = \frac{1}{0.1^2} \left[\sin(1.1) - 2\sin(1) + \sin(0.9) \right] = -0.84077.$$

The percentage of relative error is 0.083. Employing the Richardson extrapolation results in

$$f''(\text{accurate}) = \frac{(0.05/0.1)^2 \times -0.840771 - (-0.841296)}{(0.05/0.1)^2 - 1} = -0.841471.$$

8.5 Fractional Order Derivatives

The focus of this section is on the numerical evaluation of fractional order derivatives of a function. There are certain definitions of the fractional order derivative of order $\alpha > 0$ [3-6]. The widely used definitions are the Riemann–Liouville and Caputo. The Caputo's definition is often preferred in physical applications as it has the advantage that the initial condition is specified in terms of field variables and their integer order derivatives. *Why is the study of fractional derivatives important?* A fractional derivative of a function with respect to space or time variable contains memory effect, that is, it possesses details of the function at earlier space or time. The fractional derivatives are thus nonlocal operators. This feature has applications in viscoelastic materials, polymers and anomalous diffusion, a few to mention. For more details, see Chapter 16.

The most common version of the Caputo's definition of the fractional order derivative D_a^α of $f(t)$ is

$$D_a^\alpha f(t) = \frac{1}{\Gamma(m - \alpha)} \int_{a=0}^{t} (t - \tau)^{m-1-\alpha} f^{(m)}(\tau) d\tau \,, \tag{8.31}$$

where $m = [\alpha] + 1$ with $[\alpha]$ being the integer part of α and $\Gamma(\gamma)$ is the gamma function with argument γ and $\Gamma(\gamma) = (\gamma - 1)!$. The subscript a in D_a^α denotes the lower limit of the integration. For convenience and simplicity, the subscript a may be dropped. The gamma function is defined as

$$\Gamma(\gamma) = \int_0^\infty z^{\gamma-1} e^{-z} dz \,. \tag{8.32}$$

For $0 < \alpha < 1$, the value of m is 1 and for $1 < \alpha < 2$, its value is 2. Methods for evaluating numerically $D^\alpha f(t)$ are proposed and analyzed [7-11].

8.5.1 Fractional Derivative of t^β

Let us find $D^\alpha t^\beta$ for $0 < \alpha < 1$ and for positive integer values of β. For $0 < \alpha < 1$, Eq. (8.31) becomes

$$D^\alpha f(t) = \frac{1}{\Gamma(1 - \alpha)} \int_0^t (t - \tau)^{-\alpha} f'(\tau) d\tau \,, \quad t \geq 0. \tag{8.33}$$

Start with $D^\alpha t$ and obtain

$$D^\alpha t = \frac{1}{\Gamma(1 - \alpha)} \int_0^t (t - \tau)^{-\alpha} d\tau = \frac{\Gamma(2)}{\Gamma(2 - \alpha)} t^{1-\alpha}. \tag{8.34a}$$

Next, $D^{\alpha}t^2$ and $D^{\alpha}t^3$ are determined as

$$
\begin{aligned}
D^{\alpha}t^2 &= \frac{2}{\Gamma(1-\alpha)}\int_0^t (t-\tau)^{-\alpha}\tau d\tau \\
&= \frac{2}{(1-\alpha)\Gamma(1-\alpha)}\int_0^t (t-\tau)^{1-\alpha}d\tau \\
&= \frac{\Gamma(3)}{\Gamma(3-\alpha)}t^{2-\alpha}
\end{aligned}
\tag{8.34b}
$$

and

$$
\begin{aligned}
D^{\alpha}t^3 &= \frac{3}{\Gamma(1-\alpha)}\int_0^t (t-\tau)^{-\alpha}\tau^2 d\tau \\
&= \frac{3\cdot 2}{(1-\alpha)\Gamma(1-\alpha)}\int_0^t (t-\tau)^{1-\alpha}\tau d\tau \\
&= \frac{3\cdot 2\cdot 1}{(2-\alpha)(1-\alpha)\Gamma(1-\alpha)}\int_0^t (t-\tau)^{2-\alpha}d\tau \\
&= \frac{\Gamma(4)}{\Gamma(4-\alpha)}t^{3-\alpha}.
\end{aligned}
\tag{8.34c}
$$

From Eqs. (8.34) $D^{\alpha}t^{\beta}$ is written as

$$
D^{\alpha}t^{\beta} = \frac{\Gamma(\beta+1)}{\Gamma(\beta+1-\alpha)}t^{\beta-\alpha}.
\tag{8.35}
$$

8.5.2 Composite Trapezoidal Rule for $D^{\alpha}f(t)$, $0 < \alpha < 1$

For $0 < \alpha < 1$, $D^{\alpha}f(t)$ is given by Eq. (8.33). It is desired to evaluate the derivative at, say, $t = T$. *Can this integral be evaluated by applying the composite quadrature formulas?* There is a problem in applying these formulas directly to this integral. The integrand has a singularity at $\tau = T$ and for values of τ near T the term $(t-\tau)^{-\alpha}$ diverges rapidly. Interestingly, the singularity can be avoided by integrating the integral by parts once. This gives

$$
\begin{aligned}
D^{\alpha}f(t) &= \frac{t^{1-\alpha}f'(0)}{(1-\alpha)\Gamma(1-\alpha)} + \frac{1}{(1-\alpha)\Gamma(1-\alpha)}\int_0^t (t-\tau)^{1-\alpha}f''(\tau)d\tau \\
&= \frac{1}{\Gamma(2-\alpha)}\left(t^{1-\alpha}f'(0) + \int_0^t (t-\tau)^{1-\alpha}f''(\tau)d\tau\right).
\end{aligned}
\tag{8.36}
$$

The above formula is proposed and analyzed in [10]. Approximate the integral in this equation by the composite trapezoidal rule (10.35). Divide the interval $[0, t(= T)]$ into n subintervals with nodes $\tau_k = kh$, $k = 0, 1, \ldots, n$, $h = t(= T)/n$. Application of the composite trapezoidal rule leads to the equation

$$
\begin{aligned}
D^{\alpha}f(t) &= \frac{1}{\Gamma(2-\alpha)}\Bigg[t^{1-\alpha}f'(0) \\
&\quad + \frac{h}{2}\left(t^{1-\alpha}f''(0) + 2\sum_{k=1}^{n-1}(t-\tau_k)^{1-\alpha}f''(\tau_k)\right)\Bigg].
\end{aligned}
\tag{8.37}
$$

At the node $\tau_n = t$, the term $(t-\tau_n)^{1-\alpha}f''(\tau_n)$ is 0 and hence this term does not appear in the above equation.

If the analytical expressions for f' and f'' are known and determining their values is easy then the formula (8.37) can be used to compute $D^\alpha f(t)$ at $t = T$. When finding f' and f'' are difficult or only numerical values of them are known at the nodes then it is necessary to replace them by appropriate finite-difference approximations. Since $t \geq 0$ the derivatives $f'(0)$ and $f''(0)$ are approximated by the formulas (refer Eqs. (8.9) and (8.19))

$$f'(0) = \frac{1}{2h}\left[-3f(0) + 4f(h) - f(2h)\right], \tag{8.38a}$$

$$f''(0) = \frac{1}{h^2}\left[f(0) - 2f(h) + f(2h)\right]. \tag{8.38b}$$

For the nodes τ_k, $k = 1, 2, \ldots, n-1$ the three-point central-difference formula is useful and is

$$f''(\tau_k) = \frac{1}{h^2}\left[f(\tau_k - h) - 2f(\tau_k) + f(\tau_k + h)\right]. \tag{8.38c}$$

The error term in the composite trapezoidal rule is

$$E = -\frac{t}{12}h^2\frac{d^2}{d\tau^2}\left[(t - \tau)^{1-\alpha}f''(\tau)\right]\Big|_{\tau=\mu}, \tag{8.39}$$

where $0 < \mu < t$. The error $E \to 0$ as $h \to 0$. Further, the error terms in the approximations of f' and f'' also approach zero in the limit of $h \to 0$.

Example:

For the function $f(t) = t^{3+\alpha}$ the exact value of $D^\alpha f(t)$ is $\Gamma(4 + \alpha)t^3/6$ (for proof see Problem 16.8). Compute $D^\alpha f(t)$ at $t = 0.5$ numerically applying the formula (8.37), with the derivatives f' and f'' given by Eqs. (8.38), and compare the result with the exact value of it.

The formula (8.37) contains $\Gamma(2 - \alpha)$ and the exact solution contains $\Gamma(4 + \alpha)$. These values to be computed by numerically evaluating the integral in (8.31) for $0 < \alpha < 1$. Using the composite trapezoidal rule $\Gamma(\gamma)$ is approximated as

$$\Gamma(\gamma) \approx h\sum_{k=1}^{\infty} z_k^{\gamma-1}e^{-z_k}. \tag{8.40}$$

In practice, the summation can be stopped, for example, when $z > 10$ and $z_k^{\gamma-1}e^{-z_k} < 10^{-6}$. For $\alpha = 0.5$ with $h = 0.001, 0.0001$ and 0.00001 the values of $\Gamma(2 - \alpha)$ are computed as $0.886220, 0.886226$ and 0.886226, respectively. The exact value of $\Gamma(0.5) = \sqrt{\pi}$ and hence $\Gamma(2 - \alpha) = \Gamma(1.5) = 0.5\Gamma(0.5) = \sqrt{\pi}/2 = 0.886226$. Therefore, the choice $h = 0.0001$ is desirable for the calculation of $\Gamma(2 - \alpha)$. The exact value of $\Gamma(4 + \alpha) = \Gamma(4.5) = 3.5 \times 2.5 \times 1.5 \times 0.5 \times \sqrt{\pi} = 11.631728$. The numerically calculated value of $\Gamma(4.5)$ with $h = 0.0001$ is 11.631728. In the numerical calculation of $\Gamma(\gamma)$ for noninteger values of $\gamma > 0$ an appropriate value of h has to be chosen for realizing desired accuracy.

Table 8.1 presents $D^\alpha f(t)$ computed at $t = 0.5$ for $\alpha = 0.25, 0.5, 0.75$ and 1. The calculations are done for $h = 10^{-2}, 10^{-3}, 10^{-4}$ and 10^{-5}. Accuracy improves with decrease in the value of h. For $h = 10^{-4}$ the numerically computed $D^\alpha f(t)$ is in very close agreement with the exact value of it. Use of composite Simpson's rules can give more accurate results for $h = 10^{-2}$ and 10^{-3}.

In the above example, the values of $\Gamma(\gamma)$ for noninteger values of γ are computed by applying the composite trapezoidal rule. Certain other formulas are available to calculate

TABLE 8.1
The numerically computed $D^\alpha f(t)$, $f(t) = t^{3+\alpha}$, at $t = 0.5$ for four values of step size h. The exact values of $D^\alpha f(t)$ are also given. D_n^α and D_e^α denote the numerically computed and the exact values, respectively, of the derivative.

α	D_e^α	D_n^α for			
		$h = 10^{-2}$	$h = 10^{-3}$	$h = 10^{-4}$	$h = 10^{-5}$
0.25	0.17261	0.17250	0.17260	0.17261	0.17261
0.50	0.24233	0.24165	0.24231	0.24233	0.24233
0.75	0.34555	0.34221	0.34536	0.34554	0.34555
1.00	0.50000	0.48522	0.49853	0.49988	0.50001

$\Gamma(\gamma)$ for noninteger values of γ [12]. The value of $\Gamma(n + p)$ for $n \geq 2$ and $0 \leq p \leq 1$ can be calculated using the formula [12]

$$\Gamma(n + p) = n! \left[\left(n + \frac{1}{2}p \right)^2 + \frac{p(2 - p)}{12} \right]^{(p-1)/2} . \tag{8.41}$$

The error in $\Gamma(n + p)$ is never more than 1 part in $13,000$ [12]. The accuracy will increase with n. If $\Gamma(1 + p)$ value is needed then calculate $\Gamma(2 + p)$ from (8.41) and then use the relation $\Gamma(2+p) = (1+p)\Gamma(1+p)$. Similarly, for $\Gamma(p)$ use the relation $\Gamma(2+p) = (1+p)p\Gamma(p)$. Calculate the values of $\Gamma(0.5)$ and $\Gamma(1.5)$ using (8.41) and also using (8.40) and compare them with the exact values.

8.5.3 Formula for $D^\alpha f(t)$, $1 < \alpha < 2$

In the case of $1 < \alpha < 2$, Eq. (8.31) gives

$$D^\alpha f(t) = \frac{1}{\Gamma(2 - \alpha)} \int_0^t (t - \tau)^{1-\alpha} f''(\tau) d\tau , \quad t \geq 0. \tag{8.42}$$

Due to the presence of a singularity in the integrand at $\tau = t$, perform the integration by parts once. The result is

$$D^\alpha f(t) = \frac{1}{\Gamma(3 - \alpha)} \left[t^{2-\alpha} f''(0) + \int_0^t (t - \tau)^{2-\alpha} f'''(\tau) d\tau \right], \quad t \geq 0. \tag{8.43}$$

This formula is analyzed in [11]. Approximating the integral in the above equation by the composite trapezoidal rule gives

$$D^\alpha f(t) = \frac{1}{\Gamma(3 - \alpha)} \left[t^{2-\alpha} f''(0) \right.$$

$$\left. + \frac{h}{2} \left(t^{2-\alpha} f'''(0) + 2 \sum_{k=1}^{n-1} (t - \tau_k)^{2-\alpha} f'''(\tau_k) \right) \right]. \tag{8.44}$$

TABLE 8.2

The numerically computed $D^\alpha f(t)$, $f(t) = t^4$, at $t = 0.5$ and 1 for three values of step size h. The exact values of $D^\alpha f(t)$ are also given. $D^\alpha_{\rm n}$ and $D^\alpha_{\rm e}$ denote the numerically computed and the exact values, respectively, of the derivative.

t	α	$D^\alpha_{\rm e}$	$D^\alpha_{\rm n}$ for		
			$h = 10^{-2}$	$h = 10^{-3}$	$h = 10^{-4}$
0.5	1.25	0.80661	0.80800	0.80662	0.80661
0.5	1.50	1.27662	1.27620	1.27655	1.27661
0.5	1.75	1.97916	1.96857	1.97843	1.97912
0.5	2.00	3.00000	2.94322	2.99405	2.99942
1.0	1.25	5.42620	5.42836	5.42621	5.42620
1.0	1.50	7.22163	7.21939	7.22148	7.22162
1.0	1.75	9.41451	9.39103	9.41304	9.41443
1.0	2.00	12.00000	11.88326	11.98809	11.99886

The nodes are $\tau_k = kh$, $k = 0, 1, \ldots, n$, $h = t(= T)/n$. The useful approximations for the integer derivatives of f in the above equation are

$$f''(0) = \frac{1}{h^2} \left[f(0) - 2f(h) + f(2h) \right], \tag{8.45a}$$

$$f'''(0) = \frac{1}{h^3} \left[-f(0) + 3f(h) - 3f(2h) + f(3h) \right], \tag{8.45b}$$

$$f'''(\tau_k) = \frac{1}{2h^3} \left[-f(\tau_k - 2h) \right.$$
$$\left. + 2f(\tau_k - h) - 2f(\tau_k + h) + f(\tau_k + 2h) \right]. \tag{8.45c}$$

Example:

Test the applicability of the formula (8.44) for the function $f(t) = t^4$ for $1 \le \alpha \le 2$. The exact analytical expression of $D^\alpha f(t)$ is $\dfrac{24}{\Gamma(5 - \alpha)} t^{4-\alpha}$.

Choose the target values of t as 0.5 and 1 and $h = 0.01$, 0.001 and 0.00001. The computed values of $D^\alpha f(t)$ for $\alpha = 1.25$, 1.5, 1.75 and 2 are presented in Table 8.2 along with the exact values of $D^\alpha f(t)$. The convergence of the numerical result to the exact result with decrease in the value of h is clearly seen.

Formulas for $D^\alpha f(t)$ for $\alpha > 2$ can be straight-forwardly obtained following the steps used for $0 < \alpha < 1$ and $1 < \alpha < 2$.

8.6   Concluding Remarks

In the present chapter, certain formulas for computing both integer order and fractional order derivatives are derived and tested. For more accurate result the Richardson interpolation can be combined. Numerical differentiation is useful to compute the discontinuous points and the edge detection in image processing and certain inverse problems occurring in model equations. In signal processing to find the rates of change in the signals and to remove the noise present in the signal numerical differentiation is utilized. In design engineering problems numerical differentiation is applied to calculate gradients and objective functions. To identify the local maxima and minima of an analytical function or a numerical data set computation of derivatives is required.

8.7 Bibliography

[1] *What is the real-life application of numerical differentiation?* https://www.quora.com/What-is-the-real-life-application-of-numerical-differentiation (accessed on June 12, 2023).

[2] *Applications of Numerical Differentiation and Integration in Engineering.* https://www.studocu.com/in/document/amity-university/ fundamentals-of-mathematics/math-nice/59089228 (accessed on June 12, 2023).

[3] L. Debnath, *Int. J. Math. Educ. Sci. Technol.* 35:487, 2004.

[4] K.B. Oldham and J. Spanier, *The Fractional Calculus: Theory and Applications of Differentiation and Integration to Arbitrary Order.* Dover, New York, 2006.

[5] R. Hilfer, Threshold Introduction to fractional derivatives. In *Anomalous Transport: Foundations and Applications.* R. Kloges et al. (Eds.). Wiley-VCH, Weinheim, 2008.

[6] R. Herrmann, *Fractional Calculus: An Introduction for Physicists.* World Scientific, Singapore, 2014.

[7] Z. Odibat, *Appl. Math. Comput.* 178:527, 2006.

[8] E. Sousa, *How to approximate the fractional derivative of order $1 < \alpha < 2$?* In the Proceedings of the 4th IFAC Workshop Fractional Differentiation and its Applications. I. Podlubny, B.M. Vinagre Jara, Y.Q. Chen, V. Feliu Batile and I. Tejado Balsera (Eds.). 2010, Article number FDA10-019.

[9] P. Novati, *Numerische Mathematik* 127:539, 2014.

[10] R.B. Albadarneh, M. Zerqat and I.M. Batiha, *Int. J. Pure Appl. Math.* 106:859, 2016.

[11] R.B. Albadarneh, I.M. Batiha and M. Zurigat, *J. Math. Comput. Sci.* 16:103, 2016.

[12] M.S. Raff, *The Am. Stat.* 24:22, 1970.

8.8 Problems

8.1 Derive the following central-difference formulas of third derivative of $f(x)$:

$$
\begin{aligned}
f^{(3)}(x_0) &= \frac{1}{2h^3}\left[f(x_0 + 2h) - 2f(x_0 + h) + 2f(x_0 - h)\right. \\
&\quad \left. -f(x_0 - 2h)\right] + O\left(h^2\right)
\end{aligned}
$$

and

$$f^{(3)}(x_0) = \frac{1}{8h^3}[-f(x_0+3h)+8f(x_0+2h)-13f(x_0+h)$$
$$+13f(x_0-h)-8f(x_0-2h)+f(x_0-3h)]+O(h^4).$$

8.2 Compute the values of $f'(x_0)$ by the two-point and four-point formulas and $f''(x_0)$ by the three-point and five-point formulas of the following functions at the specified value of x for $h = 0.05$ and 0.1. Also, obtain the accurate values of f' and f'' by Richardson extrapolation. Compute the bounds of the truncation error for $h = 0.05$ and 0.1.

   a) $\cos x$, $x_0 = 0$, x is in radian.
   b) $\ln x$, $x_0 = 1$.
   c) e^x, $x_0 = 0.5$.
   d) xe^{-x}, $x_0 = 0$.
   e) $1/(1+x)$, $x_0 = 1$.

8.3 Obtain an expression for the optimum value of h for the computation of $f'(x_0)$ by the two-point central-difference formula.

8.4 The two-point central-difference formula for the partial derivatives $f_x(x_0, y_0)$ and $f_y(x_0, y_0)$ of $f(x, y)$ are given by

$$f_x(x_0, y_0) = \frac{1}{2h}[f(x_0+h, y_0)-f(x_0-h, y_0)]+O(h^2),$$
$$f_y(x_0, y_0) = \frac{1}{2h}[f(x_0, y_0+h)-f(x_0, y_0-h)]+O(h^2).$$

   Compute $f_x(0.2, 0.2)$ and $f_y(0.2, 0.2)$ of $f(x, y) = e^{-(x^2+2y)}$ with $h = 0.05$ and 0.1. Compare them with the exact values.

8.5 Compute $f_x(0,0)$ and $f_y(0,0)$ of $f(x, y) = (1+x)\cos y$ with $h = 0.05$ and 0.1. Compare them with the exact values.

8.6 The motion of a particle along a one-dimensional line is governed by the equation $dx/dt = v_0 + gt$, where v_0 is the initial velocity of the particle and g is the acceleration due to gravity at the surface of the earth. The position of the particle measured at three instants of time (in sec) are $x(1) = 5.4\,\text{m}$, $x(1.1) = 6.479\,\text{m}$ and $x(1.2) = 7.656\,\text{m}$. Use the two-point central-difference formula for the calculation of dx/dt at $t = 1.1$. Compute v_0. Assume that $g = 9.8\,\text{m/sec}^2$.

8.7 The equation of motion of a linear harmonic oscillator is given by (in dimensionless form) $x'' + \omega_0^2 x = 0$. Following table gives the values of x (in metre) for five values of t (in sec). Using the five-point formula compute the second derivative at $t = 0.2$. Then, determine the value of the parameter ω_0^2.

t	0	0.1	0.2	0.3	0.4
x	1	0.98	0.921	0.825	0.697

8.8 The position $x(t)$ values (in metre) of a particle executing a simple harmonic motion at five values of t (in sec) are given below. Using the four-point central-difference formula compute the first derivative x' (velocity) at $t = 0.2$. Then, determine the constant energy E of the particle, where $E = \frac{1}{2}(x')^2 + x^2$.

t	0.0	0.1	0.2	0.3	0.4
x	1.0	0.995	0.98	0.955	0.921

8.9 The state equation of an LC circuit in dimensionless form is given by $d^2v/dt^2 + (1/LC)v = 0$. The following table gives the values of $v(t)$ measured at five values of t with $C = 1$. Compute the value of L.

t	1.0	1.1	1.2	1.3	1.4
v	0.5403	0.4536	0.3624	0.2675	0.17

8.10 Obtain a six-point central-difference formula for the first-order derivative $f'(x_0)$ with sixth-order accuracy in h.

8.11 For $f(t) = t^2 - t$ the exact value of $D^{1/2}f(t) = \dfrac{8}{3}\sqrt{\dfrac{t^3}{\pi}} - 2\sqrt{\dfrac{t}{\pi}}$. Compute $D^{1/2}f(t)$ at $t = 0.5$ and compare it with the exact value of it for $h = 0.01, 0.001$ and 0.0001.

8.12 Compute $D^{1/2}f(t)$ for $f(t) = t^2$ for $t \in [0.1, 0.5]$ with $\Delta t = 0.1$ and $h = 0.0001$. The exact value of $D^{1/2}f(t)$ is $2t^{3/2}/\Gamma(5/2)$.

8.13 The exact value of $D^{1/2}(t)$ is $2\sqrt{t/\pi}$. Find the value of $D^{1/2}(t)$ for $t \in [0.1, 0.5]$ with $h = 0.0001$.

8.14 The exact value of $D^{3/2}t^2$ is $4\sqrt{t/\pi}$. Compute $D^{3/2}t^2$ for $t \in [0.1, 0.5]$ with $\Delta t = 0.1$ and for $h = 0.0001$. Compare the numerical result with the exact result.

8.15 The approximate solution of $D^{1.9}x(t) + x(t) = (2/\Gamma(3-\alpha))t^{2-\alpha} + t^3$ for $x(0) = 0$, $x'(0) = 0$ is $x(t) \approx t^2$. Verify this numerically computing the fractional order derivative.

9

Numerical Minimization of Functions

9.1 Introduction

For a function $f(x)$ a local minimum is a value of x at which the function attains a lowest value in a certain region of it. The local maximum is a value of x where $f(x)$ reaches the highest value in a certain region of it. A function can have more than one minimum and maximum. The concept of minimum and maximum of a function was introduced by Sir Issac Newton. Consider the function $f(x) = \sin x$, $0 \le x \le 2\pi$. The minimum value of $\sin x$ is -1 and this happens at $x = 3\pi/2$ while at $x = \pi/2$ it takes the maximum value 1. For every value of x in the interval $[0, 2\pi]$ the function $\sin x$ assumes a value between $[-1, 1]$.

The concept of minimum and maximum of a function has real-life applications in physics, engineering, economics and business management [1-3]. In the case of a classical particle in a potential $V(x)$ the minima of V are the stable equilibrium points and the maxima are the unstable equilibrium points. Certain coupled oscillators and networks of oscillators show multiple resonance and anti-resonance (where the amplitude A of oscillation is minimum). An anti-resonance with $A = 0$ means suppression of oscillation. It is of interest to determine the values of the parameters at which resonance and an anti-resonance occur and the conditions for their occurrence. Bubbles tend to acquire a spherical form to minimize the shape. Atoms occupy position that correspond to minimum elastic potential energy. To reduce the cost and the pump sizes the designing of piping systems requires details of minimizing pressure drops. On the other hand, maximizing the strength is the basic one in the choice of shapes of steel and concrete beams. In business management, maxima and minima are utilized to realize a maximum profit and efficiency and minimize the effort and expenses, respectively. With the knowledge of minimum and maximum values of the profit function an economist is able to derive the limits of salaries of the employees to avoid loss.

One way of finding the minima and maxima of a function is to draw a graph of the function and identify the points of minima and maxima. Mathematically, at the points of maxima and minima $f'(x) = 0$. Thus, the roots of $f'(x) = 0$ are the points of maxima or minima. At a maximum $f''(x) < 0$ while $f''(x) > 0$ at a minimum. This can be extended to multivariable functions. For some simple functions, the maxima and minima can be determined using the above conditions. When it is difficult to calculate the minima or maxima of a given function, it is desirable to use a numerical approach. This chapter is concerned with finding a minimum of single variable and multivariable functions.

9.2 Minimization of One-Dimensional Functions

This section deals with the problem of finding a minimum of one-dimensional functions. The case of two-dimensional functions will be considered in the next section. For an

DOI: 10.1201/9781032649931-9

TABLE 9.1
Result of the secant method in finding a root of $f' = e^{x-1} - e^{-x+1} = 0$.

Iteration number	x_0	x_1	x_2	$f'(x_2)$
0	0.200000	0.300000		
1	0.200000	0.300000	0.885678	$-0.229142e+00$
2	0.300000	0.885678	0.989871	$-0.202576e-01$
3	0.885678	0.989871	0.999976	$-0.479518e-04$
4	0.989871	0.999976	1.000000	$-0.821820e-09$

n−dimensional function the standard approach is the numerical implementation of conditions for a function to be minimum at a point. A straight forward (noniterative) and an iterative method for both one- and two-dimensional functions are discussed in the following. These methods can be extended to higher-dimensional functions also.

Consider a function $f(x)$ defined in the interval $x \in [a, b]$ with its value (i) decreasing (increasing) on $[a, x^*]$ and (ii) increasing (decreasing) on $[x^*, b]$, where $a < x^* < c$. The point x^* is then a *local minimum* (*maximum*) of $f(x)$. At a local minimum or maximum $f'(x)|_{x=x^*} = 0$. When $f''(x^*) > 0$ then $f(x)$ is minimum at $x = x^*$. The function $f(x)$ is maximum at x^* if $f''(x^*) < 0$. The test is inconclusive if $f''(x^*) = 0$.

9.2.1 Method by Solving $f'(x) = 0$

The point of minimum or maximum can be calculated by solving $f'(x) = 0$. That is, the roots of $f'(x) = 0$ are the minima or maxima of $f(x)$. A root of $f'(x) = 0$ can be computed numerically by employing one of the methods discussed in Chapter 3. Then, the sign of f'' can be used to identify whether x^* corresponds to a minimum or maximum of $f(x)$.

Example:

Determine numerically a local minimum of $f(x) = e^{x-1} + e^{-x+1}$.

The first derivative of the given function is $f'(x) = e^{x-1} - e^{-x+1}$ and $f'' = e^{x-1} + e^{-x+1}$ (which is always positive). To find a root of the equation $f'(x) = 0$ by the secant method (see Section 3.4) two starting values x_0 and x_1 are required and they need not enclose a root. Choose $x_0 = 0.2$ and $x_1 = 0.3$. To stop the iteration in the secant method choose the tolerances as $\delta_x = 10^{-2}$, $\delta_f = 10^{-5}$, $\delta_s = 10^{-5}$. Table 9.1 presents the successive approximate root (x_2). At the end of the 4th iteration, $|x_2 - x_1| < \delta_x$. A root of $f'(x) = 0$ is obtained as 1 which is an exact root. Since, $f''(x^* = 1) = 2 > 0$ the function f is minimum at $x = x^* = 1$.

9.2.2 Method by Enclosing the Minimum

Consider a starting value x_0, $a < x_0 < b$. The point x_0 is left to x^* if $f'(x_0) < 0$ otherwise right to x^*. Identify two more points $x_1 = x_0 + \Delta x$ and $x_2 = x_0 + 2\Delta x$ such that

$$f(x_0) > f(x_1) \text{ and } f(x_1) < f(x_2). \tag{9.1}$$

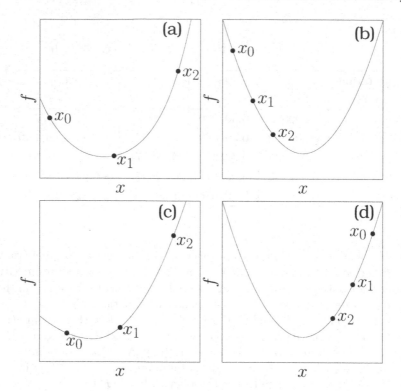

FIGURE 9.1
(a) Points x_0, x_1 and x_2 satisfying the condition (9.1). (b) x_0, x_1 and x_2 are all left to the minimum of $f(x)$ with $f(x_0) > f(x_1)$ and $f(x_1) > f(x_2)$. (c) x_0 and x_1 are such that $f(x_0) \leq f(x_1)$, x_1 is right to x^*. (d) x_0 is right to the minimum of the function in which case the initial Δx has to be set -1.

To realize the condition (9.1) a suitable Δx has to be found. Initially choose $\Delta x = 1$ if x_0 is left to x^* and analyze the following three cases.

Case (i)

The condition (9.1) is satisfied. Then, x_0, x_1 and x_2 are the desired points (Fig. 9.1a).

Case (ii)

$f(x_0) > f(x_1) > f(x_2)$. In this case $x_2 < x^*$ (Fig. 9.1b). Double the value of Δx, redefine the values of x_1 and x_2 and start again from the case (i).

Case (iii)

$f(x_0) \leq f(x_1)$. Now, x_1 is right to x^* (Fig. 9.1c) and Δx is too large. Reduce Δx by a factor of 2 and repeat the above process. When case (i) is realized, the three points x_0, x_1 and x_3 are enclosing a minimum of $f(x)$, that is, a root of $f'(x) = 0$. Then, construct a second-order Newton interpolation polynomial $P_2(x)$ for the data set (x_0, f_0), (x_1, f_1) and (x_2, f_2). The root of $P_2'(x) = 0$, denoted as x_1^*, is an approximation of x^*. Next, with $x_0 = x_1^*$ and updated Δx repeat the above process and obtain successive approximation of x^*. Stop the iterative process when, say, $|\Delta x| < 0.001$. If the initial x_0 lies right to the minimum of the function (Fig. 9.1d) then $\Delta x = -1$ and the above three cases are to be analyzed. For the data set (x_0, f_0), (x_1, f_1) and (x_2, f_2) the Newton polynomial is (see Section 6.2.3)

$$
\begin{aligned}
P_2 &= a_1 + a_2 (x - x_0) + a_3 (x - x_0)(x - x_1) \\
&= f_0 + \frac{(f_1 - f_0)}{\Delta x}(x - x_0) + \frac{f_2 - 2f_1 + f_0}{2(\Delta x)^2}(x - x_0)(x - x_0 - \Delta x). \quad (9.2)
\end{aligned}
$$

The root of the equation $P_2'(x) = 0$ (which is first-order in x) gives

$$
x_1^* = x_0 + \frac{\Delta x}{2} - \frac{(f_1 - f_0)\Delta x}{f_2 - 2f_1 + f_0} = x_0 + \frac{\Delta x}{2}\left(\frac{f_2 - 4f_1 + 3f_0}{f_2 - 2f_1 + f_0}\right). \quad (9.3)
$$

Example:

Determine numerically a local minimum of $f(x) = e^{x-1} + e^{-x+1}$ by the method of enclosing the minimum.

A Python program is developed to find a minimum of the given function. This program is used to find successive approximation of x^*. The following is the result of the execution of the program with the choice $x_0 = -0.6$.

$x_0 = -0.6$, $f'(x_0) = -4.75114$. x_0 is left to the point x^*.

First Iteration

$\Delta x = 1$, $x_0 = -0.60000$, $x_1 = 0.40000$, $x_2 = 1.40000$,

$f_0 = 5.15493$, $f_1 = 2.37093$, $f_2 = 2.16214$.

$f_0 > f_1$ and $f_1 > f_2$. Double the value of Δx.

$\Delta x = 2$, $x_0 = -0.60000$, $x_1 = 1.40000$, $x_2 = 3.40000$,

$f_0 = 5.15493$, $f_1 = 2.16214$, $f_2 = 11.11389$.

$f_0 > f_1$ and $f_1 < f_2$. x_0, x_1 and x_2 are correct choice.

$x_1^* = 0.90111$, $f^* = 2.00979$, $f^{*\prime} = -0.19810$, $f^{*\prime\prime} = 2.00979$.

Second Iteration

$\Delta x = 2$, $x_0 = 0.90111$, $x_1 = 2.90111$, $x_2 = 4.90111$,

$f_0 = 2.00979$, $f_1 = 6.84275$, $f_2 = 49.47771$. $f_0 \le f_1$. Reduce Δx.

$\Delta x = 1$, $x_0 = 0.90111$, $x_1 = 1.90111$, $x_2 = 2.90111$,

$f_0 = 2.00979$, $f_1 = 2.86846$, $f_2 = 6.84275$. $f_0 \le f_1$. Reduce Δx.

$\Delta x = 0.5$, $x_0 = 0.90111$, $x_1 = 1.40111$, $x_2 = 1.90111$,

$f_0 = 2.00979$, $f_1 = 2.16306$, $f_2 = 2.86846$. $f_0 \le f_1$. Reduce Δx.

$\Delta x = 0.25$, $x_0 = 0.90111$, $x_1 = 1.15111$, $x_2 = 1.40111$,

$f_0 = 2.00979$, $f_1 = 2.02288$, $f_2 = 2.16306$. $f_0 \le f_1$. Reduce Δx.

$\Delta x = 0.125$, $x_0 = 0.90111$, $x_1 = 1.02611$, $x_2 = 1.15111$,

$f_0 = 2.00979$, $f_1 = 2.00068$, $f_2 = 2.02288$.

$f_0 > f_1$ and $f_1 < f_2$. x_0, x_1 and x_2 are correct choice.

$x_2^* = 0.99997$, $f^* = 2.00000$, $f^{*\prime} = -0.00006$, $f^{*\prime\prime} = 2.00000$.

Third Iteration

$\Delta x = 0.125$, $x_0 = 0.99997$, $x_1 = 1.12497$, $x_2 = 1.24997$,

$f_0 = 2.000000$, $f_1 = 2.01564$, $f_2 = 2.06281$. $f_0 \le f_1$. Reduce Δx.

. . . . . .

$\Delta x = 0.00195...$, $x_0 = 0.99997$, $x_1 = 1.00193$, $x_2 = 1.00388$,

$f_0 = 2.000000$, $f_1 = 2.00000$, $f_2 = 2.00002$. $f_0 \leq f_1$. Reduce Δx.

$\Delta x = 0.000976....$ $|\Delta x| < 0.001$. Iteration is stopped.

The point of minimum is $x^* = 0.99997$.

The choice $x_0 = 1.6$ gives $f' = 1.27331$ and hence it is right to the point x^* and the initial value of $\Delta x = -1$. The first iteration gives $\Delta x = -1$, $x_1^* = 1.01110$, $f^* = 2.00012$, $f^{*\prime} = 0.02219$ and $f^{*\prime\prime} = 2.00012$. The second iteration gives $\Delta x = -0.015625$, $x_2^* = 1.0$, $f^* = 2.0$, $f^{*\prime} = 0.0$ and $f^{*\prime\prime} = 2.0$. In the third iteration before getting a desired set of points x_0, x_1 and x_2 the value of Δx becomes less than the tolerance value 0.001. Hence, the given function is minimum at $x_2^* = 1.0$ which is identical to the exact x^*.

9.3   Minimization of Two-Dimensional Functions

This section discusses two methods of minimization of two-dimensional functions. They can be extended to multivariable functions. Minimization of a function forms a major role in the numerical algorithm of computing transmission probability of scattering.

9.3.1   Using Derivatives

Let us consider a two-dimensional function $f(x,y)$ and define

$$f_x = \frac{\partial f}{\partial x}, \quad f_y = \frac{\partial f}{\partial y}, \quad f_{xx} = \frac{\partial^2 f}{\partial x^2}, \quad f_{yy} = \frac{\partial^2 f}{\partial y^2}, \quad f_{xy} = \frac{\partial^2 f}{\partial x \partial y}, \tag{9.4}$$

$$\Delta^2 = \left(f_{xx}f_{yy} - f_{xy}^2\right)|_{(x^*,y^*)}, \tag{9.5}$$

where (x^*, y^*) is a root of the system of equations

$$f_x(x,y) = 0, \quad f_y(x,y) = 0. \tag{9.6}$$

Then, (x^*, y^*) is a local minimum if

$$\Delta^2 > 0 \text{ and } f_{xx}(x^*, y^*) > 0. \tag{9.7}$$

(x^*, y^*) is a local maximum if

$$\Delta^2 > 0 \text{ and } f_{xx}(x^*, y^*) < 0. \tag{9.8}$$

When $\Delta^2 < 0$ then $f(x,y)$ does not possess a local maximum at (x^*, y^*). The test is inconclusive when $\Delta^2 = 0$. The above mathematical process of determining a minimum or a maximum of $f(x,y)$ can be used for numerical computation of the same.

Starting from an initial guess (x_0, y_0) of a minimum point of $f(x,y)$ one can find the root (x^*, y^*) of the two-coupled Eqs. (9.6), which is generally nonlinear, by the Newton–Raphson method discussed in Section 3.5.10. Then, computing Δ^2 and f_{xx} whether (x^*, y^*) is a minimum can be identified.

Example:

Compute the point (x^*, y^*) at which $f(x,y) = \sin x \cos y$ is minimum.

TABLE 9.2

Result of the Newton–Raphson method in finding a root of $f_x = \cos x \cos y = 0$ and $f_y = -\sin x \sin y = 0$ with $(x_0, y_0) = (1.2, 2.5)$.

Iteration number	x	y	f_x	f_y
0	1.200000	2.500000	——	——
1	1.861534	3.439150	0.274062	0.280881
2	1.531399	3.102196	−0.039356	−0.039356
3	1.570878	3.141674	0.000082	0.000082
4	1.570796	3.141593	0.000000	0.000000
5	1.570796	3.141593	0.000000	0.000000

For the given function

$$f_x = \cos x \cos y, \quad f_y = -\sin x \sin y, \quad f_{xx} = -\sin x \cos y,$$
$$f_{yy} = -\sin x \cos y, \quad f_{xy} = -\cos x \sin y,$$
$$\Delta^2 = \left(\sin^2 x \cos^2 y - \cos^2 x \sin^2 y\right)\big|_{(x^*, y^*)},$$

Table 9.2 shows the successive approximation of (x^*, y^*) after each iteration in the Newton–Raphson method, where $(x_0, y_0) = (1.2, 2.5)$. After 5 iterations the desired root of $f_x = 0$ and $f_y = 0$ is $(x^*, y^*) = (1.57080, 3.14159)$. At this point $\Delta^2 = 1$ and $f_{xx} = 1$. Hence, $(x^*, y^*) = (1.57080, 3.14159)$ is a point of minimum of the given function. The exact value of (x^*, y^*) is $(\pi/2, \pi)$.

9.3.2 Nelder–Mead Simplex Method

Now, outline the Nelder–Mead simplex method [4-7] for finding a minimum of a function of more than one independent variable. Let us consider a continuous function F of the form

$$F = F(x_1, x_2, \ldots, x_N). \tag{9.9}$$

The Nelder–Mead method requires only function evaluations to find a minimum of F.

For simplicity, let us describe the method for a function of two variables x, y, that is, $F(x, y)$. The method starts with a triangular simplex. A *simplex* is a geometrical figure consisting, in N-dimensions, of $N + 1$ points (or vertices) and all their interconnecting line segments. For a function of two variables, a simplex is a triangle. For a given initial three vertices of a triangle, the function values at these points are evaluated. The vertex at which $F(x, y)$ is the largest among the three vertices is called the *worst vertex*. Replace the worst vertex by a new vertex and form a new triangle. Continue this process until a minimum point is reached. The method essentially consists of six steps.

Step-1: The initial best-good-worst points

Evaluate the value of the function $F(x, y)$ at each of the three given points (initial guess), $X_k = (x_k, y_k)$, $k = 1, 2, 3$. (For a function of N variables, the number of initial points is $N + 1$.) Rearrange the points (x_k, y_k) so that $F(x_1, y_1) \leq F(x_2, y_2) \leq F(x_3, y_3)$. Use the notations B(best)$= (x_1, y_1)$, G(good)$= (x_2, y_2)$ and W(worst)$= (x_3, y_3)$.

Step-2: Mid-point of the line segment BG

Compute the mid-point M of the line segment joining the best and the good vertices:

$$M = (B + G)/2 = [(x_1 + x_2)/2, (y_1 + y_2)/2]. \tag{9.10}$$

Step-3: Computation of a test point R

Find a test point R as follows: From the mid-point M draw the line segment from W to M. The distance WM is d. Extend the last segment by a distance d through M to locate the point R as shown in Fig. 9.2a. The point R is given by $R = 2M - W$.

Step-4: Expansion using the point E

If $F(R) < F(W)$ then the movement is in the right direction towards the minimum. The minimum is faster than the point R. Therefore, a point E is formed as shown in Fig. 9.2b: $E = 2R - M$. If $F(E) < F(R)$ then E is a better vertex than R.

Step-5: Contraction using the point C

If $F(R) = F(W)$ then test another point. For this purpose, consider two points C_1 and C_2 of the line segments WM and MR as shown in Fig. 9.2c: Call the point for which F is smaller as C. The new triangle is then BGC. The contraction is along one direction only.

Step-6: Shrink towards B

A contraction along all dimensions towards the best vertex B can be achieved as follows. If $F(C) < F(W)$, the points G and W must be shrunk toward B (see Fig. 9.2d). Define a point S as the mid-point of BW. Replace the point G with M and W with S.

The logical steps to stop the process are as follows. If $F(R) < F(G)$ then analyze the following case (i) otherwise the case (ii).

Case (i)

It performs refection (step 3) or expansion (step-4). If $F(B) < F(R)$ then W is replaced with R. Otherwise, E is computed. Then W is replaced with E if $F(E) < F(B)$. If $F(E) \geq F(B)$, W is replaced with R.

Case (ii)

It performs contraction (step 5) or shrunk (step 6). Replaced W with R if $F(R) < F(W)$. Otherwise, compute C. When $F(C) < F(W)$, replace W by C. If this condition is not satisfied then compute S. Next, replace W by S and G by M.

Example:

Compute the minimum point of the function $F(x, y) = x^2 + y^2$.

Table 9.3 gives the values of the three vertices and the function values obtained successively by the Nelder–Mead method. $M = 0$ corresponds to the initial vertices. At the end of 8th iteration the termination criterion is satisfied and the process is stopped. The minimum coordinate obtained is $(x, y) = (0, 0)$ while the exact result is $(0, 0)$.

9.4   Concluding Remarks

In this chapter, numerical methods for computing minima of mathematical functions are described. For certain simple functions minima and maxima can be determined analytically.

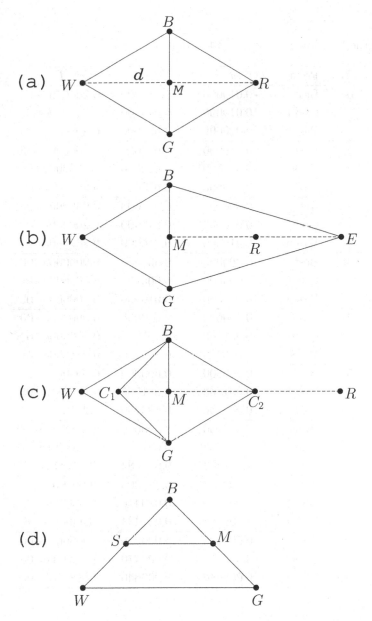

FIGURE 9.2
Possible outcomes of the Nelder–Mead simplex method. (a) Reflection away from the worst point W. (b) The triangle BGW, point R and the extended point E. This is a reflection and expansion away from the worst point. (c) A contraction along one-dimension from the worst point. (d) A contraction along all dimensions towards the best point.

When the function is complicated then it is desirable to use numerical methods to compute minima of it. In some physical phenomena minima of a quantity are beneficial. In certain events maxima of a quantity are useful. In the tunnel effect in quantum mechanics, the transmission amplitude T as a function of energy E of the incident particle shows a series of maxima and minima with $T_{\max} = 1$ corresponds to the maximum value of T. In some

TABLE 9.3
The values of the function $F = x^2 + y^2$ at various triangles.

M	Point	x	y	F
0	Best	−0.010000	−0.010000	0.200000e−03
	Good	0.010000	0.010000	0.200000e−03
	Worst	0.020000	0.020000	0.800000e−03
1	Best	−0.010000	−0.010000	0.200000e−03
	Good	0.010000	0.010000	0.200000e−03
	Worst	0.010000	0.010000	0.200000e−03
2	Best	0.000000	0.000000	0.000000e+00
	Good	0.010000	0.010000	0.200000e−03
	Worst	0.010000	0.010000	0.200000e−03
3	Best	0.000000	0.000000	0.000000e+00
	Good	−0.005000	−0.005000	0.500000e−04
	Worst	0.010000	0.010000	0.200000e−03
4	Best	0.000000	0.000000	0.000000e+00
	Good	−0.005000	−0.005000	0.500000e−04
	Worst	0.003750	0.003750	0.281250e−04
5	Best	0.000000	0.000000	0.000000e+00
	Good	−0.001563	−0.001563	0.488281e−05
	Worst	0.003750	0.003750	0.281250e−04
6	Best	0.000000	0.000000	0.000000e+00
	Good	−0.001563	−0.001563	0.488281e−05
	Worst	0.001484	0.001484	0.440674e−05
7	Best	0.000000	0.000000	0.000000e+00
	Good	−0.000410	−0.000410	0.336456e−06
	Worst	0.001484	0.001484	0.440674e−05
8	Best	0.000000	0.000000	0.000000e+00
	Good	−0.000410	−0.000410	0.336456e−06
	Worst	0.000640	0.000640	0.818300e−06

systems, occurrence of resonance is to be avoided. At resonance, a system can absorb more energy and may develop violent vibrations leading to disaster of bridges, building and airplanes. In designing such constructions the engineers wish to know the details of conditions of such resonance where the amplitude of vibrations become maximum and it is desirable to have a minimum amplitude of vibrations.

9.5   Bibliography

[1]   M. Lakshmanan and S. Rajasekar, *Nonlinear Dynamics: Integrability, Chaos and Patterns*. Springer, Berlin, 2002.

[2]   S. Rajasekar and M.A.F. Sanjuan, *Nonlinear Resonances*. Springer, Berlin, 2016.

[3]   J.R. Maddocks, *Maxima and Minima: Applications* https://science.jrank.org/pag es/4186/Maxima-Minima.html (accessed on June 18, 2023).

[4]   J.H. Mathews, *Numerical Methods for Mathematics Science and Engineering*. Prentice-Hall of India, New Delhi, 2005.

[5]   J.A. Nelder and R. Mead, *Comput. J.* 7:308, 1965.

[6]   S. Singer and J. Nelder, *Nelder-Mead Algorithm*, Scholarpedia. http://var.scholarpedia.org/article/Nelder-Mead_algorithm.

[7]   H.P. Gavin, *The Nelder-Mead algorithm in two-dimensions*. https: //people.duke. edu/~hpgavin/SystemID/CourseNotes/Nelder-Mead-2D.pdf.

9.6   Problems

9.1   Find a minimum or a maximum of $f(x) = x \sin x \cos x$ by computing the root of $f' = 0$ with the starting values a) $(x_0, x_1) = (1.5, 2.5)$, b) $(x_0, x_1) = (1.1, 1.2)$ and c) $(x_0, x_1) = (4, 5)$.

9.2   The function $f(x) = x e^{-x^2}$ has a local minimum and a maximum. Determine them by solving $f' = 0$.

9.3   Compute all the points of minima and maxima of $f(x) = x^6 + x^4 - 10x^2$ by finding the roots of $f' = 0$.

9.4   Find the minimum of the function $f(x) = 10 + x^2 + e^{-x^2}$ with the starting values $x_0 = 1$ and $x_1 = 0.8$ by solving the equation $f' = 0$.

9.5   In the method of enclosing the minimum verify that instead of Newton interpolating polynomial, if a second-order Lagrange interpolating polynomial is considered then the expression for x_1^* is identical to Eq. (9.3).

9.6   Determine a minimum of the function $f = -e^{-x^2}$ by the method of enclosing a minimum with the starting point $x_0 = 0.1$.

9.7   Determine the minimum of the function $f = x e^{-x^2}$ by the method of enclosing a minimum with the starting point $x_0 = 0.1$.

9.8   By computing a root of the first partial derivatives find the minimum of the following functions. Use the given initial guess in the Newton–Raphson method.

a) $f(x, y) = e^{-x^2} \cos y$, $(x_0, y_0) = (0.5, 2.5)$.

b) $f(x, y) = -e^{-x^2 - y^2}$, $(x_0, y_0) = (0.3, 0.3)$.

9.9   Determine the minimum or maximum of the function $f(x, y) = x^4 - x^2 + y^4 - y^2 - x^2 y^2$ with the initial guesses $(x_0, y_0) = (\pm 0.8, \pm 0.8)$, $(0.2, 0.2)$ and $(0, 0.7)$ by finding a root of its first derivatives.

9.10   a) Compute the local minimum of the function $f(x, y) = e^{-x^2} \cos y$ by the Nelder–Mead simplex method with the vertices $(0, 2.9)$, $(-0.1, 3.0)$ and $(0.2, 3.1)$.

b) Compute the local minimum of the function $f(x, y) = -e^{-x^2 - y^2}$ by the Nelder–Mead simplex method with the vertices $(0, 0.1)$, $(-0.1, 0.1)$ and $(0.2, 0.0)$.

10

Numerical Integration

10.1 Introduction

Integration of a function $f(x)$ with one variable is defined as

$$g(x) = \int f(x)\,\mathrm{d}x. \tag{10.1}$$

Similarly, integration of a multi-variable function with more than one variable can be defined. Because of $\mathrm{d}g/\mathrm{d}x = f(x)$ an integration of a function $f(x)$ is equivalent to finding the function, the derivative of which is $f(x)$. Integrals are broadly classified into two classes:

1. definite integrals and
2. indefinite integrals.

An integral of the form given by Eq. (10.1) is called an *indefinite integral*. Here, when the function $f(x)$, called the *integrand*, is integrated a new function $g(x)$ is obtained. The general form of definite integrals is

$$I = \int_a^b f(x)\,\mathrm{d}x = g(x)\big|_{x=a}^{x=b} = g(b) - g(a). \tag{10.2}$$

In this case, the final result is not a function of x but is a number. *What does the integral* $\int_a^b f(x)\mathrm{d}x$ *represent?* The integration of a function $f(x)$ in an interval $[a, b]$ represents the area under the curve in this interval.

Integrating a function or a set of data one can get an estimate of rates of change, area under the curves, volume, etc. In finite-element analysis, integrals arise in the discretization of concerned equations and are to be evaluated numerically. In engineering optimization techniques numerical integrations of objective functions are helpful to obtain optimal solutions for appropriate designs. In electrical circuits studies integration methods are used to find power dissipation in the circuits. Oscillation period of certain mechanical systems and electrical circuits, electric and magnetic potentials at a point, the centre of mass of an object and kinetic energy are expressed as integrals of certain quantities. For details see Problems 14–25 at the end of the present chapter. Calculations of average of quantities and functions over an interval, Fourier coefficients and probability of an event in an interval and normalization of quantum mechanical wave function of a system involve evaluation of certain definite integrals. In data analysis, numerical integration is employed to get an appropriate distribution of sample data.

Certain types of integrals can be evaluated employing direct methods or employing techniques such as Laplace and Fourier transforms and complex variable analysis (for example, see the refs. [1–3]). One may wish to evaluate a given integral $\int_a^b f(x)\mathrm{d}x$ numerically when

DOI: 10.1201/9781032649931-10

solving it exactly by analytical methods is not possible or difficult or $f(x)$ is known only at a finite number of values of the variable x. For example, the x-component of a magnetic field B_x at a point due to a hollow circular wire is given by

$$B_x = \int_a^b f(\theta)\,\mathrm{d}\theta = \frac{\mu_0 I}{2\pi} \int_0^{2\pi} \frac{2 - \sin\theta}{5 - 4\sin\theta}\,\mathrm{d}\theta. \tag{10.3}$$

Because of the complexity of $f(\theta)$, one may wish to evaluate it numerically.

What is numerical integration? The general formula of all the numerical integration algorithms assumes the form

$$I = \int_a^b f(x)\,\mathrm{d}x = \sum_{i=0}^n W_i f(x_i). \tag{10.4}$$

The formula given by Eq. (10.4) is called *numerical integration* or *quadrature*. In Eq. (10.4) the quantities x_i and W_i, $i = 0, 1, \ldots, n$ are called *quadrature nodes* or simply *nodes* and *weights*, respectively. I is an approximate value of the integral. The error in the approximation is given by

$$E = \int_a^b f(x)\,\mathrm{d}x - I. \tag{10.5}$$

This chapter presents some of the methods of numerical integration of definite integrals including fractional order integrations.

10.2   Newton–Cotes Methods

The quadrature formula, Eq. (10.4), is derived from Lagrange polynomial interpolation. As shown in Section 5.6 a unique polynomial $P(x)$ of degree $\leq n$ passing through the $n+1$ nodes or sample points x_i, $i = 0, 1, \ldots, n$ can be constructed. Thus, the function $f(x)$ over $[a, b]$ can be approximated by the polynomial $P(x)$. Then, the integral of $f(x)$ is approximated by the integral of $P(x)$. The resulting formula is called a *Newton–Cotes quadrature formula*.

10.2.1   Newton–Cotes Quadrature Formula

Let us assume that the integrand $f(x)$ is sufficiently smooth in the interval $[a, b]$ and $x_0, x_1, \ldots, x_n$ be $n + 1$ distinct points in the interval, say, $[a, b]$. Then, referring to Section 6.4 the Lagrange polynomial interpolation formula for the function $f(x)$ is written as

$$f(x) = P_n(x) + \sum_{i=0}^n L_i(x) f(x_i) + R_n, \tag{10.6a}$$

where

$$L_i(x) = \prod_{\substack{j=0 \\ j \neq i}}^n \frac{(x - x_j)}{(x_i - x_j)}, \quad i = 0, 1, \ldots, n. \tag{10.6b}$$

Here, R_n is the error in the polynomial approximation and is given by (refer Section 6.4)

$$R_n = \frac{1}{(n+1)!} \psi_n(x) f^{(n+1)}(x), \tag{10.6c}$$

where

$$\psi_n(x) = \prod_{i=0}^{n} (x - x_i).$$ (10.6d)

Substitution of Eq. (10.6a) for $f(x)$ in Eq. (10.4) with $I = I_n$ gives

$$I_n = \sum_{i=0}^{n} f(x_i) \int_a^b L_i(x)\,dx + \int_a^b R_n(x)\,dx.$$ (10.7)

Defining

$$W_i = \int_a^b L_i(x)\,dx, \quad E_n = \int_a^b R_n\,dx$$ (10.8)

Eq. (10.7) is rewritten as

$$I_n = \sum_{i=0}^{n} W_i f(x_i) + E_n.$$ (10.9)

Neglecting the error term E_n Eq. (10.9) becomes

$$I_n = \sum_{i=0}^{n} W_i f_i, \quad f_i = f(x_i).$$ (10.10)

When the nodes x_i are equally spaced in the interval $[a,b]$ the formula given by Eq. (10.10) is called *Newton–Cotes quadrature* and the numbers W_i are called *Cotes numbers*.

10.2.2   Error in the Newton–Cotes Formula

From Eqs. (10.6c) and (10.8) the error in the approximation given by Eq. (10.10) is obtained as

$$E_n = \frac{1}{(n+1)!} \int_a^b f^{(n+1)}(x)\psi_n(x)\,dx.$$ (10.11)

As shown below the above integral can be simplified using the intermediate-value theorem for integrals (see Section 1.6). According to this theorem if $\psi_n(x)$ in Eq. (10.11) does not change in sign on $[a,b]$ then

$$E_n = \frac{1}{(n+1)!} f^{(n+1)}(x') \int_a^b \psi_n(x)\,dx, \quad x' \in [a,b].$$ (10.12)

If the function $\psi_n(x)$ changes sign in $[a,b]$ then E_n is given by the next higher-order term of $f(x')$,

$$E_n = \frac{1}{(n+2)!} f^{(n+2)}(x') \int_a^b \psi_{n+1}(x)\,dx$$ (10.13)

provided the sign of ψ_{n+1} remains same in $[a,b]$, and so on. For polynomials of degree less than $n+2$ the function $f^{(n+2)}(x)$ is zero and hence $E_n = 0$. Therefore, the formula given by Eq. (10.10) is exact for polynomials of degree $\leq n+1$.

10.2.3   Some Newton–Cotes Formulas

Let us proceed to obtain the Newton–Cotes formulas for first few values of n. Set $x_0 = a$, $x_n = b$ and choose the spacing between the nodes as $h = (b-a)/n$. Then, x_i's are given by

$$x_i = x_0 + ih, \quad i = 0, 1, 2, \ldots, n.$$ (10.14)

Figure 10.1 shows the representation of the nodes x_i, $i = 0, 1, 2, \ldots, n$ on the interval $[a,b]$ for $n = 1, 2, 3$.

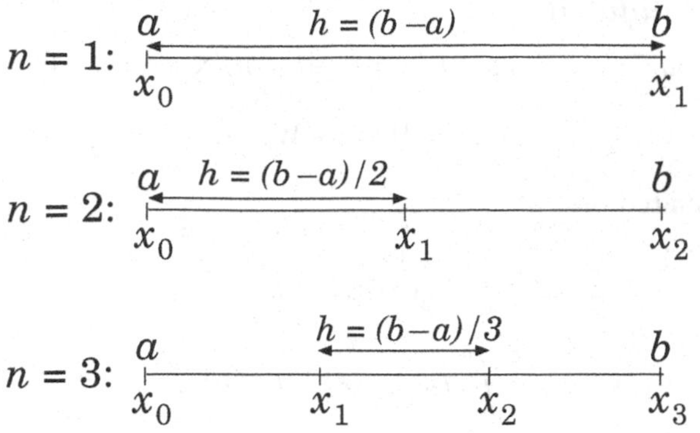

FIGURE 10.1
Representation of the nodes in the interval $[a, b]$ for the integral $\int_a^b f(x)dx$ for $n = 1, 2, 3$. n is the number of subintervals in $[a, b]$.

(i) $n = 0$: (a) Rectangle Rule

For $n = 0$, Eq. (10.10) becomes

$$I_0 = W_0 f_0 . \tag{10.15}$$

From Eqs. (10.8) and (10.6b) W_0 is evaluated as

$$W_0 = \int_a^b L_0(x)\, dx = \int_a^b dx = b - a . \tag{10.16}$$

Thus, for $x_0 = a$

$$I_0 = (b - a) f_0. \tag{10.17}$$

The quantity $(b - a)f_0$ in the above formula represents the area of rectangle of length $(b - a)$ and breadth f_0 and hence the name *rectangle rule*. To determine the error in the approximation consider the function $\psi_0(x)$. Since $\psi_0(x) = (x - x_0) = (x - a) > 0$ the error is given by Eq. (10.12). E_0 is calculated as

$$E_0 = f'(x') \int_a^b \psi_0(x)\, dx = f'(x') \int_a^b (x - x_0)\, dx = \frac{1}{2} f'(x')(b - a)^2, \tag{10.18}$$

where $x' \in [a, b]$.

(b) Mid-Point Rule

For the choice $x_0 = (a+b)/2$ the midpoint of the interval $[a, b]$, W_0, is again $(b-a)$. However,

$$I_0 = (b - a) f\left((a + b)/2\right) \tag{10.19}$$

and is called the *mid-point rule*. The function $\psi_0 = (x - x_0) = (x - (a + b)/2)$ is ≤ 0 for $a \leq x \leq (a+b)/2$ and > 0 for $(a+b)/2 < x < b$. That is, ψ changes sign in $[a, b]$. Therefore, the error term is not given by Eq. (10.12). For $x_1 = x_0$, $\psi(x) = (x - x_0)(x - x_1) = (x - x_0)^2$ does not change its sign in $[a, b]$ and hence the error is given by Eq. (10.13). E_0 is then evaluated as

$$E_0 = \frac{1}{2} f''(x') \int_a^b (x - x_0)^2\, dx = \frac{1}{6}(b - a)^3 f''(x') . \tag{10.20}$$

(ii) $n = 1$: *Trapezoidal Rule*

For $n = 1$ the nodes are $x_0 = a$, $x_1 = b$ and $h = b - a$. Now, Eq. (10.10) becomes

$$I_1 = W_0 f_0 + W_1 f_1 . \tag{10.21}$$

W_0 and W_1 are evaluated as

$$W_0 = \int_a^b L_0(x) \, dx = (b - a)/2 = h/2 , \tag{10.22a}$$

$$W_1 = \int_a^b L_1(x) \, dx = (b - a)/2 = h/2 . \tag{10.22b}$$

Then,

$$I_1 = \frac{1}{2} h \left(f_0 + f_1 \right) \tag{10.23}$$

and is called *trapezoidal rule*. The function $\psi_1(x) = (x - x_0)(x - x_1) = (x - a)(x - b)$ is always negative in the interval $[a, b]$ and therefore the error term E_1 is given by Eq. (10.12). E_1 is estimated as

$$
\begin{aligned}
E_1 &= \frac{1}{2!} f^{(2)}(x') \int_a^b \psi_1(x) \, dx \\
&= \frac{1}{2} f^{(2)}(x') \int_a^b (x - x_0) \, (x - x_1) \, dx \\
&= -\frac{1}{12} h^3 f^{(2)}(x') .
\end{aligned}
\tag{10.24}
$$

The rule is exact for linear function of x.

(iii) $n = 2$: *Simpson's* $1/3$-*Rule*

Next, consider the case $n = 2$. The nodes are $x_0 = a$, $x_1 = (a+b)/2$, $x_2 = b$ and $h = (b-a)/2$. From (10.10)

$$I_2 = W_0 f_0 + W_1 f_1 + W_2 f_2 . \tag{10.25}$$

The weight factors are worked out as

$$
\begin{aligned}
W_0 &= \int_a^b L_0(x) \, dx \\
&= \int_a^b \frac{(x - x_1) \, (x - x_2)}{(x_0 - x_1) \, (x_0 - x_2)} \, dx \\
&= \frac{2}{(a - b)^2} \int_a^b (x - (a - b)/2) \, (x - b) \, dx \\
&= \frac{1}{3} h
\end{aligned}
\tag{10.26a}
$$

and

$$W_1 = \int_a^b L_1(x) \, dx = \int_a^b \frac{(x - x_0) \, (x - x_2)}{(x_1 - x_0) \, (x_1 - x_2)} \, dx = \frac{4}{3} h , \tag{10.26b}$$

$$W_2 = \int_a^b L_2(x) dx = \int_a^b \frac{(x - x_0) \, (x - x_1)}{(x_2 - x_0) \, (x_2 - x_1)} \, dx = \frac{1}{3} h . \tag{10.26c}$$

Then,

$$I_2 = \frac{1}{3}h\left(f_0 + 4f_1 + f_2\right). \tag{10.27}$$

The above formula is known as *Simpson's 1/3-rule*. To estimate the error term first check whether the sign of $\psi_2(x)$ is changing in $[a, b]$ or not. ψ_2 is given by

$$\psi_2(x) = (x - x_0)\,(x - x_1)\,(x - x_2) = (x - a)\,(x - (a+b)/2)\,(x - b).$$

$\psi_2 > 0$ for $a < x < (a+b)/2$ while it is < 0 for $(a+b)/2 < x < b$. Therefore, consider the next higher-order term of ψ:

$$\psi_3(x) = (x - x_0)^2\,(x - x_1)\,(x - x_2). \tag{10.28}$$

In this case

$$
\begin{aligned}
E_2 &= \frac{1}{4!}f^{(4)}(x')\int_a^b \psi_3(x)\,\mathrm{d}x \\
&= \frac{1}{4!}f^{(4)}(x')\int_a^b (x - x_0)^2\,(x - x_1)\,(x - x_2)\,\mathrm{d}x \\
&= -\frac{1}{90}h^5 f^{(4)}(x').
\end{aligned}
\tag{10.29}
$$

The Simpson's 1/3-rule is exact for polynomials of degree ≤ 3.

(iv) $n = 3$: *Simpson's 3/8-Rule*

The values of the nodes for $n = 3$ are $x_0 = a$, $x_1 = (2a + b)/3$, $x_2 = (a + 2b)/3$, $x_3 = b$ and $h = (b - a)/3$. The formula for the integral $\int_a^b f(x)\mathrm{d}x$ is

$$I_3 = W_0 f_0 + W_1 f_1 + W_2 f_2 + W_3 f_3. \tag{10.30}$$

The W's are computed as (verify)

$$W_0 = \frac{3}{8}h, \quad W_1 = \frac{9}{8}h, \quad W_2 = \frac{9}{8}h, \quad W_3 = \frac{3}{8}h. \tag{10.31}$$

Then,

$$I_3 = \frac{3}{8}h\left(f_0 + 3f_1 + 3f_2 + f_3\right) \tag{10.32}$$

and is called *Simpson's 3/8-rule*. To compute the error in the above formula consider the function $\psi_3(x)$:

$$
\begin{aligned}
\psi_3(x) &= (x - x_0)\,(x - x_1)\,(x - x_2)\,(x - x_3) \\
&= [x - a]\,[x - (2a + b)/3]\,[x - (a + 2b)/3]\,[x - b].
\end{aligned}
\tag{10.33}
$$

ψ_3 is < 0 for $a < x < (2a + b)/3$ and $(a + 2b)/3 < x < b$. It is > 0 for $(2a + b)/3 < x < (a + 2b)/3$. That is, ψ_3 changes its sign in $[a, b]$. Hence,

$$
\begin{aligned}
E_3 &= \frac{1}{5!}f^{(5)}(x')\int_a^b \psi_4(x)\,\mathrm{d}x \\
&= \frac{11}{480}h^6 f^{(5)}(x').
\end{aligned}
\tag{10.34}
$$

Example 1:

Evaluate the integral $\int_0^1 (1 + x^2)\, \mathrm{d}x$ using the various formulas obtained in this section.

(i)(a) Rectangle Rule

With $x_0 = a = 0$ from Eq. (10.17) I_0 is computed as 1 while the exact value of the integral is 1.33333.

(i)(b) Mid-Point Rule

With $x_0 = (a + b)/2 = (0 + 1)/2 = 0.5$, I_0 is estimated as 1.25.

(ii) Trapezoidal Rule

The nodes are $x_0 = 0$ and $x_1 = 1$. With $h = 1$ $I = \frac{1}{2}(f_0 + f_1) = \frac{1}{2}(1 + 2) = 1.5$.

(iii) Simpson's 1/3-Rule

For the Simpson's 1/3-rule $x_0 = 0$, $x_1 = 0.5$, $x_2 = 1$ and $h = 0.5$. The function values are $f_0 = 1$, $f_1 = 1.25$ and $f_2 = 2$. Then, from Eq. (10.27)

$$I = \frac{1}{3}h(f_0 + 4f_1 + f_2) = 1.33333\ldots.$$

(iv) Simpson's 3/8-Rule

The nodes are $x_0 = 0$, $x_1 = 1/3$, $x_2 = 2/3$ and $x_3 = 1$. The step size is $h = 1/3$. The function values are $f_1 = 1$, $f_2 = 10/9$, $f_3 = 13/9$ and $f_4 = 2$. The value of the integral is

$$I = \frac{3}{8}h(f_0 + 3f_1 + 3f_2 + f_3) = 1.33333\ldots.$$

For the given integral the Simpson's rules are exact (why?).

Example 2:

Evaluate the integral $\int_0^3 (1 + e^x)\mathrm{d}x$ by the trapezoidal and the Simpson's rules and also compute the percentage of relative error, E_{rel}, using the expression

$$E_{\mathrm{rel}} = 100 \times (I_{\mathrm{exact}} - I_{\mathrm{num}})/I_{\mathrm{exact}},$$

where I_{exact} is the exact value of the integral obtained by direct integration and I_{num} is the value of the integral computed by an approximation method. The exact value of the integral is 22.085537.

(i) Trapezoidal Rule

For the trapezoidal rule $h = 3$, $x_0 = 0$, $x_1 = 3$, $f_0 = 2$ and $f_1 = 21.085537$. The value of the integral is

$$I = \frac{1}{2}h(f_0 + f_1) = 34.628305.$$

The percentage of relative error in the calculation is 56.79.

(ii) Simpson's 1/3-Rule

Now, $h = 1.5$, $x_0 = 0$, $x_1 = 1.5$, $x_2 = 3$, $f_0 = 2$, $f_1 = 5.4816891$ and $f_2 = 21.085537$. Then

$$I = \frac{1}{3} h \left(f_0 + 4f_1 + f_2 \right) = 22.506147.$$

The percentage of relative error in the result is 1.91.

(iii) Simpson's 3/8-Rule

For the Simpson's 3/8-rule $h = 1$, $x_0 = 0$, $x_1 = 1$, $x_2 = 2$, $x_3 = 3$, $f_0 = 2$, $f_1 = 3.7182818$, $f_2 = 8.3890561$, $f_3 = 21.085537$ and

$$I = \frac{3}{8} h \left(f_0 + 3f_1 + 3f_2 + f_3 \right) = 22.277832.$$

The percentage of relative error is 0.87.

In the above example 2 the error in the numerical integration of the given integral by the various methods is found to be large. *How does one improve these approximations?* This is considered in the next section.

Remark:

Can you apply the quadrature formulas to evaluate the integral $\displaystyle\int_0^1 \frac{-2x}{\sqrt{1-x^2}}\,\mathrm{d}x$ *where its exact value is* -1*?* The value of the integrand becomes $-\infty$ at the node value $x = 1$. Therefore, the quadrature formulas cannot be applied to the given integral. In general, for an integral of the form $\int_a^b f(x)/g(x)\mathrm{d}x$ if $g(x) = 0$ or $g(x) \approx 0$ and $|f(x)| \gg 0$ or $\lim_{x \to x_i} f(x_i)/g(x_i)$ is not finite at a node value then the quadrature formulas are not useful to evaluate the integral even if the exact value of the integral is finite. If $g(x) = 0$ at the end points and the value of the integral is finite then one may use the Gauss–Legendre (see Section 10.4.3) or any other suitable integration method. The integral formulas discussed in this section can be applied if the function $f(x)$ is not known or not given explicitly but its values are given for a set of points (see Problem 10.4 at the end of this chapter).

10.3 Composite Quadrature Formulas

In the quadrature formulas such as the trapezoidal and the Simpson's 1/3 and 3/8 formulas considered in the previous section the interval $[a, b]$ is divided into 1, 2 and 3 subintervals, respectively. Then, using the values of the function at the nodes the given integral can be evaluated. When the interval $[a, b]$ is large these quadrature formulas generally will not produce sufficiently accurate estimates as is the case in example 2 in the previous section. Further, some functions may have a large variation in a small interval and a small variation in a wide interval. Considering the above, it is necessary to develop more accurate formulas to evaluate the given integral. Such formulas can be obtained by dividing the interval $[a, b]$ into a large number of subintervals (conveniently equally spaced) and then applying a quadrature formula to each subinterval. An integral formula obtained in this manner is called a *composite quadrature formula*.

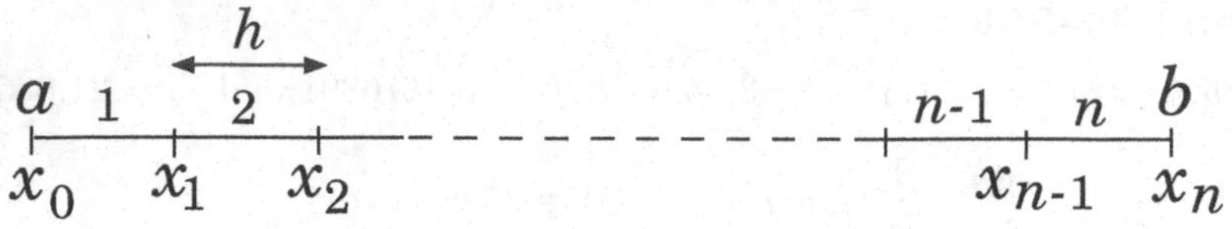

FIGURE 10.2
Dividing of the interval $[a, b]$ into n equally spaced subintervals for obtaining the composite trapezoidal rule.

10.3.1   Composite Trapezoidal Rule

Let the interval $[a, b]$ is divided into n-equally spaced subintervals as shown in Fig. 10.2. The nodes x_i are given by $x_i = a + ih$, $i = 0, 1, \ldots, n$ and $h = (b - a)/n$. Then, applying the trapezoidal rule to each interval gives

$$
\begin{aligned}
\int_a^b f(x)\, \mathrm{d}x &= \int_a^b f(x)\mathrm{d}x \\
&= \int_{x_0}^{x_1} f\mathrm{d}x + \int_{x_1}^{x_2} f\mathrm{d}x + \cdots + \int_{x_{n-2}}^{x_{n-1}} f\mathrm{d}x + \int_{x_{n-1}}^{x_n} f\mathrm{d}x \\
&= \frac{1}{2}h \left[f_0 + f_1 + f_1 + f_2 \cdots + f_{n-1} + f_{n-1} + f_n \right] \\
&= \frac{1}{2}h \left[f_0 + 2\sum_{i=1}^{n-1} f_i + f_n \right] .
\end{aligned}
\tag{10.35}
$$

Formula given by Eq. (10.35) is called the *composite trapezoidal rule.*

Referring to Eq. (10.24) the error term for ith interval is $-(h^3/12)f^{(2)}(x_i')$. The error term for the composite trapezoidal rule is obtained by summing this above error term over all the subintervals:

$$
E_{\mathrm{CT}} = -\frac{1}{12}h^3 \sum_{i=1}^{n} f^{(2)}(x_i') = -\frac{1}{12n}h^2(b - a) \sum_{i=1}^{n} f^{(2)}(x_i') ,
\tag{10.36}
$$

where x_i' is a value of x in the ith interval. Defining

$$
f^{(2)}(x') = \frac{1}{n} \sum_{i=1}^{n} f^{(2)}(x_i')
\tag{10.37}
$$

one has

$$
E_{\mathrm{CT}} = -\frac{1}{12}h^2(b - a)f^{(2)}(x') .
\tag{10.38}
$$

10.3.2   Composite Simpson's Rules

The composite Simpson's 1/3-rule is obtained by dividing the interval $[a, b]$ into $2n$ equally spaced subintervals as shown in Fig. 10.3. The nodes x_i are given by $x_i = a + ih$, $i = 0, 1, \ldots, 2n$ and $h = (b - a)/(2n)$. Then, applying the rule to each pair of subintervals gives

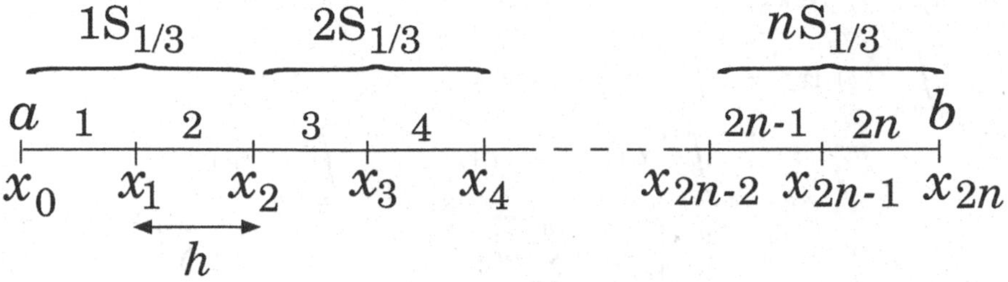

FIGURE 10.3
Dividing of the interval $[a, b]$ into $2n$ equally spaced subintervals for obtaining the composite Simpson's 1/3-rule.

$$
\begin{aligned}
\int_a^b f(x)\,\mathrm{d}x &= \int_{x_0}^{x_{2n}} f(x)\,\mathrm{d}x \\
&= \int_{x_0}^{x_2} f\,\mathrm{d}x + \int_{x_2}^{x_4} f\,\mathrm{d}x + \cdots + \int_{x_{2n-2}}^{x_{2n}} f\,\mathrm{d}x \\
&= \frac{h}{3}\left[f_0 + 4f_1 + f_2 + \cdots + f_{2n-2} + 4f_{2n-1} + f_{2n}\right] \\
&= \frac{h}{3}\left[f_0 + 4\sum_{i=1}^{n} f_{2i-1} + 2\sum_{i=1}^{n-1} f_{2i} + f_{2n}\right].
\end{aligned}
\tag{10.39}
$$

The error term for the composite Simpson's 1/3-rule is

$$
E_{CS} = -\frac{1}{180}(b-a)h^4 f^{(4)}(x'), \quad f^{(4)}(x') = \frac{1}{n}\sum_{i=1}^{n} f^{(4)}(x_i'),
\tag{10.40}
$$

where $x_i' \in [x_{2i-2}, x_{2i}]$.

To obtain the composite Simpson's 3/8-rule the interval $[a, b]$ is divided into $3n$ equally spaced subintervals (refer Fig. 10.4). The nodes are $x_i = a + ih$, $i = 0, 1, \ldots, 3n$, where

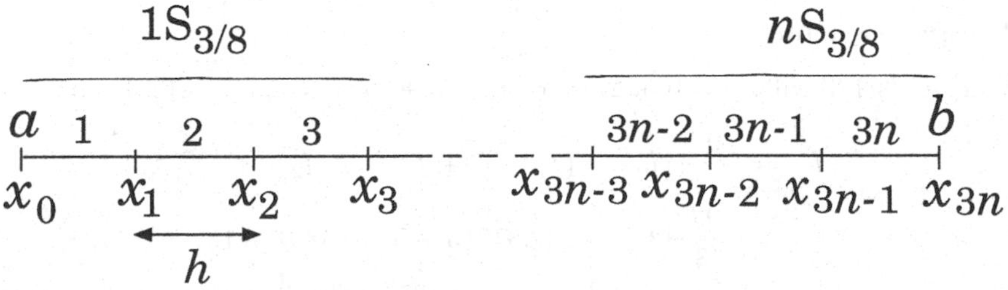

FIGURE 10.4
Dividing of the interval $[a, b]$ into $3n$ equally spaced subintervals for obtaining the composite Simpson's 3/8-rule.

$h = (b - a)/(3n)$. Now,

$$
\begin{aligned}
\int_a^b f(x)\,dx &= \int_{x_0}^{x_{3n}} f(x)\,dx \\
&= \int_{x_0}^{x_3} f\,dx + \int_{x_3}^{x_6} f\,dx + \cdots + \int_{x_{3n-3}}^{x_{3n}} f\,dx \\
&= \frac{3}{8}h\left[f_0 + 3\sum_{i=1}^{n}(f_{3i-2} + f_{3i-1}) + \cdots + 2\sum_{i=1}^{n-1} f_{3i} + f_{3n}\right]. \quad (10.41)
\end{aligned}
$$

The error term in the above formula is

$$
E_{CS} = \frac{11}{1440}h^5(b-a)f^{(5)}(x'), \quad f^{(5)}(x') = \frac{1}{n}\sum_{i=1}^{n} f^{(5)}(x'_{3i}), \quad (10.42)
$$

where $x'_{3i} \in [x_{3i-3}, x_{3i}]$.

Example:

Evaluate the integral $I = \int_0^3 (1 + e^x)\,dx$ employing the composite trapezoidal and the Simpson's rules. In all the cases choose the width of each subinterval as $h = 0.5$.

(i) Trapezoidal Rule

For the trapezoidal rule the value of n for $h = 0.5$ is 6. The nodes are $x_i = 0 + ih$, $i = 0, 1, \ldots, 6$. That is, $x_0 = 0$, $x_1 = 0.5$, $x_2 = 1$, $x_3 = 1.5$, $x_4 = 2$, $x_5 = 2.5$ and $x_6 = 3$. From Eq. (10.35)

$$
\begin{aligned}
I &= \int_0^3 (1 + e^x)\,dx \\
&= \frac{0.5}{2}\left[f_0 + 2(f_1 + f_2 + f_3 + f_4 + f_5) + f_6\right] \\
&= 0.25\left[2 + 2\{2.6487213 + 3.7182818 + 5.4816891\right. \\
&\qquad\qquad \left. + 8.3890561 + 13.182494\} + 21.085537\right] \\
&= 22.481505.
\end{aligned}
$$

The percentage of relative error in the computation is 1.79.

(ii) Simpson's Rules

For Simpson's 1/3-rule, $h = 0.5$, $2n = 6$, $x_i = 0 + ih$, $i = 0, 1, \ldots, 6$. Then,

$$
\begin{aligned}
I &= \frac{0.5}{3}\left[f_0 + 4(f_1 + f_3 + f_5) + 2(f_2 + f_4) + f_6\right] \\
&= \frac{0.5}{3}\left[2 + 4(2.6487213 + 5.4816891 + 13.182494)\right. \\
&\qquad\qquad \left. + 2(3.7182818 + 8.3890561) + 21.085537\right] \\
&= 22.091972.
\end{aligned}
$$

The percentage of error in the result is 0.03. Verify that the value of the integral by the composite Simpson's 3/8-rule with $3n = 6$ is 22.09961.

Tables 10.1 and 10.2 present the numerically computed value of the integral as a function of order of the composite quadrature rules. The percentage of relative error E_{rel} is also given.

TABLE 10.1

Computed value of the integral $\int_0^3 (1 + e^x)\, dx$ by the composite trapezoidal rule as a function of number of equally spaced subintervals n. The exact value of the integral is 22.08554.

n	h	Integral value	E_{rel}	n	h	Integral value	E_{rel}
5	0.60000	22.65470	2.58	30	0.10000	22.10144	0.07
10	0.30000	22.22846	0.65	35	0.08571	22.09722	0.05
15	0.20000	22.14911	0.29	40	0.07500	22.09448	0.04
20	0.15000	22.12131	0.16	45	0.06667	22.09261	0.03
25	0.12000	22.10843	0.10	50	0.06000	22.09126	0.03

10.3.3 Estimation of Optimum Values of n and h in the Composite Rules

What are the optimum values of n, n_{opt} and the step size h, h_{opt} so that the error in the value of an integral $\int_a^b f(x)dx$ computed by a composite quadrature rule is less than δ? For a chosen value of δ and a quadrature formula, the n_{opt} and h_{opt} can be calculated from the error term of the formula. In the following, the expressions for these quantities for the composite trapezoidal and the composite Simpson's rules are obtained. Then, they are computed for the integral $\int_0^3 (1 + e^x)\, dx$ for two values of δ.

Consider the error term given by Eq. (10.36) for the composite trapezoidal rule. Substitution of $h = (b - a)/n$ in this equation gives

$$E_{\text{CT}} = -\frac{1}{12n^2}(b - a)^3 f^{(2)}(x').$$ (10.43)

TABLE 10.2

Computed value of the integral $\int_0^3 (1 + e^x)\, dx$ by the composite Simpson's 1/3-rule as a function of number of equally spaced subintervals $2n$. The exact value of the integral is 22.08554.

$2n$	h	Integral value	E_{rel}	$2n$	h	Integral value	E_{rel}
2	1.50000	22.50615	1.90	12	0.25000	22.08595	0.00
4	0.75000	22.11696	0.14	14	0.21429	22.08576	0.00
6	0.50000	22.09197	0.03	16	0.18750	22.08567	0.00
8	0.37500	22.08760	0.01	18	0.16667	22.08562	0.00
10	0.30000	22.08639	0.00	20	0.15000	22.08559	0.00

It is required to have $|E_{CT}| < \delta$. That is,

$$\left| \frac{1}{12n^2}(b-a)^3 f^{(2)}(x') \right| < \delta . \tag{10.44}$$

This equation gives

$$n > n_{opt} \doteq \text{int.} \left[\left| \frac{1}{12\delta}(b-a)^3 f_m^{(2)}(x') \right| \right]^{1/2} , \tag{10.45}$$

where int.(y) is the integer part of y and $f_m^{(2)}(x')$ is the maximum value of $f^{(2)}(x')$ in the interval $x' \in [a,b]$. Then, $h_{opt} = (b-a)/n_{opt}$.

For the Simpson's 1/3-rule, Eq. (10.40) with the substitution $h = (b-a)/2n$ becomes

$$E_{CS} = -\frac{1}{2880n^4\delta}(b-a)^5 f^{(4)}(x') . \tag{10.46}$$

Thus,

$$n > n_{opt} = \text{int.} \left[\left| \frac{1}{2880\delta}(b-a)^5 f_m^{(4)}(x') \right| \right]^{1/4} \tag{10.47}$$

and $h_{opt} = (b-a)/2n_{opt}$.

For the Simpson's 3/8-rule n_{opt} is given by

$$n_{opt} = \text{int.} \left[\left| \frac{11}{349920\delta}(b-a)^6 f_m^{(5)}(x') \right| \right]^{1/5} \tag{10.48}$$

and h_{opt} is $(b-a)/3n_{opt}$.

Next, compute n_{opt} and h_{opt} for the integral $\int_0^3 (1 + e^x)dx$. For this integral $f^{(n)}(x') = e^{x'}$. Its maximum value in $x' \in [0,3]$ is at $x = 3$. For the composite trapezoidal rule

$$n_{opt} = \text{int.} \left(\frac{3^3 e^3}{12\delta} \right)^{1/2} . \tag{10.49}$$

For $\delta = 10^{-3}$ Eq. (10.45) gives $n_{opt} = 212$ and then $h_{opt} = (b-a)/n_{opt} = 3/212$. Thus, to have $\delta = 10^{-3}$ the value of n must be greater than 212. For $\delta = 10^{-5}$ the results are $n_{opt} = 2125$ and $h_{opt} = 3/2125$.

For the composite Simpson's 1/3-rule

$$n_{opt} = \text{int.} \left(\frac{3^5 e^3}{2880\delta} \right)^{1/4} . \tag{10.50}$$

The choice $\delta = 10^{-3}$ gives $n_{opt} = 6$ while for $\delta = 10^{-5}$ it is 20. For the composite Simpson's 3/8-rule

$$n_{opt} = \text{int.} \left(\frac{11}{349920} \frac{3^6 e^3}{\delta} \right)^{1/5} . \tag{10.51}$$

$\delta = 10^{-3}$ leads to $n_{opt} = 3$ (while for $\delta = 10^{-5}$ it is 8).

10.4   Gauss–Legendre Integration

Suppose the aim is to obtain a reasonable approximate value of an integral by evaluating the value of the function at very few nodes, say, 2 or 3 or 4. In this case, Gauss–Legendre (GL) formulas are extremely useful.

10.4.1   Gaussian Formula

Let us rewrite the integral $\int_a^b f(x)\mathrm{d}x$ as

$$I = \int_a^b g(x)W(x)\,\mathrm{d}x,\tag{10.52a}$$

where $g(x)$ is chosen as

$$g(x) = P_n(x) + g\left[x_0, x_1, \ldots, x_n, x\right]\psi_n(x)\tag{10.52b}$$

with $P_n(x)$ being a polynomial of degree $\le n$ and

$$\psi_n(x) = \sum_{i=0}^n (x - x_i)\ .\tag{10.53}$$

Now, write

$$I(g) = I(P_n) + \int_a^b g\left[x_0, \ldots, x_n, x\right]\psi_n(x)W(x)\,\mathrm{d}x\ .\tag{10.54}$$

Writing $P_n(x)$ in the Lagrange form

$$P_n = \sum_{i=0}^n g(x_i)\,L_i(x)\,,\quad L_i(x) = \prod_{\substack{j=0 \\ j\ne i}}^n \frac{(x - x_i)}{(x_i - x_j)}\,,\quad i = 0, 1, \ldots, n\tag{10.55}$$

gives

$$I(P_n) = \int_a^b P_n(x)W(x)\,\mathrm{d}x = \sum_{i=0}^n A_i g(x_i)\,,\tag{10.56a}$$

where

$$A_i = \int_a^b L_i(x)W(x)\,\mathrm{d}x\,,\quad i = 0, 1, \ldots, n.\tag{10.56b}$$

I given by Eq. (10.56) is called the *Gaussian formula*.

10.4.2   Error in the Gaussian Formula

In order to compute the error in the Gaussian formula, consider the equation

$$I(g) - I(P_n) = \int_a^b g\left[x_0, x_1, \ldots, x_n; x\right]\psi_n(x)W(x)\,\mathrm{d}x\ .\tag{10.57}$$

If

$$\int_a^b \psi(x)\,(x - x_{n-1}) \cdots (x - x_{n+1+i})\,W(x)\,\mathrm{d}x = 0, \quad i = 0, 1, \ldots, m - 1 \tag{10.58}$$

for certain $x_0, x_1, \ldots, x_{n+m}$ then for any x_{n+m+1}

$$I(g) - I(P_n) = \int_a^b g\,[x_0, x_1, \ldots, x_{n+m+1}; x]\,\psi_{n+m+1}(x)W(x)\,\mathrm{d}x. \tag{10.59}$$

One can find an orthogonal polynomial

$$P_{n+1}(x) = \alpha_{n+1}\,(x - \xi_0)\,(x - \xi_1) \cdots (x - \xi_n), \tag{10.60}$$

where $\xi_0, \xi_1, \ldots, \xi_n$ are the $(n + 1)$ distinct points in $[a, b]$ at which $P_{n+1} = 0$ such that

$$\int_a^b P_{n+1}(x)q(x)W(x)\,\mathrm{d}x = 0 \tag{10.61}$$

for all polynomials $q(x)$ of degree $\leq n$. When $m = n$ Eq. (10.59) becomes

$$I(g) - I(P_n) = \int_a^b g\,[x_0, x_1, \ldots, x_{2n+1}; x]\,\psi_{2n+1}(x)W(x)\,\mathrm{d}x. \tag{10.62}$$

The choice $x_{n+j} = \xi_{j-1}$, $j = 1, 2, \ldots n + 1$ gives

$$\psi_{2n+1}(x) = (x - x_0) \ldots (x - x_{2n+1}) = \left(\frac{P_{n+1}(x)}{\alpha_{n+1}}\right)^2. \tag{10.63}$$

$\psi_{2n+1}W(x)$ does not change the sign in $[a, b]$. Hence,

$$I(g) - I(P_n) = \frac{1}{(2n+2)!}g^{(2n+2)}(\xi)\frac{S_{n+1}}{\alpha_{n+1}^2}, \tag{10.64a}$$

where

$$S_{n+1} = \int_a^b P_{n+1}^2(x)W(x)\,\mathrm{d}x. \tag{10.64b}$$

From Eq. (10.64) one finds that the Gaussian formula (10.56) is exact for all polynomials of degree $\leq 2n + 1$ provided the points $x_0, x_1, \ldots, x_n$ in Eq. (10.56) are the zeros of $P_{n+1}(x)$ of degree $n + 1$ in $[a, b]$ and the coefficients $A_i(i = 0, 1, \ldots, n)$ in Eq. (10.56a) are chosen according to Eq. (10.56b).

10.4.3   Special Case $n = 1$: Two-Point Gauss–Legendre Integral Formula

Let us choose $W(x) = 1$ and the orthogonal polynomials $P_{n+1}(x)$ as the Legendre polynomials. Since $P_n(x)$ are defined in the interval $[-1, 1]$ the arbitrary interval $[a, b]$ has to be changed to $[-1, 1]$. This can be made by the change of variable

$$x = \frac{1}{2}(b - a)t + \frac{1}{2}(b + a). \tag{10.65}$$

TABLE 10.3

First few Legendre polynomials and their zeros.

Value of n	$P_{n+1}(x)$	Zeros of $P_{n+1}(x)$
0	$P_1(x) = x$	$x_0 = 0$
1	$P_2(x) = \left(3x^2 - 1\right)/2$	$x_0 = -1/\sqrt{3}, \ x_1 = 1/\sqrt{3}$
2	$P_3(x) = 8x^3 - 12x$	$x_0 = -\sqrt{3/2}, \ x_1 = 0, \ x_2 = \sqrt{3/2}$

Then, the integral in Eq. (10.52a) becomes

$$I = \int_a^b g(x)\,\mathrm{d}x = \frac{1}{2}(b - a) \int_{-1}^1 g\left(\frac{1}{2}t + \frac{(b+a)}{2}\right) \mathrm{d}t. \tag{10.66}$$

Since a given integral with finite limits can be re-expressed as in Eq. (10.66) consider the integral

$$I = \int_{-1}^1 g(x)\,\mathrm{d}x. \tag{10.67}$$

First few Legendre polynomials and their zeros are given in Table 10.3.

For $n = 1$ Eq. (10.56) becomes

$$I = A_0 g(x_0) + A_1 g(x_1), \tag{10.68}$$

where A_0 and A_1 are obtained from Eq. (10.56b) as

$$
\begin{aligned}
A_0 &= \int_{-1}^1 L_0(x)\,\mathrm{d}x \\
&= \int_{-1}^1 (x - x_1)/(x_0 - x_1)\,\mathrm{d}x \\
&= \int_{-1}^1 \left(x - 1/\sqrt{3}\right) / \left(-1/\sqrt{3} - 1/\sqrt{3}\right) \mathrm{d}x \\
&= 1
\end{aligned}
\tag{10.69}
$$

and

$$A_1 = \int_{-1}^1 L_1(x)\,\mathrm{d}x = 1. \tag{10.70}$$

Hence,

$$I = g(x_0) + g(x_1), \quad x_0 = -1/\sqrt{3}, \ x_1 = 1/\sqrt{3}. \tag{10.71}$$

The above formula is called a *two-point Gauss–Legendre rule*.

The error term in the formula (10.68) is obtained from Eq. (10.64) by substituting $n = 1$:

$$E = \frac{1}{4!} \frac{S_2}{\alpha_2^2} g^{(4)}(x'). \tag{10.72}$$

S_2 and α_2 are found to be $S_2 = \int_{-1}^1 (P_2(x))^2\,\mathrm{d}x = 2/5$ and $\alpha_2 = 3/2$. Then,

$$E = \frac{1}{135} g^{(4)}(x'). \tag{10.73}$$

For $n \geq 1$ the nodes are irrational.

For some other integration methods see Problems 10.1, 10.26–10.29.

Example 1:

Evaluate the integral $\int_0^3 (1 + e^x)\mathrm{d}x$ employing the two-point Gauss–Legendre formula.

First, transform the given integral into the standard form $\int_{-1}^1 g(x)\mathrm{d}x$. Substitution of $a = 0$ and $b = 3$ in Eq. (10.65) gives

$$x = 3(t+1)/2 \quad \text{and} \quad \mathrm{d}x = 3/2\,\mathrm{d}t.$$

Introduce the above change of variables in the given integral and use the Gauss–Legendre formula. Then,

$$
\begin{aligned}
I &= \frac{3}{2} \int_{-1}^1 \left(1 + e^{3(t+1)/2}\right) \mathrm{d}t \\
&= \frac{3}{2} \left[g\left(t_0\right) + g\left(t_1\right)\right], \quad t_0 = -1/\sqrt{3}, \quad t_1 = 1/\sqrt{3} \\
&= \frac{3}{2} \left[1 + e^{3\left(-1/\sqrt{3}+1\right)/2} + 1 + e^{3\left(1/\sqrt{3}+1\right)/2}\right] \\
&= 21.810071.
\end{aligned}
$$

The exact value of the integral is 22.085537. The maximum possible error computed from the formula, Eq. (10.72), is $|E| = 9\,e^3/320 = 0.5649057$. The percentage of relative error in the numerical result is 1.25. This error is lower than the error in the trapezoidal and the Simpson's 1/3-rules but slightly higher than the Simpson's 3/8-rule.

Example 2:

Evaluate the integral $\int_0^1 1/(1 + x^2)\mathrm{d}x$.

Since the limits of the integral are not $[-1, 1]$ introduce the change of variables given by Eq.(10.65) with $a = 0$ and $b = 1$. Then,

$$
\begin{aligned}
I &= \int_0^1 \frac{1}{1 + x^2}\,\mathrm{d}x \\
&= 2 \int_{-1}^1 \frac{1}{4 + (1 + t)^2}\,\mathrm{d}t \\
&= 2 \left[g\left(-1/\sqrt{3}\right) + g\left(1/\sqrt{3}\right)\right] \\
&= 2\left[0.2393127 + 0.1541299\right] \\
&= 0.7868852,
\end{aligned}
$$

where the exact value is $I = \tan^{-1} x\big|_0^1 = \pi/4 = 0.7853981$. The percentage of relative error in the result is 0.19.

10.5 Double Integration

The methods discussed so far to evaluate single integrals can be extended to solve a double integral of the form

$$I = \int_c^d \int_a^b f(x, y) \, dx \, dy. \tag{10.74}$$

This section obtains the trapezoidal rule and the composite trapezoidal rule for double integrals. Derivation of the Simpson's rules for the above integral is left as an exercise to the reader.

10.5.1 Trapezoidal Rule

Application of the trapezoidal rule, Eq. (10.23), to the inner integral gives

$$I = \frac{1}{2}(b-a) \left[\int_c^d f(a, y) \, dy + \int_c^d f(b, y) \, dy \right]. \tag{10.75}$$

Next, the application of the rule to each integral in Eq. (10.75) gives

$$\begin{aligned} I &= \frac{1}{4}(b-a)(d-c) \left[f(a,c) + f(a,b) + f(b,c) + f(b,d) \right] \\ &= \frac{1}{4}(b-a)(d-c) \left[f_{00} + f_{01} + f_{10} + f_{11} \right]. \end{aligned} \tag{10.76}$$

10.5.2 Composite Trapezoidal Rule

To obtain the composite trapezoidal rule for the double integral, in Eq. (10.74) divide the interval $[a, b]$ into N subintervals and $[c, d]$ into M subintervals. The nodes are given by

$$\begin{aligned} x_i &= x_0 + ih, \quad i = 1, 2, \dots, N \tag{10.77a} \\ y_i &= y_0 + jk, \quad j = 1, 2, \dots, M \tag{10.77b} \end{aligned}$$

where $x_0 = a$, $y_0 = b$, $h = (b-a)/N$ and $k = (d-c)/M$. Referring to the discussion in Section 10.3.1 the composite trapezoidal rule for the double integral (10.74) is obtained as

$$\begin{aligned} I = \frac{hk}{4} & \left[f_{00} + f_{N0} + f_{0M} + f_{NM} + 2 \sum_{i=1}^{N-1} (f_{i0} + f_{iM}) \right. \\ & \left. + 2 \sum_{j=1}^{M-1} (f_{0j} + f_{Nj}) + 4 \sum_{i=1}^{N-1} \sum_{j=1}^{M-1} f_{ij} \right], \end{aligned} \tag{10.78}$$

where $f_{ij} = f(x_i, y_j)$.

Example:

Evaluate the integral $\int_0^1 \int_0^1 x^2 y \, dx \, dy$ by the composite trapezoidal rule with step size $h = 0.25$ and $k = 0.25$.

$h = k = 0.25$ gives $N = M = 4$. The nodes are

$$\begin{aligned} x_i &= 0 + i \times 0.25, \quad i = 0, 1, 2, 3, 4 \\ x_j &= 0 + j \times 0.25, \quad j = 0, 1, 2, 3, 4. \end{aligned}$$

The value of the integral is obtained as

$$I = \frac{0.25^2}{4}\left[f_{00} + f_{40} + f_{04} + f_{44} + 2\sum_{i=1}^{3}(f_{i0} + f_{i4})\right.$$

$$\left. + 2\sum_{j=1}^{3} f_{0j} + 2\sum_{i=1}^{3} f_{4j} + 4\sum_{i=1}^{3}\sum_{j=1}^{3} f_{ij}\right]$$

$$= \frac{0.25^2}{4} \times 11$$

$$= 0.171875\,.$$

The exact value of the integral is $x^3 y^2/6\big|_0^1 = 1/6 = 0.16666\ldots$.

10.6   Fractional Order Integration

The generalization of ordinary integration and differentiation to noninteger is known as fractional calculus. In Section 8.5, fractional order derivatives are considered. The present section is concerned with the numerical evaluation of fractional order integration. For the numerical computation of fractional order differential equation refer Chapter 16.

The Riemann–Liouville fractional integral operator of order $\alpha \in R_+$ is defined by [4-6]

$$I^\alpha f(t) = \frac{1}{\Gamma(\alpha)}\int_0^t (t - \tau)^{\alpha-1} f(\tau)\mathrm{d}\tau\,, \quad t \geq 0,\ \alpha > 0, \tag{10.79}$$

where Γ denotes the gamma function

$$\Gamma(\gamma) = \int_0^\infty z^{\gamma-1}\mathrm{e}^{-z}\mathrm{d}z\,, \quad \Gamma(\gamma) = (\gamma - 1)!\,. \tag{10.80}$$

10.6.1   Fractional Order Exact Integration of t^β

For certain simple functions, the integral in Eq. (10.79) can be easily evaluated analytically. For example, for $f(t) = t$

$$I^\alpha t = \frac{1}{\Gamma(\alpha)}\int_0^t (t - \tau)^{\alpha-1}\tau\mathrm{d}\tau\,. \tag{10.81}$$

Integrating by parts gives

$$I^\alpha t = \frac{1}{\Gamma(\alpha)}\left[-\frac{(t-\tau)^\alpha \tau}{\alpha}\bigg|_0^t + \frac{1}{\alpha}\int_0^t (t-\tau)^\alpha \mathrm{d}\tau\right]$$

$$= \frac{1}{\alpha\Gamma(\alpha)}\int_0^t (t-\tau)^\alpha \mathrm{d}\tau$$

$$= \frac{\Gamma(2)}{\Gamma(\alpha+2)} t^{\alpha+1}\,. \tag{10.82a}$$

For $f(t) = t^2$ and t^3

$$I^\alpha t^2 = \frac{1}{\Gamma(\alpha)}\int_0^t (t-\tau)^{\alpha-1}\tau^2 \mathrm{d}\tau = \frac{\Gamma(3)}{\Gamma(\alpha+3)} t^{\alpha+2}\,, \tag{10.82b}$$

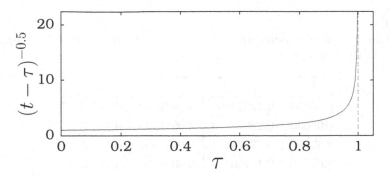

FIGURE 10.5
Plot of $(t - \tau)^{\alpha - 1}$ as a function of τ for $\alpha = 0.5$ and $t = T = 1$. The vertical dashed line represents the singularity at $\tau = T = 1$.

and

$$I^\alpha t^3 = \frac{1}{\Gamma(\alpha)} \int_0^t (t - \tau)^{\alpha-1} \tau^3 d\tau = \frac{\Gamma(4)}{\Gamma(\alpha + 4)} t^{\alpha+3}, \qquad (10.82c)$$

respectively. From Eqs. (10.82)

$$I^\alpha t^\beta = \frac{\Gamma(\beta + 1)}{\Gamma(\alpha + \beta + 1)} t^{\alpha+\beta}. \qquad (10.83)$$

10.6.2 A Modified Trapezoidal Formula

Suppose, it is desired to evaluate $I^\alpha f(t)$ over the interval $[0, T]$. The integral in Eq. (10.79) cannot be evaluated numerically for $0 < \alpha < 1$ applying the composite quadrature rules. This is because for $0 < \alpha < 1$ and for the values of τ near $t = T$ the quantity $(t - \tau)^{\alpha-1}$ diverges rapidly and it has a singularity at $\tau = T$ (refer Fig. 10.5). Numerical methods have been proposed for such integrals [7-11].

To obtain a formula for the evaluation of the integral in Eq. (10.79) divide the interval $[0, T]$ into n subintervals of equal width with the nodes $\tau_k = kh$, $k = 0, 1, \ldots, n$ and $h = T/n$. Then, Eq. (10.79) becomes

$$I^\alpha f(t)\big|_{t=T} = \frac{1}{\Gamma(\alpha)} \sum_{k=0}^{n-1} \int_{t_k}^{t_{k+1}} (T - \tau)^{\alpha-1} f(\tau) d\tau. \qquad (10.84)$$

Replace f by the piecewise linear interpolant in each interval with the nodes τ_k. This results in the formula [7]

$$\begin{aligned}
I^\alpha f(t)\big|_{t=T} = \frac{h^\alpha}{\Gamma(2 + \alpha)} &\left\{ \left[(n-1)^{\alpha+1} - (n - \alpha - 1)n^\alpha \right] f(0) \right. \\
&+ \sum_{k=1}^{n-1} \left[(n - k + 1)^{\alpha+1} - 2(n - k)^{\alpha+1} \right. \\
&\left. \left. + (n - k - 1)^{\alpha+1} \right] f(\tau_k) + f(T) \right\}.
\end{aligned} \qquad (10.85)$$

TABLE 10.4

I_{error}^{MTF} in the numerically computed value of $I^{0.5}(t^3)|_{0.5}$ by the modified trapezoidal formula as a function of the step size h.

h	I_{error}^{MTF}	h	I_{error}^{MTF}
0.10000	0.11378 e-2	0.0031250	0.12715 e-5
0.05000	0.29936 e-3	0.0015625	0.32342 e-6
0.02500	0.77347 e-4	0.0010000	0.13548 e-6
0.01250	0.19770 e-4	0.0005000	0.37243 e-7
0.00625	0.50206 e-5	0.0001000	0.56431 e-8

This is known as the *modified trapezoidal formula*. The error term is found to be $O(h^2)$ [7].

Examples:

Let us apply the modified trapezoidal formula to two integrals. The first example is $I^\alpha t^\beta$. Its exact value is given by Eq. (10.83). For $\alpha = 0.5$ and $\beta = 3$ the exact value of the integral $I^{0.5}(t^3)|_{t=0.5}$ is

$$I_{exact} = \frac{2\sqrt{2}}{35\sqrt{\pi}} \approx 0.04559. \tag{10.86}$$

Obviously, the value of a fractional order integral cannot be calculated employing the formula (10.85) using a calculator. A Python or C++ or Fortran90 program implementing the modified trapezoidal formula can be developed.

The numerical value of the integral denoted as I_{num} is computed (using a developed program) for 10 values of h and compared with I_{exact} by calculating

$$I_{error}^{MTF} = |I_{exact} - I_{num}|. \tag{10.87}$$

The value of $\Gamma(2+\alpha)$ is computed using Eq. (8.40). The exact value of $\Gamma(2.5)$ is $1.5 \times 0.5 \times \Gamma(0.5) = 0.75\sqrt{\pi} \approx 1.32934$. For the step sizes 0.1, 0.01, 0.001 and 0.0001 the values of $\Gamma(0.5)$ computed numerically are 1.32925, 1.32934, 1.32934 and 1.32934, respectively. The step size 0.0001 is chosen. This step size can also be used for other values of α. Table 10.4 shows the variation of I_{error}^{MTF} with the step size h.

The second example is with

$$f(t) = \frac{8}{3}\sqrt{\frac{t^3}{\pi}} - 2\sqrt{\frac{t}{\pi}}. \tag{10.88}$$

For $\alpha = 0.5$, $I_{exact}^{0.5} = t^2 - t$. Variation of I_{error}^{MTF} with h is given in Table 10.5 for $T = 0.5$. Tables 10.4 and 10.5 clearly illustrate the effect of h on the values of I_{num} and also the applicability of the method. The method can be applied to the case of $\alpha > 1$, for example see Problem 34 at the end of the present chapter.

10.6.3   Implementation of Composite Trapezoidal Formula

Due to the presence of singularity at $\tau = t$ in Eq. (10.79) the composite trapezoidal formula is not applicable directly to evaluate it numerically. However, the singularity can be avoided

TABLE 10.5

$I_{\text{error}}^{\text{MTF}}$ versus h for $I^{0.5}\left((8/3)\sqrt{t^3/\pi} - 2\sqrt{t/\pi}\right)\Big|_{0.5}$.

h	$I_{\text{error}}^{\text{MTF}}$	h	$I_{\text{error}}^{\text{MTF}}$
0.10000	0.74393 e-2	0.0031250	0.34285 e-4
0.05000	0.24813 e-2	0.0015625	0.11952 e-4
0.02500	0.83887 e-3	0.0010000	0.60743 e-5
0.01250	0.28668 e-3	0.0005000	0.21246 e-5
0.00625	0.98816 e-4	0.0001000	0.17955 e-6

by integrating this equation once by parts. Taking $u = f(\tau)$ and $dv = (t - \tau)^{\alpha-1}d\tau$ the relation $\int_0^t u\,dv = uv|_0^t - \int_0^t v\,du$ gives

$$I^\alpha f(t) = \frac{1}{\Gamma(1 + \alpha)}\left[t^\alpha f(0) + \int_0^t (t - \tau)^\alpha f'(\tau)d\tau\right]. \tag{10.89}$$

Then, the application of the composite trapezoidal formula to the integral in the above equation results in

$$I^\alpha f(t)|_{t=T} = \frac{1}{\Gamma(1 + \alpha)}\left\{T^\alpha f(0) + \frac{h}{2}\left[T^\alpha f'(0) + 2\sum_{k=1}^{n-1}(T - \tau_k)^\alpha f'(\tau_k)\right]\right\}, \tag{10.90}$$

where $\tau_k = kh$, $k = 0, 1, \ldots, n$ and $h = T/n$. If the analytical expression for $f'(\tau)$ is easy to determine then it can be used in the above formula. Otherwise the following finite-difference approximation can be used for $f'(0)$ and $f'(\tau_k)$:

$$f'(0) = \frac{1}{2h}\left[-3f(0) + 4f(h) - f(2h)\right], \tag{10.91a}$$

$$f''(\tau_k) = \frac{1}{2h}\left[f(\tau_k + h) - f(\tau_k - h)\right]. \tag{10.91b}$$

$I^\alpha f(t)|_{t=T}$ is computed for $f(t) = t^3$ for $\alpha = 0.5$, $T = 0.5$ and for a range of values of h. Table 10.6 displays the error $I_{\text{error}}^{\text{MTF}}$ and $I_{\text{error}}^{\text{CTF}}$, where CTF stands for composite trapezoidal formula. Desired accuracy can be obtained by choosing appropriate value of h. For a given value of h, $I_{\text{error}}^{\text{CTF}}$ is higher than $I_{\text{error}}^{\text{MTF}}$.

TABLE 10.6

Comparison of the errors in the modified trapezoidal formula and the composite trapezoidal formula in the case of $I^{0.5}(t^3)|_{0.5}$.

h	$I_{\text{error}}^{\text{MTF}}$	$I_{\text{error}}^{\text{CTF}}$	h	$I_{\text{error}}^{\text{MTF}}$	$I_{\text{error}}^{\text{CTF}}$
0.100	0.11378 e-2	0.38882 e-2	0.0100	0.12719 e-4	0.14967 e-4
0.050	0.29936 e-3	0.14089 e-2	0.0010	0.13548 e-6	0.52793 e-5
0.025	0.77347 e-4	0.53985 e-3	0.0001	0.56431 e-8	0.15665 e-6

10.7   Concluding Remarks

It is possible to perform hybrid multi-dimensional numerical integrations with a moderate precision. Methods have been introduced for numerical integration to get results with several hundreds of digits accuracy. Such techniques have applications in high-performance computations. Combination of such high-precision methods with integer relation detection algorithms is employed to obtain analytic expressions for integrals. Gaussian quadrature, tanh-sinh quadrature and highly parallel quadrature are useful for high-precision results.

10.8   Bibliography

[1]   L.A. Pipes and L.R. Harvill, *Applied Mathematics for Engineers and Physicists*. McGraw-Hill, Singapore, 1971.

[2]   A. Jeffrey, *Advanced Engineering Mathematics*. Academic Press, New Delhi, 2003.

[3]   E. Kreyszig, *Advanced Engineering Mathematics*. John Wiley, New York, 1999.

[4]   K.S. Miller and B. Ross, *An Introduction to the Fractional Calculus and Fractional Differential Equations*. John Wiley and Sons, Inc., New York, 1993.

[5]   I. Podlubny, *Fractional Differential Equations*. Academic Press, Cambridge, 1999.

[6]   A.A. Kilbas, H.H. Srivastava and J.J. Trujillo, *Theory and Applications of Fractional Differential Equations*. Elsevier, Amsterdam, 2006.

[7]   Z.M. Odibat, *Appl. Math. Comput.* 178:527, 2006.

[8]   Z.M. Odibat and S. Momani, *J. Appl. Math. & Inform.* 26:15, 2008.

[9]   T. Blaszczyk and J. Siedlecki, *J. Appl. Math. Comput. Mech.* 13:137, 2014.

[10] T. Blaszczyk, M.Ciesielski, *Fract. Calc. Appl. Anal.* 17:307, 2014.

[11] N. Pandiangan, D. Johar and S. Purwani, *World Scientific News* 153:169, 2021.

10.9   Problems

10.1 Show that (a) for $n = 5$ (Boole's rule) the integration formula given by Eq. (10.10) becomes [G. Boole and J.F. Moulton, *A Treatise on the Calculus of Finite Differences*. Dover, New York, 1960.]

$$I_5 = \frac{2h}{45} \left[7f_0 + 32f_1 + 12f_2 + 32f_3 + 7f_4 \right] - \frac{8}{945} h^7 f^{(6)}(x')$$

and (b) for $n = 6$ (Hardy's rule) [E.W. Weisstein, *Hardy's Rule*–A Wolfram Web Resource. http://mathworld.wolfram.com/HardysRule.html (accessed on May 10, 2020).]

$$I_6 = \frac{h}{100} \left[28f_0 + 162f_1 + 220f_3 + 162f_5 + 28f_6 \right].$$

10.2 Evaluate the following integrals by the trapezoidal, the Simpson's, the two-point Gauss–Legendre's, the composite trapezoidal and the composite Simpson's rules. Compare the numerical value with the exact result by computing the percentage of relative error wherever possible. For the composite rules divide the range of integration into 3 equally spaced subintervals and then apply the rules to each subinterval. That is, $n = 3$ in the formulas (10.35), (10.39) and (10.41).

(a) $\displaystyle\int_0^1 \cos x \, dx.$ (b) $\displaystyle\int_0^1 e^{-x^2} dx.$ (c) $\displaystyle\int_1^2 1/x \, dx.$

10.3 For the following problems employ the trapezoidal, the Simpson's, the two-point Gauss–Legendre, the composite trapezoidal and the Simpson's rules. For composite rules divide the range of integration into 3 equally spaced subintervals, that is $n = 3$.

(a) The $S_i(x)$ function is defined as

$$S_i(x) = \int_0^x \frac{\sin u}{u} \, du.$$

Compute $S_i(1)$ assuming $(\sin u)/u$ for $u = 0$ as 1.

(b) The beta function is given by

$$\beta(m, n) = 2 \int_0^{\pi/2} \sin^{2m-1}\theta \cos^{2n-1}\theta \, d\theta.$$

Find the value of $\beta(3, 2)$.

(c) The error function $\operatorname{erf}(x)$ is defined as

$$\operatorname{erf}(x) = \frac{2}{\sqrt{\pi}} \int_0^x e^{-t^2} dt.$$

Compute $\operatorname{erf}(0.1)$.

10.4 Evaluate the integral $\displaystyle\int_0^1 f(x) dx$, with the values of $f(x)$ at certain points given below, employing the trapezoidal and the Simpson's 1/3 rules.

x	0	0.25	0.5	0.75	1
$f(x)$	0	0.015625	0.125	0.421875	1

10.5 Compare the relative errors in the evaluation of the integral $I = \displaystyle\int_0^2 x^3 dx$ by the composite trapezoidal and Simpson's rules as a function of n, with $n = 3, 6, 12, 18, 24, 30$.

10.6 List some integrals which are not solvable by the composite trapezoidal and the Simpson's rules but solvable by the Gauss–Legendre two-point method.

10.7 For what value(s) of n the composite trapezoidal and the Simpson's rules are applicable to evaluate the integral

$$I = \int_0^\pi \frac{\sin^3 x}{\cos^2 x} \, dx.$$

10.8 Compute the absolute maximum errors possible in the trapezoidal, the Simpson's and the two-point Gauss–Legendre rules for the following integrals.

a) $\int_0^2 1/(2+x)\mathrm{d}x$.   b) $\int_0^1 x^2\mathrm{d}x$.   c) $\int_0^1 \cos x\mathrm{d}x$.

10.9 Calculate the optimum number of subintervals n_{opt} and their width h_{opt}, so that the integrals in Problem 10.8 can be evaluated using (i) the composite trapezoidal rule, (ii) the composite Simpson's 1/3-rule and (iii) the composite Simpson's 3/8-rule with accuracy 10^{-5}.

10.10 Compute the values n_{opt} and h_{opt} for the integral $\int_0^1 (e^x + \sin x)\mathrm{d}x$ (x is in radian) for a) the composite trapezoidal rule, b) the composite Simpson's 1/3-rule and c) the composite Simpson's 3/8-rule so that the error in the approximation is 10^{-3}. Then, using the calculated n_{opt} and h_{opt} evaluate the integral by the above three rules.

10.11 Write a program in Python which first calculate the optimum value of n, n_{opt}, in the composite trapezoidal rule for a given integral of the form $I = \int_a^b f(x)\mathrm{d}x$ and then evaluate the value of the integral with $n = n_{opt} + 1$. Using the program evaluate the values of the integrals of Problem 10. 8.

10.12 Do Problem 10.11 for the composite Simpson's rules.

10.13 Can one use the trapezoidal and the Simpson's rules to compute the value of the following integrals? Why?

a) $\int_0^1 x^2/\sqrt{1-x}\,\mathrm{d}x$.   b) $\int_0^{\pi/2} (\tan x/\cos x)\mathrm{d}x$.

c) $\int_0^1 1/(1-x^2)^{1/4}\mathrm{d}x$.

For the following Problems 10.14–10.30 employ the trapezoidal, the Simpson's, the two-point Gauss–Legendre, the composite trapezoidal and the Simpson's rules. For the composite rules divide the range of integration into 3 equally spaced subintervals, that is $n = 3$. Compare the numerical value with the exact result (if known) by computing percentage of relative error.

10.14 For an electron in a box of size unity the transition dipole moment integral is given by

$$I = \int_0^1 x \sin \pi x \sin 2\pi x \, \mathrm{d}x .$$

Compute the value of I.

10.15 The oscillation period T of a Josephson junction is given by

$$T = \frac{\hbar}{2er} \int_0^{2\pi} \frac{1}{I_b - I \sin \theta} \, \mathrm{d}\theta .$$

For $I_b = 2\,\mathrm{mA}$ and $I = 1\,\mathrm{mA}$ evaluate the integral.

10.16 The time period of oscillation of a pendulum is given by

$$T(h) = \frac{4}{\sqrt{2h}} \int_0^{\pi/2} \frac{1}{\sqrt{1 - \frac{2}{h} \sin^2 \theta}} \, d\theta,$$

where h is the energy of the system. Compute $T(4)$.

10.17 The calculation of total dipole moment of a dielectric material involves the integral $\int_{-1}^{1} x e^x dx$. Compute the value of this integral.

10.18 The vector potential at a point r, due to a current (I) carrying wire is given by

$$\Lambda = \mathbf{K} \frac{\mu_0 I}{4\pi} \int_{-L}^{L} \frac{1}{(r_1^2 + z^2)^{1/2}} \, dz,$$

where $2L$ is the length of the wire. Compute the integral for $L = 1$ metre and $r_1 = 1$ metre.

10.19 The quantum mechanical probability P_{QM} for a particle to be found in a classically allowed region of a linear harmonic oscillator potential is given by

$$P_{\mathrm{QM}} = \frac{2}{\sqrt{\pi}} \int_0^{0.5} e^{-x^2} \, dx.$$

Compute P_{QM}.

10.20 Consider a spherical shell of radius R, carrying a uniform surface charge density σ which is set to spin at an angular velocity ω. The expression for the vector potential at a point P produced by the spherical shell involves the integral

$$\int_{-1}^{1} \frac{u}{\sqrt{R^2 + s^2 - 2Rsu}} \, du,$$

where s is the distance of the point P from the centre of the sphere. Compute the value of the above integral for $R = 1$ and $s = 2$.

10.21 The magnetic field at the origin due to a current (I) carrying elliptic wire centered at origin with the lengths of major and minor axes $2a$ and $2b$, respectively, is given by

$$B = \frac{\mu_0 I}{\pi a} \int_0^{\pi/2} \sqrt{1 - k^2 \sin^2 \theta} \, d\theta,$$

where $k = \sqrt{1 - (a^2/b^2)}$. Compute the value of the integral in B for $a = 2b$ (that is, $k^2 = -3$).

10.22 The magnetic flux ϕ flowing through a surface due to a linear conductor of infinite length carrying a current I_0 is given by

$$\phi = \frac{3}{\sqrt{2}} \frac{\mu_0 I_0}{\pi} \int_{-1}^{1} \frac{(x+1)}{(x^2 + 2x + 2)} \, dx.$$

Evaluate the above integral.

10.23 The centre of mass of a uniform solid hemisphere of radius R and mass M is given by

$$Z = \frac{3}{2R^3} \int_0^R x(R^2 - x^2)\, dx.$$

Compute Z for $R = 0.2$.

10.24 The x-component of a magnetic field B_x at a point due to a hollow circular wire is given by

$$B_x = \frac{\mu_0 I}{2\pi} \int_0^{2\pi} \frac{2 - \sin\theta}{5 - 4\sin\theta}\, d\theta.$$

Compute the value of the above integral.

10.25 A rod of 0.3 m length (L) has a nonuniform density λ. The mass per unit length of the rod varies as $\lambda = \lambda_0(s/L)$, where λ_0 is a constant and s is the distance from the end. The centre of the mass is given by the integral

$$R = \frac{2}{\lambda_0 L} \int_0^L x\lambda_0 \frac{x}{L}\, dx = \frac{2}{L^2} \int_0^L x^2\, dx.$$

Calculate R.

10.26 Show that the three-point Gauss–Legendre formula for the integral $I = \int_{-1}^{1} g(x)dx$ is

$$I = \frac{5}{9}g(-\sqrt{3/5}) + \frac{8}{9}g(0) + \frac{5}{9}g(\sqrt{3/5}) + \frac{1}{15750}g^{(6)}(x').$$

Then evaluate the integrals

a) $\int_0^1 e^{-x^2}\, dx$ b) $\int_1^2 1/x\, dx$ c) $\int_0^1 (\sin x)/x\, dx.$

10.27 Derive a four-point Gauss–Legendre rule for the integrals of the form $\int_{-1}^{1} f(x)dx.$ Then find the values of the following integrals given below.

a) $\int_0^1 1/(2 - x)\, dx.$ b) $\int_0^1 1/(2 + x^2)\, dx.$ c) $\int_0^3 (1 + e^x)\, dx.$

10.28 A three-point Lobatto integration rule is given by

$$\int_{-1}^{1} g(x)\, dx = \frac{1}{3}\left[g(-1) + 4g(0) + g(1)\right] - \frac{1}{90}g^{(4)}(x').$$

Compute the values of the integrals in Problem 10.26 by this rule.

10.29 A three-point Radan integration rule is

$$\int_{-1}^{1} g(x)\, dx = \frac{2}{9}g(-1) + \frac{16 + \sqrt{6}}{18} g\left(\frac{1 - \sqrt{6}}{5}\right)$$

$$+ \frac{16 - \sqrt{6}}{18} g\left(\frac{1 + \sqrt{6}}{5}\right).$$

Compute the values of the integrals in Problem 10.26 by this rule.

10.30 Evaluate the integral

$$I = \int_0^1 \int_0^1 \frac{1}{1+x+y} \, dx \, dy$$

using the composite trapezoidal rule with step size $h = k = 0.25$.

10.31 Develop composite Simpson's 1/3-rule to evaluate the double integral

$$\int_c^d \int_a^b f(x, y) \, dx \, dy \,.$$

10.32 The exact value of $I^{1.5} \left(4\sqrt{t/\pi} \right) \Big|_2$ is t^2. Compute the value of this integral for $h = 0.1, 0.01$ and 0.001 by the modified trapezoidal formula.

10.33 Develop a program for evaluating $I^{0.75} \left(t^3 \right) \Big|_{0.5}$. Compute the value of the integral for $h = 0.1, 0.01$ and 0.001.

10.34 Determine the value of $I^{1.5} \left(4\sqrt{t/\pi} \right) \Big|_{t=1}$ by the composite trapezoidal formula given by Eq. (10.91) for $h = 0.01, 0.001$ and 0.0001.

10.35 Develop a program to evaluate the integral $I^\alpha t^\beta$ by the fractional Simpson's rule given in [9]. Compare the value of the integral for $\alpha = 0.5$ with the exact value of the integral.

11

Ordinary Differential Equations – Initial-Value Problems

11.1 Introduction

Differential equations arise in the mathematical modelling of many physical, chemical and biological systems. A differential equation is said to be a *linear* if the degree or the power of dependent variables and their derivatives is zero or one and no product of them occurs. A differential equation which is not linear is called a *nonlinear*. A differential equation with only one independent variable is called an *ordinary differential equation*, otherwise a *partial differential equation*. Given a differential equation our prime aim is to find its solution. Particular interest is to find an exact analytical solution expressed in terms of well-known functions. This is because, from the solution, the state of the system can be determined immediately for any time for a given initial condition. General analytical methods are available to find exact solution for linear differential equations with constant coefficients [1–4]. For other types of equations, general methods are not available. When the given equation is not solvable exactly by the existing analytical methods then it can be solved by an appropriate numerical integration scheme. Then, proceed to analyze the behaviour of the solution.

The problem of solving differential equations of the form

$$\frac{\mathrm{d}x_i}{\mathrm{d}t} = f_i\left(x_1, x_2, \ldots, x_n, t\right) , \quad i = 1, 2, \ldots, n \tag{11.1}$$

is generally classified into the following three main classes:

1. Initial-Value Problem

Solving a differential equation of the form (11.1) with the values of x_i given at a single starting value of t is known as an *initial-value problem*.

2. Boundary-Value Problem

Solving a differential equation with the value of x_i given at the starting and ending values of t is known as a *boundary-value problem*.

3. Eigenvalue Problem

For certain differential equations of the form $Ly = \lambda y$, where L is a differential operator, bounded solution exists only for a specific set of values of the parameter λ. These specific values of λ are called *eigenvalues* and the corresponding solutions are called *eigenfunctions*. The eigenvalue problem of equations of the form $Ly = \lambda y$ involves determination of those specific values of the parameter λ and then the corresponding solution.

DOI: 10.1201/9781032649931-11

Any nth order differential equation of the form

$$\frac{\mathrm{d}^n x}{\mathrm{d}t^n} + a_1(t)\frac{\mathrm{d}^{n-1}x}{\mathrm{d}t^{n-1}} + \cdots + a_{n-1}(t)\frac{\mathrm{d}x}{\mathrm{d}t} + a_n(t)x = 0 \qquad (11.2)$$

can always be rewritten in the form (11.1) as $x_1 = x$,

$$\begin{aligned}
\dot{x}_1 &= x_2, \\
\dot{x}_2 &= x_3, \\
&\;\;\vdots \\
\dot{x}_n &= -a_1(t)x_{n-1} + \cdots + a_{n-1}(t)x_2 + a_n(t)x_1,
\end{aligned} \qquad (11.3)$$

where overdot refers to differentiation with respect to the independent variable t.

The basic idea of any method of solving the initial-value problem is the following. For simplicity consider a first-order equation of the form

$$\dot{x} = f(x,t). \qquad (11.4)$$

The goal is to find the solution of Eq. (11.4) for $t \in [a,b]$ with $x(a) = x_a$ or x_0. The interval $[a,b]$ is divided into m intervals by the points $t_0 = a < t_1 = a + h < t_2 = a + 2h < \cdots < t_m = a + mh = b$, where $h = (b-a)/m$. Write $t_n = a + nh$, $n = 0, 1, \ldots, m$. The points t_n are called *grid points* and h is called *step size*. The exact solution $x(t)$ at $t = t_n$ is approximated by a number and is denoted as x_n. The sequence $x_0, x_1, \ldots, x_m$ is called a *numerical solution*. In general, the numerical value x_{n+1} at t_{n+1} is obtained from the formula

$$x_{n+1} = x_n + h\phi(x_n, x_{n-1}, \ldots, x_{n-k}, t_n, t_{n-1}, \ldots, t_{n-k}). \qquad (11.5)$$

When $k = 0$, the value of x_{n+1} is found by using (x_n, t_n) only. In this case, the method is called a *single-step method*. For $k > 0$, the method is a *multi-step method*. If the error in a method is $O(h^{N+1})$ then it is called Nth *order method*. Some of the single-step methods are the Euler method and the Runge–Kutta methods. An example of a multi-step method is the Adams–Bashforth–Moulton method. These and other methods essentially discretize the differential equation and produce a difference equation generally of the form (11.5). Different methods give rise to different difference equations for the same differential equation. However, their aim is to produce a solution which should correspond closely to the exact solution of the given equation.

The present and the next two chapters concentrate on numerical integration of ordinary differential equations. Some methods for solving partial differential equations will be described in Chapters 14 and 15. In the present chapter, for the initial-value problem certain single-step methods such as the Euler and the Runge–Kutta methods and the multi-step methods such as the Adams–Bashforth–Moulton, the Milne–Simpson and the Hamming methods are considered. The features of these methods are discussed. Also, the methods to solve stiff equations are presented. Numerical methods for Hamiltonian equations will be addressed in Chapter 12. Methods for boundary-value problems will be considered in Chapter 13.

11.2 Euler Method

In this section let us obtain the Euler formula for the numerical integration of an ordinary differential equation, discuss the error in the formula and analyze the stability of the method.

11.2.1   Euler Formula

For simplicity and illustrative purpose consider the first-order equations of the form (11.4). Assume that $x(t_0 = a)$ is given and denote it as x_0. If $\dot{x} = dx/dt$ (rate of change of x) remains as $f(x_0, t_0)$ for all time then integration of Eq. (11.4) gives

$$x(t) = f(x_0, t_0) t + C, \tag{11.6}$$

where C is an integration constant. With $x(t) = x_0$ at $t = t_0$ the above equation is rewritten as

$$C = x_0 - f(x_0, t_0) t_0 . \tag{11.7}$$

Then,

$$x(t) = x_0 + f(x_0, t_0) (t - t_0) . \tag{11.8}$$

When $\dot{x}$ changes with time, it is reasonable to expect that it remains close to $f(x_0, t_0)$ for t close t_0. In this case

$$x(t) \approx x_0 + f(x_0, t_0) (t - t_0) + \frac{1}{2} x''(c_1) (t - t_0)^2 , \tag{11.9}$$

where $t_0 < c_1 < t$.

Define $t_1 = t_0 + h$, where h is a small number then at $t = t_1$

$$x(t_1) = x_1 = x_0 + hf(x_0, t_0) + \frac{1}{2} h^2 x''(c_1). \tag{11.10}$$

Dropping the h^2 term gives

$$x_1 = x_0 + hf(x_0, t_0) . \tag{11.11}$$

Like-wise, define

$$t_2 = t_1 + h = t_0 + 2h \tag{11.12a}$$

and write at $t = t_2$

$$x(t_2) = x_2 = x_1 + hf(x_1, t_1) . \tag{11.12b}$$

In general, with

$$t_{n+1} = t_0 + (n+1)h , \quad n = 0, 1, \ldots \tag{11.13a}$$

x_{n+1} is given by

$$x_{n+1} = x_n + hf(x_n, t_n) . \tag{11.13b}$$

The above algorithm of determining $x(t)$ from $x(t_0)$ is called *Euler method*. For an mth-order equation the Euler formula is

$$t_{n+1} = t_0 + (n+1)h , \quad n = 0, 1, \ldots \tag{11.14a}$$
$$\mathbf{x}_{n+1} = \mathbf{x}_n + h\mathbf{F}(\mathbf{x}_n, t_n) . \tag{11.14b}$$

In order to understand the geometric picture of the Euler method consider the direct integration of Eq. (11.4) from t_n to t_{n+1}:

$$\int_{t_n}^{t_{n+1}} \frac{dx}{dt} \, dt = x(t_{n+1}) - x(t_n) = \int_{t_n}^{t_{n+1}} f(x(t), t) \, dt . \tag{11.15}$$

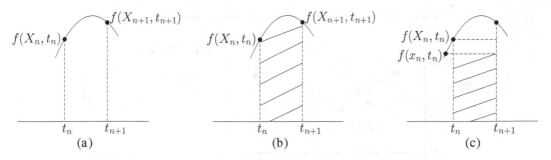

FIGURE 11.1

Geometrical description of the Euler method. For details see the text.

The above is exact therefore replace x by X, the exact value of x, and write

$$X\left(t_{n+1}\right) = X\left(t_n\right) + \int_{t_n}^{t_{n+1}} f(X(t), t)\, dt. \tag{11.16}$$

Because the exact value X is not known a numerical approximation can be used for the value of the integral in Eq. (11.16). Denoting x_n as the approximation to $X(t_n)$ Eq. (11.16) becomes

$$x_{n+1} = x_n + \text{an approximate value for } \int_{t_n}^{t_{n+1}} f(X(t), t)\, dt. \tag{11.17}$$

Comparison of Eqs. (11.17) and (11.13b) indicates that in the Euler method the second term in the right-side of (11.17) is $hf(x_n, t_n)$. Geometrically, *what does $hf(x_n, t_n)$ represent?* Assume that the form of $f(x, t)$ over the interval $[t_n, t_{n+1}]$ is as shown in Fig. 11.1a. The value of the integral in Eq. (11.16) is the area of the stripped portion in Fig. 11.1b. This area is, say, approximated by the area of the stripped rectangle shown in Fig. 11.1c and is $(t_{n+1} - t_n)f(x_n, t_n)$, that is $hf(x_n, t_n)$.

Example:

The equation of motion of the overdamped linear harmonic oscillator driven by the periodic external driving force $\sin t$ is the first-order equation

$$\dot{x} = -x + \sin t, \quad x(0) = 1. \tag{11.18}$$

The exact solution of this equation is

$$x(t) = (x(0) + 0.5)\, e^{-t} + 0.5(\sin t - \cos t). \tag{11.19}$$

(a) Calculate $x(0.1)$ by the Euler method with step size $h = 0.1$.

(b) Solve the equation in the interval $t \in [0, 3]$ with $h = 0.01, 0.1, 0.25, 0.5$.

(a) The Euler formula for $x(0.1)$ is

$$x(0.1) = x(0) + hf(x(0), t = 0).$$

Then,

$$x(0.1) = x(0) + h(-x(0) + \sin 0) = 1.0 + 0.1(-1.0 + 0) = 0.9$$

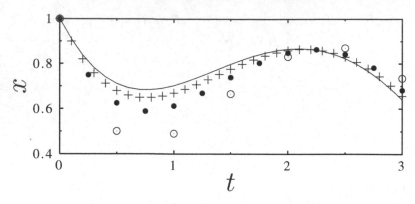

FIGURE 11.2
Comparison of the numerical solution of the equation $\dot{x} = -x + \sin t$ obtained with the step sizes $h = 0.1$ (marked by $+$), 0.25 (marked by the painted circles) and 0.5 (marked by the open circles) by the Euler method. The continuous curve represents the exact solution. The initial condition used is $x(0) = 1$.

while the exact solution is 0.9096707. The difference is 0.0096707 and the percentage of relative error ($= 100 \times |(x_{\text{exact}} - x_{\text{numer}})/x_{\text{exact}}|$) is 1.063.

(b) Figure 11.2 shows the plot of $x(t)$ with $h = 0.10$, 0.25 and 0.50 along with the exact solution. Table 11.1 gives the numerical values of x at a few values of t for different values of h.

11.2.2   Error Analysis

Generally, there are three types of numerical errors in a numerical integration algorithm. They are round-off error, local truncation error and final global error or simply global error. The round-off error is due to the finite-precision arithmetic operations. It depends on the

TABLE 11.1
Comparison of the numerical solution obtained by the Euler method with different step sizes for the equation $\dot{x} = -x + \sin t$, $x(0) = 1$. The exact solution is $x(t) = (x(0) + 0.5)\,\mathrm{e}^{-t} + 0.5(\sin t - \cos t)$.

| t | Numerically computed x with | | | | |
	$h = 0.01$	$h = 0.10$	$h = 0.25$	$h = 0.50$	Exact x
0.0	1.0000000	1.0000000	1.0000000	1.0000000	1.0000000
0.5	0.7077484	0.6793987	0.6243510	0.5000000	0.7107175
1.0	0.6991978	0.6688721	0.6114994	0.4897128	0.7024035
1.5	0.7959364	0.7759006	0.7389904	0.6655919	0.7980741
2.0	0.8650723	0.8590772	0.8487089	0.8315434	0.8657251
2.5	0.8236080	0.8300115	0.8424103	0.8704204	0.8229354
3.0	0.6417220	0.6556509	0.6814846	0.7344463	0.6402369

type of arithmetical operations and the number of operations used in a step. A consequence of this is that a very small step size cannot be chosen since this involves a greater number of arithmetical operations leading to an increase in the round-off error. Since the round-off error depends on the computer on which the method is implemented, it is generally not considered. In a numerical integration of ordinary differential equations, the other two errors are important.

The local truncation error is the error in a single-step due to the approximation used. It is present even with an infinite-precision arithmetic. It depends on the step size as well as the order of the method. The global error is the sum of the local truncation error over the interval say $t_0 = a$ to $t_N = b$ and is due to the repeated application of the algorithm.

Denoting $X(t)$ as the exact solution, the local truncation error is given by

$$
\begin{aligned}
E_{n+1} &= X_{n+1} - x_{n+1} \\
&= X_{n+1} - x_n - h\phi(x_n, t_n; h).
\end{aligned}
\tag{11.20}
$$

For the Euler method the local truncation error E_{n+1} is given by

$$
E_{n+1} = X_{n+1} - x_{n+1}, \quad n = 0, 1, \ldots, m - 1.
\tag{11.21}
$$

Use of Eq. (11.13b) in the above equation gives

$$
E_{n+1} = X_{n+1} - x_n - hf(x_n, t_n).
\tag{11.22}
$$

Since, the exact solution is not known replace X_{n+1} by $= x_{n+1}$. Then,

$$
E_{n+1} = \frac{1}{2}h^2 x''(c_1),
\tag{11.23}
$$

where $t_n < c_1 < t_{n+1}$. The local truncation error is $O(h^2)$. If

$$
\max_{|t_0, t_m|} |E_{n+1}| = T
\tag{11.24a}
$$

and

$$
\max_{|t_0, t_m|} |x''(c)| = M_1, \quad t_0 < c < t_N
\tag{11.24b}
$$

then $T \leq h^2 M_1/2$. The error E_{n+1} given by Eq. (11.23) is the error made in the single-step from t_n to t_{n+1}. Therefore, the global error committed in repeating the above formula from $t_0 = a$ to $t_m = b$ with step size h is

$$
\begin{aligned}
E(x(b), h) &= \frac{1}{2}h^2 \sum_{n=1}^{m} x''(c_n) \\
&= \frac{1}{2}mx''(c)h^2 \\
&= \frac{1}{2}(b - a)x''(c)h \\
&= O(h).
\end{aligned}
\tag{11.25}
$$

The global error in the Euler method is $O(h)$.

TABLE 11.2

The final global error ($|x_{\text{exact}} - x_{\text{numer}}|$) and the percentage of relative error at $t = 1$ for Eq. (11.18) with the Euler method. The exact solution at $t = 1$ is 0.7024035.

h	x by the Euler method	Final global error at $t = 1$	% of relative error
0.001	0.7020843	0.0003192	00.045
0.010	0.6991978	0.0032057	00.456
0.025	0.6943304	0.0080731	01.149
0.050	0.6860566	0.0163469	02.327
0.100	0.6688721	0.0335314	04.774
0.250	0.6114994	0.0909041	12.942
0.500	0.4897128	0.2126907	30.280

Example:

Compare the final global error in the solution of Eq. (11.18) obtained at $t = 1$ by the Euler method with $h = 0.001, 0.01, 0.025, 0.05, 0.1, 0.25$ and 0.5 by calculating the percentage of relative error.

Table 11.2 presents the numerically computed final global error at $t = 1$ for different values of h. The final global and the relative errors decrease with decrease in the value of h.

11.2.3 Prefactor in Global Error

The global error $E(x(b), h) = \frac{1}{2}(b - a)x''(c)h$ in the previous subsection is simply obtained by summing all local truncation errors from $t = a$ to $t = b$. In [5] it has been shown that the prefactor $\frac{1}{2}(b - a)x''(c)$ violates the actual error.

Denote the global error from $t = a$ to $t = b$ as $E_m = \mathcal{E}$. It is to be noted that $E_m = \mathcal{E}$ and $E_m = E = \frac{1}{2}(b - a)x''(c)h$ (obtained in the previous subsection) are not the same. From Eq. (11.21) E_{n+1} is written as

$$
\begin{aligned}
E_{n+1} &= X_{n+1} - x_{n+1} \\
&= X_n + hf(X_n, t_n) - x_n - hf(x_n, t_n) - \frac{1}{2}h^2 x''(c_n) \\
&= E_n + h\left[f(X_n, t_n) - f(x_n, t_n)\right] - \frac{1}{2}h^2 x''(c_n).
\end{aligned}
\tag{11.26}
$$

To determine $\mathcal{E}$ this formula has to be used with a care. One cannot simply sum all the single-step local truncation errors to get the global error (why?). It has been pointed out in [5] that

$$
|\mathcal{E}| \leq \frac{hM}{2L}\left(e^{L(b-a)} - 1\right),
\tag{11.27}
$$

where L is a constant and $x''(c_n)$ are bounded by M. The errors given by both (11.25) and (11.27) are $O(h)$. However, the prefactors in E and $\mathcal{E}$ are different.

For further understanding, let us take the problem of solving the equation [5]

$$
x' = 1 - t^2 + x, \quad x(0) = 0.5, \quad t \in [0, 2]
\tag{11.28}
$$

TABLE 11.3

The variation of $|x_{\text{exact}} - x_{\text{numer}}|$ and $x''(t)$ with t for the Eq. (11.28) with time step $h = 0.1$.

t	$\begin{array}{c}\|x_{\text{exact}} \\ -x_{\text{numer}}\|\end{array}$	$x''(t)$	t	$\begin{array}{c}\|x_{\text{exact}} \\ -x_{\text{numer}}\|\end{array}$	$x''(t)$
0.1	0.00741	1.44741	1.1	0.10979	0.49792
0.2	0.01530	1.38930	1.2	0.12300	0.33994
0.3	0.02367	1.32507	1.3	0.13671	0.16535
0.4	0.03255	1.25409	1.4	0.15090	−0.02760
0.5	0.04195	1.17564	1.5	0.16550	−0.24084
0.6	0.05188	1.08894	1.6	0.18047	−0.47652
0.7	0.06235	0.99312	1.7	0.19571	−0.73697
0.8	0.07338	0.88723	1.8	0.21113	−1.02482
0.9	0.08497	0.77020	1.9	0.22660	−1.34295
1.0	0.09710	0.64086	2.0	0.24197	−1.69453

by the Euler method. The exact solution of this equation is (derivation of the solution is left as an exercise to the readers)

$$x(t) = 1 + 2t + t^2 - 0.5e^t. \tag{11.29}$$

Its second derivative is $x''(t) = 2 - 0.5e^t$. For the chosen equation $L = 1$ and $x''(c_n)$ is bounded by $M = 0.5e^2 - 2 = \frac{1}{2}\left(e^2 - 4\right)$. Then,

$$|\mathcal{E}| \le \frac{h}{4}\left(e^2 - 4\right)\left(e^2 - 1\right). \tag{11.30}$$

For the choice $h = 0.1$ the error bound is $|\mathcal{E}| \le 0.54132$. The error cannot be larger than $|\mathcal{E}|$.

What is the (global) error predicted by the Euler method? What is the actual difference between the exact value of $x(t)$ and the solution computed by the Euler method at $t = 2$? Since the exact solution of Eq. (11.28) is known, these two quantities can be easily computed. Table 11.3 presents $|x_{\text{exact}} - x_{\text{numer}}|$ and $x''(t)$ as a function of t with the time step $h = 0.1$ for Eq. (11.28). Here, x_{exact} is the exact solution given by Eq. (11.29), x_{numer} is the Euler method predicted solution of Eq. (11.28) and $x''(t)$ is the second derivative of the exact solution given by Eq. (11.29).

From Table 11.3 the global error at $t = 2$ is $d = |x_{\text{exact}} - x_{\text{numer}}| = 0.24197$. Compare this error with the global error $|\mathcal{E}|$ given by Eq. (11.30) and $|E| \le |\frac{1}{2}(b-a)x''(c)h|$ (obtained by adding the local errors). d is 2.24 times less than the error bound $|\mathcal{E}|$. For E choose the maximum value of $|x''(c)|$ where $c \in [0, 2]$. From Table 11.3 the maximum value of $|x''|$ is 1.69453 and hence $|E| \le 0.16945$. The actual global error $d = 0.24197$ is higher than $|E|$. That is, d is not $\le |E|$ but roughly 1.4 times higher than $|E|$. Note that $d \le |\mathcal{E}| = 0.54312$. The point is that simply summing the local truncation errors does not give an appropriate global error bound. This is because the single-step errors are not independent from each other [5].

11.2.4 Stability Analysis

In a numerical integration method if the step size is too small then an enormous amount of computation time is required and round-off error result. Computation time can be reduced by taking a higher step size. When the step size is above a critical value a method numerically produces a solution which no longer falls close to the exact solution. Then, the method is said to be *numerically unstable*. In general, a method is said to be *stable* if

$$\lim_{h \to 0} x_n = X_n, \quad n = 0, 1, \ldots, m. \tag{11.31}$$

Here, $mh = b - a$ is kept constant so that t_n is always the same point. It is possible to find the upper bound on the step size of a method for its stability.

To study the stability of a method it is necessary to analyze the solution of Eq. (11.4) in the neighbourhood of a point say, t' (denote the corresponding value of x as $x'(t')$). The behaviour of the solution of Eq. (11.4) near t' can be determined from the linearization of it. Replacing f by its Taylor series expansion about $x'(t')$ and keeping only the first-order terms leads to

$$\begin{aligned}
\dot{x} &= f(x', t') + (x - x') \left.\frac{\partial f}{\partial x}\right|_{t'} + (t - t') \left.\frac{\partial f}{\partial t}\right|_{t'} \\
&= \lambda x + c,
\end{aligned} \tag{11.32a}$$

where

$$\lambda = \left.\frac{\partial f}{\partial x}\right|_{t'}, \quad c = f(x', t') + (t - t') \left.\frac{\partial f}{\partial t}\right|_{t'} - x' \left.\frac{\partial f}{\partial x}\right|_{t'}. \tag{11.32b}$$

The change of variable $x = x' - c/\lambda$ and dropping the prime bring the above equation into the linear equation

$$\dot{x} = \lambda x. \tag{11.33}$$

The analytical solution (X) of Eq. (11.33) is

$$X(t) = \alpha e^{\lambda t}, \tag{11.34}$$

where $\alpha = X(0)$. From Eq. (11.34)

$$x_1 = X(h) = \alpha e^{\lambda h}. \tag{11.35}$$

Writing the Taylor series approximation of $e^{\lambda h}$ gives

$$\begin{aligned}
x_1 &= X(h) \\
&= \alpha e^{\lambda h} \\
&= \alpha \left[1 + \lambda h + \frac{1}{2}(\lambda h)^2 + \cdots\right]
\end{aligned} \tag{11.36a}$$

and

$$x_{n+1} = x((n+1)h) = x_n E(\lambda h), \quad n = 0, 1, \ldots \tag{11.36b}$$

where $E(\lambda h)$ is some approximation of $e^{\lambda h}$ and is called *stability function*. Defining

$$e_n = X_n - x_n \tag{11.37}$$

Eq. (11.36b) can be written as

$$X_{n+1} - e_{n+1} = (X_n - e_n) E(\lambda h). \tag{11.38}$$

From Eq. (11.35) write

$$x_{n+1} = x_1 e^{\lambda n h} = x_2 e^{\lambda(n-1)h} = \cdots = x_n e^{\lambda h} = X_n e^{\lambda h}. \tag{11.39}$$

The above equation becomes

$$e_{n+1} = \left(e^{\lambda h} - E(\lambda h)\right) X_n + E(\lambda h)e_n. \tag{11.40}$$

The first part in Eq. (11.40) is the local truncation error. By choosing higher-order methods the error can be made as small as possible. The second term is the propagation error from time t_n to t_{n+1}. This term will not diverge when $|E(\lambda h)| \leq 1$. Now, the following two cases to be considered.

(1)   If $|E(\lambda h)| \leq 1$ the method is said to be *absolutely stable*.

(2)   If $|E(\lambda h)| \leq e^{\lambda h}$ the method is said to be *relatively stable*.

When the real part of λ is negative ($\lambda_R < 0$) then the exact solution decays to zero as $t \to \infty$ and the important condition is (1). The region of absolute stability for a method is the set of values of h (which is real and positive) and λ (complex) for which $x_n \to 0$ as $n \to \infty$. Thus, for a method to be stable the condition is $|E(\lambda h)| \leq 1$. That is, the values of h and λ are to be set in such a way that $|E(\lambda h)| \leq 1$.

The above treatment is for a first-order Eq. (11.4). For a system of nonlinear equations $\dot{\mathbf{x}} = F(\mathbf{x}, t)$, where F is nonlinear, its linearized system of equations $\dot{\mathbf{x}} = A\mathbf{x}$, where A is $m \times m$ constant matrix and $\mathbf{x}$ has m components, the condition for a solution of the nonlinear system to be stable is the real part of all the eigenvalues of the matrix A should be negative.

For the Euler method

$$
\begin{aligned}
x_{n+1} &= x_n + hf_n \\
&= (1 + \lambda h)x_n \\
&= E(\lambda h)x_n
\end{aligned}
\tag{11.41}
$$

giving $E(\lambda h) = 1 + \lambda h$. Write $\lambda h = \bar{h} = \bar{h}_R + i\bar{h}_I$. Then,

$$E(\lambda h) = 1 + \bar{h}_R + i\bar{h}_I, \tag{11.42a}$$

$$|E| = \sqrt{\left(1 + \bar{h}_R\right)^2 + \bar{h}_I^2}. \tag{11.42b}$$

The requirement is $|E| \leq 1$. When λ is real then $|E| = |1 + \bar{h}_R| \leq 1$. This condition is satisfied for $-2 < \lambda h < 0$. For complex λ the condition $|E| \leq 1$ is possible if λh lies inside a circle of radius unity centered at $\lambda h = (-1, 0)$.

11.2.5   Improved Euler Method

In the Euler method, the value of the integral in Eq. (11.17) is approximated into the area of the stripped rectangle of Fig. 11.1c. In the improved Euler method, a better approximation for this integral is realized by means of the trapezoid (stripped region in Fig. 11.3) rather than the rectangle in Fig. 11.1c. The exact area of this trapezoid is the length h of the base multiplied by the average of the heights of the two sides:

$$h\frac{[f\left(X_n, t_n\right) + f\left(X_{n+1}, t_{n+1}\right)]}{2}. \tag{11.43}$$

The integral in (11.17) is approximated as

$$\int_{t_n}^{t_{n+1}} f(X, t)\,\mathrm{d}t = \frac{1}{2}h\left[f\left(X_n, t_n\right) + f\left(X_{n+1}, t_{n+1}\right)\right]. \tag{11.44}$$

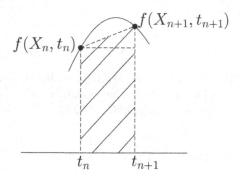

FIGURE 11.3
Approximation of the stripped area of Fig. 11.1b.

Since X is not known replace X_n by x_n and write

$$\int_{t_n}^{t_{n+1}} f(X,t)\, dt = \frac{1}{2}h\left[f(x_n,t_n) + f(x_n + hf(x_n,t_n),t_{n+1})\right]. \tag{11.45}$$

Then,

$$x_{n+1} = x_n + \frac{1}{2}h\left[f(x_n,t_n) + f(x_n + hf_n,t_{n+1})\right]. \tag{11.46}$$

The result is the improved Euler algorithm

$$p_{n+1} = x_n + hf(x_n,t_n), \tag{11.47a}$$

$$x_{n+1} = x_n + \frac{1}{2}h\left[f(x_n,t_n) + f(p_{n+1},t_{n+1})\right]. \tag{11.47b}$$

Defining

$$k_1 = hf(x_n,t_n), \tag{11.48a}$$

$$k_2 = hf(x_n + k_1,t_n + h) \tag{11.48b}$$

the improved Euler algorithm takes the form

$$x_{n+1} = x_n + \frac{1}{2}(k_1 + k_2). \tag{11.48c}$$

In Subsection 11.3.2, the improved Euler algorithm is shown to be exactly the second-order Runge–Kutta method. Solving Eq. (11.18) by this method is taken up in the next section.

Equation (11.47a) is simply the Euler formula and the last term in Eq. (11.47b) is the trapezoidal rule. At a point t_{n+1} the Euler method is used to predict the value of x_{n+1} (and hence p_{n+1} is called *predictor*) while the trapezoidal rule is employed to get a correction to the final value. The corrected value of x_{n+1} is re-substituted in Eq. (11.47b) to get a re-correction. In this way, successive re-correction can be obtained. The above method is also called the *Euler predictor-corrector method*.

The approximation of the integral in Eq. (11.17) involves trapezoidal rule. Hence, the error term for this approximation is simply the error term for the trapezoidal rule and is

$$E_{n+1} = -\frac{1}{12}h^3 x''(c_n). \tag{11.49}$$

The final global error after m steps is

$$
\begin{aligned}
E &= -\sum_{n=1}^{m} \frac{1}{12} h^3 x''(c_n) \\
&\approx -\frac{1}{12}(b-a)h^2 x''(c) \\
&\approx O\left(h^2\right).
\end{aligned}
\tag{11.50}
$$

11.3 Runge–Kutta Methods

The Euler method is usually not recommended for practical purposes because it is not very accurate when compared to other methods and further it is not very stable. The German mathematicians Runge and Kutta derived integration formulas for which the truncation error is $O(h^{N+1})$ while the final global error is $O(h^N)$. The Runge–Kutta methods can be constructed for any order p. These methods are closely related to the Taylor series expansion of $f(x, t)$ but differentiation of f is not necessary in the use of the methods. Essentially, they construct a solution over an interval by combining the information from a few Euler-style steps, each of which needs only one calculation of the function f and then equate the obtained result to a Taylor series expansion up to some higher order. The present section discusses the features of the Runge–Kutta methods. First, the algorithm for the Nth order Runge–Kutta method is derived. Then, the second-order and fourth-order methods are treated.

11.3.1 Nth Order Runge–Kutta Method

Let us first obtain the Taylor series approximation of the solution of Eq. (11.4). Denote $X(t) = X(t_n)$, $t_n = a + nh$, $n = 0, 1, \ldots$ as the exact solution of Eq. (11.4) with the initial condition $X(a) = x(a) = x_0$. The Taylor series expansion of the approximate solution $x(t)$ about $t = t_n$ is written as

$$
\begin{aligned}
x(t_n + h) &= x(t_n) + hx^{(1)}(t_n) + \frac{1}{2}h^2 x^{(2)}(t_n) + \cdots \\
&\quad + \frac{1}{N!} h^N x^{(N)}(t_n) + O\left(h^N\right),
\end{aligned}
\tag{11.51a}
$$

where

$$
x^{(N)}(t_n) = \frac{\mathrm{d}^N}{\mathrm{d}t^N} x(t_n).
\tag{11.51b}
$$

Defining $x_n = x(t_n)$ the above equation is rewritten as

$$
\begin{aligned}
x_{n+1} &= x_n + hx_n^{(1)} + \frac{1}{2}h^2 x_n^{(2)} + \cdots + \frac{1}{N!}h^N x_n^{(N)} + O\left(h^{N+1}\right) \\
&= x_n + \sum_{i=1}^{N} \frac{1}{i!} h^i x_n^{(i)} + O\left(h^{N+1}\right).
\end{aligned}
\tag{11.52}
$$

The derivatives of x can be computed recursively as

$$
\begin{aligned}
x^{(1)}(t) &= f, \\
x^{(2)}(t) &= f_t + f_x x' = f_t + f_x f, \\
x^{(3)}(t) &= f_{tt} + 2f_{tx}f + f^2 f_{xx} + ff_x(f_t + f_x), \\
x^{(4)}(t) &= \left(f_{ttt} + 3f_{ttx} + 3f^2 f_{txx} + f^3 f_{xxx}\right) + f_t\left(f_{tt} + 2ff_{tx} + f^2 f_{xx}\right) \\
&\quad +3\left(f_t + ff_x\right)\left(ff_{xx} + f_{tx}\right) + f_x^2\left(f_t + ff_x\right)
\end{aligned}
\tag{11.53}
$$

and in general

$$
x^{(N)}(t) = \left(\frac{\partial}{\partial t} + f\frac{\partial}{\partial x}\right)^{N-1} f(x,t).
\tag{11.54}
$$

Equation (11.52) can be considered as an approximate solution to Eq. (11.4). The global error in the approximation is $O(h^{N+1})$ if N terms in Eq. (11.52) are taken into account. A drawback of the Taylor method is the computation of higher derivatives. However, from the Taylor method, it is possible to obtain methods with global error of the order of h^N but avoiding the computation of derivatives. Such methods are called the *Runge–Kutta* methods of order N.

To obtain a N-stage Runge–Kutta method write

$$
x_{n+1} = x_n + h\phi(x_n, t_n; h), \quad \phi(x_n, t_n; h) = \sum_{i=1}^{N} \omega_i k_i.
\tag{11.55}
$$

It can be written as

$$
x_{n+1} = x_n + h\sum_{i=1}^{N} \omega_i k_i,
\tag{11.56a}
$$

where

$$
k_i = f\left(x_n + h\sum_{j=1}^{i-1} \alpha_{ij}k_j, t_n + h\beta_i\right)
\tag{11.56b}
$$

and $\beta_1 = 0$. For Nth order method determine the values of α_{ij}, β_i and ω_i for $1 \le (i,j) \le N$ so that $N+1$ terms in Eq. (11.56a) and (11.52) are identical. For this purpose write the Taylor expansion of Eq. (11.56a) about (x_n, t_n) assuming $x(t_n) = X(t_n)$ and compare this with Eq. (11.52).

When $N = 1$

$$
x_{n+1} = x_n + h\omega_1 k_1 = x_n + hf(x_n, t_n)
\tag{11.57}
$$

which is the Euler formula. The following subsections discuss the cases with $N = 2$ and 4.

11.3.2   Second-Order Runge–Kutta Method

The choice $N = 2$ in Eq. (11.51) gives

$$
x_{n+1} = x_n + hf + \frac{1}{2}h^2 \frac{df}{dt}\bigg|_{(x_n, t_n)} + O\left(h^3\right),
\tag{11.58}
$$

where $f = f(x_n, t_n)$. Inserting the Taylor expansion of f' gives

$$
x_{n+1} = x_n + hf + \frac{1}{2}h^2\left[f_t + f_x f\right] + O\left(h^3\right),
\tag{11.59}
$$

where f and its partial derivatives are evaluated at (x_n, t_n). From Eqs. (11.56) write (with $\beta_1 = 0$)

$$x_{n+1} = x_n + h\omega_1 k_1 + h\omega_2 k_2, \tag{11.60a}$$

where

$$
\begin{aligned}
k_1 &= f(x_n, t_n), & \text{(11.60b)}\\
k_2 &= f(x_n + h\alpha_{21}k_1, t_n + h\beta_2)\\
&= f(x_n + h\alpha_{21}k_1, t_n) + h\beta_2 \frac{\partial}{\partial t} f(x_n + h\alpha_{21}k_1, t_n) + O(h^2)\\
&= f(x_n, t_n) + h\alpha_{21}k_1 f_x + h\beta_2 f_t + O(h^2)\\
&= f + h\alpha_{21} f f_x + h\beta_2 f_t + O(h^2) & \text{(11.60c)}
\end{aligned}
$$

with f_t and f_x being evaluated at $(x, t) = (x_n, t_n)$. Now, Eq. (11.60a) becomes

$$
\begin{aligned}
x_{n+1} &= x_n + h\omega_1 f + h\omega_2 (f + h\beta_2 f_t + h\alpha_{21} f f_x) + O(h^3)\\
&= x_n + h(\omega_1 + \omega_2) f + h^2 \omega_2 \beta_2 f_t + h^2 \omega_2 \alpha_{21} f f_x. & \text{(11.61)}
\end{aligned}
$$

Comparison of Eqs. (11.59) and (11.61) gives

$$\omega_1 + \omega_2 = 1, \quad \omega_2 \beta_2 = 1/2, \quad \omega_2 \alpha_{21} = 1/2. \tag{11.62}$$

For the four unknowns ω_1, ω_2, α_{21} and β there are only three equations. Therefore, solve Eqs. (11.62) in terms of ω_2 and find $\omega_1 = 1 - \omega_2$, $\beta_2 = \alpha_{21} = 1/(2\omega_2)$. Then, the choice $\omega_2 = 1/2$ gives

$$
\begin{aligned}
k_1 &= f(x_n, t_n), & \text{(11.63a)}\\
k_2 &= f(x_n + hk_1, t_n + h), & \text{(11.63b)}\\
x_{n+1} &= x_n + \frac{1}{2} h(k_1 + k_2). & \text{(11.63c)}
\end{aligned}
$$

Equations (11.63) can be rewritten as

$$
\begin{aligned}
k_1 &= hf(x_n, t_n), & \text{(11.64a)}\\
k_2 &= hf(x_n + k_1, t_n + h), & \text{(11.64b)}\\
x_{n+1} &= x_n + \frac{1}{2}(k_1 + k_2). & \text{(11.64c)}
\end{aligned}
$$

This is called the *second-order Runge–Kutta* method. The algorithm given by Eqs. (11.64) is identical to the improved Euler formula given by Eqs. (11.48).

The local truncation error of the method is $O(h^3)$. It can be shown that the second-order Runge–Kutta method reduces to the trapezoidal rule when $f(x, t)$ is independent of x. In fact, each Runge–Kutta method reduces to a quadrature formula. Application of the trapezoidal rule to evaluate the integral in $x = \int f dt$ from t_n to t_{n+1} gives

$$
\begin{aligned}
x(t_{n+1}) - x(t_n) &= \int_{t_n}^{t_{n+1}} f(x, t)\, dt\\
&= \frac{1}{2} h[f(x(t_n), t_n) + f(x(t_n + h), t_n + h)]. & \text{(11.65)}
\end{aligned}
$$

Comparison of Eqs. (11.65) and (11.64) gives

$$f(x(t_n), t_n) + f(x(t_n + h), t_n + h) = k_1 + k_2$$

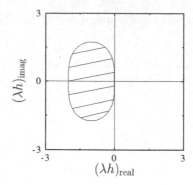

FIGURE 11.4
Stability boundary of the second-order Runge–Kutta method in the complex λh plane. The method is stable in the stripped region.

or

$$f\left(x\left(t_n + h\right), t_n + h\right) = k_2. \tag{11.66}$$

When $f(x, t_n + h)$ is independent of x then $f(x, t_n + h) = f(t_n + h) = k_2$ and Eq. (11.65) becomes

$$x\left(t_{n+1}\right) - x\left(t_n\right) = \frac{1}{2} h\left[f\left(t_n\right) + f\left(t_{n+1}\right)\right]. \tag{11.67}$$

The right-side of Eq. (11.67) is identified as the trapezoidal rule. The formula (11.64) can be viewed as the trapezoidal rule when the value of k_2 is used instead of $f(x(t_n + h), t_n + h)$. Therefore, the local truncation error in the second-order Runge–Kutta method is approximately the error involved in the trapezoidal rule and is $-h^3 x''(c_1)/12$ where $t_n < c_1 < t_{n+1}$ and the global error is $O(h^2)$.

Next, proceed to study the stability of the second-order Runge–Kutta method. Write

$$
\begin{aligned}
x_{n+1} &= x_n + \frac{1}{2} h\left(k_1 + k_2\right) \\
&= x_n + \frac{1}{2} h\left[f\left(x_n, t_n\right) + f\left(x_n + h k_1, t_n + h\right)\right] \\
&= x_n + \frac{1}{2} h f\left(x_n, t_n\right) + \frac{1}{2} f\left(x_n + h k_1, t_n + h\right) \\
&= x_n + \frac{1}{2} h f + \frac{1}{2} h\left[f + h k_1 f_x + h f_t\right] \\
&= x_n + h f + \frac{1}{2} h^2 k_1 f_x + \frac{1}{2} h^2 f_t \\
&= x_n + h \lambda x_n + \frac{1}{2} h^2 f f_x + \frac{1}{2} h^2 f_t \\
&= x_n + h \lambda x_n + \frac{1}{2} h^2 \lambda x_n f_x + \frac{1}{2} h^2 f_t \\
&= x_n + h \lambda x_n + \frac{1}{2} h^2 \lambda^2 x_n \\
&= x_n\left(1 + \lambda h + \frac{1}{2}(\lambda h)^2\right) \\
&= x_n E(\lambda h), \tag{11.68}
\end{aligned}
$$

where

$$E(\lambda h) = 1 + \lambda h + \frac{1}{2}(\lambda h)^2. \tag{11.69}$$

The stability condition is $|E| \leq 1$ with λh being generally complex. Figure 11.4 depicts the region of absolute stability for the second-order Runge–Kutta method. For real λ the method is absolutely stable for $-2 < \lambda h < 0$.

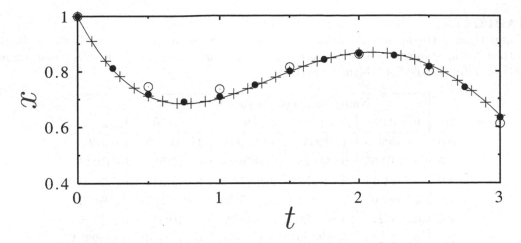

FIGURE 11.5

Comparison of the numerical solution of $\dot{x} = -x + \sin t$ obtained with the step sizes $h = 0.1$ (marked by +), 0.25 (marked by solid circles) and 0.5 (marked by open circles) by the second-order Runge–Kutta method. The continuous curve represents the exact solution. The initial condition used is $x(0) = 1$.

Example:

Consider the overdamped linear harmonic oscillator Eq. (11.18).

(a) Calculate $x(0.1)$ by the second-order Runge–Kutta method with step size $h = 0.1$.

(b) Solve the equation in the interval $t \in [0, 3]$ with $h = 0.01, 0.10, 0.25$ and 0.5.

(c) Compare the final global error at $t = 1$ for $h = 0.001, 0.01, 0.025, 0.05, 0.1, 0.25$ and 0.5 by calculating percentage of relative error.

(a) For $h = 0.1$

$$
\begin{aligned}
k_1 &= hf(x(0), t = 0) = -0.10, \\
k_2 &= hf(x(0) + k_1, 0 + h) = f(0.9, 0.1) = -0.0800167, \\
x(0.1) &= x(0) + \frac{1}{2}(k_1 + k_2) \\
&= 1.0 + \frac{1}{2}(-0.10 - 0.0800167) \\
&= 0.9099917
\end{aligned}
$$

while the exact solution is 0.9096707. The difference is 0.0003210 and the percentage of relative error is 0.035 which is much lower than the value obtained for the Euler method.

(b) Figure 11.5 shows the plot of $x(t)$ with $h = 0.1, 0.25$ and 0.5 along with the exact solution. Table 11.4 gives the numerical values of x at certain values of t for different values of h.

(c) Table 11.5 presents the numerically computed final global error at $t = 1$ for different values of h. The final global error and the relative error decrease with decrease in the value of h.

TABLE 11.4

Comparison of the numerical solution obtained by the second-order Runge–Kutta method with different step sizes for the equation $\dot{x} = -x + \sin t$, $x(0) = 1$. The exact solution is $x(t) = (x(0) + 0.5)\,e^{-t} + 0.5(\sin t - \cos t)$.

t	Numerically computed x with				Exact x
	$h = 0.01$	$h = 0.10$	$h = 0.25$	$h = 0.50$	
0.0	1.0000000	1.0000000	1.0000000	1.0000000	1.0000000
0.5	0.7107265	0.7116925	0.7176344	0.7448564	0.7107175
1.0	0.7024114	0.7032733	0.7087773	0.7358312	0.7024035
1.5	0.7980765	0.7983631	0.8005629	0.8144521	0.7980741
2.0	0.8657212	0.8653533	0.8636868	0.8610438	0.8657251
2.5	0.8229270	0.8220796	0.8174886	0.8014326	0.8229354
3.0	0.6402270	0.6392140	0.6334812	0.6109844	0.6402369

11.4　Fourth-Order Runge–Kutta Method

In the second-order Runge–Kutta method the local truncation error is $O(h^3)$. The present section considers a Runge–Kutta algorithm with the local truncation error $O(h^5)$.

11.4.1　Integration Formula

For $N = 4$, Eq. (11.52) takes the form

$$x_{n+1} = x_n + hf + \frac{1}{2}h^2 f' + \frac{1}{6}h^3 f'' + \frac{1}{24}h^4 f''' + O\left(h^5\right), \tag{11.70}$$

TABLE 11.5

Final global error ($|x_{\text{exact}} - x_{\text{numer}}|$) and the percentage of relative error at $t = 1$ for Eq. (11.18) with the second-order Runge–Kutta (RK) method. The exact solution at $t = 1$ is 0.7024035.

h	x by the second-order RK method	Final global error at $t = 1$	% of relative error
0.001	0.7024036	0.0000001	0.000
0.010	0.7024114	0.0000079	0.001
0.025	0.7024538	0.0000503	0.007
0.050	0.7026100	0.0002065	0.029
0.100	0.7032733	0.0008698	0.124
0.250	0.7087773	0.0063738	0.907
0.500	0.7358312	0.0334277	4.759

whereas Eq. (11.56) becomes

$$x_{n+1} = x_n + h\omega_1 k_1 + h\omega_2 k_2 + h\omega_3 k_3 + h\omega_4 k_4, \tag{11.71a}$$

where

$$k_1 = f(x_n, t_n), \tag{11.71b}$$
$$k_2 = f(x_n + h\alpha_{21}k_1, t_n + h\beta_2), \tag{11.71c}$$
$$k_3 = f(x_n + h\alpha_{31}k_1 + h\alpha_{32}k_2, t_n + h\beta_3), \tag{11.71d}$$
$$k_4 = f(x_n + h\alpha_{41}k_1 + h\alpha_{42}k_2 + h\alpha_{43}k_3, t_n + h\beta_4). \tag{11.71e}$$

Expanding k_2, k_3 and k_4, substituting them in Eq. (11.71a) and then comparing the resultant equation with Eq. (11.70) one obtains the following system of nonlinear algebraic equations:

$$\alpha_{21} = \beta_2, \tag{11.72a}$$
$$\alpha_{31} + \alpha_{32} = \beta_3, \tag{11.72b}$$
$$\alpha_{41} + \alpha_{42} + \alpha_{43} = \beta_4, \tag{11.72c}$$
$$\omega_1 + \omega_2 + \omega_3 + \omega_4 = 1, \tag{11.72d}$$
$$\omega_2\beta_2 + \omega_3\beta_3 + \omega_4\beta_4 = \frac{1}{2}, \tag{11.72e}$$
$$\omega_2\beta_2^2 + \omega_3\beta_3^2 + \omega_4\beta_4^2 = \frac{1}{3}, \tag{11.72f}$$
$$\omega_3\beta_2^3 + \omega_3\beta_3^3 + \omega_4\beta_4^3 = \frac{1}{4}, \tag{11.72g}$$
$$\omega_3\beta_2\alpha_{32} + \omega_4(\beta_2\alpha_{42} + \beta_3\alpha_{43}) = \frac{1}{6}, \tag{11.72h}$$
$$\omega_3\beta_2\beta_3\alpha_{32} + \omega_4\beta_4(\beta_2\alpha_{42} + \beta_3\alpha_{43}) = \frac{1}{18}, \tag{11.72i}$$
$$\omega_3\beta_2^2\alpha_{32} + \omega_4(\beta_2^2\alpha_{42} + \beta_3^2\alpha_{43}) = \frac{1}{12}, \tag{11.72j}$$
$$\omega_4\beta_2\alpha_{32}\alpha_{43} = \frac{1}{24}. \tag{11.72k}$$

For 13 unknowns there are only 11 equations. Therefore, two unknowns can be fixed. The most useful choice is $\beta_2 = 1/2$ and $\alpha_{31} = 0$. Then, the solution of Eqs. (11.72) is

$$\alpha_{21} = 1/2, \; \alpha_{32} = 1/2, \; \alpha_{41} = 0, \; \alpha_{42} = 0, \; \alpha_{43} = 1, \tag{11.73a}$$
$$\beta_3 = 1/2, \; \beta_4 = 1, \; \omega_1 = 1/6, \; \omega_2 = 1/3, \tag{11.73b}$$
$$\omega_3 = 1/3, \; \omega_4 = 1/6. \tag{11.73c}$$

With these values of the parameters, the fourth-order Runge–Kutta formula for the first-order differential equation is given by

$$k_1 = f(x_n, t_n), \tag{11.74a}$$
$$k_2 = f\left(x_n + \frac{1}{2}hk_1, t_n + \frac{1}{2}h\right), \tag{11.74b}$$
$$k_3 = f\left(x_n + \frac{1}{2}hk_2, t_n + \frac{1}{2}h\right), \tag{11.74c}$$
$$k_4 = f(x_n + hk_3, t_n + h), \tag{11.74d}$$
$$x_{n+1} = x_n + \frac{1}{6}h(k_1 + 2k_2 + 2k_3 + k_4). \tag{11.74e}$$

The above formula can also be written as

$$k_1 = hf(x_n, t_n), \tag{11.75a}$$

$$k_2 = hf\left(x_n + \frac{1}{2}k_1, t_n + \frac{1}{2}h\right), \tag{11.75b}$$

$$k_3 = hf\left(x_n + \frac{1}{2}k_2, t_n + \frac{1}{2}h\right), \tag{11.75c}$$

$$k_4 = hf(x_n + k_3, t_n + h), \tag{11.75d}$$

$$x_{n+1} = x_n + \frac{1}{6}(k_1 + 2k_2 + 2k_3 + k_4). \tag{11.75e}$$

For simplicity use the form given by Eqs. (11.75). The fourth-order method is the most popular and widely recommended method. It is quite accurate, stable and easy to program. It is not necessary to go for higher-order methods because accuracy can be improved by choosing either a small step size or using an adaptive size (see next section).

Another fourth-order Runge–Kutta method which uses the slope f at 5 different points is the *Runge–Kutta–Merson method*. Its algorithm is the following:

$$k_1 = hf(x_n, t_n), \tag{11.76a}$$

$$k_2 = hf\left(x_n + \frac{1}{3}k_1, t_n + \frac{1}{3}h\right), \tag{11.76b}$$

$$k_3 = hf\left(x_n + \frac{1}{6}k_1 + \frac{1}{6}k_2, t_n + \frac{1}{3}h\right), \tag{11.76c}$$

$$k_4 = hf\left(x_n + \frac{1}{8}k_1 + \frac{3}{8}k_3, t_n + \frac{1}{2}h\right), \tag{11.76d}$$

$$k_5 = hf\left(x_n + \frac{1}{2}k_1 - \frac{3}{2}k_3 + 2k_4, t_n + h\right), \tag{11.76e}$$

$$x_{n+1} = x_n + \frac{1}{6}(k_1 + 4k_4 + k_5) + O\left(h^5\right). \tag{11.76f}$$

11.4.2   Geometric Description

Let us briefly discuss the geometric meaning of the components in the formula (11.75) [6-8]. Essentially, the Runge–Kutta method iterates the t values by adding a step size h at each iteration. *What about x-iteration formula?* Equation (11.75) implies that x-iteration is a weighted average of the four values k_1, k_2, k_3 and k_4. k_1 and k_4 are given a weight of $1/6$ in the weighted average whereas k_2 and k_3 are weighted by the factor $1/3$ which is twice as heavily as k_1 and k_4.

What are k_1, k_2, k_3 and k_4?   The quantity k_1 is h times $x'(t) = f(x, t)$ evaluated at t_n. It is the vertical jump from the current time t_n to the next Euler computed point along the numerical solution. Next, consider k_2. For k_2, the value of t at which f is calculated is $t_n + h/2$ and is the midpoint of the interval h between the times t_n and $t_{n+1} = t_n + h$. The x-value corresponding to k_2 is $x_n + k_1/2$ and is the current x-value and half of the Euler calculated value of x. $x_n + k_1/2$ is vertically half-way up from the current point to the Euler calculated next point. $f(x_n + k_1/2, t_n + h/2)$ is an estimation of the slope x' of the solution at $h/2$. Therefore, $hf(x_n + k_1/2, t_n + h/2)$ gives an estimation of the jump in x in the actual solution over the full length of the interval.

The meaning of k_3 is similar to k_2, however, instead of k_1 used in k_2 now k_2 is used in k_3. The f value is another estimation of the interval. The x-value of the midpoint is based on the jump in x predicted with k_2. This slope multiplied by h gives another estimation of the jump in x made by the actual solution across the full length of the interval.

For k_4 the value of f is at $t = t_n + h$. This time value is the extreme right of the interval considered. Here, f is evaluated with $x = x_n + k_3$ and is an estimate of the value of x at the right end of the interval, considering the jump in x found by k_3. The calculated f value is then multiplied by h.

Note that k_1 uses the Euler method; k_2 and k_3 use estimates of the slope x' of the solution x at the middle point; k_4 uses an estimate of the slope at the right extreme point. Each k_i uses the k_{i-1} as a basis for its calculation of the jump in x.

Use of the Simpson's 1/3-rule with step size $h/2$ to evaluate the integral in the equation

$$x = \int_{t_n}^{t_{n+1}} f \, dt \text{ gives}$$

$$\int_{t_n}^{t_{n+1}} f(x,t) \, dt \approx \frac{1}{3}\frac{h}{2} \left[f(x_n, t_n) + 4f(x(t_n + h/2), t_n + h/2) \right.$$
$$\left. + f(x(t_{n+1}), t_{n+1}) \right]. \tag{11.77}$$

Comparison of Eqs. (11.77) and (11.75) gives

$$f(x(t_n + h/2), t_n + h/2) = (k_2 + k_3)/2. \tag{11.78}$$

Therefore, Eq. (11.75) can be treated as the Simpson's rule, where instead of $f(x(t_n + h/2), t_n + h/2)$ the average of k_2 and k_3 is considered.

11.4.3   Error Versus Step Size and Stability

For the fourth-order Runge–Kutta method, referring to Eq. (11.70), the local truncation error is $T_{n+1} \approx O(h^5)$. Equation (11.68) can be viewed as the Simpson's 1/3-rule with step size $h/2$. So, take T_{n+1} as the error in the Simpson's rule. It is simply $-x^{(4)}(c_1)(h/2)^5/90$ (refer Eq. (10.30)). If $t \in [a,b]$ then the error (global) at the end point $t = b$ with $h = (b-a)/m$ is $O(h^4)$. For two different time steps, say h and $h/2$,

$$E(x(b), h) \approx \alpha h^4, \tag{11.79a}$$

$$E(x(b), h/2) \approx \frac{1}{16}\alpha h^4 \approx \frac{1}{16}E(x(b), h). \tag{11.79b}$$

Thus, if the step size is reduced by a factor of $h/2$ then the overall global error will be reduced by a factor of $1/16$.

For the test Eq. (11.33) the stability condition is

$$|E(\lambda h)| = \left| 1 + \lambda h + \frac{1}{2}(\lambda h)^2 + \frac{1}{3!}(\lambda h)^3 + \frac{1}{4!}(\lambda h)^4 \right| \le 1. \tag{11.80}$$

The stability region is shown in Fig. 11.6. For real λ the stability condition is $-2.795 < \lambda h < 0$. For pure imaginary λ the condition is $0 < |\lambda h| \le 2\sqrt{2}$.

Example:

Consider the first-order Eq. (11.18).

(a)   Calculate $x(0.1)$ by the fourth-order Runge–Kutta method with step size $h = 0.1$.

(b)   Solve it in the interval $t \in [0,3]$ with $h = 0.01, 0.10, 0.25$ and 0.5.

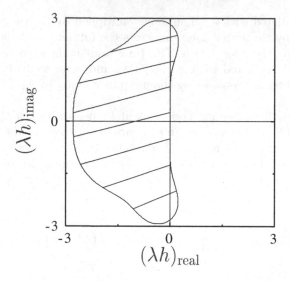

FIGURE 11.6
Stability boundary of the fourth-order Runge–Kutta method in the complex λh plane. The method is stable in the stripped region.

(c)   Compare the final global error at $t = 1$ for $h = 0.001, 0.01, 0.025, 0.05, 0.1, 0.25$ and 0.5 by calculating percentage of relative error.

(a) The fourth-order Runge–Kutta algorithm for $x(0.1)$ is

$$
\begin{aligned}
k_1 &= hf(x(0), t = 0), \\
k_2 &= hf(x(0) + k_1/2, 0.05), \\
k_3 &= hf(x(0) + k_2/2, 0.05), \\
k_4 &= hf(x(0) + k_3, 0.1), \\
x(0.1) &= x(0) + (k_1 + 2k_2 + 2k_3 + k_4)/6.
\end{aligned}
$$

Then,

$$
\begin{aligned}
k_1 &= 0.1 \times f(1, 0) = -0.1, \\
k_2 &= 0.1 \times f(0.95, 0.05) = 0.1 \times -0.9000208 = -0.090002, \\
k_3 &= 0.1 \times f(0.954999, 0.05) = 0.1 \times -0.9050198 = -0.0905019, \\
k_4 &= 0.1 \times f(0.909498, 0.1) = 0.1 \times -0.8096646 = -0.0809664, \\
x(0.1) &= 0.9096709
\end{aligned}
$$

while the exact solution is 0.9096707. The difference is -0.0000002 and the percentage of relative error is 2×10^{-5} which is negligible.

(b) Figure 11.7 shows the plot of $x(t)$ with $h = 0.1, 0.25$ and 0.5 along with the exact solution. Table 11.6 presents the numerical values of x at a few values of t for different values of h.

(c) Table 11.7 gives the numerically computed final global error at $t = 1$ for different values of h. The final global error and the relative error can be compared with the values obtained for the Euler and the second-order Runge–Kutta methods (Tables 11.2 and 11.5).

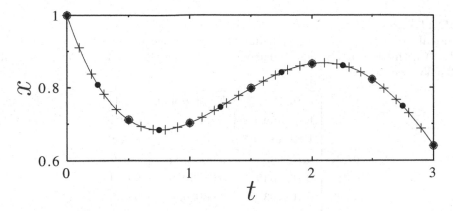

FIGURE 11.7
Comparison of the numerical solution of $\dot{x} = -x + \sin t$ with $x(0) = 1$ obtained for $h = 0.1$ (marked by +), 0.25 (solid circles) and 0.5 (open circles) by the fourth-order Runge–Kutta method. The continuous curve is the exact solution.

11.5   Convergence of Runge–Kutta Methods

An integration scheme is said to be *convergent* if the numerical solution converges to the exact solution as the step size h is decreased. That is,

$$\lim_{\substack{h \to 0 \\ nh = t - a}} x_n = X_n , \tag{11.81}$$

where X_n is the exact solution at $t = t_n$. Here, nh is kept constant so that t_n is always the same point. A method is said to be *consistent* if

$$\phi(x_n, t_n; h = 0) = f(x_n, t_n) . \tag{11.82}$$

TABLE 11.6
Comparison of the numerical solution obtained by the fourth-order Runge–Kutta method with different step sizes for the equation $\dot{x} = -x + \sin t$, $x(0) = 1$. The exact solution is $x(t) = (x(0) + 0.5)\, e^{-t} + 0.5(\sin t - \cos t)$.

	Numerically computed x with				
t	$h = 0.01$	$h = 0.10$	$h = 0.25$	$h = 0.50$	Exact x
0.0	1.0000000	1.0000000	1.0000000	1.0000000	1.0000000
0.5	0.7107175	0.7107179	0.7107387	0.7111511	0.7107175
1.0	0.7024035	0.7024040	0.7024248	0.7028600	0.7024035
1.5	0.7980741	0.7980744	0.7980863	0.7983662	0.7980741
2.0	0.8657251	0.8657250	0.8657260	0.8658017	0.8657251
2.5	0.8229354	0.8229352	0.8229274	0.8228304	0.8229354
3.0	0.6402369	0.6402366	0.6402247	0.6400323	0.6402369

TABLE 11.7
Final global error ($|x_{\text{exact}} - x_{\text{numer}}|$) and the percentage of relative error at $t = 1$ for Eq. (11.18) with the fourth-order Runge–Kutta (RK) method. The exact solution at $t = 1$ is 0.7024035.

h	x by the fourth-order RK method	Final global error at $t = 1$	% of relative error
0.001	0.7024035	0.0000000	0.000
0.010	0.7024035	0.0000000	0.000
0.025	0.7024035	0.0000000	0.000
0.050	0.7024035	0.0000000	0.000
0.100	0.7024039	0.0000005	0.000
0.250	0.7024248	0.0000213	0.003
0.500	0.7028600	0.0004565	0.065

This section shows that the Runge–Kutta methods are convergent by mainly following the ref. [9].

Let Eq. (11.4) has a unique solution $x(t)$ such that $f(x,t) = \phi(x,t;0)$ and write $x(t_n) \approx (x_{n+1} - x_n)/h$. When $x(t)$ is continuous on the closed interval $[a,b]$ then at some point c with $a < c < b$

$$\dot{x} = x'(c) = \frac{(x(b) - x(a))}{(b-a)} . \tag{11.83}$$

This is known as the mean-value theorem for the derivative x'. Choosing $a = t_n$ and $b = t_{n+1} = t_n + h$, the above equation is written as

$$x'(c) = (x(t_n + h) - x(t_n))/h$$

or

$$x(t_n + h) - x(t_n) = hx'(c), \tag{11.84}$$

where $t_n < c < t_{n+1}$. Define $c = t_n + \delta h$ with $0 < \delta < 1$. Then,

$$x(t_n + h) - x(t_n) = hx'(t_n + \delta h) . \tag{11.85}$$

Therefore,

$$\begin{aligned} x_{n+1} - x_n &= x(t_n + h) - x(t_n) \\ &= hx'(t_n + \delta h) \\ &= hf(x(t_n + \delta h), t_n + \delta h) . \end{aligned} \tag{11.86}$$

For the exact solution $X(t)$ with $X(a) = X_0$

$$X_{n+1} = X_n + h\phi(X_n, t_n; h) \tag{11.87}$$

and the global truncation error is

$$e_n = |x_n - X_n| . \tag{11.88}$$

Then,

$$
\begin{aligned}
e_{n+1} &= |x_{n+1} - X_{n+1}| \\
&= |x_n + hf\left(x\left(t_n + \theta h\right), t_n + \theta h\right) - X_n - h\phi\left(X_n, t_n; h\right)| \\
&\leq e_n + h\left[|f\left(x\left(t_n + \theta h\right), t_n + \theta h\right) - \phi\left(X_n, t_n; h\right)|\right] \\
&\leq e_n + h|f\left(x\left(t_n + \theta h\right), t_n + \theta h\right) - f\left(x_n, t_n\right) + \phi\left(x_n, t_n; 0\right) \\
&\quad - \phi\left(x_n, t_n; h\right) + \phi\left(x_n, t_n; h\right) - \phi\left(X_n, t_n; h\right)| .
\end{aligned}
\tag{11.89}
$$

Define

$$
\chi(h) = \max|f\left(x(t + \theta h), t + \theta h\right) - f(x, t)| \tag{11.90}
$$

and

$$
\xi(h) = \max|\phi(x, t; 0) - \phi(x, t; h)| , \tag{11.91}
$$

where $t \in [a, b]$ and $\theta \in [0, 1]$. In the limit $h \to 0$, $\chi(h) = 0$ and $\xi(h) = 0$.

If $\phi(x, t; h)$ is a continuous function and satisfies the Lipschitz condition then

$$
|\phi\left(x_n, t_n; h\right) - \phi\left(X_n, t_n; h\right)| \leq L|x_n - X_n|. \tag{11.92}
$$

Therefore,

$$
\begin{aligned}
e_{n+1} &\leq e_n + h\left(\chi(h) + \xi(h) + Le_n\right) \\
&\leq (1 + hL)e_n + h(\chi(h) + \xi(h))
\end{aligned}
$$

or

$$
e_n \leq (1 + hL)e_{n-1} + h(\chi(h) + \xi(h)) . \tag{11.93}
$$

Replacing e_{n-1} in terms of e_{n-2}; e_{n-2} in terms of e_{n-3} and so on lead to

$$
e_n \leq (1 + hL)^n e_0 + \frac{1}{L}\left[(1 + hL)^n - 1\right]\left[\chi(h) + \xi(h)\right] . \tag{11.94}
$$

Substitutions of $x_0 = X_0$ and $e_0 = 0$ give

$$
e_n \leq \frac{1}{L}\left[(1 + hL)^n - 1\right]\left[\chi(h) + \xi(h)\right] . \tag{11.95}
$$

In the limit $h \to 0$

$$
\lim_{\substack{h \to 0 \\ nh = t - a}} e_n \leq \lim_{\substack{h \to 0 \\ nh = t - a}} \frac{1}{L}\left[(1 + hL)^n - 1\right]\left[\chi(h) + \xi(h)\right] . \tag{11.96}
$$

Because of $\chi(h)$ and $\xi(h) \to 0$ as $h \to 0$

$$
\lim_{\substack{h \to 0 \\ nh = t - a}} e_n = 0 . \tag{11.97}
$$

In the limit $h \to 0$, for the Runge–Kutta methods, Eq. (11.56b) becomes

$$
k_i = f\left(x_n, t_n\right) \tag{11.98}
$$

and hence

$$
\phi\left(x_n, t_n; 0\right) = \sum_{i=1}^{N} \omega_i k_i = \sum_{i=1}^{n} \omega_i f\left(x_n, t_n\right) . \tag{11.99}
$$

Because of $\phi(x, t; 0) = f(x, t)$

$$\sum_{i=1}^{N} \omega_i = 1.$$

(11.100)

This equation is satisfied in deriving the Runge–Kutta methods. That is, consistency condition is satisfied for all the Runge–Kutta methods. Equation (11.100) is the *necessary and sufficient condition* for the Runge–Kutta methods to be consistent. Since, the consistency is the necessary and sufficient condition for convergence of the Runge–Kutta methods the point is that *all Runge–Kutta methods are convergent.*

11.6   Adaptive Step Size and Runge–Kutta–Fehlberg Method

How does one make sure that the solution obtained by a Runge–Kutta method is sufficiently accurate? One approach is comparison of solutions at the end of a time interval computed with two different time steps, for example, with h and $h/2$. If the difference between the two solutions is slightly changed or less than a preassumed tolerance then the result is accepted. If the difference is higher than the tolerance then the step size must be halved again until a satisfactory result is obtained. However, to apply this procedure with the fourth-order Runge–Kutta method an additional seven function evaluations are required. That is, totally 11 function evaluations to go from x_n to x_{n+1}.

Another approach to get some preassumed accuracy in the numerical solution is to introduce a suitable adaptive control on step size. Suppose, for the fourth-order Runge–Kutta method find the solution at $t = t_0 + 2h$ from t_0 using two different step sizes h and $2h$. Denoting x_1 and x_2 are the two solutions obtained at $t = t_0 + 2h$ with step sizes $2h$ and h, respectively, then

$$X(t_0 + 2h) = x_1 + (2h)^5 \phi + O\left(h^6\right),$$

(11.101a)

$$X(t_0 + 2h) = x_2 + 2h^5 \phi + O\left(h^6\right).$$

(11.101b)

Since, the error is $h^5 \phi$ when the step size is h Eq. (11.101b) has the error term $2h^5 \phi$. Now, the difference between the two numerical estimations is

$$\Delta = x_2 - x_1.$$

(11.102)

It is desirable to keep this quantity lower than the preassumed accuracy by adjusting the value of h. Ignoring the terms $O(h^6)$ in Eqs. (11.101) they can be solved for h^5. The result is

$$h^5 = (x_2 - x_1)/30$$

(11.103)

and

$$X(t_0 + 2h) = x_2 + \frac{\Delta}{15} + O\left(h^6\right).$$

(11.104)

This estimate has fifth-order accuracy, however, there is no way of monitoring its truncation error [7]. Another disadvantage of this method is the calculation of the solution with two different step sizes h and $2h$.

An alternative approach developed by Fehlberg uses two Runge–Kutta methods of different orders. For example, the fourth-order and fifth-order methods can be used to go from x_n to x_{n+1}, compare the solution at x_{n+1} and obtain the error. If the error is larger than the preassumed tolerance then compute a new step size so that the error is less than the tolerance. The method which implements the above is the Runge–Kutta–Fehlberg method. It requires six function evaluation and has an estimate of the error. The following is the algorithm:

$$k_1 = hf(x_n, t_n), \tag{11.105a}$$

$$k_2 = hf\left(x_n + \frac{1}{4}k_1, t_n + \frac{1}{4}h\right), \tag{11.105b}$$

$$k_3 = hf\left(x_n + \frac{3}{32}k_1 + \frac{9}{32}k_2, t_n + \frac{3}{8}h\right), \tag{11.105c}$$

$$k_4 = hf\left(x_n + \frac{1}{2197}(1932k_1 - 7200k_2 + 7296k_3), t_n + \frac{12}{13}h\right), \tag{11.105d}$$

$$k_5 = hf\left(x_n + \frac{439}{216}k_1 - 8k_2 + \frac{3680}{513}k_3 - \frac{845}{4104}k_4, t_n + h\right), \tag{11.105e}$$

$$k_6 = hf\left(x_n - \frac{8}{27}k_1 + 2k_2 - \frac{3544}{2565}k_3 - \frac{1859}{4104}k_4 - \frac{11}{40}k_5, t_n + \frac{1}{2}h\right). \tag{11.105f}$$

The fourth-order and fifth-order solutions are given by

$$\hat{x}_{n+1} = x_n + \left(\frac{25}{216}k_1 + \frac{1408}{2565}k_3 + \frac{2197}{4104}k_4 - \frac{1}{5}k_5\right) \tag{11.106}$$

and

$$x_{n+1} = x_n + \left(\frac{16}{135}k_1 + \frac{6656}{12825}k_3 + \frac{28561}{56430}k_4 - \frac{9}{50}k_5 + \frac{2}{55}k_6\right), \tag{11.107}$$

respectively.

The error in the approximation $(x_{n+1} - \hat{x}_{n+1})$ is

$$E = \frac{1}{360}k_1 - \frac{128}{4275}k_3 - \frac{2197}{75240}k_4 + \frac{1}{50}k_5 + \frac{2}{55}k_6. \tag{11.108}$$

Note that $\hat{x}_{n+1}$ and x_{n+1} are calculated using the same k's. The global error in $\hat{x}_{n+1}$ and x_{n+1} are $O(h^4)$ and $O(h^5)$, respectively. If the error E is less than the tolerance chosen then x_{n+1} is accepted. If the error E is higher than the preassumed tolerance then the time step h can be reduced to meet the requirement. The time step can be increased depending upon the error.

Let us denote h as the step size giving the error E and E_0 is the desired accuracy. The time step, say h_{new}, which would produce the error E_0 is given by

$$h_{\text{new}} = h\,|E_0/E|^{1/5}. \tag{11.109}$$

This equation points out how much h to be decreased (increased) if $E_1 > E_0$ $(E_1 < E_0)$ to have the error $\approx E_0$. In this way, the adaptive strategy continuously controls the value of h and helps to maintain the accuracy E_0.

Example:

Compare the fourth-order Runge–Kutta (RK4) method and the Runge–Kutta–Fehlberg (RKF) method with reference to the equation

$$\dot{x} = 3x - 3t + 1, \quad x(0) = 1.$$

The exact solution of this equation is $x(t) = x(0)\,e^{3t} + t$.

TABLE 11.8

The numerical solution of $x' = 3x - 3t + 1$, $x(0) = 1$ obtained using the RK4 method with $h = 0.1$. PER denotes the percentage of relative error.

t	x_{numer}	x_{exact}	$|x_{exact} - x_{numer}|$	PER
0.0	1.0000000	1.0000000	0.0000000	0.000
0.1	1.4498375	1.4498588	0.0000213	0.001
0.2	2.0220613	2.0221188	0.0000575	0.003
0.3	2.7594866	2.7596031	0.0001165	0.004
0.4	3.7199073	3.7201169	0.0002096	0.006
0.5	4.9813354	4.9816891	0.0003537	0.007
0.6	6.6490745	6.6496475	0.0005730	0.009
0.7	8.8652676	8.8661699	0.0009023	0.010
0.8	11.8217844	11.8231763	0.0013919	0.012

Table 11.8 gives the numerical solution obtained by the RK4 method with the fixed step size $h = 0.1$. Table 11.9 presents the numerical solution and the relative error with the RKF method. At time $t = 0$ the step size is chosen as 0.1. The tolerance E_0 is chosen as 10^{-5}. At $t = 0.6$ the error $|E|$ (Eq. (11.108)) is 1.124e−4 which is higher than E_0. The new step size h_0 (Eq. (11.109)) is computed as 0.09580074 and rounded it to 0.09. Then, using the value of $x(0.5)$ the values of $x(0.59)$ and $x(0.68)$ are obtained. At $t = 0.77$ the error in the approximation is $|E| = 1.273$e−4. As this error is higher than E_0 the new step size is calculated and is 0.08.

Table 11.10 gives a comparison of the various numerical integration methods. The methods discussed so far are called *single-step* methods because they use only (x_n, t_n) to obtain (x_{n+1}, t_{n+1}). The next section is devoted to the multi-step methods.

TABLE 11.9

The numerical approximation of the solution of $x' = 3x - 3t + 1$, $x(0) = 1$ by the RKF method. PER denotes the percentage of relative error.

t	h	x_{numer}	$|E|$	$|x_{exact} - x_{numer}|$	PER
0.00	––	1.0000000	––	0.0000000	0.000000
0.10	0.10	1.4498581	0.277e−5	0.0000007	0.000048
0.20	0.10	2.0221169	0.373e−5	0.0000019	0.000094
0.30	0.10	2.7595993	0.503e−5	0.0000038	0.000138
0.40	0.10	3.7201100	0.680e−5	0.0000069	0.000185
0.50	0.10	4.9816774	0.919e−5	0.0000117	0.000235
0.59	0.09	6.4608363	0.742e−5	0.0000171	0.000265
0.68	0.09	8.3705846	0.971e−5	0.0000246	0.000294
0.76	0.08	10.5366478	0.717e−5	0.0000326	0.000309

TABLE 11.10
Comparison of various single-step numerical integration schemes.

Method	Estimation of f	Number of calculation of f per step	Local truncation error	Global error
Euler	Initial value	1	$O\left(h^2\right)$	$O(h)$
Improved Euler	Average of initial and final predicted values of f	2	$O\left(h^3\right)$	$O\left(h^2\right)$
Second-order Runge–Kutta	Average of two values of f	2	$O\left(h^3\right)$	$O\left(h^2\right)$
Fourth-order Runge–Kutta	Weighted average of four values of f	4	$O\left(h^5\right)$	$O\left(h^4\right)$
Runge–Kutta– Merson	Weighted average of three values of f	5	$O\left(h^5\right)$	$O\left(h^4\right)$
Runge–Kutta– Fehlberg	Weighted average of six values of f	6	$O\left(h^6\right)$	$O\left(h^5\right)$

11.7 Multi-Step Methods

In the single-step methods (x_0, t_0) is alone required to compute (x_1, t_1) and (x_1, t_1) alone is needed to compute (x_2, t_2) and so on. In general, knowing (x_n, t_n) the single-step methods determine (x_{n+1}, t_{n+1}). Once (x_n, t_n) is known for a first few values of n then they can be used for further calculation. The methods which use the information at two or more prior points are called *multi-step* methods. An advantage of a multi-step method is that it is possible to determine the local truncation error. Further, a correction term can be included to improve the accuracy of the result at each step. This section mainly focuses on the Adams–Bashforth–Moulton four-step method. It requires x_{n-3}, x_{n-2}, x_{n-1} and x_n to compute x_{n+1}. The formulas of a few other methods are given at the end of this section.

11.7.1 Derivation of Adams–Bashforth–Moulton Four-Step Method

Consider that for Eq. (11.4) the values of (x_{n-3}, t_{n-3}), (x_{n-2}, t_{n-2}), (x_{n-1}, t_{n-1}) and (x_n, t_n) are given. If (x_{n-3}, t_{n-3}) alone given then the other three sets can be calculated using a suitable single-step method, for example, a fourth-order Runge–Kutta method. From the above four sets, the values of f_{n-3}, f_{n-2}, f_{n-1} and f_n can be computed. Now, consider the equation

$$
\begin{aligned}
x_{n+1} &= x_n + \int_{t_n}^{t_{n+1}} \frac{dx}{dt}\, dt \\
&= x_n + \int_{t_n}^{t_{n+1}} f(x(t), t)\, dt.
\end{aligned}
\tag{11.110}
$$

Adams' methods are based on the use of the Newton polynomial approximation for f in the interval (t_n, t_{n+1}). Since, f is known at four points it is desirable to choose the third-order Newton polynomial $p_3(x)$. Using the divided-differences write $p_3(t)$ as (refer the Section 6.2.4)

$$
\begin{aligned}
p_3(t) \;=\; & f_{n-3} + f\,[t_{n-3}, t_{n-2}]\,(t - t_{n-3}) \\
& + f\,[t_{n-3}, t_{n-2}, t_{n-1}]\,(t - t_{n-3})\,(t - t_{n-2}) \\
& + f\,[t_{n-3}, t_{n-2}, t_{n-1}, t_n]\,(t - t_{n-3})\,(t - t_{n-2})\,(t - t_{n-1}) \;, \quad (11.111a)
\end{aligned}
$$

where

$$
f\,[t_{n-3}, t_{n-2}] \;=\; \frac{1}{h}\,(f_{n-2} - f_{n-3}) \;, \tag{11.111b}
$$

$$
f\,[t_{n-3}, t_{n-2}, t_{n-1}] \;=\; \frac{1}{2h^2}\,(f_{n-1} - 2f_{n-2} + f_{n-3}) \;, \tag{11.111c}
$$

$$
f\,[t_{n-3}, t_{n-2}, t_{n-1}, t_n] \;=\; \frac{1}{6h^3}\,(f_n - 3f_{n-1} + 3f_{n-2} - f_{n-3}) \;. \tag{11.111d}
$$

Then,

$$
x_{n+1} = x_n + \int_{t_n}^{t_{n+1}} p_3(t)\,\mathrm{d}t + \int_{t_n}^{t_{n+1}} E(t)\,\mathrm{d}t \;. \tag{11.112}
$$

The integrals

$$
I_0 \;=\; \int_{t_n}^{t_{n+1}} \mathrm{d}t \;, \tag{11.113a}
$$

$$
I_1 \;=\; \int_{t_n}^{t_{n+1}} (t - t_{n-3})\,\mathrm{d}t \;, \tag{11.113b}
$$

$$
I_2 \;=\; \int_{t_n}^{t_{n+1}} (t - t_{n-3})\,(t - t_{n-2})\,\mathrm{d}t \;, \tag{11.113c}
$$

$$
I_3 \;=\; \int_{t_n}^{t_{n+1}} (t - t_{n-3})\,(t - t_{n-2})\,(t - t_{n-3})\,\mathrm{d}t \tag{11.113d}
$$

occurring in the second term are to be evaluated. I_0 is found to be h. The other integrals can be evaluated by introducing the change of variable $t = t_0 + h\tau$ with τ varying from 0 to 1. Consider the integral I_1. In this integral $t - t_{n-3} = h(\tau + 3)$, $\mathrm{d}t = h\mathrm{d}\tau$ and hence

$$
I_1 = h^2 \int_0^1 (\tau + 3)\mathrm{d}\tau = \frac{7}{2}h^2. \tag{11.114}
$$

Similarly, the other two integrals are worked out as

$$
I_2 = \frac{53}{6}h^3 \;\text{ and }\; I_3 = \frac{55}{4}h^4. \tag{11.115}
$$

Then,

$$
\begin{aligned}
\int_{t_n}^{t_{n+1}} \mathrm{d}t \;=\; & hf_{n-3} + \frac{7}{2}h\,(f_{n-2} - f_{n-3}) + \frac{53}{12}h\,(f_{n-1} - 2f_{n-2} + f_{n-3}) \\
& + \frac{55}{24}h\,(f_n - 3f_{n-1} + 3f_{n-2} - f_{n-3}) \\
\;=\; & \frac{1}{24}h\,[55f_n - 59f_{n-1} + 37f_{n-2} - 9f_{n-3}] \;. \tag{11.116}
\end{aligned}
$$

The numerical estimation

$$p_{n+1} = x_n + \int_{t_n}^{t_{n+1}} p_3(t) \, dt$$

$$= x_n + \frac{1}{24} h \left[55 f_n - 59 f_{n-1} + 37 f_{n-2} - 9 f_{n-3} \right] \qquad (11.117a)$$

is called the *Adams–Bashforth predictor.*

Next, the integral $\int p_3(t) dt$ is evaluated by constructing the Newton polynomial of f at t_{n-2}, t_{n-1}, t_n and t_{n+1}. Here, p_{n+1} is used for t_{n+1}, that is, $f_{n+1} = f(p_{n+1}, t_{n+1})$ while x's are used for other t's. The result is

$$x_{n+1} = x_n + \frac{1}{24} h \left[f_{n-2} - 5 f_{n-1} + 19 f_n + 9 f_{n+1} \right] \qquad (11.117b)$$

and is called the *Adams–Moulton corrector.* Redefining the obtained x_{n+1} as p_{n+1} and using it in Eq. (11.117b) obtain a re-correction. In this way, successive re-correction can be achieved.

The error term in the predictor is worked out as

$$\int_{t_n}^{t_{n+1}} E(t) \, dt = \frac{1}{24} f^{(4)} \int_{t_n}^{t_{n+1}} (t - t_{n-3})(t - t_{n-2})(t - t_{n-1})(t - t_n) \, dt$$

$$= \frac{1}{24} x^{(5)} \int_{t_n}^{t_{n+1}} (t - t_{n-3})(t - t_{n-2})(t - t_{n-1})(t - t_n) \, dt$$

$$= \frac{1}{24} x^{(5)} \int_{t_n}^{t_{n+1}} (u + 3h)(u + 2h)(u + h) \, du$$

$$= \frac{251}{720} x^{(5)} (c_{n+1}), \qquad (11.118)$$

where $f^{(4)} = f^{(4)}(c_{n+1})$, $u = t - t_n$, $t - t_{n-3} = t - (t_n - 3h) = u + 3h$ and so on. The error given above is the local truncation error. The error term in the corrector is $-(19/720) h^5 x^{(5)} (d_{n+1})$. Thus, the error terms in the predictor and corrector are $O(h^5)$.

Example:

Compute the numerical solution of the first-order Eq. (11.18) by the Adams–Bashforth–Moulton four-step method over the interval $[0, 3]$ with different step sizes. Compare the numerical solution with the exact solution $x(t) = (x(0) + 0.5) e^{-t} + 0.5 (\sin t - \cos t)$.

In the Adams–Bashforth–Moulton method x_{n-3}, x_{n-2}, x_{n-1} and x_n are used to calculate x_{n+1}. For the given equation only $x(0)$ is given. The method requires $x(0)$, $x(h)$, $x(2h)$ and $x(3h)$ to calculate $x(4h)$. The unknown quantities $x(h)$, $x(2h)$ and $x(3h)$ can be calculated using a method which requires only one previous value of x. Using the fourth-order Runge–Kutta method the values of $x(h)$, $x(2h)$ and $x(3h)$ are found. Then, the values of $x(0)$ to $x(3h)$ are used to calculate $x(4h)$ and so on. Table 11.11 gives the numerical solution for different values of h. The final global error and the percentage of relative error at $t = 1$ for different values of h are given in Table 11.12.

11.7.2 Other Methods

Some other multi-step methods for Eq. (11.4) are given below:

TABLE 11.11
Comparison of the numerical solution obtained for the equation $\dot{x} = -x + \sin t$, $x(0) = 1$ by the Adams–Bashforth–Moulton four-step method.

t	Numerically computed x with				Exact x
	$h = 0.01$	$h = 0.10$	$h = 0.25$	$h = 0.50$	
0.0	1.0000000	1.0000000	1.0000000	1.0000000	1.0000000
0.5	0.7107175	0.7107172	0.7107387	0.7111511	0.7107175
1.0	0.7024035	0.7024030	0.7024112	0.7028600	0.7024035
1.5	0.7980741	0.7980743	0.7980822	0.7983662	0.7980741
2.0	0.8657251	0.8657256	0.8657681	0.8662206	0.8657251
2.5	0.8229354	0.8229356	0.8229858	0.8238864	0.8229354
3.0	0.6402369	0.6402363	0.6402523	0.6411582	0.6402369

(a) Adams–Bashforth–Moulton Three-Step Method

$$p_{n+1} = x_n + \frac{1}{12}h\left(5f_{n-2} - 16f_{n-1} + 23f_n\right), \tag{11.119a}$$

$$x_{n+1} = x_n + \frac{1}{12}h\left(-f_{n-1} + 8f_n + 5f_{n+1}\right). \tag{11.119b}$$

(b) Milne–Simpson Method

$$p_{n+1} = x_{n-3} + \frac{4}{3}h\left(2f_{n-2} - f_{n-1} + 2f_n\right), \tag{11.120a}$$

$$x_{n+1} = x_{n-1} + \frac{1}{3}h\left(f_{n-1} + 4f_n + f_{n+1}\right). \tag{11.120b}$$

TABLE 11.12
Comparison of the numerical solutions obtained at $t = 1$ for the equation $x' = -x + \sin t$, $x(0) = 1$ by the Adams–Bashforth–Moulton four-step method with different values of h.

h	x_{numer}	Final global error	% of relative error
0.001	0.7024035	0.0000000	0.000
0.010	0.7024035	0.0000000	0.000
0.025	0.7024035	0.0000000	0.000
0.050	0.7024035	0.0000000	0.000
0.100	0.7024030	0.0000005	0.000
0.250	0.7024112	−0.0000077	0.001
0.500	0.7028600	−0.0004565	0.065

TABLE 11.13

The numerical solution of the equation $\dot{x} = -x + \sin t$, $x(0) = 1$ at $t = 1$ obtained by the Adams–Bashforth–Moulton (ABM) three- and four-step methods, the Milne–Simpson (MS) method and the Hamming method. The exact solution is $x(1) = 0.7024035$.

h	Numerically computed x by the methods			
	ABM three-step	ABM four-step	MS	Hamming
0.001	0.7024035	0.7024035	0.7024035	0.7024035
0.010	0.7024035	0.7024035	0.7024035	0.7024035
0.025	0.7024039	0.7024035	0.7024035	0.7024035
0.050	0.7024069	0.7024035	0.7024035	0.7024035
0.100	0.7024362	0.7024030	0.7024032	0.7024030
0.250	0.7029729	0.7024112	0.7024105	0.7024185
0.500	0.7029729	0.7028600	0.7628600	0.7028600

(c) Hamming Method

$$p_{n+1} = x_{n-3} + \frac{4}{3}h\left(2f_{n-2} - f_{n-1} + 2f_n\right), \tag{11.121a}$$

$$x_{n+1} = \frac{1}{8}\left(-x_{n-2} + 9x_n\right) + \frac{3}{8}h\left(-f_{n-1} + 2f_n + f_{n+1}\right). \tag{11.121b}$$

Example:

Compare the numerical solutions obtained at $t = 1$ by the various multi-step methods with reference to the first-order Eq. (11.18).

The solution at $t = 1$ is calculated by the methods, namely, Adams–Bashforth–Moulton three- and four-step, Milne–Simpson and Hamming and compared with the exact solution by calculating the percentage of the relative error. This is done for the various values of h. The results are presented in Tables 11.13 and 11.14.

TABLE 11.14

Computed percentage of relative errors for the Adams–Bashforth–Moulton (ABM) three- and four-step methods, the Milne–Simpson (MS) method and the Hamming method for the numerical solution of the equation $\dot{x} = -x + \sin t$, $x(0) = 1$ at $t = 1$.

h	ABM three-step	ABM four-step	MS	Hamming
0.001	0.000	0.000	0.000	0.000
0.010	0.000	0.000	0.000	0.000
0.025	0.000	0.000	0.000	0.000
0.050	0.000	0.000	0.000	0.000
0.100	0.005	0.000	0.000	0.000
0.250	0.065	0.001	0.001	0.002
0.500	0.081	0.065	0.065	0.065

11.7.3   Convergence Criteria – Right Step

A predictor-corrector method becomes unstable if

1. $f_x(x, t) < 0$ and

2.   the step size h is too large.

Now, obtain a condition on h for the stability and the convergence of the three predictor-corrector methods.

Let the value of x_{n+1} from the predictor and the corrector formulas as x_p and x_c, respectively, and define $D = x_c - x_p$. The successive re-corrections are denoted as x_{cc}, x_{ccc} and so on. x^* is the value to which the successive re-corrections converge. First, obtain the convergence criteria for the Adams–Bashforth–Moulton method. The correction x_c and the re-correction x_{cc} are

$$
\begin{aligned}
x_c &= x_n + \frac{1}{24} h \left(f_{n-2} - 5 f_{n-1} + 19 f_n + 9 f_{n+1} \right) \\
&= x_n + \frac{1}{24} h \left(x'_{n-2} - 5 x'_{n-1} + 19 x'_n + 9 x'_p \right)
\end{aligned}
\tag{11.122a}
$$

and

$$
x_{cc} = x_n + \frac{1}{24} h \left(x'_{n-2} - 5 x'_{n-1} + 19 x'_n + 9 x'_c \right) .
\tag{11.122b}
$$

Then,

$$
x_{cc} - x_c = \frac{3}{8} h \left(x'_c - x'_p \right) .
\tag{11.123}
$$

The difference quantity $x'_c - x'_p$ is

$$
\begin{aligned}
x'_c - x'_p &= f(x_c, t_{n+1}) - f(x_p, t_{n+1}) \\
&= \frac{1}{(x_c - x_p)} \left[f(x_c, t_{n+1}) - f(x_p, t_{n+1}) \right] (x_c - x_p) \\
&= f_x(\xi_1) D , \quad x_c < \xi < x_p \text{ and } D = x_c - x_p .
\end{aligned}
\tag{11.124}
$$

Substitution of Eq. (11.124) in Eq. (11.123) gives

$$
x_{cc} - x_c = \frac{3}{8} h D f_x(\xi_1) .
\tag{11.125}
$$

Further,

$$
\begin{aligned}
x_{ccc} - x_{cc} &= \frac{3}{8} h \left(x'_{cc} - x'_c \right) \\
&= \frac{3}{8} h f_x(\xi_2) (x_{cc} - x_c) \\
&= \frac{3}{8} h f_x(\xi_2) \cdot \frac{3}{8} h D f_x(\xi_1) \\
&= \left[\frac{3}{8} h f_x(\xi) \right]^2 D ,
\end{aligned}
\tag{11.126}
$$

where $x_c < \xi < x_{cc}$. The sum of such successive corrections gives

$$
\begin{aligned}
x^* &= x_p + (x_c - x_p) + (x_{cc} - x_c) + (x_{ccc} - x_{cc}) + \cdots \\
&= x_p + D + rd + r^2 D + r^3 D + \cdots ,
\end{aligned}
\tag{11.127}
$$

where $r = 3hf_x(\xi)/8$. Rewrite Eq. (11.127) as

$$x^* = x_p + D(1 - r)^{-1}. \qquad (11.128)$$

The right-side of Eq. (11.128) is a convergent series provided $|r| < 1$, that is

$$h < \frac{8}{3}\frac{1}{|f_x(x_n, t_n)|}. \qquad (11.129)$$

The above is the first criterion for the convergence. It is desired to have x_c and x^* the same within one in the mth decimal place. Then,

$$x^* - x_c = x_p + D(1 - r)^{-1} - (d + x_p) = \frac{r}{1 - r}D < 10^{-m}. \qquad (11.130)$$

For $r \ll 1$, $r/(1 - r)$ is $\approx r$ and the second criterion for the convergence is $rD < 10^{-m}$, that is,

$$10^m D < |1/r| = \frac{8}{3}\frac{1}{|f_x(x_n, t_n)|}. \qquad (11.131)$$

Note that the above criteria are for a first-order equation. Similar analysis can be performed for a system of first-order equations but the analysis is tedious. For the other two multi-step methods the convergence criteria are given in Problems 11.3 and 11.4.

11.8   Second-Order Equations

The methods discussed so far for first-order ordinary differential equations can be extended to second-order and higher-order equations. For example, consider a system of two first-order equations of the form

$$\dot{x} = f(x, y, t), \quad \dot{y} = g(x, y, t). \qquad (11.132)$$

Any second-order equation can always be rewritten in the above form. The fourth-order Runge–Kutta algorithm for Eqs. (11.132) is

$$X1 = hf(x_n, y_n, t_n), \qquad (11.133a)$$

$$Y1 = hg(x_n, y_n, t_n), \qquad (11.133b)$$

$$X2 = hf\left(x_n + \frac{1}{2}X1, y_n + \frac{1}{2}Y1, t_n + \frac{1}{2}h\right), \qquad (11.133c)$$

$$Y2 = hg\left(x_n + \frac{1}{2}X1, y_n + \frac{1}{2}Y1, t_n + \frac{1}{2}h\right), \qquad (11.133d)$$

$$X3 = hf\left(x_n + \frac{1}{2}X2, y_n + \frac{1}{2}Y2, t_n + \frac{1}{2}h\right), \qquad (11.133e)$$

$$Y3 = hg\left(x_n + \frac{1}{2}X2, y_n + \frac{1}{2}Y2, t_n + \frac{1}{2}h\right), \qquad (11.133f)$$

$$X4 = hf(x_n + X3, y_n + Y3, t_n + h), \qquad (11.133g)$$

$$Y4 = hg(x_n + X3, y_n + Y3, t_n + h), \qquad (11.133h)$$

$$x_{n+1} = x_n + \frac{1}{6}(X1 + 2X2 + 2X3 + X4), \qquad (11.133i)$$

$$y_{n+1} = y_n + \frac{1}{6}(Y1 + 2Y2 + 2Y3 + Y4). \qquad (11.133j)$$

TABLE 11.15
The numerical solution of Eq. (11.134) by the fourth-order Runge–Kutta method with step size $h = 0.1$. The exact solution is given by Eq. (11.135).

t	x_{numer}	x_{exact}	y_{numer}	y_{exact}
0.0	0.0000000	0.0000000	0.0000000	0.0000000
0.1	0.0048333	0.0048333	0.0950000	0.0950001
0.2	0.0186666	0.0186668	0.1800024	0.1800026
0.3	0.0405007	0.0405010	0.2550190	0.2550192
0.4	0.0693384	0.0693387	0.3200794	0.3200797
0.5	0.1041864	0.1041868	0.3752384	0.3752387
0.6	0.1440587	0.1440593	0.4205830	0.4205832
0.7	0.1879798	0.1879804	0.4562370	0.4562373
0.8	0.2349885	0.2349892	0.4823667	0.4823669
0.9	0.2841425	0.2841433	0.4991834	0.4991836
1.0	0.3345232	0.3345241	0.5069468	0.5069469

Example:

Using the fourth-order Runge–Kutta method compute the solution of

$$\ddot{x} + x = e^{-t}, \quad x(0) = 0, \ \dot{x}(0) = 0 \tag{11.134}$$

for $t \in [0, 1]$ with time step $h = 0.1$.

The exact solution of the given equation is

$$x(t) = (\dot{x}(0) + 0.5) \sin t + (x(0) - 0.5) \cos t + 0.5e^{-t}. \tag{11.135}$$

In order to use the Runge–Kutta method rewrite the given equation as

$$\dot{x} = y, \quad \dot{y} = -x + e^{-t}, \quad x(0) = 0, \ y(0) = 0. \tag{11.136}$$

Table 11.15 gives the solution $x(t)$ and $y(t)$, where $h = 0.1$.

Programs can be developed to solve Eq. (11.134) by the Euler, second-order Runge–Kutta, fourth-order Runge–Kutta and Adams–Bashforth–Moulton four-step methods.

11.9 Stiff Equations

There is a class of ordinary differential equations which are numerically unstable when solved by certain methods unless the time step is extremely small.

11.9.1 What are Stiff Equations?

Consider a system of N first-order differential equations of the form

$$\dot{\mathbf{x}} = \mathbf{f}(\mathbf{x}, t), \tag{11.137}$$

where $\mathbf{x} = (x_1, x_2, \ldots, x_N)$, $\mathbf{f} = (f_1, f_2, \ldots, f_N)$. The system (11.137) is said to be *stiff* if the real part of all the eigenvalues of the Jacobian matrix

$$J = \begin{pmatrix} \partial f_1/\partial x_1 & \partial f_1/\partial x_2 & \cdots & \partial f_1/\partial x_N \\ & & \vdots & \\ \partial f_N/\partial x_1 & \partial f_N/\partial x_2 & \cdots & \partial f_N/\partial x_N \end{pmatrix} \tag{11.138}$$

for every value of t are all negative and they differ greatly in magnitude. That is, for a stiff equation

1. $\mathrm{Re}\lambda_i < 0$ and
2. $\max.|\mathrm{Re}\lambda_i| \gg \min.|\mathrm{Re}\lambda_i|, \quad i = 1, 2, \ldots, N$.

In stiff equations, the dependent variables $\mathbf{x}$ depend on independent variables with two or more greatly differing scales. That is, a stiff differential equation is one in which there are both *slow and fast-changing components*.

11.9.2 Examples

The analytical solution of the equation

$$\dot{x} = -100x - 99e^{-t}, \quad x(0) = 0 \tag{11.139}$$

is

$$x(t) = e^{-100t} - e^{-t}. \tag{11.140}$$

Let us calculate the value of $x(0.02)$ with $h = 0.02$ by the Euler method. The result is $x(0.02) = -1.98$ while the exact value is -0.8448634. The error is very large. For $h = 0.001(\ll 1/100)$ the numerical solution is $x(0.001) = -0.099$ while the exact solution is -0.0941631. That is, the time step h should be $\ll 1/100$.

Next, consider the equation

$$\dot{x} = -1998x + 1996y = f_1, \tag{11.141a}$$
$$\dot{y} = -998x + 996y = f_2. \tag{11.141b}$$

Its Jacobian matrix is

$$\begin{aligned} J &= \begin{pmatrix} \partial f_1/\partial x & \partial f_1/\partial y \\ \partial f_2/\partial x & \partial f_2/\partial y \end{pmatrix} \\ &= \begin{pmatrix} -1998 & 1996 \\ -998 & 996 \end{pmatrix}. \end{aligned} \tag{11.142}$$

Its eigenvalues are the roots of the equation

$$\begin{aligned} |J - \lambda I| &= \begin{vmatrix} -1998 - \lambda & 1996 \\ -998 & 996 - \lambda \end{vmatrix} \\ &= \lambda^2 + 1002\lambda + 2000 = 0. \end{aligned} \tag{11.143}$$

Solving of Eq. (11.143) gives $\lambda_1 = -2$, $\lambda_2 = -1000$. Here, $|\lambda_1| \ll |\lambda_2|$ and $\mathrm{Re}\lambda_1$, $\mathrm{Re}\lambda_2 < 0$. The exact solution of Eq. (11.141) is

$$x(t) = Ae^{-2t} + Be^{-1000t}, \tag{11.144a}$$
$$y(t) = Ae^{-2t} + \frac{B}{2}e^{-1000t}, \tag{11.144b}$$

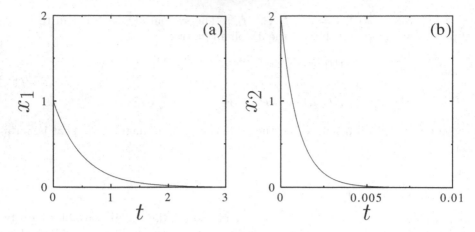

FIGURE 11.8
Variation of the two particular solutions of Eq. (11.141).

where

$$A = -x(0) + 2y(0), \quad B = 2x(0) - 2y(0).$$
(11.144c)

The two particular solutions $x_1 = e^{-2t}$ and $x_2 = e^{-1000t}$ have time dependence with greatly differing scales. Figure 11.8 depicts the variation of these two solutions. The solution x_2 decays to zero much faster than the solution x_1.

In order to solve Eq. (11.141) by the Euler or the Runge–Kutta method the step size h must be $\ll 2/1000$ so that these methods will be stable. This can be easily verified. With $x(0) = 3$ and $y(0) = 2$ solve Eq. (11.141) by the fourth-order Runge–Kutta method with different step sizes, namely, $h = 0.0001$, 0.001 and 0.002. Figure 11.9 depicts the numerical solution $x(t)$ obtained with these three step sizes. For $h = 0.0001$ the numerical solution (represented by solid circles in Fig. 11.9) is in good agreement with the exact solution (continuous curve). But the solutions obtained with $h = 0.001$ and 0.002 deviate largely

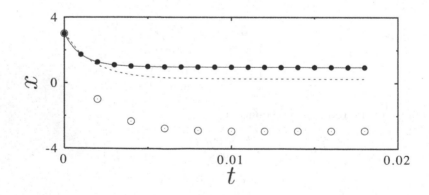

FIGURE 11.9
Comparison of the numerical solutions of Eq. (11.141) by the fourth-order Runge–Kutta method with the step sizes $h = 0.0001$ (solid circles), $h = 0.001$ (dashed curve) and $h = 0.002$ (open circles). Continuous curve is the exact solution.

from the exact solution. Similar result is observed when the Euler method is used. For example, the Euler method gives

$$x(0.005) \ = \ x(0) + hf_1 = -7.05, \quad x_{\text{exact}} = 1.00353, \tag{11.145}$$
$$y(0.005) \ = \ y(0) + hf_2 = -3.01, \quad y_{\text{exact}} = 0.9967877, \tag{11.146}$$

and

$$x(0.01) \ = \ 33.3397, \quad x_{\text{exact}} = 0.9802894, \tag{11.147}$$
$$y(0.01) \ = \ 17.1797, \quad y_{\text{exact}} = 0.9802440. \tag{11.148}$$

The numerical solution diverges as t increases whereas the exact solution decays to zero.

To understand the problem with Eq. (11.141) consider the equation

$$\dot{x} = -cx, \quad c > 0. \tag{11.149}$$

Its exact solution is $x(t) = x(0) \, \text{e}^{-ct}$. The explicit Euler method gives

$$
\begin{aligned}
x_{n+1} \ &= \ x_n + hf \\
&= \ x_n - hcx_n \\
&= \ (1 - ch)x_n \, .
\end{aligned}
\tag{11.150}
$$

If $ch > 2$ (that is $h > 2/c$) then $|x_n|$ diverges to infinity as $n \to \infty$. The method is thus unstable for $h > 2/c$.

11.9.3 Implicit Euler Method

To overcome the problem with the explicit Euler method consider an implicit differencing:

$$
\begin{aligned}
x_{n+1} \ &= \ x_n + hf\,(x_{n+1}) \\
&= \ x_n - hcx_{n+1}
\end{aligned}
$$

or

$$x_{n+1} = \frac{x_n}{1 + hc} \, . \tag{11.151}$$

This is called an *implicit backward Euler scheme*. It is to be noted that $x_{n+1} \to 0$ (as is the case of the exact solution) even for $h \to \infty$. That is, in the long time limit, $t \to \infty$, the implicit method converges to the exact fixed point solution for large step sizes [7].

Let us derive an implicit Euler method for a system of differential equations. For simplicity consider a system of linear equations of the form

$$\dot{\mathbf{x}} = -C\mathbf{x}, \tag{11.152}$$

where $\mathbf{x} = (x_1, x_2, \ldots, x_N)$ and C is a $N \times N$ definite matrix. It is easy to show that for an explicit Euler method, the numerical solution x_n is bounded only if $h < 1/\lambda_{\text{max}}$, where λ_{max} is the largest eigenvalue of the matrix C [7].

The implicit differencing gives

$$\mathbf{x}_{n+1} = \mathbf{x}_n + hf\,(\mathbf{x}_{n+1}), \tag{11.153}$$

that is,

$$\mathbf{x}_{n+1} = (I + Ch)^{-1}\mathbf{x}_n \, . \tag{11.154}$$

If f is nonlinear then Eqs. (11.153) for known $\mathbf{x}_n$ are nonlinear equations. These can be solved iteratively for each n, for example, using the Newton–Raphson method. In this case

$$\mathbf{x}_{n+1} = \mathbf{x}_n + h\left[f\left(\mathbf{x}_n\right) + \left.\frac{\partial \mathbf{f}}{\partial \mathbf{x}}\right|_{\mathbf{x}_n} \left(\mathbf{x}_{n+1} - \mathbf{x}\right)\right]$$

or

$$\mathbf{x}_{n+1} = \mathbf{x}_n + h\left[I - h\frac{\partial \mathbf{f}}{\partial \mathbf{x}}\right]^{-1} \cdot \mathbf{f}\left(\mathbf{x}_n\right) . \tag{11.155}$$

This is called the *semi-implicit Euler method*.

Example:

Develop an implicit Euler algorithm to solve a second-order linear stiff equation of the form

$$\dot{x} = -a_{11}x - a_{12}y, \tag{11.156a}$$

$$\dot{y} = -a_{21}x - a_{22}y, \quad x(0) = x_0 = 3, \quad y(0) = y_0 = 2. \tag{11.156b}$$

Then, apply the algorithm to Eq. (11.141) with $h = 0.0001$, 0.001 and 0.01.

Since the given system (11.156) is linear the implicit Euler algorithm is given by Eq. (11.154). Explicitly it is

$$\left(\begin{array}{c} x_{n+1} \\ y_{n+1} \end{array}\right) = \left(\begin{array}{cc} 1 + a_{11}h & a_{12}h \\ a_{21}h & 1 + a_{22}h \end{array}\right)^{-1} \left(\begin{array}{c} x_n \\ y_n \end{array}\right) . \tag{11.157}$$

Equation (11.157) can be rewritten as

$$\left(\begin{array}{c} x_{n+1} \\ y_{n+1} \end{array}\right) = \left(\begin{array}{cc} b_{11} & b_{12} \\ b_{21} & b_{22} \end{array}\right) \left(\begin{array}{c} x_n \\ y_n \end{array}\right)$$

or

$$x_{n+1} = b_{11}x_n + b_{12}y_n, \tag{11.158a}$$

$$y_{n+1} = b_{21}x_n + b_{22}y_n, \tag{11.158b}$$

where

$$b_{11} = (1 + a_{22}h)/\det, \quad b_{12} = -a_{12}h/\det, \tag{11.159a}$$

$$b_{21} = -a_{21}h/\det, \quad b_{22} = (1 + a_{11}h)/\det, \tag{11.159b}$$

$$\det = (1 + a_{11}h)(1 + a_{22}h) - a_{12}a_{21}h^2. \tag{11.159c}$$

The quantities b_{11}, b_{12}, b_{21} and b_{22} are constants and do not change at every iteration. Equations (11.158) appear as a two-dimensional linear map.

Now, apply the above algorithm to Eq. (11.141). For this equation

$$a_{11} = 1998, \quad a_{12} = -1996, \quad a_{21} = 998, \quad a_{22} = -996.$$

For $h = 0.001$

$$b_{11} = 0.0019960, \quad b_{12} = 0.9960079,$$
$$b_{21} = -0.4980039, \quad b_{22} = 1.4960080, \quad \det = 2.004$$

and

$$x(0.001) = 1.9980038, \quad y(0.001) = 1.4980043 .$$

Using Eqs. (11.158) the values of x and y for $t = ih$, $i = 2, 3, \ldots$ are easily computed. Table 11.16 presents the numerical solution over the time interval $[0, 5]$ for different values of h. For large t the numerical solution converges to the exact solution even for a large time step $(h > 1/1000)$.

TABLE 11.16
The numerical solution of Eq. (11.141) is obtained by the implicit Euler method with different step sizes.

t	Numerically computed x with			Exact x
	$h = 0.0001$	$h = 0.001$	$h = 0.01$	
0.000	3.0000000	3.0000000	3.0000000	3.0000000
0.002	1.2932956	1.4960120	$- - - -$	1.2666786
0.004	1.0362226	1.1170398	$- - - -$	1.0286632
0.006	0.9946414	1.0193336	$- - - -$	0.9930292
0.008	0.9851053	0.9919555	$- - - -$	0.9847982
0.010	0.9803458	0.9821714	1.1622103	0.9802895
0.100	0.8187471	0.8188943	0.8203483	0.8187308
1.000	0.1353623	0.1356059	0.1380330	0.1353353
5.000	0.0000454	0.0000459	0.0000501	0.0000454

11.9.4 Rosenbrock Method

Another method of solving the stiff equations of order $N \leq 10$ is Rosenbrock method. Let us review the description of this method given in ref. [7]. For the equation of the form

$$\dot{\mathbf{x}} = \mathbf{f}(\mathbf{x}) \tag{11.160}$$

the method looks for a solution of the form

$$\mathbf{x}(t_0 + h) = \mathbf{x}_0 + \sum_{i=1}^{s} c_i \mathbf{k}_i, \tag{11.161}$$

where $\mathbf{k}_i$ is to be determined by solving the following equations which are the general form of Eq. (11.155):

$$(I - \gamma h \mathbf{f}') \cdot \mathbf{k}_i = h \mathbf{f}\left(\mathbf{x}_0 + \sum_{j=1}^{i-1} a_{ij} \mathbf{k}_j\right) + h \mathbf{f}' \cdot \sum_{j=1}^{i-1} c_{ij} \mathbf{k}_j, \quad i = 1, 2, \ldots, s. \tag{11.162}$$

Here, $\mathbf{f}'$ is the Jacobian matrix of $\mathbf{f}$ and γ, c_i, a_{ij} and c_{ij} are fixed constants and s is chosen as 4. To minimize the matrix vector multiplications define

$$\mathbf{g}_i = \sum_{j=1}^{i-1} c_{ij} \mathbf{k}_j + \gamma \mathbf{k}_i. \tag{11.163}$$

Then, Eqs. (11.162) are rewritten as

$$\left(\frac{I}{h\gamma} - \mathbf{f}'\right) \cdot \mathbf{g}_i = \mathbf{f}\left(\mathbf{x}_0 + \sum_{j=1}^{i-1} a_{ij} \mathbf{g}_j\right) + \frac{1}{h} \sum_{j=1}^{i-1} c_{ij} \mathbf{g}_j, \quad i = 1, 2, \ldots, s \tag{11.164}$$

The above set of equations is solved by first calculating the matrix $(I/(h\gamma) - \mathbf{f}')$, next LU decomposition of it and then $\mathbf{g}_i$ are determined by back-substitution. The result is

$$\mathbf{x}(t_0 + h) = \mathbf{x}_0 + \sum_{i=1}^{s} c_i \mathbf{g}_i. \tag{11.165}$$

TABLE 11.17

The numerical solution of Eq. (11.141) is obtained by the Rosenbrock method with different step sizes.

	Numerically computed x with			
t	$h = 0.0001$	$h = 0.001$	$h = 0.002$	Exact x
0.000	3.0000000	3.0000000	3.0000000	3.0000000
0.002	1.2684000	1.2490727	1.4323740	1.2666786
0.004	1.0280742	1.0214135	0.9249354	1.0286632
0.006	0.9928845	0.9918221	1.0367018	0.9930292
0.008	0.9848135	0.9850506	0.9637871	0.9847982
0.010	0.9803393	0.9808123	0.9919037	0.9802895
0.100	0.8187754	0.8191774	0.8196245	0.8187308
1.000	0.1353369	0.1353513	0.1353674	0.1353353
5.000	0.0000454	0.0000453	0.0000452	0.0000454

A Fortran code algorithm for the Rosenbrock method is given in ref. [7] with step-size control for the equations which have explicit time dependence also. The values of the parameters in the method are given by

$$c_1 = 19/9, \quad c_2 = 1/2, \quad c_3 = 25/108, \quad c_4 = 125/108, \tag{11.166a}$$

$$\gamma = 1/2, \quad A_{21} = 2, \quad A_{31} = 48/25, \quad A_{32} = 6/25, \tag{11.166b}$$

$$c_{21} = -8, \quad c_{31} = 372/25, \quad c_{32} = 12/5, \tag{11.166c}$$

$$c_{41} = -112/125, \quad c_{42} = -54/125, \quad c_{43} - 2/5. \tag{11.166d}$$

Equation (11.141) is solved by the Rosenbrock method with $h = 0.0001$, 0.001 and 0.002. The result is presented in Table 11.17.

An adaptive Runge–Kutta integration algorithm for stiff systems like a linear harmonic oscillator subjected to nonlinear thermal constraints (Nosé–Hoover equations) has been described in the ref. [10].

11.10   Solving a Differential Equation with a Noise Term

This section points out how to solve numerically a differential equation in the presence of a noise term. As an example, consider the first-order system

$$\frac{dx}{dt} = F(x) + \eta(t), \tag{11.167}$$

where $\eta(t)$ is, say, a continuous time white noise. $\eta(t)$ is specified by its mean and correlations. Assume that

$$\langle \eta(t) \rangle = 0, \quad \langle \eta(t)\eta(t+\tau) \rangle = D\delta(\tau), \tag{11.168}$$

where $\langle \cdot \rangle$ denotes an averaging over an ensemble of realizations of the random variable η at time t and D is the strength of the noise at any time t.

Because the numerical methods such as the Euler and the Runge–Kutta solve a differential equation at discrete times (independent variable) $\eta(t)$ has to be treated as a discrete process. Let $\eta(t_i) = \eta_i$ denote the noise at time $t = t_i$. Choose to represent the continuous time white noise at discrete times: $\{t_i : i = 1, 2, \ldots\}$. Thus, $\{\eta_i : i = 1, 2, \ldots\}$ denotes the discrete time noise process (random numbers) equivalent to the continuous time noise process $\eta(t)$.

What are the statistical properties of the discrete time process $\{\eta_i : i = 1, 2, \ldots\}$? The answer is

$$\langle \eta_i \rangle = 0 \text{ for all } i = 1, 2, \ldots, \quad \langle \eta_i \eta_j \rangle = D' \delta_{ij}, \tag{11.169}$$

where the scaled noise strength D' (of the discrete process) and noise strength D (of the continuous process) are related to each other as follows:

$$D' = \frac{D}{\Delta t}, \tag{11.170}$$

where $\Delta t = t_{i+1} - t_i$. To prove this consider

$$\int_0^t \eta(t')\, dt' = \lim_{\Delta t \to 0} \sum_i \eta(t_i) \Delta t. \tag{11.171}$$

As $\langle \eta(t_i) \eta(t') \rangle = D\delta(t_i, t')$

$$\int_0^t \langle \eta(t_i) \eta(t') \rangle\, dt' = \int_0^t D\delta(t_i, t')\, dt' = D. \tag{11.172}$$

Evaluating the above integral in discrete time results in

$$\int_0^t \langle \eta(t_i) \eta(t') \rangle\, dt' = \sum_j \langle \eta(t_i) \eta(t_j) \rangle \Delta t = \sum_j \langle \eta_i \eta_j \rangle \Delta t = D' \Delta t. \tag{11.173}$$

Integrals (11.172) and (11.173) are the same and hence $D' = D/\Delta t$.

Let $\eta(t)$ is a Gaussian noise with zero mean and unit variance. *How does one sample* η_i? First, generate a set of Gaussian random numbers of zero mean and unit variance employing a standard sampling technique, for example, the Box–Muller algorithm and call it ξ. Then,

$$\eta_i = \frac{\sqrt{D}}{\sqrt{\Delta t}} \xi_i. \tag{11.174}$$

Finite differencing of Eq. (11.167) with $x_i = x(t_i)$ and $\eta(t_i) = \eta_i$ yields

$$x_{i+1} = x_i + \Delta t f(x_i) + \Delta t\, \eta_i = x_i + \Delta t f(x_i) + \sqrt{D\Delta t}\, \xi_i. \tag{11.175}$$

In other words, given $x(t_i)$ integrate Eq. (11.167) numerically without $\eta(t)$ and obtain $x(t_{i+1}) = x(t_i + \Delta t)$. Then, the solution of Eq. (11.167) with $\eta(t)$ at $t = t_i + \Delta t$ is given by $x(t_{i+1}) \to x(t_{i+1}) + \sqrt{D\Delta t}\, \xi_i$.

11.11 Dynamical Systems with Coloured Noise

A white noise is a uniform mixture of random energy at every frequency. Its Fourier spectrum contains all frequencies. For a white noise $\epsilon(t)$, $\langle \epsilon(t) \rangle = 0$ and $\langle \epsilon(t) \epsilon(t') \rangle = \delta(t - t')$. The term white is by analogy with white light which is a mixture of all different possible colours. Using a filter a white colour can be transformed into a desired colour. Like-wise one can use filters on signals to change the balance of frequency components. The resulting noise is no longer has the quality of white but has some other quality with certain frequencies are more prominent.

In contrast to δ-correlation of white noise, in certain events fluctuation is time-correlated. For example, fluctuations associated with the opening and closing of calcium channels are slow in comparison with those associated with the opening and closing of sodium and potassium channels. To model such fluctuations δ-correlated noise terms are inappropriate. In such a case coloured noise, for example, time-correlated Gaussian noise with zero mean and an exponential correlation function, that is

$$\langle \epsilon(t) \rangle = 0, \quad \langle \epsilon(t)\epsilon'(t) \rangle = D\lambda e^{-\lambda|t-t'|}, \tag{11.176}$$

where D is the strength of the noise and λ is the inverse of the correlation time τ, is desirable. The variance of $\epsilon(t)$ with the properties given by Eq. (11.176) is $\sigma^2 = \langle \epsilon^2 \rangle = D\lambda$. In the limit $\lambda \to \infty$, the correlation function approaches the δ-function and thus $\epsilon(t)$ is a white noise if D is kept constant.

The noise with the property (11.176) can be generated from the linear Ornstein–Uhlenbeck process given by

$$\dot{\epsilon} = h(\epsilon) + \lambda\psi_w, \quad h(\epsilon) = -\lambda\epsilon, \tag{11.177}$$

where the Gaussian white noise ψ_w has the properties

$$\langle \psi_w(t) \rangle = 0, \quad \langle \psi_w(t)\psi_w(t') \rangle = 2D\delta(t - t'). \tag{11.178}$$

$\psi(t)$ can be generated by the Box–Muller method.

As Eq. (11.177) for the coloured noise is an ordinary differential equation the numerical integration algorithms such as the Euler and the Runge–Kutta methods can be used. Let us now describe these algorithms.

1. The Euler Integration Scheme

The Euler algorithm is

$$\epsilon(t_0 + \Delta t) = \epsilon(t_0) + \Delta t\, h(\epsilon(t_0)) + (2D\lambda^2\Delta t)^{1/2}\psi, \tag{11.179a}$$

where

$$\psi = (-2\ln a)^{1/2} \cos(2\pi b) \tag{11.179b}$$

wth a and b are two uniform random numbers in the interval $[0, 1]$ and ψ is a Gaussian noise with mean zero and unit variance. To start the simulation, an initial value for ϵ is needed and is given by

$$\epsilon_0 = \epsilon(t_0) = (-2D\lambda \ln c)^{1/2} \cos(2\pi d), \tag{11.180}$$

where c and d are uniform random numbers in the interval $[0, 1]$. Using the $\epsilon(t_0)$ given by Eq. (11.180) compute $\epsilon(t_0 + \Delta t)$ employing Eq. (11.179a). Next, with the calculated $\epsilon(t_0 + \Delta t)$ find $\epsilon(t_0 + 2\Delta t)$ by replacing t_0 by $t_0 + \Delta t$. In this way by iterating Eq. (11.179a) it is easy to generate $\epsilon(t_0 + i\Delta t)$, $i = 0, 1, 2, \ldots$.

The problem is to solve numerically the one-variable and additive coloured noise equation

$$\dot{x} = f(x) + \epsilon = f(x, \epsilon) \tag{11.181}$$

with given initial condition $x(t_0) = x_0$. Since $\epsilon(t)$ is described by the ordinary differential Eq. (11.2) it is evident that

$$\dot{x} = f(x, \epsilon) = f(x) + \epsilon, \tag{11.182a}$$

$$\dot{\epsilon} = h(\epsilon) + \lambda\psi = -\lambda\epsilon + \lambda\psi. \tag{11.182b}$$

2. The Second-Order Runge–Kutta Scheme

Honeycutt [11] developed a second- and fourth-orders Runge–Kutta algorithms for coloured noise equations. The second-order Runge–Kutta scheme for $\epsilon(t_0 + \Delta t)$ from t_0 to $t_0 + \Delta t$ is given by

$$\alpha = \lambda(2D\Delta t)^{1/2}\psi, \tag{11.183a}$$

$$H_1 = \Delta t\, h(\epsilon(t_0)), \tag{11.183b}$$

$$H_2 = \Delta t\, h\left(\epsilon(t_0) + H_1 + \alpha\right), \tag{11.183c}$$

$$\epsilon(t_0 + \Delta t) = \epsilon(t_0) + \frac{1}{2}\left(H_1 + H_2\right) + \alpha, \tag{11.183d}$$

where ψ is a Gaussian random variable with zero mean and unit variance. Initial value of ϵ, $\epsilon(t_0)$, is given by Eq. (11.180).

For the system (11.181) driven by the coloured noise ϵ the stochastic Runge–Kutta second-order scheme is

$$\alpha = \lambda(2D\Delta t)^{1/2}\psi, \tag{11.184a}$$

$$H_1 = \Delta t\, h(\epsilon(t_0)), \tag{11.184b}$$

$$F_1 = \Delta t\, f(x(t_0), \epsilon(t_0)), \tag{11.184c}$$

$$H_2 = \Delta t\, h\left(\epsilon(t_0) + H_1 + \alpha\right), \tag{11.184d}$$

$$F_2 = \Delta t\, f\left(x(t_0) + F_1, \epsilon(t_0) + H_1 + \alpha\right), \tag{11.184e}$$

$$x(t_0 + \Delta t) = x(t_0) + \frac{1}{2}\left(F_1 + F_2\right), \tag{11.184f}$$

$$\epsilon(t_0 + \Delta t) = \epsilon(t_0) + \frac{1}{2}\left(H_1 + H_2\right) + \alpha. \tag{11.184g}$$

3. The Fourth-Order Runge–Kutta Scheme

Define

$$a_0 = 1, \quad b_0 = 1, \tag{11.185a}$$

$$a_1 = \frac{1}{4} + \frac{\sqrt{3}}{6}, \quad b_1 = \frac{1}{4} - \frac{\sqrt{3}}{6} + \frac{\sqrt{6}}{12}, \tag{11.185b}$$

$$a_2 = \frac{1}{4} + \frac{\sqrt{3}}{6}, \quad b_2 = \frac{1}{4} - \frac{\sqrt{3}}{6} - \frac{\sqrt{6}}{12}, \tag{11.185c}$$

$$a_3 = \frac{1}{2} + \frac{\sqrt{3}}{6}, \quad b_2 = \frac{1}{2} - \frac{\sqrt{3}}{6}, \tag{11.185d}$$

$$a_4 = \frac{5}{4} + \frac{\sqrt{3}}{6}, \quad b_4 = \frac{5}{4} - \frac{\sqrt{3}}{6} + \frac{\sqrt{6}}{12}, \tag{11.185e}$$

and ψ_1 and ψ_2 as Gaussian random numbers with $\langle \psi_i \rangle = 0$, $\langle \psi_i \psi_j \rangle = \delta_{ij}$, $i = 1, 2$. The stochastic fourth-order Runge–Kutta scheme is given by

$$\alpha_i = \lambda (D \Delta t)^{1/2} (a_i \psi_1 + b_i \psi_2), \quad i = 0, 1, 2, 3, 4 \tag{11.186a}$$

$$H_1 = \Delta t \, h (\epsilon(t_0) + \alpha_1), \tag{11.186b}$$

$$H_2 = \Delta t \, h \left(\epsilon(t_0) + \frac{1}{2} H_1 + \alpha_2 \right), \tag{11.186c}$$

$$H_3 = \Delta t \, h \left(\epsilon(t_0) + \frac{1}{2} H_2 + \alpha_3 \right), \tag{11.186d}$$

$$H_4 = \Delta t \, h (\epsilon(t_0) + H_3 + \alpha_4), \tag{11.186e}$$

$$\epsilon(t_0 + \Delta t) = \epsilon(t_0) + \frac{1}{6} (H_1 + 2H_2 + 2H_3 + H_4) + \alpha_0. \tag{11.186f}$$

The algorithm for the system (11.181) is

$$H_1 = \Delta t \, h (\epsilon(t_0) + \alpha_1), \tag{11.187a}$$

$$F_1 = \Delta t \, f (x(t_0), \epsilon(t_0) + \alpha_1), \tag{11.187b}$$

$$H_2 = \Delta t \, h \left(\epsilon(t_0) + \frac{1}{2} H_1 + \alpha_2 \right), \tag{11.187c}$$

$$F_2 = \Delta t \, f \left(x(t_0) + \frac{1}{2} F_1, \epsilon(t_0) + \frac{1}{2} H_1 + \alpha_2 \right), \tag{11.187d}$$

$$H_3 = \Delta t \, h \left(\epsilon(t_0) + \frac{1}{2} H_2 + \alpha_3 \right), \tag{11.187e}$$

$$F_3 = \Delta t \, f \left(x(t_0) + \frac{1}{2} F_2, \epsilon(t_0) + \frac{1}{2} H_2 + \alpha_3 \right), \tag{11.187f}$$

$$H_4 = \Delta t \, h (\epsilon(t_0) + H_3 + \alpha_4), \tag{11.187g}$$

$$F_4 = \Delta t \, f (x(t_0) + F_3, \epsilon(t_0) + H_3 + \alpha_4), \tag{11.187h}$$

$$\epsilon(t_0 + \Delta t) = \epsilon(t_0) + \frac{1}{6} (H_1 + 2H_2 + 2H_3 + H_4) + \alpha_0, \tag{11.187i}$$

$$x(t_0 + \Delta t) = x(t_0) + \frac{1}{6} (F_1 + 2F_2 + 2F_3 + F_4). \tag{11.187j}$$

11.12   Concluding Remarks

Various phenomena, processes and events in science, engineering and social science are modelled by appropriated mathematical model equations. These equations are generally nonlinear differential equations. In the present chapter, certain methods for numerical computation of solutions of ordinary differential equations are presented and applied to equations for which exact solutions are known. Developing methods for computing numerical solutions with higher accuracy is important since exactly solvable systems are very limited. Numerical methods of solving differential equations are playing a key role in the study of dynamical systems. Features of novel phenomena such as complicated regular solutions, chaotic solutions, different routes to chaotic behaviour, synchronization of chaotic systems, various resonances, etc. are investigated through numerical solutions. In this chapter ordinary differential equations with integer order derivatives are considered. The case of ordinary differential equations with fractional order derivatives is dealt in Chapter 16.

11.13　Bibliography

[1] E. Kreyszig, *Advanced Engineering Mathematics*. John Wiley, New York, 1999. 8th edition.

[2] G.F. Simmons, *Differential Equations*. Tata McGraw–Hill, New Delhi, 2001.

[3] E.A. Coddington, *An Introduction to Ordinary Differential Equations*. Printice-Hall of India, New Delhi, 2002.

[4] A. Jeffrey, *Advanced Engineering Mathematics*. Academic Press, San Diego, 2003. Indian reprint.

[5] A. Gezerlis and M. Williams, *Am. J. Phys.* 89:51, 2021.

[6] E. Suli and D. Mayers, *An Introduction to Numerical Analysis*. Cambridge University Press, Cambridge, 2003.

[7] W.H. Press, S.A. Teukolsky, W.T. Vetterling, B.P. Flannery, *Numerical Recipes in Fortran*. Foundation Books, New Delhi, 1993. Indian edition.

[8] J.H. Mathews, *Numerical Methods for Mathematics Science and Engineering*. Prentice-Hall of India, New Delhi, 2005.

[9] J. Cartwright, Convergence of Runge-Kutta Methods (1995); https://www.iact.ugr-csic.es/personal/julyan_cartwright/papers/rkpaper/node13.html (accessed on June 20, 2020).

[10] W.G. Hoover, J.C. Sprott and C.G. Hoover, *Am. J. Phys.* 84:786, 2016.

[11] R.L. Honeycutt, *Phys. Rev. A* 45:600, 604, 1992.

11.14　Problems

11.1 Obtain the second-order Runge–Kutta methods with $\omega_2 = 3/4$ and 0. Compare them with $\omega_2 = 1/2$.

11.2 Show that for the linear differential equation $x' = x$, $x(0) = A$ the truncation error in the fourth-order Runge–Kutta method is $\approx (A/120)h^5 + O(h^6)$.

11.3 Show that for a first-order equation the convergence criteria for the Milne–Simpson method are

$$h < \frac{3}{|f_x(x_n, t_n)|} \quad \text{and} \quad 10^m D < \frac{3}{h\,|f_x(x_n, t_n)|}\,.$$

11.4 Show that for a first-order equation the convergence criteria for the Hamming method are

$$h < \frac{8}{3\,|f_x(x_n, t_n)|} \quad \text{and} \quad 10^m D < \frac{8}{3h\,|f_x(x_n, t_n)|}\,.$$

11.5 The error function erf(x) is defined as

$$\text{erf}(x) = \frac{2}{\sqrt{\pi}} \int_0^x e^{-u^2}\,\mathrm{d}u\,.$$

Solve the differential equation for erf(x) with the initial condition erf(0) = $2/\sqrt{\pi}$ by the Euler, the second- and the fourth-order Runge–Kutta methods with step size $h = 0.1$ and obtain the table of erf(x) in the interval $x \in [0, 1]$.

11.6 The Fresnel integral is given by

$$C(x) \;=\; \int_0^x \cos\left(\pi u^2/2\right) du$$

Solve the differential equation for $C(x)$ with the initial condition $C(0) = 1$ by the Euler, second and fourth-order Runge–Kutta methods with step size $h = 0.1$ and obtain the table of $C(x)$ in the interval $x \in [0, 1]$.

For the following problems find the numerical solution using Euler, second-order and fourth-order Runge–Kutta methods at time (independent variable) $t = 0.1$ with $h = 0.1$. Also, find the numerical solution in the interval $t \in [0, 1]$ with $h = 0.01, 0.05, 0.1$ and 0.2 by the above methods and compare the numerical solution with the exact solution.

11.7 Find the numerical solution of the equation $\dot{x} = 1 - x^2$ with $x(0) = 0$. The exact solution of the above equation is $x(t) = \tan\left(t + \tan^{-1} x(0)\right)$.

11.8 Find the numerical solution of the equation $\dot{x} = 2x - x^2$ with $x(0) = 1$. The exact solution of the above equation is $x(t) = 1 + \tanh t$.

11.9 The exact solution of the equation $\dot{x} = -x^3$ is $x(t) = x(0)/\sqrt{1 + 2x(0)^2 t}$. Obtain the numerical solution for $x(0) = 1$.

11.10 Find the numerical solution of the equation $\dot{x} = \mu x - x^3$ for $\mu = 1$ and $x(0) = 0.5$ whose exact solution is

$$x(t) = \left[\frac{1}{\mu} + \left(\frac{1}{x(0)^2} - \frac{1}{\mu}\right) e^{-2\mu t}\right]^{-1/2}.$$

11.11 Find the numerical solution of the equation $\dot{x} = -\mu x + x^2$ for $\mu = 2$ and $x(0) = 1$ whose exact solution is

$$x(t) = \mu \frac{A e^{-\mu t}}{A e^{-\mu t} - 1}, \quad A = \frac{x(0)}{x(0) - \mu}, \quad \mu \neq 0.$$

11.12 Find the numerical solution of the Bernoulli equation $\dot{x} = -x + x^2 t$ for the initial condition $x(0) = 0.5$. Its exact solution is $x(t) = 1/(A e^t + 1 + t)$, where $A = (1 - x(0))/x(0)$.

11.13 A radioactive substance disintegrate at a rate proportional to the amount present. The equation governing the process is $\dot{x} = -\lambda x$, where $x(t)$ is the amount of the substance present at time t and λ is a constant. Find the numerical solution for $\lambda = 0.5$ and $x(0) = 5\,\text{gm}$.

11.14 The equation of motion of an overdamped linear harmonic oscillator driven by a periodic force is given by the equation $\dot{x} = -x + \cos t$. Find the numerical solution for $x(0) = 0.1$.

11.15 Hormone secretion is modelled by the equation $\dot{x} = a - b\cos(2\pi t/24) - cx$, where $x(t)$ is the amount of a certain hormone in the blood, a is the average secretion rate, cx models the removal rate of the hormone from the blood and $b\cos(2\pi t/24)$

represents the daily secretion cycles. Find the solution for $a = b = c = 1$ and $x(0) = 3.0$. The exact solution is given by

$$x(t) = \frac{a}{c} + Ae^{-ct} - \frac{b\omega}{\omega^2 + c^2} \sin \omega t - \frac{bc}{\omega^2 + c^2} \cos \omega t,$$

where

$$A = \left(x(0) + \frac{bc}{\omega^2 + c^2} - \frac{a}{c} \right), \quad \omega = \frac{2\pi}{24}.$$

11.16 The equation for the velocity of a parachutist is given by $\dot{v} = g - (a/m)v^2$, where v is the velocity and m is the mass of the person plus the equipment. For simplicity choose $m = 1$ and $a = 1$. g is 9.8 m/sec^2. If the initial velocity is $v(0) = 0.5$ m/sec calculate $v(t)$.

11.17 The evolution of population density x of species in a place is governed by the logistic equation $\dot{x} = kx(1-x)$, where k is a constant parameter. Its exact solution is

$$x(t) = \frac{x(0)}{x(0) + (1 - x(0))e^{-kt}}.$$

Find the numerical solution for $k = 0.5$ and $x(0) = 0.5$.

11.18 Find the numerical solution of the RC circuit equation

$$\dot{q} + q/(RC) - V/R = 0, \quad R = 10\,\Omega, \ C = 0.1\,\text{farad}, \ V = 4\,\text{V}$$

with the initial condition $q(0) = 0$. The exact solution of the equation is $q(t) = CV + (q(0) - CV)e^{-t/RC}$.

11.19 The RL circuit equation is given by

$$i' + iR/L - V/L = 0, \quad R = 5\,\Omega, \ L = 10\,\text{henry}, \ V = 2\,\text{volts}.$$

Solve the equation with $i(0) = 0$. Its exact solution is

$$i(t) = \frac{V}{R} + \left(i(0) - \frac{V}{R} \right) e^{-t/(L/R)}.$$

11.20 In a chemical reaction cane sugar and water in the presence of the catalyst H$^+$ react to form glucose. The rate equation describing the time variation of concentration $x(t)$ of glucose is given by $\dot{x} = k(\alpha - x)$, where k is a positive constant for reaction and α is the initial concentration of cane sugar. Find the numerical solution for $k = 1$, $\alpha = 0.5$ and $x(0) = 5$. The exact solution is given by $x(t) = \alpha + (x(0) - \alpha)e^{-kt}$.

11.21 The intensity $I(x)$ of sound waves travelling through a medium is given by the equation $dI/dx = -I$, where x is the distance travelled. If $I(0) = 3$ units find the numerical solution. The exact solution is $I(x) = I(0)e^{-x}$.

11.22 A hot object of temperature $T_0 = 60°$C is put in an environment of temperature $T_1 = 30°$C. The variation of temperature of the object is governed by the Newton's cooling law given by $\dot{T} = -(T - T_1)$. Find the numerical solution. The exact solution is $T(t) = T_1 + (T_0 - T_1)e^{-t}$.

11.23 The Hermite differential equation is given by $\ddot{x} - 2t\dot{x} + 2x = 0$. Find the numerical solution for $x(0) = 0$, $\dot{x}(0) = 2$. The exact solution is $x = 2t$.

11.24 A particle of unit mass moves in a vertical direction under the force of gravity. If the downward direction is positive the equation of motion is $\ddot{x} - g = 0$, $g = 9.8\,\text{m/sec}^2$. For the initial conditions $x(0) = 0$, $\dot{x}(0) = v_0 = 1$ solve the equation. The exact solution of the equation is

$$x(t) = x(0) + v_0 t + \frac{1}{2}gt^2 \,.$$

11.25 Find the numerical solution of the equation $\dot{N}_1 = -\lambda_1 N_1$, $\dot{N}_2 = \lambda_1 N_1 - \lambda_2 N_2$ for $\lambda_1 = 0.7\,\text{hour}^{-1}$, $\lambda_2 = 0.14\,\text{hour}^{-1}$ with the initial conditions $N_1(0) = N_0 = 100$ and $N_2(0) = 0$. Its exact solution is

$$N_1 = N_0 e^{-\lambda_1 t}, \quad N_2 = \frac{\lambda_1}{\lambda_2 - \lambda_1} N_0 \left(e^{-\lambda_1 t} - e^{-\lambda_2 t}\right) \,.$$

12

Symplectic Integrators for Hamiltonian Systems

12.1 Introduction

Hamiltonian systems of ordinary differential equations describing non-dissipative phenomena occur in many fields of physics, chemistry and other branches of science. Numerical methods such as the Euler and the Runge–Kutta discussed in the previous chapter are not appropriate for numerically solving Hamiltonian systems. These methods do not conserve the total energy (H). Further, such systems are not structurally stable against non-Hamiltonian perturbations. The numerical approximation to the differential equations of Hamiltonian systems essentially introduce non-Hamiltonian perturbations. Consequently, the long-time behaviour obtained from an ordinary numerical method will be completely different from the actual behaviour of the original Hamiltonian system and the Hamiltonian is not conserved, that is, the Hamiltonian associated with the numerical solution is *not a constant of motion*. To overcome this problem special types of methods called *symplectic integrators* are developed for Hamiltonian systems. Symplectic integrators are a special type of *geometric integrators*. Geometric integration methods are the numerical methods developed for Hamiltonian systems that preserve at least any one of the geometric properties [1–7] of the flow of the Hamiltonian systems. *Why are the symplectic methods called so?* Because when these methods are employed to Hamiltonian systems they preserve the so-called *linear symplectic structure* that are inherent in the phase space.

The first symplectic method was proposed by de Vogelaere [8] and later by Ruth [9]. Symplectic integration algorithms are important in many branches of science. For instance, in the simulation of particle accelerators and in the numerical fluid analysis the conservation of symplectic structure is very important. Symplectic integration methods are very useful for accurate computation of singularities, long-term integration of solar systems, analysis of highly oscillatory systems, study of molecular dynamics and astronomy [7,10]. Highly accurate determination of track of dynamical variables over a long time is very important in accelerators like the Large Hadron Collider, interplanetary spacecraft and near-earth satellites for which symplectic integrators are helpful.

Symplectic methods of Hamiltonian systems have many features concerning long-time integration. Particularly, they have

1. no divergence of error in the energy (H),

2. linear error growth in the canonically conjugate dependent variables instead of quadratic growth and

3. correct qualitative behaviour.

The present chapter, first briefly introduces some basic ideas of Hamiltonian systems and shows the failure of Euler method in solving them. Then, presents a few symplectic integration methods. For more details on symplectic integrators the refs. [1–19] can be referred.

DOI: 10.1201/9781032649931-12

12.2   Hamiltonian Systems and Hamilton's Equation of Motion

The *Lagrangian* of a classical mechanical system is defined as

$$L = T - V, \tag{12.1}$$

where T and V are kinetic and potential energies, respectively. Assume that V is a function of generalized coordinates $\mathbf{q}$ only while T is a function of $\mathbf{q}$ and $\dot{\mathbf{q}}$. The *Euler–Lagrange equations* describing the motion of the system are

$$\frac{\mathrm{d}}{\mathrm{d}t}\left(\frac{\partial L}{\partial \dot{\mathbf{q}}}\right) - \frac{\partial L}{\partial \mathbf{q}} = 0. \tag{12.2}$$

Hamilton introduced the coordinates

$$\mathbf{p} = \frac{\partial L}{\partial \dot{\mathbf{q}}} \tag{12.3}$$

and defined the *Hamiltonian*

$$H(\mathbf{q}, \mathbf{p}) = T + V. \tag{12.4}$$

H represents the total energy of the system and $\mathbf{p}$ is called the *conjugate generalized momentum*. Hamilton showed that Eqs. (12.2) are equivalent to the *Hamilton's equations*

$$\dot{\mathbf{q}} = \frac{\partial H}{\partial \mathbf{p}}, \tag{12.5a}$$

$$\dot{\mathbf{p}} = -\frac{\partial H}{\partial \mathbf{q}}. \tag{12.5b}$$

Examples:

1. Harmonic Oscillator

The Hamiltonian of a one-dimensional particle in the potential $V = q^2/2$ is

$$H = \frac{1}{2}\left(p^2 + q^2\right) = c, \quad \text{a constant.} \tag{12.6}$$

The associated equations of motion are

$$\dot{q} = p, \quad \dot{p} = -q. \tag{12.7}$$

2. Pendulum System

The Hamiltonian of a pendulum system is

$$H = \frac{1}{2}p^2 - \cos q. \tag{12.8}$$

The Hamilton's equations of motion are

$$\dot{q} = p, \quad \dot{p} = -\sin q. \tag{12.9}$$

The *phase space* of the system (12.5) is an abstract space formed by the coordinates q_i and p_i. For a one-dimensional particle, the coordinates axes of the phase space are q_x and p_x

only and hence its dimension is 2. For a particle moving in a three-dimensional real space q and p have components q_x, q_y, q_z and p_x, p_y, p_z, and hence its phase space dimension is 6. Each point in a phase space represents a state of the system at a given time. As time changes, the state of the system evolves and the particle traces out a trajectory in the phase space.

What is the rate of change of phase space volume under time evolution? To determine this let us consider the equation of motion of the form

$$\frac{d\mathbf{X}}{dt} = \mathbf{F}(\mathbf{X}), \tag{12.10}$$

where $\mathbf{X} = (x_1, x_2, \ldots, x_n)$ and $\mathbf{F} = (F_1, F_2, \ldots, F_n)$. The Hamilton's Eqs. (12.5) can be rewritten in the above form. The rate of change of a volume is given by

$$\Lambda = \sum \frac{\partial F_i}{\partial x_i} = \nabla \cdot \mathbf{F}. \tag{12.11}$$

The conditions $\nabla \cdot \mathbf{F} = 0$, < 0 and > 0 represent conservative, dissipative and volume expanding systems, respectively. For details about the various types of solutions and the behaviours exhibited by these systems, a reader may refer to the refs. [20–23].

For a Hamiltonian system of the form

$$\dot{q}_i = \frac{\partial H}{\partial p_i}, \quad \dot{p}_i = -\frac{\partial H}{\partial q_i}, \quad i = 1, 2, \ldots, n \tag{12.12}$$

define $\mathbf{X} = (x_1, x_2, \ldots, x_{2n}) = (q_1, q_2, \ldots, q_n, p_1, p_2, \ldots, p_n)$. Then,

$$\begin{aligned}
\nabla \cdot \mathbf{F} &= \sum_{k=1}^{2n} \frac{\partial F_k}{\partial x_k} \\
&= \sum_{i=1}^{n} \left[\frac{\partial}{\partial q_i} \left(\frac{\partial H}{\partial p_i} \right) + \frac{\partial}{\partial p_i} \left(-\frac{\partial H}{\partial q_i} \right) \right] \\
&= 0.
\end{aligned} \tag{12.13}$$

At $t = 0$ choose a closed $(N-1)$ dimensional surface S_0 in the N-dimensional phase space and then evolve each point on the surface S_0 forward in time by using them as a set of initial conditions. S_0 evolves to a closed surface S_t at some later time t. The N-dimensional volume V_0 of the region enclosed by S_0 and $V(t)$ of the region enclosed by S_t are the same. That is, $V(t) = V(0)$. Therefore, the system is a volume preserving in phase space and is conservative.

From Eq. (12.5) notice that

$$\frac{dH}{dt} = \frac{\partial H}{\partial q}\frac{dq}{dt} + \frac{\partial H}{\partial p}\frac{dp}{dt} = -\dot{p}\dot{q} + \dot{q}\dot{p} = 0. \tag{12.14}$$

Further, for $\omega = d\mathbf{p} \wedge d\mathbf{q}$ (called *symplectic manifold*), where $\wedge$ represents wedge product,

$$\begin{aligned}
\frac{d\omega}{dt} &= d\dot{\mathbf{p}} \wedge d\mathbf{q} + d\mathbf{p} \wedge d\dot{\mathbf{q}} \\
&= -(\nabla_{\mathbf{qq}} H)\,d\mathbf{q} \wedge d\mathbf{q} + d\mathbf{p} \wedge (\nabla_{\mathbf{pp}} H)\,d\mathbf{p} \\
&= 0.
\end{aligned} \tag{12.15}$$

Equation (12.5) conserves the energy and the symplectic form $\omega = d\mathbf{p} \wedge d\mathbf{q}$.

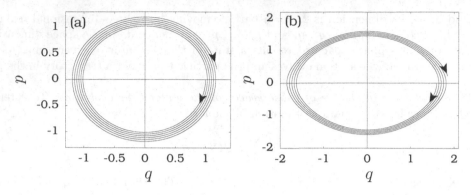

FIGURE 12.1
The numerical solution of (a) the linear harmonic oscillator Eq. (12.7) and (b) the pendulum
Eq. (12.9) obtained by the explicit Euler method with step size $h = 0.01$. For Eq. (12.7)
the energy is $H = 0.5$ and the initial conditions are $q(0) = 1$ and $p(0) = 0$. For Eq. (12.9)
$q(0) = 1.6$ and $p(0) = 0$.

12.3   Application of the Explicit Euler and the Runge–Kutta Methods

First, consider the harmonic oscillator Eq. (12.7). Its exact solution is

$$q(t) \;=\; q(0)\cos t + p(0)\sin t \,, \tag{12.16a}$$
$$p(t) \;=\; -q(0)\sin t + p(0)\cos t \,. \tag{12.16b}$$

The Hamiltonian is a conserved quantity $(\mathrm{d}H/\mathrm{d}t = 0)$, that is, H is a constant of time. The
solutions for different initial conditions are bounded and periodic.

A forward Euler discretization of Eq. (12.7) gives

$$\frac{q_{n+1} - q_n}{h} \;=\; p_n, \quad \longrightarrow \quad q_{n+1} = q_n + hp_n, \tag{12.17a}$$

$$\frac{p_{n+1} - p_n}{h} \;=\; -q_n \,, \quad \longrightarrow \quad p_{n+1} = p_n - hq_n. \tag{12.17b}$$

Now, obtain

$$\frac{1}{2}\left(p_{n+1}^2 + q_{n+1}^2\right) = \frac{1}{2}\left(1 + h^2\right)\left(p_n^2 + q_n^2\right) . \tag{12.18}$$

That is, $H_{n+1} = (1 + h^2)H_n$. H increases with n and H is not a conserved quantity.
Consequently, the solution obtained by the Euler method is unbounded and nonperiodic.
Figure 12.1a shows the plot of the numerical solution obtained by the Euler method with $h = 0.01$. The initial conditions used are $q(0) = 1$, $p(0) = 0$ and $H(0) = 0.5$. The solution spirals
outward and diverges to infinity. Similar result is obtained for the pendulum Eq. (12.9)
(Fig. 12.1b).

The fourth-order Runge–Kutta method is also found to be unsuitable for Eqs. (12.7)
and (12.9). However, H is conserved for much longer time compared to Euler method.
Figure 12.2a depicts the variation of the error in the Hamiltonian (H_{e}) computed from the
numerical solution (q, p) for different values of h in the case of the Euler method. Here,
$H_{\mathrm{e}} = H_{\mathrm{exact}} - H_{\mathrm{numerical}}$ with $H_{\mathrm{exact}} = 0.5$. H_{e} diverges rapidly with time. Figure 12.2b

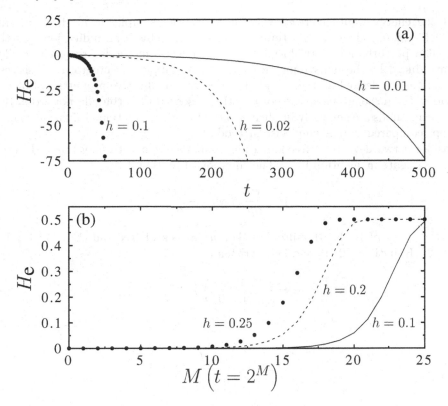

FIGURE 12.2
Variation of the error associated with the Hamiltonian computed from the numerical solution of the linear harmonic oscillator. (a) Solution is obtained by the explicit Euler method with three different step sizes. (b) Solution is obtained by the fourth-order Runge–Kutta method. In both the methods $q(0) = 1$, $p(0) = 0$ and $H(0) = 0.5$.

shows the result for the fourth-order Runge–Kutta method. The Euler method produced solutions which are diverging to $\pm\infty$. In Fig. 12.2b as $t \to \infty$, $H_e \to 0.5$, that is $H_{\mathrm{numer}} = 0$, which implies $q(t)$, $p(t) \to 0$ as $t \to \infty$. The solution approached the equilibrium point $(q, p) = (0, 0)$.

From the above examples, it is clear that ordinary numerical methods are not suitable for numerically integrating Hamiltonian systems. A reason is that the numerical approximation to the ordinary differential equation of the equation of motion of a Hamiltonian system introduces a non-Hamiltonian perturbation. Consequently, the Hamiltonian H is not conserved. Numerical methods are developed for Hamiltonian systems which preserve their features and are called *symplectic methods* or *symplectic integrators*. The basic idea of symplectic methods is presented in the next section.

12.4 Basic Idea of Symplectic Methods

In symplectic methods the Hamiltonian is not strictly conserved, however, it undergoes bounded oscillations. In contrast, as shown above, in nonsymplectic methods like the Euler method the energy would increase without limit. The symplectic methods produce behaviour which looks like that of a Hamiltonian system. This is an indication that the non-Hamiltonian perturbations introduced by the symplectic methods are much smaller than those introduced by the nonsymplectic methods. A symplectic integrator preserves phase space volume to a desired accuracy and also preserves qualitative properties of phase space trajectories. Even though energy is not exactly conserved the orbits do not cross. However, symplectic methods are generally implicit and require more function evaluations and smaller time steps compared to nonsymplectic methods.

Consider a two degrees of freedom Hamiltonian system with coordinate denoted by q and the conjugate momentum by p. The equation of motion is

$$\dot{q} = \frac{\partial H}{\partial p}, \quad \dot{p} = -\frac{\partial H}{\partial q}, \tag{12.19}$$

where $H = H(q,p)$ is the Hamiltonian. For n-degrees of freedom, Eq. (12.19) becomes Eq. (12.12). Equation (12.19) can be rewritten as

$$\begin{pmatrix} \dot{q} \\ \dot{p} \end{pmatrix} = \begin{pmatrix} 0 & 1 \\ -1 & 0 \end{pmatrix} \begin{pmatrix} \partial H/\partial q \\ \partial H/\partial p \end{pmatrix}. \tag{12.20}$$

or

$$\dot{\mathbf{Z}} = J\frac{\partial H}{\partial Z}, \quad \mathbf{Z} = \begin{pmatrix} q \\ p \end{pmatrix}, \quad J = \begin{pmatrix} 0 & 1 \\ -1 & 0 \end{pmatrix}, \tag{12.21a}$$

and

$$\frac{\partial H}{\partial \mathbf{Z}} = \begin{pmatrix} \partial H/\partial q \\ \partial H/\partial p \end{pmatrix}. \tag{12.21b}$$

The matrix J is called *symplectic matrix* (which means interwined matrix) [24].

The transformation

$$\mathbf{Z}(t_0) = \mathbf{Z}(t) \tag{12.22}$$

conserves both the energy (H) and

$$S(\mathbf{Z}_1, \mathbf{Z}_2) = \mathbf{Z}_1^T J \mathbf{Z}_2, \tag{12.23}$$

where S is the area of the parallelogram defined by the two initial vectors $\mathbf{Z}_1$ and $\mathbf{Z}_2$.

Let us write

$$\mathbf{Z}(t) = A\mathbf{Z}(0). \tag{12.24}$$

Then, Eq. (12.23) becomes

$$S(\mathbf{Z}_1, \mathbf{Z}_2) = \mathbf{Z}_1^T(0)A^T J A\mathbf{Z}_2(0) = \mathbf{Z}_1^T(0)J\mathbf{Z}_2(0), \tag{12.25}$$

where $A^T H A = J$. The solution of Eq. (12.19) induces a transformation A on the phase space and the associated Jacobian is A. The associated map is called *symplectic*.

Example:

Consider the linear harmonic oscillator Hamiltonian (Eq. (12.6)) and its equation of motion (Eq. (12.7)). Its solution (12.16) can be rewritten in the form of Eq. (12.24) as

$$\mathbf{Z}(t) = \begin{pmatrix} \cos t & \sin t \\ -\sin t & \cos t \end{pmatrix} \begin{pmatrix} q(0) \\ p(0) \end{pmatrix} = A\mathbf{Z}(0). \tag{12.26}$$

For the linear harmonic oscillator Hamiltonian

$$A^T J A = \begin{pmatrix} \cos t & -\sin t \\ \sin t & \cos t \end{pmatrix} \begin{pmatrix} 0 & 1 \\ -1 & 0 \end{pmatrix} \begin{pmatrix} \cos t & \sin t \\ -\sin t & \cos t \end{pmatrix}$$

$$= \begin{pmatrix} 0 & 1 \\ -1 & 0 \end{pmatrix}$$

$$= J. \tag{12.27}$$

Thus, the transformation (12.22) not only conserves H but also preserves the symplectic form given by Eq. (12.23).

Suppose, a numerical method approximates the solution of a differential equation, for example Eq. (12.19), and the time step h is a constant. Let the exact solution of the equation is $\psi(h)$ and the approximation is ψ_h. The map ψ_h to be *symplectic* if it satisfies the equation

$$\psi_h^T J \psi_h = J. \tag{12.28}$$

ψ_h has to be symplectic if $\psi(h)$ is symplectic. For a symplectic $\psi(h)$ certain numerical methods produce nonsymplectic ψ_h. An example is the Euler method. The numerical methods which give rise to symplectic map are called *symplectic numerical methods*.

12.5 Explicit Euler Method is Not a Symplectic Method

In this section, let us apply the Euler method to the linear harmonic oscillator and show that it is nonsymplectic.

The Euler algorithm for Eq. (12.7) is given by Eq. (12.17). A discrete map of the map

$$\mathbf{Z}_{n+1} = \mathbf{F}(\mathbf{Z}_n) \tag{12.29}$$

is said to be *conservative* if the determinant of the Jacobian is 1:

$$|J| = |\partial \mathbf{F} / \partial \mathbf{Z}|$$

$$= 1. \tag{12.30}$$

For the map (12.17)

$$|J| = \begin{vmatrix} 1 & h \\ -h & 1 \end{vmatrix}$$

$$= 1 + h^2. \tag{12.31}$$

Since $|J| \neq 1$, the map (12.17) is not conservative or area-preserving. Further, as $h^2 > 0$ and $1 + h^2 > 1$ the map is area expanding.

Consider

$$q(h) = q(0) + hp(0), \tag{12.32a}$$
$$p(h) = p(0) - hq(0). \tag{12.32b}$$

Or

$$\mathbf{Z}(h) = \begin{pmatrix} 1 & h \\ -h & 1 \end{pmatrix} \begin{pmatrix} q(0) \\ p(0) \end{pmatrix}$$

$$= A\mathbf{Z}(\mathbf{0}). \tag{12.33}$$

The symplectic form (12.23) or (12.25) is

$$
\begin{aligned}
S &= \mathbf{Z}_1^T(0) A^T J A \mathbf{Z}_2(0) \\
&= \mathbf{Z}_1^T(0) \begin{pmatrix} 1 & -h \\ h & 1 \end{pmatrix} \begin{pmatrix} 0 & 1 \\ -1 & 0 \end{pmatrix} \begin{pmatrix} 1 & h \\ -h & 1 \end{pmatrix} \mathbf{Z}_2(0) \\
&= \mathbf{Z}_1^T(0) \begin{pmatrix} h & 1 \\ -1 & h \end{pmatrix} \begin{pmatrix} 1 & h \\ -h & 1 \end{pmatrix} \mathbf{Z}_2(0) \\
&= \mathbf{Z}_1^T(0) \begin{pmatrix} 0 & 1+h^2 \\ -1-h^2 & 0 \end{pmatrix} \mathbf{Z}_2(0) \\
&= \mathbf{Z}_1^T(0) \left(1+h^2\right) \begin{pmatrix} 0 & 1 \\ -1 & 0 \end{pmatrix} \mathbf{Z}_2(0) \\
&= \left(1+h^2\right) \mathbf{Z}_1^T(0) J \mathbf{Z}_2(0) .
\end{aligned}
\tag{12.34}
$$

That is, the symplectic form is increased by a factor $(1 + h^2)$ at each time step. When Eq. (12.17) is viewed as a discrete version of Eq. (12.7) the following can be noticed:

1. Equation (12.7) is area-preserving (conservative) while Eq. (12.17) is area expanding.

2. The (exact)solution of Eq. (12.7) preserves the symplectic form. But the solution of Eq. (12.17) not preserve the symplectic form.

The point is that the explicit Euler algorithm is not a symplectic algorithm.

12.6   First-Order Symplectic Algorithms

For the equation $\dot{x} = f(x)$, $x(t_0) = x_0$, an integration scheme is said to be nth-order if $x_1 - x(t_0 + h) = O(h^{n+1})$ as $h \to 0$, where x_1 is the result obtained by applying the numerical scheme to the initial state x_0 for one time step. Let us begin with simple first-order methods.

The simplest symplectic method is the leapfrog method. Restrict to the equations of the form

$$
\dot{\mathbf{q}} = \mathbf{p}, \quad \dot{\mathbf{p}} = \mathbf{F}(\mathbf{q}),
\tag{12.35}
$$

where $\mathbf{q} = (q_1, q_2, \ldots, q_N)$, $\mathbf{p} = (p_1, p_2, \ldots, p_N)$ and $\mathbf{F} = (f_1, f_2, \ldots, f_N)$. The leapfrog method looks like one of the following two ways.

1. One approach is the discretization of Eq. (12.35) in the form

$$
\begin{aligned}
\mathbf{q}_{n+1} &= \mathbf{q}_n + h\mathbf{p}_n, \tag{12.36a} \\
\mathbf{p}_{n+1} &= \mathbf{p}_n + h\mathbf{F}\left(\mathbf{q}_{n+1}\right). \tag{12.36b}
\end{aligned}
$$

In the above algorithm, first $\mathbf{q}$ is updated using $\mathbf{p}_n$ and $\mathbf{q}_n$ and then $\mathbf{p}$ is updated using the current $\mathbf{p}_n$ and the updated latest $\mathbf{q}$, $\mathbf{q}_{n+1}$.

2. The second approach first determines $\mathbf{p}_{n+1}$ using $\mathbf{q}_n$ and $\mathbf{p}_n$ and then determines $\mathbf{q}_{n+1}$ using the current $\mathbf{q}_n$ and the updated $\mathbf{p}$, namely, $\mathbf{p}_{n+1}$. This gives

$$
\begin{aligned}
\mathbf{p}_{n+1} &= \mathbf{p}_n + h\mathbf{F}\left(\mathbf{q}_n\right), \tag{12.37a} \\
\mathbf{q}_{n+1} &= \mathbf{q}_n + h\mathbf{p}_{n+1}. \tag{12.37b}
\end{aligned}
$$

Because of the alternating process in the algorithms (12.36) and (12.37) these methods are called *leapfrog methods*. The above two algorithms are accurate to first-order in h.

For the linear harmonic oscillator Eq. (12.7) the algorithm (12.36) becomes

$$q_{n+1} = q_n + hp_n, \tag{12.38a}$$
$$p_{n+1} = \left(1 - h^2\right) p_n - hq_n. \tag{12.38b}$$

Then,

$$|J| = \begin{vmatrix} 1 & h \\ -h & 1 - h^2 \end{vmatrix} = 1. \tag{12.39}$$

The map (12.38) is area-preserving. The discrete evolutionary operator ψ_h for (q, p) from Eq. (12.38) is written as [11]

$$\psi_h \begin{pmatrix} q \\ p \end{pmatrix} = \begin{pmatrix} 1 & h \\ h & 1 - h^2 \end{pmatrix} \begin{pmatrix} q \\ p \end{pmatrix}. \tag{12.40}$$

Note that $\det(\psi_h) = 1$ and ψ_h is a symplectic.

Next, the symplectic form (12.25) is

$$
\begin{aligned}
S &= \mathbf{Z}_1^{\mathrm{T}}(0) \begin{pmatrix} 1 & h \\ -h & 1 - h^2 \end{pmatrix} \begin{pmatrix} 0 & 1 \\ -1 & 0 \end{pmatrix} \begin{pmatrix} 1 & -h \\ h & 1 - h^2 \end{pmatrix} \mathbf{Z}_2(0) \\
&= \mathbf{Z}_1^{\mathrm{T}}(0) \begin{pmatrix} -h & 1 \\ -1 + h^2 & -h \end{pmatrix} \begin{pmatrix} 1 & -h \\ h & 1 - h^2 \end{pmatrix} \mathbf{Z}_2(0) \\
&= \mathbf{Z}_1^{\mathrm{T}}(0) \begin{pmatrix} 0 & 1 \\ -1 & 0 \end{pmatrix} \mathbf{Z}_2(0) \\
&= \mathbf{Z}_1^{\mathrm{T}}(0) J \mathbf{Z}_2(0).
\end{aligned}
\tag{12.41}
$$

Thus, the formula (12.38) preserves the symplectic form and hence it is a symplectic integrator. Similarly, it can be shown that the formula (12.37) for the linear harmonic oscillator is also a symplectic integrator. The algorithms (12.36) and (12.37) are called *implicit Euler algorithms* or *symplectic Euler algorithms. Why is the symplectic Euler method called a first-order method?* [5].

What is $p^2 + q^2$?

For the continuous time dynamical system (12.7)

$$p^2 + q^2 = 2H = C^2 \tag{12.42}$$

which is an equation for a circle Γ in $p - q$ phase space. The solutions of the ordinary differential equation of the linear harmonic oscillator with different values of H form circles in $p - q$ plane. Let us compute $p^2 + q^2$ for the map (12.38). Solving the map (12.38) for q_n and p_n gives

$$q_n = \left(1 - h^2\right) q_{n+1} - hp_{n+1}, \tag{12.43a}$$
$$p_n = p_{n+1} + hq_{n+1}. \tag{12.43b}$$

Then,

$$
\begin{aligned}
C^2 &= 2H \\
&= p_n^2 + q_n^2 \\
&= p_{n+1}^2 + h^2 q_{n+1}^2 + 2hp_{n+1}q_{n+1} + \left(1 - h^2\right)^2 q_{n+1}^2 \\
&\quad + h^2 p_{n+1}^2 - 2h\left(1 - h^2\right) p_{n+1}q_{n+1} \\
&= \left(1 + h^2\right) p_{n+1}^2 + \left(1 + h^4 - h^2\right) q_{n+1}^2 + 2h^3 p_{n+1}q_{n+1}
\end{aligned}
\tag{12.44}
$$

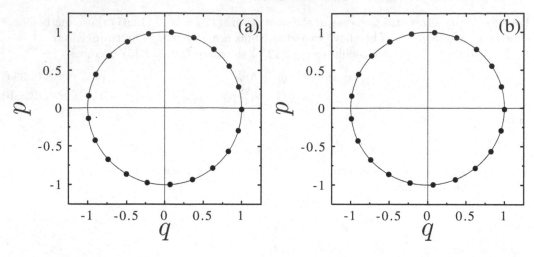

FIGURE 12.3
Comparison of the numerical solution with the exact solution (at some selected values of t) for the linear harmonic oscillator Eq. (12.7) with $h = 0.01$. The solid circles represent the exact solution. Continuous curve is the numerical solution. Here, $q(0) = 1$, $p(0) = 1$ and the exact $H = 0.5$. (a) The numerical solution is obtained by the implicit first-order Euler algorithm given by Eq. (12.45). (b) The numerical solution is obtained by the algorithm given by Eq. (12.46).

which is an equation for an ellipse γ. Under one iteration the circle Γ is given by $p_n^2 + q_n^2 = 2H = C^2$ which is mapped into the ellipse γ given by Eq. (12.44). However, both Γ and γ have same area. Thus, the qualitative features of closed orbits and bound solutions (nonperiodic) are preserved by the symplectic Euler methods while they are not preserved in the ordinary Euler method.

Example:

For the linear harmonic oscillator Eq. (12.7) an implicit Euler algorithm (refer Eq. (12.36)) is

$$q_{n+1} = q_n + hp_n, \qquad (12.45a)$$
$$p_{n+1} = p_n - hq_{n+1}. \qquad (12.45b)$$

In Eq. (12.45) q is first updated and this updated q is used to update p. An alternative algorithm is

$$p_{n+1} = p_n - hq_n, \qquad (12.46a)$$
$$q_{n+1} = q_n + hp_{n+1}. \qquad (12.46b)$$

In Eq. (12.46) p is first updated and then this p is used to update q. Equation (12.7) is solved by the above two methods with time step $h = 0.01$ and with $q(0) = 1$ and $p(0) = 0$ and $H = 0.5$. Figure 12.3 presents the numerical approximation along with the exact solution. The obtained numerical solution not diverge in both methods. Figure 12.4 depicts the variation of H associated with the numerical solution. H is not constant but it oscillates and is bounded. It remains near its true value. Table 12.1 gives the numerical solution for a few values of t.

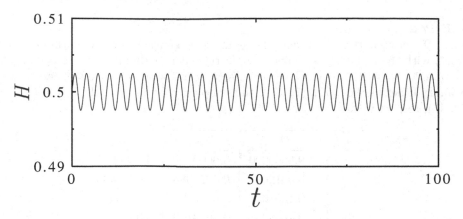

FIGURE 12.4
Variation of Hamiltonian (energy) H of linear harmonic oscillator computed using the numerical solution obtained by the symplectic Euler algorithm (12.45). Here, $h = 0.01$, $q(0) = 1$, $p(0) = 1$ and the exact $H = 0.5$.

To compare the numerical solutions of different methods use the distance D between the exact and the numerical solutions. It is defined as

$$D = \sqrt{(q_{\text{exact}} - q_{\text{numer}})^2 + (p_{\text{exact}} - p_{\text{numer}})^2}. \qquad (12.47)$$

Table 12.2 presents D versus t with the nonsymplectic and the symplectic Euler methods.

TABLE 12.1
Comparison of the numerical solution with the exact solution of Eq. (12.7). The symplectic Euler method, Eq. (12.45), with time step $h = 0.01$ is used. Here, $q(0) = 1$, $p(0) = 0$ and the exact $H = 0.5$.

t	q_{numer}	q_{exact}	p_{numer}	p_{exact}	H_{numer}
10	-0.8417693	-0.8390715	0.5440629	0.5440211	0.5022899
20	0.4125709	0.4080821	-0.9129907	-0.9129453	0.5018834
30	0.1494348	0.1542514	0.9880247	0.9880316	0.4992618
40	-0.6633372	-0.6669381	-0.7450113	-0.7451132	0.4975290
50	0.9637098	0.9649660	0.2621771	0.2623749	0.4987367
60	-0.9538620	-0.9524130	0.3050525	0.3048106	0.5014549
70	0.6369639	0.6333192	-0.7740850	-0.7738907	0.5024653
80	-0.1150256	-0.1103872	0.9939378	0.9938887	0.5005716
90	-0.4439396	-0.4480736	-0.8938397	-0.8939967	0.4980159
100	0.8599997	0.8623189	0.5060126	0.5063656	0.4978241
200	0.4835508	0.4871877	0.8729019	0.8732973	0.4978895
500	-0.8852209	-0.8838493	0.4696180	0.4677718	0.5020786

TABLE 12.2

Distance D between the exact and the numerical solutions of the harmonic oscillator Eq. (12.7) with the nonsymplectic and symplectic Euler methods. Here, $H = 0.5$, $q(0) = 1$, $p(0) = 0$ and $h = 0.01$. D_{NSE}, D_{SEQ} and D_{SEP} correspond to the distances obtained by the ordinary Euler method (Eq. (12.17)), the symplectic Euler algorithms (Eq. (12.45)) and Eq. (12.46), respectively.

t	D_{NSE}	D_{SEQ}	D_{SEP}
0	0.0000000	0.0000000	0.0000000
10	0.0512696	0.0026980	0.0027433
20	0.1051677	0.0044891	0.0046413
30	0.1618291	0.0048166	0.0050636
40	0.2213954	0.0036023	0.0038506
50	0.2840156	0.0012717	0.0013798
100	0.6486942	0.0023459	0.0027636
200	1.7181811	0.0036584	0.0051074
300	3.4814173	0.0037495	0.0062488
400	6.3884201	0.0029748	0.0057465
500	11.1811226	0.0023000	0.0038028
1000	147.3766221	0.0024475	0.0079505

12.7   A Second-Order Symplectic Algorithm

A second-order algorithm can be obtained by taking the time step as $h/2$ instead of h and making use of the formulas (12.36) and (12.37). Use the algorithm Eq. (12.37) from t to $t + h/2$ and the algorithm (12.36) from $t + h/2$ to $t + h$. These give the following set of four equations:

$$\mathbf{p}(t + h/2) = \mathbf{p}(t) + \frac{1}{2}h\mathbf{F}(\mathbf{q}(t)), \tag{12.48a}$$

$$\mathbf{q}(t + h/2) = \mathbf{q}(t) + \frac{1}{2}h\mathbf{p}(t + h/2), \tag{12.48b}$$

$$\mathbf{q}(t + h) = \mathbf{q}(t + h/2) + \frac{1}{2}h\mathbf{p}(t + h/2), \tag{12.48c}$$

$$\mathbf{p}(t + h) = \mathbf{p}(t + h/2) + \frac{1}{2}h\mathbf{F}(\mathbf{q}(t + h)). \tag{12.48d}$$

Replacing $\mathbf{p}(t + h/2)$ and $\mathbf{q}(t + h/2)$ in Eqs. (12.48c–d) by Eqs. (12.48a–b), respectively, leads to

$$\mathbf{q}(t + h) = \mathbf{q}(t) + h\mathbf{p}(t) + \frac{1}{2}h^2\mathbf{F}(\mathbf{q}(t)), \tag{12.49a}$$

$$\mathbf{p}(t + h) = \mathbf{p}(t) + \frac{1}{2}h\left[\mathbf{F}(\mathbf{q}(t)) + \mathbf{F}(\mathbf{q}(t + h))\right]. \tag{12.49b}$$

Rewrite Eqs. (12.49) as

$$\mathbf{q}_{n+1} = \mathbf{q}_n + h\mathbf{p}_n + \frac{1}{2}h^2\mathbf{F}(\mathbf{q}_n), \tag{12.50a}$$

$$\mathbf{p}_{n+1} = \mathbf{p}_n + \frac{1}{2}h\left[\mathbf{F}(\mathbf{q}_n) + \mathbf{F}(\mathbf{q}_{n+1})\right]. \tag{12.50b}$$

The above method is called *Störmer–Verlet method*. Comparison of this method with the other symplectic algorithms is discussed at the end of the next section.

12.8   Runge–Kutta Type Algorithms

For separable Hamiltonians of the form

$$H(\mathbf{q}, \mathbf{p}) = T(\mathbf{p}) + V(\mathbf{q}), \tag{12.51}$$

where $\mathbf{q} = (q_1, q_2, \ldots, q_N)$, $\mathbf{p} = (p_1, p_2, \ldots p_N)$ explicit Runge–Kutta type algorithms exist which preserve the symplectic structure [1,14,15,25]. This section describes the symplectic method of Candy and Rozmus [25] for separable Hamiltonians.

Consider the Hamiltonian of the form (12.51). The aim is to find a series of difference equations which preserve $\omega^2 = d\mathbf{q} \wedge d\mathbf{p}$ and give the approximation of $(\mathbf{q}(t = h), \mathbf{p}(t = h))$ for a given $(\mathbf{q}(0), \mathbf{p}(0)) = (\mathbf{q}_0, \mathbf{p}_0)$. (If the error in the difference approximation is $O(h^n)$, where h is the step size then the approximation is said to be an *nth-order symplectic method*.) That is, the goal is to find a map of the form

$$(\mathbf{q}, \mathbf{p}) \longrightarrow (\widetilde{\mathbf{q}}_0, \widetilde{\mathbf{p}}_0) + O\left(h^{n+1}\right), \tag{12.52}$$

where $\widetilde{\mathbf{q}}_0$ and $\widetilde{\mathbf{p}}_0$ represent approximations to $\mathbf{q}_0$ and $\mathbf{p}_0$, respectively. If it is possible to find a set of transformations which leave the Hamiltonian with the final form

$$H(\widetilde{\mathbf{q}}_0, \widetilde{\mathbf{p}}_0) = \sum_{i=n}^{\infty} H_i(\widetilde{\mathbf{q}}_0, \widetilde{\mathbf{p}}_0) h^i \tag{12.53}$$

such that $(\widetilde{\mathbf{q}}_0, \widetilde{\mathbf{p}}_0) \longrightarrow (\mathbf{q}_0, \mathbf{p}_0)$ as $h \to 0$ then it can be proved that Eq. (12.52) is satisfied [25].

12.8.1   Generating Function

In order to find an nth-order integration algorithm, introduce the series of l (canonical) transformations

$$(\mathbf{q}_l, \mathbf{p}_l) \xrightarrow{K_l} (\mathbf{q}_{l-1}, \mathbf{p}_{l-l}) \xrightarrow{K_{l-1}} \cdots \xrightarrow{K_l} (\mathbf{q}_0, \mathbf{p}_0). \tag{12.54}$$

In Eq. (12.54) $(\mathbf{q}_0, \mathbf{p}_0)$ is the initial condition at $t = 0$, $(\mathbf{q}_l, \mathbf{p}_l)$ is the numerical approximation at $t = t_0 + lh$ and the other set $(\mathbf{q}_1, \mathbf{p}_1), \ldots, (\mathbf{q}_{l-1}, \mathbf{p}_{l-1})$ are intermediate points. The choice

$$\begin{aligned}
K_i &= K_i(\mathbf{q}_{i-1}, \mathbf{p}_i, t) \\
&= -\mathbf{q}_{i-1} \cdot \mathbf{p}_i - h\left[a_i T(\mathbf{p}_i) + b_i V(\mathbf{q}_{i-1})\right], \quad i = 1, 2, \ldots, l \tag{12.55}
\end{aligned}$$

gives

$$\mathbf{p}_{i-1} \quad = \quad -\nabla_{\mathbf{q}_{i-1}} K_i = \mathbf{p}_i + h b_i \nabla_{\mathbf{q}_{i-1}} V \left(\mathbf{q}_{i-1}\right) , \tag{12.56a}$$

$$\mathbf{q}_i \quad = \quad -\nabla_{\mathbf{p}_i} K_i \quad = \mathbf{q}_{i-1} + h a_i \nabla_{\mathbf{p}_i} T \left(\mathbf{p}_i\right) \tag{12.56b}$$

and

$$H_{i-1} = H_i + \partial_t K_i = H_i - \left[a_i T \left(\mathbf{p}_i\right) + b_i V \left(\mathbf{q}_{i-1}\right) \right] . \tag{12.57}$$

Define $\mathbf{F}(\mathbf{q}) = -\nabla_{\mathbf{q}} V(\mathbf{q})$ and $\mathbf{P}(\mathbf{p}) = -\nabla_{\mathbf{p}} T(\mathbf{p})$. Then, from Eqs. (12.56)

$$\mathbf{q}_i \quad = \quad \mathbf{q}_0 + h \sum_{m=1}^{i} a_m P \left(\mathbf{p}_m\right) , \tag{12.58a}$$

$$\mathbf{p}_i \quad = \quad \mathbf{p}_0 + h \sum_{m=1}^{i} b_m F \left(\mathbf{q}_{m-1}\right) , \quad i = 1, 2, \ldots, l. \tag{12.58b}$$

To determine the unknown coefficients a_i and b_i expand $\mathbf{q}_i$ and $\mathbf{p}_i$ so that

$$H_0 \left(\mathbf{q}_0, \mathbf{p}_0\right) = \sum_{m=1}^{n-1} H_m \left(a_i, b_i, \mathbf{q}_0, \mathbf{p}_0\right) h^m + O \left(h^n\right) . \tag{12.59}$$

a_i and b_i are found by setting $H_m = 0$ for $m = 0, 1, \ldots, n-1$. Generally, for an equation of the form $\dot{x} = f(x(t))$ a K-stage Runge–Kutta method is

$$y_i \quad = \quad x_0 + h \sum_{m=1}^{K} a_{im} f \left(y_m\right) , \quad i = 1, 2, \ldots, K \tag{12.60a}$$

$$x_1 \quad = \quad x_0 + h \sum_{m=1}^{K} b_m f \left(y_m\right), \tag{12.60b}$$

where x_0, x_1 are the values of x at time $t = t_0$ and $t = t_0 + h$. Since Eqs. (12.58) are of the form of Eqs. (12.60) the method is called *Runge–Kutta type* symplectic method since Eqs. (12.58) are in the form of the Runge–Kutta algorithms.

12.8.2 H_m for $n \leq 4$

For $n \leq 4$

$$\mathbf{q}_i \quad = \quad \mathbf{q}_0 + \sum_{j=1}^{3} \alpha_{ji} h^j + O \left(h^4\right) , \tag{12.61a}$$

$$\mathbf{p}_i \quad = \quad \mathbf{p}_0 + \sum_{j=1}^{3} \beta_{ji} h^j + O \left(h^4\right) , \tag{12.61b}$$

where

$$\alpha_{1i} \quad = \quad \mathbf{P} \sum_{m=1}^{i} a_m , \quad \alpha_{21} = (\mathbf{F} \cdot \nabla_{\mathbf{p}}) \mathbf{P} \sum_{m=1}^{i} a_m \sum_{r=1}^{m} b_r , \tag{12.61c}$$

$$\alpha_{3i} \quad = \quad \frac{1}{2} (\mathbf{F} \cdot \nabla_{\mathbf{p}})^2 \mathbf{P} \sum_{m=1}^{i} a_m \left[\sum_{r=1}^{m} b_r \right]^2$$

$$+ \left[((\mathbf{P} \cdot \nabla_{\mathbf{q}}) \mathbf{F}) \cdot \nabla_{\mathbf{p}} \right] \mathbf{P} \sum_{m=2}^{i} a_m \sum_{r=2}^{m} b_r \sum_{s=1}^{r-1} a_s , \tag{12.61d}$$

and

$$\beta_{1i} = F \sum_{m=1}^{i} b_m, \quad \beta_{2i} = (\mathbf{P} \cdot \nabla_{\mathbf{p}}) \mathbf{F} \sum_{m=2}^{i} b_m \sum_{r=1}^{m-1} a_i, \tag{12.61e}$$

$$\beta_{3i} = \frac{1}{2} (\mathbf{P} \cdot \nabla_{\mathbf{q}})^2 \mathbf{F} \sum_{m=2}^{i} b_m \left[\sum_{r=1}^{m-1} a_r \right]^2$$

$$+ [((\mathbf{F} \cdot \nabla_{\mathbf{p}}) \mathbf{P}) \cdot \nabla_{\mathbf{q}}] \mathbf{F} \sum_{m=2}^{i} b_m \sum_{r=1}^{m-1} b_r \sum_{s=1}^{r} b_s. \tag{12.61f}$$

Next, expand $V(\mathbf{q}_i)$ and $T(\mathbf{p}_i)$ as

$$V(\mathbf{q}_i) = V(\mathbf{q}_0) + h [\alpha_{1i} \cdot \nabla_{\mathbf{q}}] V(\mathbf{q})$$

$$+ h^2 \left[\alpha_{2i} \cdot \nabla_{\mathbf{q}} + \frac{1}{2} (\alpha_{1i} \cdot \nabla_{\mathbf{q}})^2 \right] V(\mathbf{q}) + h^3 \left[\alpha_{3i} \cdot \nabla_{\mathbf{q}} + \frac{1}{6} (\alpha_{1i} \cdot \nabla_{\mathbf{q}})^3 \right.$$

$$\left. + (\alpha_{1i} \cdot \nabla_{\mathbf{q}})(\alpha_{2i} \cdot \nabla_{\mathbf{q}}) \right] V(\mathbf{q}) + O(h^4) \tag{12.62a}$$

and

$$T(\mathbf{p}_i) = T(\mathbf{p}_0) + h [\beta_{1i} \cdot \nabla_{\mathbf{p}}] T(\mathbf{p})$$

$$+ h^2 \left[\beta_{2i} \cdot \nabla_{\mathbf{p}} + \frac{1}{2} (\beta_{1i} \cdot \nabla_{\mathbf{p}})^2 \right] T(\mathbf{p}) + h^3 \left[\beta_{3i} \cdot \nabla_{\mathbf{p}} + \frac{1}{6} (\beta_{1i} \cdot \nabla_{\mathbf{p}})^3 \right.$$

$$\left. + (\beta_{1i} \cdot \nabla_{\mathbf{p}})(\beta_{2i} \cdot \nabla_{\mathbf{p}}) \right] T(\mathbf{p}) + O(h^4). \tag{12.62b}$$

In Eqs. (12.61) and (12.62) the derivatives are evaluated at $(\mathbf{q}_0, \mathbf{p}_0)$. Substitution of Eqs. (12.61) and (12.62) in Eq. (12.59) results in

$$H_0(\mathbf{q}_0, \mathbf{p}_0) = \sum_{i=0}^{3} h^i H_i + O(h^4), \tag{12.63}$$

where H_0 and H_1 are

$$H_0 = T(\mathbf{p}_0) \left[1 - \sum_{i=1}^{n} a_i \right] + V(\mathbf{q}_0) \left[1 - \sum_{i=1}^{n} b_i \right], \tag{12.64a}$$

$$H_1 = (\mathbf{F}_0 \cdot \mathbf{P}_0) \left[\sum_{m=1}^{n} b_m - \sum_{i=1}^{n} a_i \sum_{m=1}^{i} b_m - \sum_{m=1}^{n} a_m + \sum_{i=2}^{n} b_i \sum_{m=1}^{i-1} a_m \right]. \tag{12.64b}$$

H_2 is

$$H_2 = (\mathbf{P}_0 \cdot \nabla_{\mathbf{q}})(\mathbf{F} \cdot \mathbf{P}_0) \left[\sum_{m=2}^{n} b_m \sum_{r=1}^{m-1} a_r - \sum_{i=2}^{n} a_i \sum_{m=2}^{i} b_m \sum_{r=1}^{m-1} a_r \right.$$

$$\left. - \frac{1}{2} \left(\sum_{m=1}^{n} a_m \right)^2 + \frac{1}{2} \sum_{i=2}^{n} b_i \left(\sum_{m=1}^{i-1} a_m \right)^2 \right]$$

$$- (\mathbf{F}_0 \cdot \nabla_{\mathbf{p}})(\mathbf{P} \cdot \mathbf{F}_0) \left[\sum_{m=1}^{n} a_m \sum_{r=1}^{m} b_r - \sum_{i=2}^{n} b_i \sum_{m=1}^{i-1} a_m \sum_{r=1}^{m} b_r \right.$$

$$\left. - \frac{1}{2} \left(\sum_{m=1}^{n} b_m \right)^2 + \frac{1}{2} \sum_{i=1}^{n} a_i \left(\sum_{m=1}^{i} b_m \right)^2 \right]. \tag{12.64c}$$

H_3 is worked out as

$$
\begin{aligned}
H_3 \;=\;& (\mathbf{P}_0 \cdot \nabla_{\mathbf{q}})^2 \, (\mathbf{F} \cdot \mathbf{P}_0) \left[\frac{1}{2} \sum_{m=2}^{n} b_m \left(\sum_{r=1}^{m-1} a_r \right)^2 - \frac{1}{6} \left(\sum_{m=1}^{n} a_m \right)^3 \right. \\
& \left. - \frac{1}{2} \sum_{i=2}^{n} a_i \sum_{m=2}^{i} b_m \left(\sum_{r=1}^{m-1} a_r \right)^2 + \frac{1}{6} \sum_{i=2}^{n} b_i \left(\sum_{m=1}^{i-1} a_m \right)^3 \right] \\
& - (\mathbf{F}_0 \cdot \nabla_{\mathbf{p}})^2 \, (\mathbf{P} \cdot \mathbf{F}_0) \left[\frac{1}{2} \sum_{m=1}^{n} a_m \left(\sum_{r=1}^{m} b_r \right)^2 - \frac{1}{6} \left(\sum_{m=1}^{n} b_m \right)^3 \right. \\
& \left. - \frac{1}{2} \sum_{i=2}^{n} b_i \sum_{m=1}^{i-1} a_m \left(\sum_{r=1}^{m} b_r \right)^2 + \frac{1}{6} \sum_{i=1}^{n} a_i \left(\sum_{m=1}^{i} b_m \right)^3 \right] \\
& + (\mathbf{P}_0 \cdot \nabla_{\mathbf{q}}) \, (\mathbf{F}_0 \cdot \nabla_{\mathbf{p}}) \, (\mathbf{P} \cdot \mathbf{F}) \left[\sum_{m=2}^{n} b_m \sum_{r=1}^{m-1} a_r \sum_{s=1}^{r} b_s \right. \\
& - \sum_{m=2}^{n} a_m \sum_{r=2}^{m} b_r \sum_{s=1}^{r-1} a_s - \sum_{i=2}^{n} a_i \sum_{m=2}^{i} b_m \sum_{r=1}^{m-1} a_r \sum_{s=1}^{r} b_s \\
& + \sum_{i=3}^{n} b_i \sum_{m=2}^{i-1} a_m \sum_{r=2}^{m-1} b_r \sum_{s=1}^{r-1} a_s + \sum_{m=1}^{n} b_m \sum_{m=2}^{n} b_m \sum_{r=1}^{m-1} a_r \\
& - \sum_{i=2}^{n} a_i \sum_{m=1}^{i} b_m \sum_{m=2}^{i} b_m \sum_{r=1}^{m-1} a_r - \sum_{m=1}^{n} a_m \sum_{m=1}^{n} a_m \sum_{r=1}^{n} b_r \\
& \left. + \sum_{i=2}^{n} b_i \sum_{m=1}^{i-1} a_m \sum_{r=1}^{m} b_r \right] . \tag{12.64d}
\end{aligned}
$$

12.8.3   Second-Order Algorithm

To obtain an algorithm which is second-order in h ($n = 2$) equate H_0 and H_1 to zero and obtain

$$
H_0 \;=\; T(\mathbf{p}_0)(1 - a_1 - a_2) + V(\mathbf{q}_0)(1 - b_1 - b_2) = 0, \tag{12.65a}
$$

$$
\begin{aligned}
H_1 \;=\;& (\mathbf{F}_0 \cdot \mathbf{P}_0)(b_1 + b_2 - a_1 b_1 - a_2 b_1 - a_2 b_2 - a_1 - a_2 + a_1 b_2) \\
\;=\;& 0. \tag{12.65b}
\end{aligned}
$$

Since $T(\mathbf{p}_0)$, $V(\mathbf{q}_0)$ and $(\mathbf{F}_0 \cdot \mathbf{P}_0)$ need not be zero, choose

$$
1 - a_1 - a_2 = 0, \quad 1 - b_1 - b_2 = 0, \tag{12.66a}
$$

$$
b_1 + b_2 - a_1 b_1 - a_2 b_1 - a_2 b_2 - a_1 - a_2 + a_1 b_2 = 0 \tag{12.66b}
$$

and obtain

$$
a_1 + a_2 = 1, \quad b_1 + b_2 = 1, \quad a_1 b_1 + a_2 - a_1 b_2 = 0. \tag{12.67}
$$

There are three equations for the four unknowns a_1, a_2, b_1 and b_2. The choice $b_2 = 1$ gives $b_1 = 0$. Then, the last equation in (12.67) gives $a_1 = a_2$. Since $a_1 + a_2 = 1$ the result is $a_1 = a_2 = 1/2$. Thus, a solution of Eq. (12.67) is

$$
a_1 = a_2 = 1/2, \quad b_1 = 0, \quad b_2 = 1. \tag{12.68}
$$

12.8.4   Third-Order Algorithm

For a third-order algorithm

$$a_1 = 2/3, \quad a_2 = -2/3, \quad a_3 = 1, \quad b_1 = 7/4 \ b_2 = 3/4, \quad b_3 = -1/24. \tag{12.69}$$

12.8.5   Fourth-Order Algorithm

For $n = 4$, equating H_0, H_1, H_2 and H_3 separately to zero leads to

$$a_1 + a_2 + a_3 + a_4 - 1 = 0, \tag{12.70a}$$

$$b_1 + b_2 + b_3 + b_4 - 1 = 0, \tag{12.70b}$$

$$a_1 b_2 + (a_1 + a_2) b_3 + (1 - a_4) b_4 - \frac{1}{2} = 0, \tag{12.70c}$$

$$a_1 b_1^2 + a_2 (b_1 + b_2)^2 + a_3 (1 - b_4)^2 + a_4 - \frac{1}{3} = 0, \tag{12.70d}$$

$$a_1^2 b_2 + (a_1 + a_2)^2 b_3 + (1 - a_4)^2 b_4 - \frac{1}{3} = 0, \tag{12.70e}$$

$$a_1 b_1^3 + a_2 (b_1 + b_2)^3 + a_3 (1 - b_4)^3 + a_4 - \frac{1}{4} = 0, \tag{12.70f}$$

$$b_2 a_1^3 + (a_1 + a_2)^3 b_3 + (1 - a_4)^3 b_4 - \frac{1}{4} = 0, \tag{12.70g}$$

$$a_1 b_2 + (a_1 + a_2) b_3 \left[a_1 b_1 + a_2 (b_1 + b_2) \right]$$
$$+ (a_1 - a_4)(1/2 - a_4) b_4 - (b_1 + b_2) a_1 a_2 b_2$$
$$- a_3 (1 - b_4) \left[\frac{1}{2} - b_4 (1 - a_4) \right] - \frac{1}{2} a_4 = 0. \tag{12.70h}$$

An analytic solution of the above set of equations is

$$a_1 = \frac{1}{6} \left(2 + 2^{1/3} + 2^{-1/3} \right), \tag{12.71a}$$

$$a_2 = \frac{1}{6} \left(1 - 2^{1/3} - 2^{-1/3} \right), \tag{12.71b}$$

$$a_3 = a_2, \quad a_4 = a_1, \tag{12.71c}$$

$$b_1 = 0, \quad b_2 = 1 / \left(2 - 2^{1/3} \right), \tag{12.71d}$$

$$b_3 = 1 / \left(1 - 2^{2/3} \right), \quad b_4 = b_2. \tag{12.71e}$$

12.8.6   General Algorithm for nth Order Method

From Eqs. (12.58) the general scheme for the calculation of $(\mathbf{q}(h), \mathbf{p}(h)) = (\mathbf{q}_n, \mathbf{p}_n)$ at $t = h$ from the given $(\mathbf{q}(0), \mathbf{p}(0)) = (\mathbf{q}_0, \mathbf{p}_0)$ is

$$\mathbf{p}_i = \mathbf{p}_{i-1} + h b_i \mathbf{F} (\mathbf{q}_{i-1}), \tag{12.72a}$$

$$\mathbf{q}_i = \mathbf{q}_{i-1} + h a_i \mathbf{P} (\mathbf{p}_i), \quad i = 1, 2, \ldots, n. \tag{12.72b}$$

The above algorithm is called nth-order *Candy–Rozmus* method. Since $b_1 = 0$ in the second-order solution set (12.68) and the fourth-order set (12.71) only one evaluation of $\mathbf{F}$ and three evaluations of $\mathbf{P}$ are necessary in the second-order and fourth-order algorithms, respectively. For a detailed analysis of the Candy–Rozmus method one may refer to ref. [25].

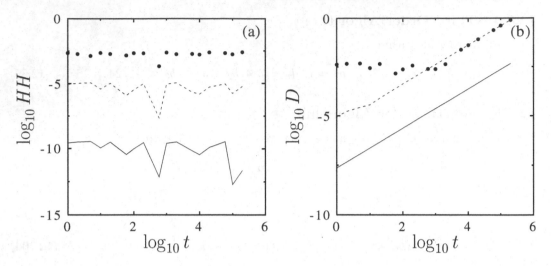

FIGURE 12.5

(a) Variation of HH, the absolute difference between the exact value of the Hamiltonian and the numerically calculated value of the Hamiltonian of the harmonic oscillator obtained by the three symplectic methods with $h = 0.01$. (b) Plot of the error D, Eq. (12.47), versus time in $\log_{10}$ scale for the three symplectic methods. In both the subplots, solid circles, dashed curve and continuous curve represent the result obtained by the implicit Euler, Störmer–Verlet and the fourth-order methods, respectively.

For a time-dependent potential $V = V(\mathbf{q}, t)$ the algorithm is

$$
\begin{aligned}
\mathbf{p}_i &= \mathbf{p}_{i-1} + hb_i\mathbf{F}\left(\mathbf{q}_{i-1}, t_{i-1}\right), & (12.73a) \\
\mathbf{q}_i &= \mathbf{q}_{i-1} + ha_i\mathbf{P}\left(\mathbf{p}_i\right), & (12.73b) \\
t_i &= t_{i-1} - ha_i, \quad i = 1, 2, \ldots, n. & (12.73c)
\end{aligned}
$$

The linear harmonic oscillator Eq. (12.7) is solved by the Störmer–Verlet and the fourth-order algorithms with

$$
H = 0.5, \quad q(0) = 1, \quad p(0) = 0 \text{ and } h = 0.01. \tag{12.74}
$$

Figure 12.5a shows the variation of the Hamiltonian H against time t in $\log_{10} - \log_{10}$ scale for the three symplectic methods considered so far. As expected the variation of H is relatively very small and of the order of 10^{-10} for the fourth-order method. In all the methods H is not diverging. Figure 12.5b presents the error D, Eq. (12.47), versus t for the three methods. For large t, the error D grows linearly for all the three methods.

Example:

Find the numerical solution of the harmonic oscillator Eq. (12.17) at $t = 1$ subjected to the initial conditions $q(0) = 1$ and $p(0) = 0$ by the explicit Euler method (Eq. (12.17)), the symplectic Euler method (Eq. (12.45)), the Störmer–Verlet method (Eq. (12.50)) and the fourth-order Candy–Rozmus symplectic method (Eq. (12.72)) with $h = 0.01$. Also, compute the error D in each method.

The exact solution at $t = 1$ is

$$q(1) = 0.5403023, \quad p(1) = -0.8414710.$$

By the explicit Euler method:

$q(1) = 0.5430387, \quad p(1) = -0.8456706, \quad D = 0.50124\mathrm{e}{-2}.$

By the symplectic Euler method:

$q(1) = 0.5445062, \quad p(1) = -0.8414838, \quad D = 0.42039\mathrm{e}{-2}.$

By the Störmer–Verlet method:

$q(1) = 0.5402988, \quad p(1) = -0.8414627, \quad D = 0.89800\mathrm{e}{-5}.$

By the Candy–Rozmus method:

$q(1) = 0.5403023, \quad p(1) = -0.8414710, \quad D = 0.87679\mathrm{e}{-9}.$

12.9   Other Methods

The symplectic algorithms considered so far are of order ≤ 4 and are for separable Hamiltonian systems. Much interest has been focused on developing higher-order symplectic integrators and integrators for certain special problems. Higher-order Runge–Kutta–Nyström methods are constructed [26–28]. Methods preserving first integrals (constants of motion) have also been proposed [29]. For general Hamiltonians the symplectic Runge–Kutta methods are implicit [30,31], where one has to solve the implicit algebraic equations for the intermediate stage values using an iterative approximation method like the Newton–Raphson.

If the potential $V(\mathbf{q})$ of a Hamiltonian system has a singularity then the numerical integration of equations of motion would produce large errors when the trajectory approaches the singularity. Symplectic integration methods have been proposed and analyzed for this problem [10,32–35]. Symplectic variational integrators with energy and momentum conserving are developed [12].

Yoshida and McLachlan introduced splitting methods [36,37] for the equations of the form

$$\frac{\mathrm{d}\mathbf{Y}}{\mathrm{d}t} = \mathbf{f}_1 + \mathbf{f}_2. \tag{12.75}$$

In their approaches the equation is decomposed into

$$\frac{\mathrm{d}\mathbf{y}_1}{\mathrm{d}t} = \mathbf{f}_1, \quad \frac{\mathrm{d}\mathbf{y}_2}{\mathrm{d}t} = \mathbf{f}_2 \tag{12.76}$$

so that both the equations in (12.76) can be integrated exactly or approximately using a discrete method. Such methods have been used very successfully in the study of solar system and molecular dynamics [10,38].

Numerical experiments [39,40] have shown that variable time step symplectic algorithms lose their favourable properties for a long-time integration and are not superior to standard nonsymplectic methods. Several methods have been suggested to overcome this difficulty [13,41–45]. In an approach of Hairer [13], small perturbations are added to the discretization such that the qualitative behaviour for long-time integration is re-stabilized. These algorithms are theoretically justified and the practical considerations are discussed. In the following, these algorithms [13] are briefly outlined.

The step size is chosen as $h = \epsilon s(\mathbf{q}, \mathbf{p}, \epsilon)$, where $s > 0$ is a state-dependent given function and $\epsilon > 0$ is a small parameter. For a fixed initial value $(\mathbf{q}_0, \mathbf{p}_0)$ with $H_0 = H(\mathbf{q}_0, \mathbf{p}_0)$ the new Hamiltonian is

$$H_{\mathrm{new}} = s(\mathbf{q}, \mathbf{p}, \epsilon)\,(H(\mathbf{q}, \mathbf{p}) - H_0)\,. \tag{12.77}$$

The Hamilton's equations

$$q' = H_{\mathbf{p}}, \quad \mathbf{p}' = -H_{\mathbf{q}}. \tag{12.78}$$

become

$$\mathbf{q}' = sH_{\mathbf{p}} + s_{\mathbf{p}}(H - H_0), \quad \mathbf{p}' = -sH_{\mathbf{q}} - s_{\mathbf{q}}(H - H_0). \tag{12.79}$$

The introduced perturbation vanishes on the solution of Eqs. (12.78) passing through $(\mathbf{p}_0, \mathbf{q}_0)$ but makes the system Hamiltonian. Transformations such as in Eq. (12.77) are already used in classical mechanics for an analytic treatment of Hamiltonian systems.

12.9.1   Symplectic Euler Method with q-Dependent Step Size Function

Consider the Hamiltonian systems of the form

$$H = \frac{1}{2}\mathbf{p}^{\mathrm{T}}M^{-1}\mathbf{p} + V(\mathbf{q}), \tag{12.80}$$

where M is a constant symmetric matrix. Application of the Euler method with step size function $h = \epsilon s$ to Eq. (12.79) with the Hamiltonian (12.80) gives the approximation at $t_1 = t_0 + \epsilon s(\mathbf{q}_0, \epsilon)$ as

$$\mathbf{p}_1 = \mathbf{p}_0 - \epsilon s(\mathbf{q}_0, \epsilon) V_{\mathbf{q}}(\mathbf{q}_0) - \epsilon s_{\mathbf{q}}(\mathbf{q}_0, \epsilon)\left(\frac{1}{2}\mathbf{p}_1^{\mathrm{T}}M^{-1}\mathbf{p}_1 + V(\mathbf{q}_0) - H_0\right), \tag{12.81a}$$

$$\mathbf{q}_1 = \mathbf{q}_0 + \epsilon s(\mathbf{q}_0, \epsilon) M^{-1}\mathbf{p}_1. \tag{12.81b}$$

A quadratic equation in $\mathbf{p}_1$ is obtained. To solve this let us consider the norm $\|\mathbf{p}\|_M^2 := \mathbf{p}^{\mathrm{T}}M^{-1}\mathbf{p}$ and calculate the scalar $\beta = \mathbf{p}_1^{\mathrm{T}}M^{-1}\mathbf{p}_1 = \|\mathbf{p}_1\|_M^2$ from

$$\beta = \|\mathbf{p}_0 - \epsilon s V_{\mathbf{q}} - \epsilon s_{\mathbf{q}}(\beta/2 + V - H_0)\|_M^2. \tag{12.82}$$

After solving the above equation for β the numerical solutions $\mathbf{p}_1$ and $\mathbf{q}_1$ can be obtained from Eqs. (12.81).

12.9.2   Verlet Scheme with q-Dependent Step Size Function

The Störmer–Verlet algorithm for the system (12.79) with H given by Eq. (12.80) and with p-dependent function s is

$$\mathbf{p}_{n+1/2} = \mathbf{p}_n - \frac{1}{2}\epsilon s(\mathbf{q}_n, \epsilon) V_{\mathbf{q}}(\mathbf{q}_n) - \frac{1}{2}\epsilon s_{\mathbf{q}}(\mathbf{q}_n, \epsilon)\left(H(\mathbf{p}_{n+1/2}, \mathbf{q}_n) - H_0\right), \tag{12.83a}$$

$$\mathbf{q}_{n+1} = \mathbf{q}_n + \frac{1}{2}\epsilon(s(\mathbf{q}_n, \epsilon) + s(\mathbf{q}_{n+1}, \epsilon)) M^{-1}\mathbf{p}_{n+1/2}, \tag{12.83b}$$

$$\mathbf{p}_{n+1} = \mathbf{p}_{n+1/2} - \frac{1}{2}\epsilon s(\mathbf{q}_{n+1}, \epsilon) V_{\mathbf{q}}(\mathbf{q}_{n+1})$$
$$\qquad - \frac{1}{2}\epsilon s_{\mathbf{q}}(\mathbf{q}_{n+1}, \epsilon)\left(H(\mathbf{p}_{n+1/2}, \mathbf{q}_{n+1}) - H_0\right), \tag{12.83c}$$

$$t_{n+1} = t_n + \frac{1}{2}\epsilon(s(\mathbf{q}_n, \epsilon) + s(\mathbf{q}_{n+1}, \epsilon)). \tag{12.83d}$$

Equation (12.83b) is implicit in $\mathbf{q}_{n+1}$. Solve the scalar equation

$$\gamma = s\left(\mathbf{q}_n + \frac{1}{2}\epsilon(s(\mathbf{q}_n, \epsilon) + \gamma) M^{-1}\mathbf{p}_{n+1/2}, \epsilon\right) \tag{12.84}$$

for $\gamma := s(\mathbf{s}_{n+1}, \epsilon)$. The Newton–Raphson iterations can be used because the derivative $s_{\mathbf{q}}$ is available.

12.9.3 An Adaptive Störmer–Verlet Method

A variable step size method proposed by Huang and Leimkuhler [42] is

$$\mathbf{p}_{n+1/2} = \mathbf{p}_n - \frac{1}{2}\epsilon\sigma_{n+1/2}V_{\mathbf{q}}(\mathbf{q}_n), \tag{12.85a}$$

$$\mathbf{q}_{n+1} = \mathbf{q}_n + \epsilon\sigma_{n+1/2}M^{-1}\mathbf{p}_{n+1/2}, \tag{12.85b}$$

$$\mathbf{p}_{n+1} = \mathbf{p}_{n+1/2} - \frac{1}{2}\epsilon\sigma_{n+1/2}V_{\mathbf{q}}(\mathbf{q}_{n+1}), \tag{12.85c}$$

where $\sigma_{n+1/2}$ is defined by the recursion

$$\sigma_{1/2} = s(\mathbf{q}_0, \epsilon), \quad \frac{1}{\sigma_{n+1/2}} + \frac{1}{\sigma_{n-1/2}} = \frac{2}{s(\mathbf{q}_n, \epsilon)}. \tag{12.85d}$$

12.10 A Symplectic Integrator for Spin Systems

Classical spin systems are essentially noncanonical Hamiltonian systems. Examples of spin systems include point vortices on a sphere, the Landau–Lifshitz equation of micromagnetics, reduced motion of a spinning top and the classical limit of Heisenberg spin chains [46–48]. Such systems have phase space $(S^2)^N$. Moreover, symplectic is said to form the sum of the standard area element on each sphere. If the spheres are realized as $||\mathbf{s}_i||^2 = 1$, $\mathbf{s}_i \in \mathbb{R}^3$ then the Hamiltonian H is on $(S^2)^N$. The equations of motion are given by

$$\dot{\mathbf{s}}_i = \mathbf{s}_i \times \nabla_{\mathbf{s}_i} H(\mathbf{s}_1, \ldots, \mathbf{s}_N) = f_i(\mathbf{s}_1, \ldots, \mathbf{s}_N). \tag{12.86}$$

McLachlan, Modin and Verdier [49] proposed the integrating scheme

$$\mathbf{s}_{i,n+1} = \mathbf{s}_{i,n} + hf_i(\mathbf{u}_1, \ldots, \mathbf{u}_N), \quad \mathbf{u}_i = \frac{\mathbf{s}_{i,n} + \mathbf{s}_{i,n+1}}{||\mathbf{s}_{i,n} + \mathbf{s}_{i,n+1}||}. \tag{12.87}$$

This scheme is found to preserve certain structural properties associated with the exact flow of the system.

12.11 Concluding Remarks

The symplectic integrators are shown to be successful in determining qualitative features of conservative systems over a very long-time integration including 1-million-year, 5-million-year and 210-billion-year integration. The simulation study of motion of Pluto has predicted chaotic behaviour of it. Numerical simulation of dynamics of a large system called molecular dynamics requires solving a system of nonlinear differential equations. Such systems are conservative and highly sensitive to discretization and round-off errors. Symplectiic integrators are appropriate for molecular dynamics applications.

Symplectic methods are proposed for stochastic Hamiltonian systems governed by stochastic ordinary differential equations and Hamiltonian systems subjected to non-dissipative forces. Symplectic integrators in curved spacetimes and for inseparable Hamiltonian systems are developed. Symplectic methods are applicable to eigenvalue problems, harmonic analysis, optimization and control problems, robotic dynamics and control and quantum cosmology have been reported. [50-54].

12.12 Bibliography

[1] E. Hairer, C. Lubich and G. Wanner, *Geometric Numerical Integration: Structure Preserving Algorithms for Ordinary Differential Equations*. Springer, Berlin, 2002.

[2] J.M. Sanz-Serna, *Acta Numerica* 1:243, 1992.

[3] R.D. Skeel, G. Zhang and T. Schlick, *SIAM J. Sci. Comput.* 18:203, 1997.

[4] S. Blanes, F. Casas and J. Ros, *SIAM J. Sci. Comput.* 21:711, 1999.

[5] D. Donnelly and E. Rogers, *Am. J. Phys.* 73:938, 2005.

[6] R. McLachlan and G.R.W. Quispel, *Six Lectures on the Geometric Integration of Ordinary Differential Equations* in *Foundations of Computational Mathematics*, R. DeVore, A. Iserles and E. Suli (Eds). Cambridge University Press, Cambridge, 2001. pp.155-210.

[7] C.J. Budd and D. Piggott, *Geometric Integration and Its Applications* in *Handbook of Numerical Analysis* 11:35, 2003.

[8] R. de Vogelaere, *Methods of integration which preserve the contact transformation property of the Hamiltonian equations*, Report No. 4, Dept. Math., Univ. of Notre Dame, Notre Dame, Ind., 1956.

[9] R.D. Ruth, *IEEE Trans. Nuclear Science* 30:2669, 1983.

[10] B. Leimkuhler, *Phil. Trans. Roy. Soc. London A* 357:1101, 1999.

[11] J.M. Sanz-Serna and M.P. Calvo, *Numerical Hamiltonian Problems*. Chapman and Hall, London, 1994.

[12] C. Kane, J.E. Marsden and M. Ortiz, *J. Math. Phys.* 40:3353, 1999.

[13] E. Hairer, *Appl. Numer. Math.* 25:219, 1997.

[14] R.I. McLachlan and P. Atela, *Nonlinearity* 5:541, 1992.

[15] E. Forest and R.D. Ruth, *Physica D* 43:105, 1990.

[16] P.J. Channel and C. Scovel, *Nonlinearity* 3:231, 1990.

[17] P.J. Channell and C. Scovel, Los Alamos National Laboratory LA–VR–88–1828, 1988.

[18] M.H. Holmes, *Am. J. Phys.* 88:60, 2019.

[19] T. Itoh and K. Abe, *J. Comput. Phys.* 77:85, 1988.

[20] A.M. Ozorio de Almeida, *Hamiltonian Systems*. Cambridge Univ. Press, Cambridge, 1988.

[21] R.S. MacKay and J.D. Meiss, *Hamiltonian Dynamical Systems*. Adams Hilger, Bristol, 1987.

[22] A.J. Lichtenberg and M.A. Lieberman, *Regular and Stochastic Motion*. Springer, New York, 1983.

[23] M. Lakshmanan and S. Rajasekar, *Nonlinear Dynamics: Integrability, Chaos and Patterns*. Springer, Berlin, 2003.

[24] F.J. Vesely, *Computational Physics: An Introduction*. Springer, New York, 1994.

[25] J. Candy and W. Rozmus, *J. Comput. Phys.* 92:230, 1991.

[26]   M. Qin and W.J. Zhu, *Computing* 47:309, 1992.

[27]   S. Blanes, F. Casas and J. Ros, *New families of symplectic Runge–Kutta–Nyström integration methods*. In the *Proceedings of the Second International Conference on Numerical Analysis and Its Applications*. L. Vulkov, J. Wasniewski and P. Yalamov (Eds.). Springer, London, 2001. pp.102-109.

[28]   S. Blanes, F. Casas and J. Ros, *App. Num. Math.* 39:245, 2001.

[29]   G.R.W. Quispel and G.S. Turner, *J. Phys. A* 29:L341, 1996.

[30]   J.M. Sanz-Serna, *BIT* 28:877, 1988.

[31]   X. Tan, *J. Comput. Phys.* 203:250, 2005.

[32]   C.J. Budd, W. Huang and R.D. Russel, *SIAM J. Sci. Comp.* 17:305, 1996.

[33]   C.J. Budd, B. Leimkuhler and M.D. Piggott, *Appl. Numer. Math.* 39:261, 2001.

[34]   H. Arakida and T.Fukushima, *Astron. J.* 120:3333, 2000.

[35]   C.J. Budd and M.D. Piggott, *J. Comp. Appl. Math.* 128:399, 2001.

[36]   H. Yoshida, *Phys. Lett. A* 150:262, 1990.

[37]   R.I. McLachlan, *SIAM J. Sci. Comput.* 16:15, 1995.

[38]   Ch. Schlier and A. Seiter, *J. Chem. Phys. A* 102:9399, 1998.

[39]   M.P. Calvo and J.M.Sanz-Serna, *SIAM J. Sci. Comput.* 14:936, 1993.

[40]   B. Gladman, M. Duncan and J. Candy, *Celestial Mech. Dyn. Astro.* 52:221, 1991.

[41]   E. Hairer and D. Stoffer, *SIAM J. Sci. Comput.* 18:257, 1997.

[42]   W. Huang and B. Leimkuhler, *SIAM J. Sci. Comput.* 18:239, 1997.

[43]   P. Hut, J. Makino and S. McMillan, *Astrophys. J.* 443:L193, 1995.

[44]   R.D. Skeel and J.J. Biesiadecki, *Ann. Numer. Math.* 1:191, 1994.

[45]   D. Stoffer, *Computing* 55:1, 1995.

[46]   M. Lakshmanan, *Philos. Trans. R. Soc. A: Math. Phys. Eng. Sci.* 369:1280, 2011.

[47]   S. Pekarsky and J.E. Marsden, *J. Math. Phys.* 39:5894, 1998.

[48]   J.E. Marsden and T.S. Ratiu, *Introduction to Mechanics and Symmetry: A Basic Exposition of Classical Mechanical Systems*. Springer, Berlin, 1999.

[49]   R.I. McLachlan, K. Modin and O. Verdier, *Phys. Rev. E* 89:061301, 2014.

[50]   H. Fassbender, *Symplectic Methods for the Symplectic Eigenproblems*. Kluwer Academic Publishers, New York, 2002.

[51]   M.A. de Gosson, *Symplectic Methods in Harmonic Analysis in Mathematical Physics*. Birkhauser, Berlin, 2011.

[52]   A. Agrachev and G. Revaz, *Pure Appl. Math.* 207:19, 1998.

[53]   Z. Xu, L. Du, H. Wang and Z. Deng, *Appl. Math. Mech.* 40:111, 2019.

[54]   E.V. Correa Silva, G.A. Monerat, G. Olivera-Neto, C. Neves and L.G. Ferreira Filtio, *Phys. Rev. D* 80:047302, 2009.

12.13   Problems

12.1 Obtain a second-order symplectic Euler algorithm using (12.37) from t to $t+h/2$ and Eq. (12.36) from $t+h/2$ to $t+h$.

12.2 From Eq. (12.64) obtain the conditions on a_i and b_i for a third-order method. Solve the obtained equations for $b_3 = 1$ analytically.

12.3 Obtain a solution of Eqs. (12.70) with $a_1 = 0$.

12.4 Find the numerical solution of the harmonic oscillator Eq. (12.7) at $t = 0.1$ subjected to $q(0) = 0$ and $p(0) = 1$ by the explicit Euler, the symplectic Euler, the Störmer–Verlet and the Candy–Rozmus methods with $h = 0.1$. Compute the error D in each method.

12.5 Find the numerical solution of the anharmonic oscillator equation $\dot{q} = p$, $\dot{p} = q - q^3$ at $t = 0.1$ subjected to the conditions $q(0) = 0.9$ and $p(0) = 0$ by the explicit Euler, symplectic Euler, Störmer–Verlet and Candy–Rozmus methods with $h = 0.1$. Compute the Hamiltonian at $t = 0.1$ (the exact value of it is -0.25) in each method.

12.6 Find the numerical solution of the anharmonic oscillator equation $\dot{q} = p$, $\dot{p} = -q - q^3$ at $t = 0.1$ subjected to $q(0) = 0$ and $p(0) = 1$ by the explicit Euler, symplectic Euler, Störmer–Verlet and Candy–Rozmus methods with $h = 0.1$. Compute the Hamiltonian at $t = 0.1$ (the exact value of it is 0.5) in each method.

12.7 Find the numerical solution of the anharmonic oscillator equation $\dot{q} = p$, $\dot{p} = -q + q^3$ at $t = 0.1$ subjected to the initial conditions $q(0) = 0.0$ and $p(0) = 0.5$ by the explicit Euler, symplectic Euler, Störmer–Verlet and Candy–Rozmus methods with $h = 0.1$. Compute the Hamiltonian at $t = 0.1$ (the exact value of the Hamiltonian is 0.125) in each method.

12.8 Find the numerical solution of the pendulum equation $\dot{q} = p$, $\dot{p} = -\sin q$ at $t = 0.1$ subjected to the conditions $q(0) = 2$ and $p(0) = 0$ by the explicit Euler, symplectic Euler, Störmer–Verlet and Candy–Rozmus methods with $h = 0.1$. Compute the Hamiltonian at $t = 0.1$ (the exact value of it is 0.4161468) in each method.

12.9 The Hamiltonian of a linear harmonic oscillator in the presence of an applied external periodic force is

$$H = \frac{1}{2}p^2 + \frac{1}{2}q^2 - qf\sin\omega t\,.$$

Fix $\omega = 2$ and $f = 0.2$. If the initial conditions are $q(0) = 1$ and $p(0) = 1$ compute $q(0.01)$ and $p(0.01)$ with time step 0.01 by the fourth-order method. Also, develop a program to solve the above equation, where the potential has explicit time-dependence.

12.10 The Hamiltonian of a particle in the field of a standing wave is

$$H = \frac{1}{2}p^2 - f\cos q\cos t\,.$$

Write its equation of motion. Calculate $q(0.1)$ and $p(0.1)$ if $q(0) = 0.3$, $p(0) = 0.5$, $f = 1$ and $h = 0.1$ by the fourth-order method.

13

Ordinary Differential Equations – Boundary-Value Problems

13.1 Introduction

In Chapter 11 the problem of numerically solving ordinary differential equations with specified initial conditions is discussed. To find the numerical solution of an nth-order equation of the form $\dot{\mathbf{X}} = \mathbf{F}(\mathbf{X}, t)$ by a single-step method over the interval $t \in [a, b]$ the values of the components of $\mathbf{X}$ at time $t = a$ are need to be specified. However, there are instances, where the conditions on the solution are specified at more than one value of time (independent variable), often, at two different values of t. Usually, these time values are at the end points a and b. Such problems are called *boundary-value problems*.

Let the problem is to determine the electrostatic potential $V(r)$ between two concentric spheres of radii $r_1 = 2\,\text{cm}$ and $r_2 = 10\,\text{cm}$ kept at potentials $V_1 = 2\,\text{V}$ and $V_2 = 0\,\text{V}$, respectively. The equation for V is

$$V'' + \frac{2}{r} V' = 0. \tag{13.1}$$

The solution of this equation is subjected to the conditions $V(r_1) = V(2) = 2\,\text{V}$ and $V(r_2) = V(10) = 0\,\text{V}$. The exact solution of (13.1) is

$$V(r) = c_1 + c_2 \frac{1}{r}. \tag{13.2}$$

Determination of the integration constants c_1 and c_2 by the values of V and V' at a particular value of r is known as the *initial-value problem*. In the present case, c_1 and c_2 are to be determined by the values of V given at $r = 2$ and $r = 10$. This is an example of a boundary-value problem. The solution $V(r)$ should satisfy the values of V given at the two values of r, namely, at $r = 2$ and 10. Boundary-value problems have applications in boundary layer flow, population balance, transport phenomena, vibration analysis and so on.

Consider the vibration of a string whose ends are fixed to rigid objects so that displacements of the end points are zero. If the string is displaced from the rest state then its vibration must be such that at the end points the displacement is always zero. The mathematical model equation for the vibration of the string is now subjected to the boundary conditions. That is, this is another boundary-value problem.

The numerical methods for solving the boundary-value problems are of two types: difference methods and shooting methods. In a difference method derivatives are replaced by finite-difference approximations. In this case, the solution is found by solving a set of simultaneous equations. In the shooting method sufficient conditions at the starting point, say, at $t = a$ are introduced and adjusted them suitably so that the conditions at the end points are satisfied. First-order equations are essentially initial-value problem. For simplicity restrict the discussion and examples on second-order linear equations only.

DOI: 10.1201/9781032649931-13

13.2   Shooting Method – Reduction to Two Initial-Value Problems

A general form of a second-order linear ordinary differential equation is given by

$$y'' - p(x)y' - q(x)y = r(x).$$ (13.3)

Assume that the boundary conditions are

$$y(a) = y_a \ \text{ and } \ y(b) = y_b.$$ (13.4)

Since the solutions of linear equations satisfy linear superposition principle one can show that the solution of the boundary-value problem is a suitable linear combination of solutions of two initial-value problems.

Suppose that u and v are two solutions of the boundary-value problem. It is easy to show that

$$y(x) = \frac{c_1 u + c_2 v}{c_1 + c_2}$$ (13.5)

is also a solution. Substitution of this solution (13.5) in the left-hand side (LHS) of Eq. (13.3) yields

$$
\begin{aligned}
\text{LHS} &= \frac{c_1 u'' + c_2 v''}{c_1 + c_2} - p(x)\left(\frac{c_1 u' + c_2 v'}{c_1 + c_2}\right) - q(x)\left(\frac{c_1 u + c_2 v}{c_1 + c_2}\right) \\
&= \frac{1}{c_1 + c_2}\left[c_1\left(u'' - pu' - qu\right) + c_2\left(v'' - pv' - qv\right)\right] \\
&= \frac{1}{c_1 + c_2}\left(c_1 r + c_2 r\right) \\
&= r.
\end{aligned}
$$ (13.6)

Thus, y given by Eq. (13.5) is also a solution of Eq. (13.3).

From Eq. (13.5)

$$
\begin{aligned}
y(a) &= \frac{c_1 u(a) + c_2 v(a)}{c_1 + c_2}, \\
y(b) &= \frac{c_1 u(b) + c_2 v(b)}{c_1 + c_2}.
\end{aligned}
$$
(13.7a)

(13.7b)

Choose $u(a) = v(a) = y(a)$ and $u'(a)$ and $v'(a)$ as our own choice. Then, by suitably adjusting c_1 and c_2 it is possible to make $y(b) = y_b$. The suitably chosen values of c_1 and c_2 can be used in Eq. (13.5). Because the solutions u and v are solutions of Eq. (13.3) corresponding to the initial conditions $(u(a), u'(a))$ and $(v(a), v'(a))$ they can be obtained by solving Eq. (13.3) by using, say, the fourth-order Runge–Kutta method for $a < x \le b$. Then, Eq. (13.5) can be used to compute $y(x)$. However, it cannot be determined immediately after knowing $u(x)$ and $v(x)$. This is because Eq. (13.5) contains both $u(b)$ and $v(b)$.

A simpler and a variant of the above method is the following: Consider the two initial-value problems given by

$$
\begin{aligned}
u'' - p(x)u' - q(x)u - r &= 0, & u(a) = y_a, \ u'(a) = 0, \\
v'' - p(x)v' - q(x)v &= 0, & v(a) = 0, \ v'(a) = 0.
\end{aligned}
$$
(13.8)

(13.9)

If $u(x)$ and $v(x)$ are the solutions of Eqs. (13.8) and (13.9), respectively, then the solution of Eq. (13.3) is given by

$$y(x) = u(x) + cv(x).$$ (13.10)

Differentiation of Eq. (13.10) with respect to x twice gives

$$
\begin{aligned}
y'' &= u'' + cv'' \\
&= p(x)\left[u' + cv'\right] + q(x)\left[u + cv\right] + r(x) \\
&= p(x)y' + q(x)y + r(x)
\end{aligned}
\tag{13.11}
$$

which is Eq. (13.3). The unknown constant c in Eq. (13.10) can be determined by imposing the conditions (13.4) to the solution (13.10). At $x = a$ the solution $y(x)$ is y_a and the left-side of Eq. (13.10) is $u(a) + cv(a) = y_a$. Equation (13.10) is satisfied at $x = a$. On the other hand, when $x = b$

$$
y(b) = y_b = u(b) + cv(b)
\tag{13.12}
$$

which gives

$$
c = \frac{y_b - u(b)}{v(b)}.
\tag{13.13}
$$

Now,

$$
y(x) = u(x) + \frac{y_b - u(b)}{v(b)} v(x).
\tag{13.14}
$$

Example:

Solve the Hermite differential equation

$$
y'' - 2xy' + 6y + 6 = 0.
\tag{13.15}
$$

subjected to the boundary conditions $y(0) = -1$, $y(1) = -5$ by the shooting method with step size $h = 0.1$ and $h = 0.01$.

Equations (13.8) and (13.9) for the above Hermite equation are

$$
\begin{aligned}
u'' - 2xu' + 6u + 6 &= 0, \quad u(0) = -1, \quad u'(0) = 0, \tag{13.16} \\
v'' - 2xv' + 6v &= 0, \quad v(0) = 0, \quad v'(0) = 1. \tag{13.17}
\end{aligned}
$$

The above two initial-value problems are solved using the fourth-order Runge–Kutta method with $h = 0.1$ and $h = 0.01$. Then, y is obtained from

$$
y(x) = u(x) + \frac{y(1) - u(1)}{v(1)} v(x).
\tag{13.18}
$$

The obtained solutions for $h = 0.1$ and $h = 0.01$ are given in Table 13.1 along with the exact solution

$$
y(x) = 8x^3 - 12x - 1.
\tag{13.19}
$$

Does it possible to solve the equation $y'' - 2xyy' = 0$, $y(0) = 0$, $y(1) = 1$ by the linear shooting method?

13.3 Finite-Difference Method

Another method of solving boundary-value problems of linear second-order differential equations is approximating the derivatives by finite-difference quotients. The differential equation

TABLE 13.1

Comparison of the numerical solution of Eq. (13.15) obtained by the shooting method with the exact solution given by Eq. (13.19).

x	y_{numer} with $h = 0.1$	y_{numer} with $h = 0.01$	y_{exact}
0.0	-1.0000000	-1.0000000	-1.0000000
0.1	-2.1943810	-2.1920024	-2.1920000
0.2	-3.3404645	-3.3360044	-3.3360000
0.3	-4.3901570	-4.3840061	-4.3840000
0.4	-5.2953662	-5.2880073	-5.2880000
0.5	-6.0080070	-6.0000079	-6.0000000
0.6	-6.4799962	-6.4720079	-6.4720000
0.7	-6.6632590	-6.6560071	-6.6560000
0.8	-6.5097209	-6.5040056	-6.5040000
0.9	-5.9713210	-5.9680033	-5.9680000
1.0	-5.0000000	-5.0000000	-5.0000000

of the form (13.3) can be converted into a difference equation by replacing the derivative by finite-difference quotients. This results in a system of algebraic equations whose solution is an approximation to the solution of the differential equation. For a nonlinear differential equation this method gives rise to a set of nonlinear equations which are generally difficult to solve. The set of obtained nonlinear equations can be solved by iterative methods.

Consider Eq. (13.3) over the interval $[a, b]$ with $y(a) = y_a$ and $y(b) = y_b$. Divide the interval $[a, b]$ into N equal intervals with grid points $x_0(= a) < x_1 < x_2 < \cdots < x_N(= b)$, where $x_i = a + ih$, $i = 0, 1, \ldots, N$ and $h = (b - a)/N$ and denote y_i as the solution at x_i. The central-difference approximations for the first and second derivatives are given by

$$y'(x_i) = \frac{1}{2h}[y(x_{i+1}) - y(x_{i-1})] + O(h^2),\tag{13.20a}$$

$$y''(x_i) = \frac{1}{h^2}[y(x_{i+1}) - 2y(x_i) + y(x_{i-1})] + O(h^2).\tag{13.20b}$$

Substitution of y' and y'' given by Eqs. (13.20) in Eq. (13.3) gives

$$\frac{1}{h^2}(y_{i+1} - 2y_i + y_{i-1}) = \frac{1}{2h}(y_{i+1} - y_{i-1})p_i + q_i y_i + r_i.\tag{13.21}$$

Rearrange Eq. (13.21) as

$$\left(-\frac{h}{2}p_i - 1\right)y_{i-1} + \left(2 + h^2 q_i\right)y_i + \left(\frac{h}{2}p_i - 1\right)y_{i+1}$$
$$= -h^2 r_i, \quad i = 1, 2, \ldots, N - 1\tag{13.22}$$

with $y_0 = y_a$ and $y_N = y_b$. When $N = 4$, Eq. (13.22) takes the form

$$\left(-\frac{h}{2}p_1 - 1\right) y_0 + \left(2 + h^2 q_1\right) y_1 + \left(\frac{h}{2}p_1 - 1\right) y_2 = -h^2 r_1, \tag{13.23a}$$

$$\left(-\frac{h}{2}p_2 - 1\right) y_1 + \left(2 + h^2 q_2\right) y_2 + \left(\frac{h}{2}p_2 - 1\right) y_3 = -h^2 r_2, \tag{13.23b}$$

$$\left(-\frac{h}{2}p_3 - 1\right) y_2 + \left(2 + h^2 q_3\right) y_3 + \left(\frac{h}{2}p_3 - 1\right) y_4 = -h^2 r_3, \tag{13.23c}$$

where $y_0 = y_a$, $y_4 = y_b$, $x_0 = a$, $x_1 = a + h$, $x_2 = a + 2h$, $x_3 = a + 3h$ and $x_4 = a + 4h = b$. In matrix form Eq. (13.23) is written as

$$\begin{pmatrix} 2 + h^2 q_1 & \frac{h}{2}p_1 - 1 & 0 \\ -\frac{h}{2}p_2 - 1 & 2 + h^2 q_2 & \frac{h}{2}p_2 - 1 \\ 0 & -\frac{h}{2}p_3 - 1 & 2 + h^2 q_3 \end{pmatrix} \begin{pmatrix} y_1 \\ y_2 \\ y_3 \end{pmatrix} = \begin{pmatrix} -h^2 r_1 + B_0 \\ -h^2 r_2 \\ -h^2 r_3 + B_4 \end{pmatrix},$$

$$\tag{13.24a}$$

where

$$B_0 = \left(\frac{h}{2}p_1 + 1\right) y_a, \quad B_4 = \left(1 - \frac{h}{2}p_3\right) y_b. \tag{13.24b}$$

An interesting observation is that the system (13.24) is a tridiagonal. This is true for arbitrary values of N. This is because the approximation of derivative of y involves only points to the left, to the right and the central point.

For an arbitrary N Eq. (13.23) can be written in matrix form as

$$\begin{pmatrix} Q_1 & P_1 - 1 & & & \\ -P_2 - 1 & Q_2 & P_2 - 1 & & \\ & & \vdots & & \\ & & -P_{N-2} - 1 & Q_{N-2} & P_{N-1} - 1 \\ & & & -P_{N-1} - 1 & Q_{N-1} \end{pmatrix}$$

$$\times \begin{pmatrix} y_1 \\ y_2 \\ y_3 \\ \vdots \\ y_{N-2} \\ y_{N-1} \end{pmatrix} = \begin{pmatrix} -h^2 r_1 + B_0 \\ -h^2 r_2 \\ \vdots \\ -h^2 r_{N-2} \\ -h^2 r_{N-1} + B_N \end{pmatrix}, \tag{13.25a}$$

where

$$Q_i = 2 + h^2 q_i, \tag{13.25b}$$

$$P_i = \frac{h}{2}p_i, \quad i = 1, 2, \ldots, N - 1 \tag{13.25c}$$

and

$$B_0 = \left(1 + \frac{h}{2}p_1\right) y_a, \quad B_N = \left(1 - \frac{h}{2}p_{N-1}\right) y_b. \tag{13.25d}$$

TABLE 13.2
The numerical solution of the Hermite differential Eq. (13.15) over the interval $x \in [0, 1]$ by the finite-difference method.

x	Numerical solution			Exact solution
	$h = 0.025$	$h = 0.05$	by Richardson extrapolation	
0.0	-1.0000000	-1.0000000	$- - - - -$	-1.0000000
0.1	-2.1924953	-2.1939850	-2.1919987	-2.1920000
0.2	-3.3369606	-3.3398496	-3.3359976	-3.3360000
0.3	-4.3853659	-4.3894737	-4.3839999	-4.3840000
0.4	-5.2896811	-5.2947368	-5.2879958	-5.2880000
0.5	-6.0018762	-6.0075188	-5.9999953	-6.0000000
0.6	-6.4739212	-6.4796992	-6.4719952	-6.4720000
0.7	-6.6577861	-6.6631579	-6.6559955	-6.6560000
0.8	-6.5054409	-6.5097744	-6.5039964	-6.5040000
0.9	-5.9688555	-5.9714286	-5.9679978	-5.9680000
1.0	-5.0000000	-5.0000000	-5.0000000	-5.0000000

The error in the numerical approximation is $O(h^2)$. Accuracy can be improved by using the Richardson extrapolation.

Let $y(x, h)$ and $y(x, 2h)$ are two numerical approximations of $y(x)$ at a point x computed with two different step sizes h and $2h$. Denoting $y(x)$ as the exact value of y then

$$
\begin{aligned}
y(x) &= y(x, h) + ch^2, & \text{(13.26a)} \\
y(x) &= y(x, 2h) + c(2h)^2, & \\
&= y(x, 2h) + 4ch^2. & \text{(13.26b)}
\end{aligned}
$$

Multiplying Eq. (13.26a) by 4 and subtracting the resulting equation from (13.26b) we have

$$
y(x) = [4y(x, h) - y(x, 2h)] / 3. \tag{13.27}
$$

Example:

Solve the Hermite differential Eq. (13.15) subjected to the boundary conditions $y(0) = -1$, $y(1) = -5$ over the interval $x \in [0, 1]$ by the finite-difference method with the step sizes $h = 0.025$ and 0.05. Applying the Richardson extrapolation for the numerical solutions obtained with $h = 0.025$ and 0.05 determines the improved solution. The exact solution of the given equation is given by Eq. (13.19).

Table 13.2 presents the numerical solutions with $h = 0.025$, 0.05 and by the Richardson extrapolation.

13.4 Solving Time-Independent Schrödinger Equation

This section deals with the problem of finding the eigenvalues and the eigenfunctions of the time-independent Schrödinger equation

$$\psi''(x) + \frac{2m}{\hbar^2}[E - V(x)]\psi(x) = 0, \quad x \in [a, b] \tag{13.28}$$

of a quantum mechanical particle of mass m and subjected to the boundary conditions $\psi(a) = 0$ and $\psi(b) = 0$. In Eq. (13.28) $\hbar$ is $h/(2\pi)$ and h is the Planck constant. For simplicity set $m = 1$ and $\hbar = 1$. Solving the above equation is the starting point of study of quantum mechanical systems [1-3] with the potential $V(x)$. For most of the systems, the boundary conditions are $\psi \to 0$ as $x \to \pm\infty$. In numerical computation choose these boundary conditions as $\psi = 0$ for $x = a$ and b, where $|a|$ and $|b|$ are sufficiently large. $\psi(x)$ is zero outside the interval $[a, b]$. There is a simple procedure developed by van der Maelen Uria et al [4] to solve Eq. (13.28) for arbitrary numerical or analytical potentials $V(x)$. Their method is illustrated in the following.

13.4.1 Discrete Equivalent of Schrödinger Equation

Discretize the space variable x in the interval $[a, b]$ by choosing a grid of points x_0, x_1, x_2, ..., x_N with $x_{i+1} - x_i = h$, where h is a constant. The solutions of Eq. (13.28) are called *eigenfunctions*. The eigenfunctions will take values only in the grid of points. That is, ψ will be a set of numbers ψ_0, ψ_1, ..., ψ_N, where $\psi_i = \psi(x_i)$. Usually $\psi_0 = \psi(a) = 0$ and $\psi_N = \psi(b) = 0$. Now, Eq. (13.28) becomes the following discrete equation

$$\psi_i'' + 2[E - V(x_i)]\psi_i = 0, \quad i = 1, 2, \dots N - 1. \tag{13.29}$$

Equation (13.29) is converted into a difference equation by the substitution

$$\psi''|_{x=x_i} = \frac{1}{h^2}[\psi_{i+1} - 2\psi_i + \psi_{i-1}]. \tag{13.30}$$

The result is the equation

$$\psi_{i-1} + 2\left[h^2 E - h^2 V(x_i) - 1\right]\psi_i + \psi_{i+1} = 0, \quad i = 1, 2, \dots, N - 1. \tag{13.31}$$

Thus, the Schrödinger Eq. (13.28) is transformed into a linear finite-difference equation. The above $N - 1$ equations can be written in a matrix form as

$$S\psi = 0 \tag{13.32a}$$

with

$$\begin{aligned}
S_{ii} &= 2\left[h^2 E - h^2 V(x_i) - 1\right], \quad i = 1, 2, \dots, N - 1 \\
S_{ii+1} &= S_{i+1i} = 1, \quad i = 1, 2, \dots, N - 2 \\
S_{ij} &= 0, \quad j \neq i - 1, i, i + 1, \quad i = 1, 2, \dots, N - 1.
\end{aligned} \tag{13.32b}$$

Equations (13.31) can be written as

$$(A - \lambda I)\psi = 0, \tag{13.33a}$$

where

$$
\begin{aligned}
a_{ii} &= 2\left[1 + h^2 V\left(x_i\right)\right], \quad i = 1, 2, \ldots, N-1 \\
a_{ii+1} &= a_{i+1i} = -1, \quad i = 1, 2, \ldots, N-2 \\
a_{ij} &= 0, \quad j \neq i-1, i, i+1, \quad i = 1, 2, \ldots, N-1 \\
\lambda &= 2h^2 E
\end{aligned} \tag{13.33b}
$$

and I is the unit matrix. The matrix A is explicitly written as

$$
A = \begin{pmatrix}
a_{11} & -1 & 0 & 0 & & & & \\
-1 & a_{22} & -1 & 0 & & & & \\
0 & -1 & a_{33} & -1 & & & & \\
& & & \vdots & \vdots & \vdots & & \\
& & & & & -1 & a_{N-2} & -1 \\
& & & & & & -1 & a_{N-1}
\end{pmatrix}. \tag{13.33c}
$$

Thus, the original differential eigenvalue problem is transformed into an algebraic eigenvalue problem. The values of λ for which nontrivial solutions to the system of linear Eqs. (13.33) exist are called *(energy) eigenvalues* and the corresponding solutions are *eigenfunctions*. So, the calculation of E and ψ of the Schrödinger Eq. (13.28) for a given potential is reduced now to the calculation of λ and ψ of the matrix A.

13.4.2 Eigenvalues of the Matrix A

Now, choose a suitable technique to compute the eigenvalues and the eigenfunctions of the matrix A. Interestingly, the matrix A is a tridiagonal symmetric matrix. For such a matrix, efficient algorithms are available to calculate λ. A simple and efficient technique is the QL algorithm discussed in Chapter 7.

The truncation error in the finite-difference approximation of the second derivative term, Eq. (13.30), is proportional to h^2. The error can be made proportional to h^4 using the Richardson's extrapolation. Suppose $E_0^{(1)}, E_1^{(1)}, \ldots, E_m^{(1)}$ and $E_0^{(2)}, E_1^{(2)}, \ldots, E_m^{(2)}$ are two sets of eigenvalues computed with two values of h, say h_1 and h_2. Then, a more accurate estimation is given by

$$
E_i(\text{more accurate}) = \frac{1}{h_2^2 - h_1^2}\left[h_2^2 E_i^{(1)} - h_1^2 E_i^{(2)}\right]. \tag{13.34}
$$

The eigenvalues of Eq. (13.28) may be infinite in number. Since, A is $(N-1) \times (N-1)$ matrix the numerical computation gives only $N-1$ eigenvalues.

13.4.3 Eigenfunction of the Matrix A

The eigenfunction of the matrix A is obtained by solving the algebraic eigenvalue problem, Eq. (13.33), where λ is a given eigenvalue, $\lambda = 2h^2 E$. Equation (13.33a) is written as (after redefining $N-1$ as N)

$$
\begin{pmatrix}
A_1 & -1 & 0 & 0 & \ldots & 0 & 0 & 0 \\
-1 & A_2 & -1 & 0 & \ldots & 0 & 0 & 0 \\
0 & -1 & A_3 & -1 & \ldots & 0 & 0 & 0 \\
& & & \vdots & & & & \\
0 & 0 & 0 & 0 & \ldots & -1 & A_{N-1} & -1 \\
0 & 0 & 0 & 0 & \ldots & 0 & -1 & A_N
\end{pmatrix}
\begin{pmatrix}
\psi_1 \\
\psi_2 \\
\psi_3 \\
\vdots \\
\psi_{N-1} \\
\psi_N
\end{pmatrix} = 0. \tag{13.35}
$$

It is necessary to normalize the eigenfunction. The normalization condition is $\int_a^b \psi^*\psi \mathrm{d}x = 1$. Let $C\psi$ is the normalized eigenfunction, where C is the normalization constant. This constant can be determined from the condition

$$C^2 \int_a^b \psi^*\psi \mathrm{d}x = 1 . \tag{13.36}$$

Keeping this in mind, choose $\psi_1 = 1$ and calculate other ψ_i's from Eq. (13.35). The first row gives

$$\psi_2 = A_1 \psi_1 . \tag{13.37}$$

That is, row 1 can be used to find ψ_2, row 2 can be used to find ψ_3 and so on. $(N-2)$th row can be used to calculate ψ_{N-1}. Now, $(N-1)$th and Nth rows are not yet used whereas only ψ_N to be determined. The last two rows can be added to reduce the two equations into one equation. Using this new equation find ψ_N as

$$\psi_N = \frac{1}{1 - A_N} \left[-\psi_{N-2} + (A_{N-1} - 1)\psi_{N-1} \right] . \tag{13.38}$$

Inspection of the expressions for $\psi_3, \psi_4, \ldots, \psi_{N-1}$ leads to the following general expression

$$\psi_i = -\psi_{i-2} + A_{i-1}\psi_{i-1} , \quad i = 3, 4, \ldots, N-1 . \tag{13.39}$$

Here, $\psi_1 = 1$, $\psi_2 = A_1 \psi_1$ and ψ_N is given by Eq. (13.38). After computing an unnormalized eigenfunction, the normalization constant C is determined from Eq. (13.36). The integral in Eq. (13.36) can be evaluated, for example, using the composite trapezoidal rule. That is,

$$
\begin{aligned}
\frac{1}{C^2} &= \int_a^b |\psi|^2 \, \mathrm{d}x \\
&= h \left[\frac{1}{2}\psi_1^2 + \psi_2^2 + \psi_3^2 + \cdots + \psi_{N-1}^2 + \frac{1}{2}\psi_N^2 \right] .
\end{aligned} \tag{13.40}
$$

Example:

For the linear harmonic oscillator potential $V(x) = x^2/2$ the exact eigenvalues are [1, 2]

$$E_{\text{exact}} = \left(n + \frac{1}{2} \right) , \quad n = 0, 1, 2, \ldots . \tag{13.41}$$

Compute numerically the energy eigenvalues and the eigenfunctions.

A Python program implementing the QL algorithm is developed to compute the energy eigenvalues for a given value of h. The program is instructed to store only the eigenvalues less than, for example, 10 in an external output file. In order to get a more accurate result, eigenvalues are calculated for two values of h, namely, h_1 and h_2. Then, using the Richardson extrapolation more accurate values of E are obtained. Table 13.3 presents first few eigenvalues calculated with $h_1 = 0.1$ and $h_2 = 0.05$, by extrapolation and the exact values. Here, $a = -10$ and $b = 10$. Accuracy can be improved by reducing the values of h_1 and h_2 and choosing large values of $|a|$ and $|b|$.

The exact eigenfunctions $\psi_n(x)$ are given by

$$\psi_n(x) = N_n H_n(x) \, \mathrm{e}^{-x^2/2} , \tag{13.42}$$

where

$$N_n = \left(\frac{1}{\sqrt{\pi}\, 2^n n!} \right)^{1/2} , \quad H_n(x) = (-1)^n \, \mathrm{e}^{x^2} \frac{\partial^n}{\partial x^n} \mathrm{e}^{-x^2} . \tag{13.43}$$

TABLE 13.3
The numerically computed first few eigenvalues of the one-dimensional linear harmonic oscillator with the potential $V(x) = x^2/2$. Here, $a = -10$ and $b = 10$.

Computed E with		Extra-	Exact
$h = 0.1$	$h = 0.05$	polated E	E
9.44309	9.48584	9.50009	9.5
8.45444	8.48868	8.50007	8.5
7.46452	7.49116	7.50004	7.5
6.47333	6.49335	6.50002	6.5
5.48087	5.49523	5.50002	5.5
4.48715	4.49679	4.50000	4.5
3.49217	3.49805	3.50001	3.5
2.49593	2.49898	2.50000	2.5
1.49844	1.49961	1.50000	1.5
0.49969	0.49992	0.50000	0.5

It is easy to develop a Python program to compute the normalized eigenfunction for a given value of energy E. For the lowest energy $E = 0.5$ the exact eigenfunction is given by

$$\psi(x) = \left(\frac{1}{\pi}\right)^{1/4} e^{-x^2/2}. \tag{13.44}$$

Table 13.4 gives the numerically computed lowest energy (ground state) eigenfunction values at some selected grid points for three values of h, where $a = -5$ and $b = 5$. The numerically computed values of $\psi(x)$ approach the exact result as h decreases. Very good agreement with exact eigenfunction is observed for other values of E also. The numerical solution is symmetric with respect to the origin. This solution is called an *even-parity solution*. An odd-parity solution can be obtained by choosing $\psi(a) = -1$. Figure 13.1 shows the numerically computed eigenfunctions for $E = 0.5, 1.5, 2.5$ and 3.5. For $E = 0.5$ and 2.5 the value of $\psi(a)$ is set as 1 whereas for $E = 1.5$ and 3.5 the value of $\psi(a)$ is set as -1.

The problem of solving the time-dependent Schrödinger equation is discussed in Section 14.9.

13.5   Concluding Remarks

In the present chapter, only two methods for the boundary-value problems of ordinary differential equations are described. There are many methods that exist, for example, multiple shooting methods [5] and wavelet approaches are developed [6]. Numerical methods for ordinary differential equations with singular coefficients [7], deviated arguments [8] and on an infinite domain with solutions having a slowly decaying tail [8] have been reported. Higher-order boundary-value problems occur in astrophysics, astronomy, fluid dynamics, viscoelastic flows, coating flows, etc. For the numerical methods for higher-order equations, one may refer to the refs. [9-15].

TABLE 13.4

The numerically computed lowest energy eigenfunction values at some grid points for three values of h. The last column is the exact result.

x	Numerical ψ with			Exact ψ
	$h = 0.02$	$h = 0.01$	$h = 0.001$	
± 5.0	0.00000	0.00000	0.00000	0.00000
± 4.0	0.00024	0.00025	0.00025	0.00025
± 3.0	0.00803	0.00832	0.00834	0.00833
± 2.5	0.03175	0.03292	0.03300	0.03300
± 2.0	0.09779	0.10138	0.10165	0.10165
± 1.5	0.23459	0.24321	0.24385	0.24385
± 1.0	0.43828	0.45438	0.45558	0.45558
± 0.5	0.63770	0.66112	0.66287	0.66287
0.0	0.72261	0.74915	0.75113	0.75113

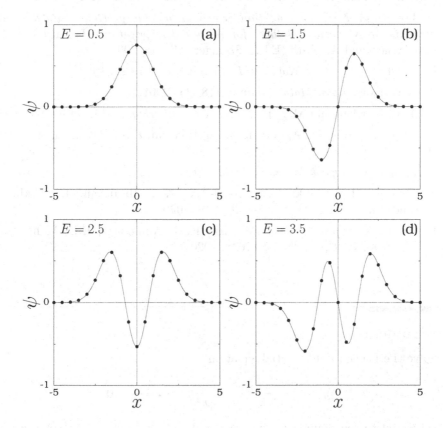

FIGURE 13.1

The numerically computed eigenfunctions of the linear harmonic oscillator (continuous curve) for first few values of E. The exact values of the eigenfunctions for selected values of x are represented by solid circles.

13.6 Bibliography

[1] L.I. Schiff, *Quantum Mechanics*. McGraw Hill, Singapore, 1968.

[2] P.M. Mathews and K. Venkatesan, *A Text Book of Quantum Mechanics*. Tata McGraw Hill, New Delhi, 2006.

[3] S. Rajasekar and R. Velusamy, *Quantum Mechanics I: The Fundamentals*. CRC Press, Boca Raton, 2022.

[4] J.F. van der Maelen Uria, S. Garcia-Granda and A. Menendez-Velazzquez, *Am. J. Phys.* 64:327, 1996.

[5] U.M. Aischer and L.R. Petzold, *Computational Methods for Ordinary Differential Equations and Differential Algebraic Equations*, SIAM, Philadelphia, 1998.

[6] S. Ul Arifeen, S. Haq, A. Ghafoor, A. Ullah, P. Kumam and P. Chatpanya, *Adv. Diff. Eqs.* 2021:347, 2021.

[7] R.D. Russell and L.F. Shampine, *SIAM J. Numer. Anal.* 12:13, 1975.

[8] T. Jankowski, *Numerical solution of boundary-value problems with deviated arguments*. In *Numerical Methods for Advanced Applications*. F. Brezzi, A. Buffa, S. Corasaro and A. Murli (Eds.). Springer, Milano, 2003.

[9] M.M. Chawla and C.P. Katti, *BIT Numer. Math.* 19:27, 1979.

[10] A.M. Wazwaz, *Appl. Math. Comput.* 118:311, 2001.

[11] I. Ullah, H. Khan and M.T. Rahim, *J. Comput. Eng.*, 2014, article ID 286039.

[12] M.J. Iqbal, S. Rehman, A. Pervaiz and A. Hakeem, *Proc. Pak. Acad. Sci.* 182:389, 2015.

[13] J. Ahmed, *Kyungpook Math. J.* 57:651, 2017.

[14] S. Owyed, M.A. Abdou, A. Abdel-Aty, A.A. Ibraheem, R. Nekhile and D. Baleanu, *J. Intell. Fuzzy Syst.* 38:2859, 2020.

[15] R. Amin, K. Shah, I. Khan, M. Arif, K.M. Abualnaja, E.E. Mahmoud and A.H. Abdel-Aty, *Open Phys.* 18:1048, 2020.

13.7 Problems

13.1 Solve the Legendre differential equation

$$y'' - \frac{2x}{1-x^2}y' + \frac{2}{1-x^2}y + \frac{4}{1-x^2} = 0$$

subjected to the boundary conditions $y(0) = -2$, $y(1) = -1$ over the interval $x \in [0,1]$ by the step sizes $h = 0.025$ and 0.05. Applying the Richardson extrapolation for the numerical solution obtained with $h = 0.025$ and 0.05 determine the improved solution. The exact solution of the given equation is $y = x - 2$.

13.2 Solve the differential equation

$$y'' + 4y' + 4y - 4e^{-2x} = 0$$

subjected to the boundary conditions $y(0) = -1$, $y(2) = 11e^{-2}$ over the interval $x \in [0, 2]$ by the step sizes $h = 0.025$ and 0.05. Applying the Richardson extrapolation for the numerical solution obtained with $h = 0.025$ and 0.05 determines the improved solution. The exact solution of the given equation is

$$y = e^{-2x} \left(2x^2 + 2x - 1\right).$$

13.3 Consider the differential equation

$$\left(x^2 - 1\right)^2 y'' + 2(x - 1)(x - 2)y' + 4y + 4 = 0$$

subjected to the boundary conditions $y(0) = -1$, $y(1) = 1$. Find the solution in the interval $x \in [0, 1]$. The exact solution of the given equation is $y = (3x - 1)/(x + 1)$.

13.4 Consider a metal bar of length $1\,\mathrm{m}$ and of uniform cross-section. At a certain time t, the temperature y of the bar at the ends is found to be $y(0) = 50°\mathrm{C}$ and $y(1\mathrm{m}) = 50°\mathrm{C}$. The temperature variation is governed by the equation

$$y'' - y = 0.$$

Find $y(x)$ in the interval $0 < x < 1\mathrm{m}$. The exact solution is $y(x) = Ae^x + Be^{-x}$, where $A = 4.9378898$ and $B = 45.06211$.

13.5 A violin string is stretched a little and then its ends are fixed at $x = 0$, π units. It is allowed to vibrate. Assume that the deflection of the string is given by $u(x, t) = y(x)W(t)$. If y is the solution of the equation

$$y'' + 4y = 0, \quad y(0) = 0, \quad y(\pi) = 0$$

find $y(x)$ in the interval $0 < x < \pi$. The exact solution is $y(x) = \sin 2x$.

13.6 The variation of the electrostatic potential $V(r)$ between two concentric spheres of radii $r_1 = 2\,\mathrm{cm}$ and $r_2 = 10\,\mathrm{cm}$ kept at the potentials $V_1 = 2$ volts and $V_2 = 0$ volt, respectively, is governed by the equation

$$V'' + 2V'/r = 0.$$

The boundary conditions are $V(r_1) = V(2) = 2$ volts and $V(r_2) = V(10) = 0$ volt. Find $V(r)$ in the interval $2\,\mathrm{cm} < r < 10\,\mathrm{cm}$. The exact solution is $V(r) = -0.5 + 5/r$ volts.

13.7 The temperature distribution $y(x, t)$ of a solid remains the same for all values of t. That is, $y(x, t)$ is in equilibrium with respect to time. The variation of temperature with respect to space variable is governed by the equation

$$y'' + 2y' + 2y = 0.$$

If the temperature at $x = 0$ and $5\pi/2$ units are $30°\mathrm{C}$ and $1°\mathrm{C}$ calculate the temperature distribution $y(x)$ for $0 < x < 5\pi/2$ units. The exact solution is $y(x) = Ae^{-x}\sin(x + B)$, where $A = 2576.1605$ and $B = 0.0116455$.

13.8 The wave function of a quantum mechanical particle in the potential $V(x) = x^{2p}$, for large values of p is the solution of the Schrödinger equation $\psi_{xx} + \lambda^2\psi = 0$, $\lambda = \pi/2$. ψ is subjected to the boundary conditions $\psi(0) = 1$, $\psi(1) = 0$. Find $\psi(x)$ for $0 < x < 1$. Exact solution is $\psi(x) = \cos(\pi x/2)$.

13.9 For the shifted harmonic oscillator potential $V(x) = \frac{1}{2}x^2 + m$, $m = 0, 1, 2, \ldots$ the exact energy eigenvalues are $E_n^{(m)} = (m + n + \frac{1}{2})$.

  (a) Compute numerically the energy eigenvalues for several values of m and compare them with the exact values.

  (b) When $n = 0$, $E_0^{(m)} = (m + \frac{1}{2})$, $m = 0, 1, \ldots$. Verify that the eigenfunctions $\psi_m(x)$, $m = 0, 1, \ldots$ all are identical even though the eigenvalues are distinct.

13.10 The eigenvalues and the ground state eigenfunction of some interesting potentials are given below. Obtain the first few lowest energy eigenvalues and the corresponding eigenfunctions.

  (a) *Box Potential:*

$$V(x) = \begin{cases} 0 & \text{for } |x| < a \\ \infty & \text{for } |x| > a \end{cases}$$

$$E_n = \frac{(n+1)^2 \pi^2}{8a^2}, \qquad \psi_0 = \sqrt{\frac{2}{a}} \sin(\pi x/a).$$

  (b) *Coulomb Potential:*

$$V(r) = -\frac{b}{r}, \quad E_n = -\frac{b^2}{2(n+1)^2}, \quad n = 0, 1, \ldots, \quad \psi_0 \sim re^{-br}.$$

  (c) *Perturbed Coulomb Potential:*

$$V(r) = -\frac{b}{r} + \frac{1}{r^2},$$

$$E_n = -\frac{b^2}{2(n+2)^2}, \quad n = 0, 1, \ldots, \quad \psi_0 \sim r^2 e^{-br/2}.$$

14

Linear Partial Differential Equations

14.1 Introduction

When the number of independent variables of a differential equation is more than one then it is said to be a *partial differential equation*. In a partial differential equation all the independent variables may be space variables or one of them may be a time variable. When a physical property of a system or a process or an event depends on more than one independent variable then it is generally described by a partial differential equation. Partial differential equations occur in many branches of engineering, physics, chemistry, biology and so on. Some of the ubiquitous partial differential equations are the heat equation, the wave equation, the Laplace equation, the Poisson equation and the Maxwell equations. For certain linear partial differential equations exact analytical solutions can be obtained by the well-known variable separable method.

Simple and efficient numerical methods for partial differential equations can be obtained by the finite-difference methods. In 1908, Runge studied the numerical solution of the Poisson equation $u_{tt} + u_{xx} = $ constant (where $u_{tt} = \partial^2 u/\partial t^2$ and $u_{xx} = \partial^2 u/\partial x^2$) using finite-difference methods. Bindes and Schmidt proposed a finite-difference formula for the heat equation. The work of Courant and Friedrichs and Lévy in 1928 was considered as the beginning of numerical methods for partial differential equations. In the case of ordinary differential equations, a particular method, for example, Runge–Kutta method, can be applied to a variety of equations. In contrast, one has to develop a suitable formula separately for each of the partial differential equations.

Partial differential equations are generally classified according to their linearity, order and boundary conditions. In this chapter, restrict ourselves to linear equations and take up the problem of nonlinear equations in the next chapter. The present chapter first gives a brief account of classification of partial differential equations into linear, quasilinear and nonlinear. Many partial differential equations of practical applications are of second-order in form. They are classified into hyperbolic, parabolic and elliptic. Next, certain useful boundary conditions are defined. Some of the partial derivatives are expressed in terms of finite-differences. Then, finite-difference formulas are developed for certain partial differential equations. Particular emphasize is on the numerical methods to solve hyperbolic, parabolic and elliptic equations.

14.2 Classification of Partial Differential Equations

The order of a partial differential equation is simply determined by the highest-order partial derivative occurring in the given equation. Let us consider a physical variable u a function of two variables x and t. The partial derivatives $\partial u/\partial t$ and $\partial u/\partial x$ are called *first-order partial*

DOI: 10.1201/9781032649931-14

derivatives. The derivatives $\partial^2 u/\partial t^2$, $\partial^2 u/\partial x^2$, $\partial^2 u/(\partial x \partial t)$ and $\partial^2 u/(\partial t \partial x)$ are *second-order partial derivatives.* In $(\partial u/\partial x)^n$ or $(\partial^2 u/\partial t^2)^n$ n is the degree or power of the respective derivatives.

If each of the terms of a given partial differential equation, after rationalization, has a total degree either 0 or 1 in the dependent variables and their partial derivatives then it is said to be a *linear partial differential equation.* Otherwise, it can be either quasilinear or nonlinear. For example, consider the following second-order partial differential equation

$$a\frac{\partial^2 u}{\partial t^2} + b\frac{\partial^2 u}{\partial t \partial x} + c\frac{\partial^2 u}{\partial x^2} + d = 0. \tag{14.1}$$

If all the coefficients a, b, c and d are constants then Eq. (14.1) is a linear second-order partial differential equation with constant coefficients. If at least one of the coefficients is a function of the independent variables x and/or t then it is a *variable coefficient linear equation.* When any one of the coefficients is a function of the dependent variable u and/or any of its derivatives of lower order than that of the differential equation then the equation is said to be a *quasilinear* (often called *nonlinear*). For example, if b has a term u or $\partial u/\partial x$ or $\partial u/\partial t$ then Eq. (14.1) is a quasilinear. If any of the coefficients a, b, c and d is a function of the dependent variable u and/or any of its derivatives of the same order as that of the equation then the equation is *nonlinear.* For example, if c has a term $\partial^2 u/\partial t^2$, $\partial^2 u/\partial x^2$ or $\partial^2 u/\partial t \partial x$ then it is a nonlinear. Examples of partial differential equations with different order and linearity are given below:

1. **First-order linear equation:**

$$\frac{\partial u}{\partial t} = \alpha \frac{\partial u}{\partial x}. \tag{14.2}$$

2. **First-order quasilinear equation:**

$$\frac{\partial u}{\partial t} + u\frac{\partial u}{\partial x} = 0. \tag{14.3}$$

3. **First-order nonlinear equation:**

$$\frac{\partial u}{\partial t} + \frac{\partial u}{\partial t}\frac{\partial u}{\partial x} = 0. \tag{14.4}$$

4. **Second-order linear equation:**

$$\frac{\partial^2 u}{\partial x^2} + \frac{\partial^2 u}{\partial y^2} + u = 0. \tag{14.5}$$

5. **Second-order quasilinear equation:**

$$\frac{\partial^2 u}{\partial t^2} + \frac{\partial u}{\partial x} + u^2 = 0. \tag{14.6}$$

6. **Second-order nonlinear equation:**

$$\frac{\partial^2 u}{\partial t^2} + \frac{\partial^2 u}{\partial t \partial x}\frac{\partial^2 u}{\partial x^2} = 0. \tag{14.7}$$

Linear second-order partial differential equations with two independent variables are found to describe many real physical systems and are classified into hyperbolic, parabolic and elliptic. Consider the following second-order linear partial differential equation

$$a\frac{\partial^2 u}{\partial t^2} + 2b\frac{\partial^2 u}{\partial t \partial x} + c\frac{\partial^2 u}{\partial x^2} + d\frac{\partial u}{\partial t} + e\frac{\partial u}{\partial x} + fu + g = 0, \tag{14.8}$$

where $u = u(x, t)$ and the coefficients are either constants or functions of the independent variables only. Equation (14.8) is classified into the following types:

$$\text{Hyperbolic} \quad : \quad b^2 - ac > 0. \tag{14.9a}$$
$$\text{Parabolic} \quad : \quad b^2 - ac = 0. \tag{14.9b}$$
$$\text{Elliptic} \quad : \quad b^2 - ac < 0. \tag{14.9c}$$

Examples:

1. Wave equation (hyperbolic):

$$\frac{\partial^2 u}{\partial t^2} = c^2 \frac{\partial^2 u}{\partial x^2}. \tag{14.10}$$

2. Heat equation (parabolic):

$$\frac{\partial u}{\partial t} = c^2 \frac{\partial^2 u}{\partial x^2}. \tag{14.11}$$

3. Laplace equation (elliptic):

$$\frac{\partial^2 u}{\partial x^2} + \frac{\partial^2 u}{\partial y^2} = 0. \tag{14.12}$$

Partial differential equations of physical systems are further classified into 1) steady-state problem, 2) propagation problem and 3) eigenvalue problem [1]. In a closed domain, the solution, say, $u(x, y)$ of a steady state problem is described by an elliptic type differential equation along with appropriate boundary conditions. Propagation problems are essentially initial-value problems, that is, the time evolutions of the systems are subjected to the initial conditions in an open interval. The eigenvalue problems correspond to the cases where bounded meaningful solutions exist only for certain specific values of the parameters called, *eigenvalues*.

14.3 Initial and Boundary Conditions

Partial differential equations are solved with the given initial and boundary conditions. Space variables generally extends from $-\infty$ to ∞. However, real systems are finite in extension or periodic in the independent variables. Therefore, the solution must be subjected to appropriate boundary conditions. A given equation with different boundary conditions give rise to qualitatively different solutions.

The following are the initial/boundary conditions often employed in solving the second-order partial differential equations.

1. Dirichlet Conditions

Consider a second-order partial differential equation with dependent variable $u(x, t)$ and the independent variables t and x with $a_1 \leq x \leq a_2$ and $b_1 \leq t \leq b_2$. In the Dirichlet conditions, the values of u are given at $x = a_1$ and a_2 and $t = b_1$ and b_2.

Example:

Initial Condition:

An initial condition specifies the value of u and u_t at the initial value of time t, say, at b_1 for the range of x considered:

$$u(x, t = b_1) \quad = \quad f_1(x), \quad a_1 \leq x \leq a_2 \tag{14.13a}$$
$$u_t(x, t = b_1) \quad = \quad f_2(x), \quad a_1 \leq x \leq a_2. \tag{14.13b}$$

Boundary Condition:

A boundary condition usually specifies the solution $u(x, t)$ at the left- and the right-boundaries of the space variable x, namely at $x = a_1$ and a_2, respectively. An example is

$$u(x = a_1, t) \quad = \quad g_1(t), \quad b_1 \leq t \leq b_2 \tag{14.14a}$$
$$u(x = a_2, t) \quad = \quad g_2(t), \quad b_1 \leq t \leq b_2. \tag{14.14b}$$

2. Neumann Condition

In certain problems, it is required to fix the value of, say, u_x at $x = a_1$ or a_2 for all t. An example is

$$u_x(x = a_2, t) = 0, \quad b_1 \leq t \leq b_2. \tag{14.15}$$

This Neumann boundary condition specifies that the partial derivatives u_x at right-boundary is zero. In the heat conduction problem, this boundary condition is realized if the right-boundary is attached to a perfect insulator.

3. Cauchy Condition

Cauchy condition is a combination of both the Dirichlet condition and the Neumann condition.

14.4   Finite-Difference Approximations of Partial Derivatives

In Chapter 8, ordinary derivatives are approximated in terms of finite-differences and demonstrated the evaluation of first and second derivatives. The method of finite-differences can be applied for expressing partial derivatives. In a given partial differential equation each partial derivative can be replaced by the corresponding finite-difference expression. Rearranging the terms in the resulting expression gives an approximate formula for the computation of numerical solution of the equation. Therefore, this section expresses a few partial derivatives in terms of finite-differences.

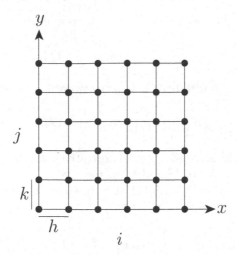

FIGURE 14.1
A two-dimensional finite-difference 6 × 6 grid.

For an ordinary derivative, the number of independent variables is one and hence the one-dimensional axis of that variable is divided into number of grid points. The derivatives are then approximated by the value of the dependent variable on either side of the grid points. Since a partial differential equation involves more than one independent variable it is desirable to consider grid points on a two-dimensional plane or a three-dimensional volume for systems with two or three independent variables, respectively. The equations that are going to consider in this chapter and in the next chapter have only two independent variables. Therefore, obtain the finite-differences of partial derivatives of a dependent variable with two independent variables.

Let the variable $u(x, y)$ is a function of two variables x and y. The two-dimensional $x - y$ plane is divided into number of rectangles of sides say $\Delta x = h$ and $\Delta y = k$ by drawing a set of equally spaced lines parallel to the x and y axes as shown in Fig. 14.1. The points of intersection of these lines are the *grid points*. They are also called *lattice points* or *mesh points*. Use the notation (i, j) for the grid points (x_i, y_i) where i and j are the counters in the x and y directions, respectively.

Express first, second and mixed partial derivatives in terms of finite-differences at a grid point (x_i, y_i), that is at (i, j). Because the partial derivatives of u with respect to x implies that y is kept constant one can write

$$\left.\frac{\partial u}{\partial x}\right|_{i,j} = \left.\frac{du}{dx}\right|_{i,j}. \tag{14.16}$$

Then, use of Eq. (8.12), which is the two-point central-difference formula of the first derivative, gives

$$\begin{aligned}
\left.\frac{\partial u}{\partial x}\right|_{i,j} &= \frac{\partial}{\partial x} u(x_i, y_j) \\
&= \frac{1}{2h}\left[u(x_i + h, y_j) - u(x_i - h, y_j)\right] + O(h^2) \\
&= \frac{1}{2h}\left[u(x_{i+1}, y_j) - u(x_{i-1}, y_j)\right] + O(h^2). \tag{14.17}
\end{aligned}$$

Writting $u(x_i, y_j)$ as $u_{i,j}$ the above equation can be rewritten as

$$\frac{\partial}{\partial x} u_{i,j} = \frac{1}{2h} \left(u_{i+1,j} - u_{i-1,j} \right) + O\left(h^2\right) . \tag{14.18a}$$

For the partial derivative of u with respect to y the variable x is held constant and at (x_i, y_j)

$$\frac{\partial}{\partial y} u_{i,j} = \frac{1}{2k} \left(u_{i,j+1} - u_{i,j-1} \right) + O\left(k^2\right) . \tag{14.18b}$$

The truncation errors in the first-order partial derivatives $\partial u / \partial x$ and $\partial u / \partial y$ approximations in Eqs. (14.18) are of the order of h^2 and k^2, respectively.

In a similar manner, using Eq. (8.18) the three-point central-difference formulas of second-order partial derivatives are obtained as

$$\frac{\partial^2}{\partial x^2} u\left(x_i, y_j\right) = \frac{1}{h^2} \left(u_{i+1,j} - 2u_{i,j} + u_{i-1,j} \right) + O\left(h^2\right) , \tag{14.19a}$$

$$\frac{\partial^2}{\partial y^2} u\left(x_i, y_j\right) = \frac{1}{k^2} \left(u_{i,j+1} - 2u_{i,j} + u_{i,j-1} \right) + O\left(k^2\right) . \tag{14.19b}$$

For the mixed derivative

$$\frac{\partial^2 u}{\partial x \partial y} = \frac{\partial^2 u}{\partial y \partial x} \tag{14.20}$$

the finite-difference approximation is

$$\frac{\partial^2}{\partial x \partial y} u\left(x_i, y_j\right) = \frac{1}{4hk} \left(u_{i+1,j+1} - u_{i-1,j+1} - u_{i+1,j-1} + u_{i-1,j-1} \right)$$
$$+ O\left(h^2 + k^2\right) . \tag{14.21}$$

In the above finite-difference approximations of derivatives the truncation error is second-order in the grid sizes. In Chapter 8, the finite-difference approximations with higher-order accuracies have also been developed. However, such more accurate formulas are not commonly employed to solve partial differential equations because they involve a larger number of terms in the resulting formula and require more extensive computation times.

Here onwards for simplicity denote $\partial u / \partial t$ as u_t, $\partial^2 u / \partial t^2$ as u_{tt}, $\partial u / \partial x$ as u_x, $\partial^2 u / \partial x^2$ as u_{xx}, $\partial^2 u / \partial t \partial x$ as u_{tx} and so on.

14.5   Hyperbolic Equations

Hyperbolic equations occur in the fields of light waves, acoustic waves, gravitational waves and vibrations of certain musical instruments. Specifically, torsional oscillations of a rod, vibrations of a stretched string, sound waves in a pipe, water waves in a narrow canal, predator-prey mute system on oranges, oil pipe lines system, gravitational instability of nebulae, aerodynamic and atmospheric flows in normal conditions, flows of contaminants through a porus medium are governed by or well approximated by (linear) hyperbolic partial differential equations.

In this section, explicit and implicit finite-difference formulas are set up for the ubiquitous wave equation, their stabilities are analyzed and the numerical solutions are compared with the exact solutions.

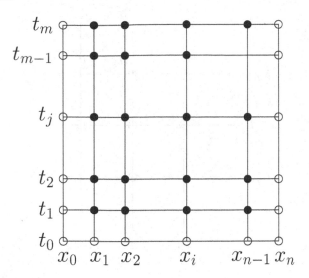

FIGURE 14.2

The $(n+1) \times (m+1)$ grid points for solving the wave equation. The values of u at the grid points marked by open circles are given by the conditions (14.23). The values of u at the grid points marked by solid circles are to be determined.

14.5.1   The Wave Equation

The wave equation is

$$u_{tt} = c^2 u_{xx}, \quad 0 \le x \le a, \ 0 \le t \le b. \tag{14.22}$$

This partial differential equation models the displacement $u(x,t)$ of a stretched elastic string with the ends fixed at $x = 0$ and a. Let Eq. (14.22) is subjected to the following initial and boundary conditions:

$$
\begin{aligned}
u(x,0) &= f(x), && 0 \le x \le a & \text{(14.23a)} \\
u_t(x,0) &= g(x), && 0 \le x \le a & \text{(14.23b)} \\
u(0,t) &= v_1(t), && 0 \le t \le b & \text{(14.23c)} \\
u(a,t) &= v_2(t), && 0 \le t \le b. & \text{(14.23d)}
\end{aligned}
$$

Divide the region $R = \{(x,t) : 0 \le x \le a, \ 0 \le t \le b\}$ into $(n+1) \times (m+1)$ rectangles with sides $\Delta x = h$ and $\Delta t = k$ as shown in Fig. 14.2. The values of x_i and t_i at the grid points are given by

$$x_i = ih, \quad i = 0,1,\ldots,n; \quad t_j = jk, \quad j = 0,1,\ldots,m. \tag{14.24}$$

The values of u at the bottom-row and the left-edge and right-edge are given by the boundary conditions. The solution at the grid points $(i, j = 0)$ in the bottom-row namely $u(i,0)$, $i = 0,1,\ldots,n$ is given by the initial condition (14.23a). The values of the partial derivative u_t at these grid points are given by the initial condition (14.23b). The values of u at the grid points on the left-edge, that is the values of $u(0,j)$, $j = 0,1,\ldots,m$ are given by Eq. (14.23c). The solution at the right-side grid points is given by Eq. (14.23d). The solution at the grid points marked by solid circles has to be determined while that at the points marked by open circles is known from the given boundary conditions.

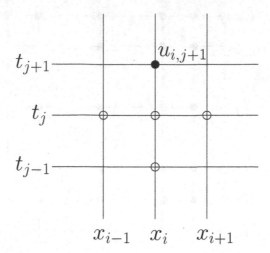

FIGURE 14.3
The known grids (marked by open circles) are used to compute the unknown solution at the grid $(i, j + 1)$ (marked by a solid circle) for the wave equation.

14.5.2   Derivation of Finite-Difference Formula

The central-difference approximations for u_{tt} and u_{xx} from Eq. (14.19) are

$$u_{tt}(x_i, t_j) = \frac{1}{k^2}(u_{i,j+1} - 2u_{i,j} + u_{i,j-1}) + O(k^2), \tag{14.25a}$$

$$u_{xx}(x_i, t_j) = \frac{1}{h^2}(u_{i+1,j} - 2u_{i,j} + u_{i-1,j}) + O(h^2). \tag{14.25b}$$

Substitution of (14.25) in (14.22) after dropping $O(k^2)$ and $O(h^2)$ gives

$$\frac{1}{k^2}(u_{i,j+1} - 2u_{i,j} + u_{i,j-1}) = \frac{c^2}{h^2}(u_{i+1,j} - 2u_{i,j} + u_{i-1,j}). \tag{14.26}$$

Define $s = ck/h$ and rewrite Eq. (14.26) as

$$u_{i,j+1} - 2u_{i,j} + u_{i,j-1} = s^2(u_{i+1,j} - 2u_{i,j} + u_{i-1,j}). \tag{14.27}$$

In the above equation assume that the value of u in the rows j and $j - 1$ are known. This results in

$$u_{i,j+1} = (2 - 2s^2)u_{i,j} + s^2(u_{i+1,j} + u_{i-1,j}) - u_{i,j-1},$$
$$i = 1, 2, \ldots, n - 1, \quad j = 1, 2, \ldots, m. \tag{14.28}$$

The values of u at the grid points (i, j), $(i+1, j)$, $(i-1, j)$ and $(i, j-1)$ are used to determine the value of u at the grid point $(i, j+1)$. This is depicted in Fig.14.3.

14.5.3   Calculation of u at $t = k$

In the formula (14.28) to calculate the value of $u_{i,j+1}$ the values of u at the grid points (i, j), $(i+1, j)$, $(i-1, j)$ and $(i, j-1)$ must be known. Set the $(j-1)$th row as the bottom-row (zeroth row) then the calculation of u begins from the grid points in the $(j+1)$th row, that is from the second row. The values of u at the grid points in the zeroth row are given by the initial condition (14.23a). In the second row, the values of u at the grid points $(0,1)$ and $(n,1)$ alone are known from the boundary conditions (14.23c) and (14.23d), respectively. The u values at other grid points in the first row are also required to compute u at the grid points in the second row. Therefore, before using the formula (14.28) it is necessary to determine the solution corresponding to the first row in Fig. 14.2. This can be done as follows.

To know $u(x_i, t_1)$ consider the Taylor series of $u(x_i, t_1)$ about $u(x_i, t_0)$:

$$u(x_i, t_1) = u(x_i, t_0) + k u_t(x_i, t_0) + O(k^2).$$ (14.29)

In other words

$$u_{i,1} = u_{i,0} + k(u_t)_{i,0} + O(k^2).$$ (14.30)

$(u_t)_{i,0}$ is known from the condition (14.23b). Neglecting the truncation error term and substituting $(u_t)_{i,0} = g_i = g(x_i)$ and $u_{i,0} = f_i$ in Eq. (14.30) yield

$$u_{i,1} = f_i + k g_i.$$ (14.31)

This numerically computed $u_{i,1}$ is not identical to the exact $u_{i,1}$ but is only an approximation. The point is that the error in (14.31) will propagate throughout the grid points and will not be dampened out. Therefore, to reduce the error in (14.31) include the next higher-order term in (14.29).

Equation (14.29) with the inclusion of one more term in the series is

$$u_{i,1} = u_{i,0} + k(u_t)_{i,0} + \frac{1}{2} k^2 (u_{tt})_{i,0} + O(k^2).$$ (14.32)

For finding $(u_{tt})_{i,0}$ consider the wave equation at the zeroth row:

$$
\begin{aligned}
(u_{tt})_{i,0} &= c^2 (u_{xx})_{i,0} \\
&= c^2 (f_{xx})_i \\
&= \frac{c^2}{h^2} (f_{i+1} - 2f_i + f_{i-1}) + O(h^2).
\end{aligned}
$$ (14.33)

Use (14.33) in (14.32) and obtain

$$
\begin{aligned}
u_{i,1} &= f_i + k g_i + \frac{c^2 k^2}{2h^2} (f_{i+1} - 2f_i + f_{i-1}) + O(h^2) O(k^2) + O(k^2) \\
&= (1 - s^2) f_i + k g_i + \frac{1}{2} s^2 (f_{i+1} + f_{i-1}), \quad i = 1, 2, \ldots, n-1
\end{aligned}
$$ (14.34)

where in the last step the error term is neglected.

The truncation error of the formula (14.28) is $O(h^2 + k^2)$. It is the error in the finite-difference equation and not of the solution. The independent variables x and t actually assume continuous values. In the finite-difference methods, they take only discrete values. That is, they are *discretized*. The discretization of a continuous problem gives rise to an error in the solution and is termed as *discretization error*. An additional error called round-off error will occur if the finite-difference equation is not solved exactly. Further, the mesh sizes h and k affect the discretization and round-off errors. Decrease in h or k decreases the discretization error but increases the round-off error since more number of iterations are used. This means that one cannot conclude that accuracy can always be improved by decreasing h and k.

14.5.4   Stability Condition

The stability of the method, that is the condition for decaying of an error made at one stage of computation, can be investigated using the so-called Fourier method (also known as von Neumann analysis) [2]. To obtain the stability condition let us assume a solution of the given equation as

$$u(x,t) = e^{i\alpha x} e^{\beta t}. \tag{14.35}$$

Replacing x by ih and t by jk in the solution gives

$$u_{i,j} = e^{i\alpha ih} e^{\beta jk}. \tag{14.36}$$

Define $\xi = e^{\beta k}$. Then,

$$u_{i,j} = e^{i\alpha ih} \xi^j. \tag{14.37}$$

The finite-difference formula will be stable if $|u_{i,j}|$ remain bounded as h, $k \to 0$ for $jk \le b$, $0 \le t \le b$, where b is finite. Suppose, the exact solution is bounded. Then, the necessary and sufficient condition for stability is $|\xi| \le 1$, $-1 \le \xi \le 1$.

For the wave equation the substitution of

$$u_{i,j} = e^{i\alpha ih} \xi^j \tag{14.38}$$

in (14.28) gives

$$\xi = \left(2 - 2s^2\right) + s^2 \left(e^{i\alpha h} + e^{-i\alpha h}\right) - \xi^{-1}. \tag{14.39}$$

Substitute

$$e^{i\alpha h} + e^{-i\alpha h} = 2\cos\alpha h = 2 - 4\sin^2(\alpha h/2) \tag{14.40}$$

and obtain

$$\xi^2 - 2\xi \left(1 - 2s^2 \sin^2(\alpha h/2)\right) + 1 = 0. \tag{14.41}$$

The above equation can be rewritten as

$$\xi^2 - 2\delta\xi + 1 = 0, \quad \delta = 1 - 2s^2 \sin^2(\alpha h/2). \tag{14.42}$$

The two roots of the above equation are

$$\xi_+ = \delta + \sqrt{\delta^2 - 1}, \quad \xi_- = \delta - \sqrt{\delta^2 - 1}. \tag{14.43}$$

$|\xi_+|$ and $|\xi_-|$ are ≤ 1 only if $|\delta| \le 1$. That is,

$$-1 \le 1 - 2s^2 \sin^2(\alpha h/2) \le 1. \tag{14.44}$$

The right-side inequality gives $-s^2 \sin^2(\alpha h/2) \le 0$. This is satisfied for all values of s (since s is real). The left-side inequality gives $s^2 \sin^2(\alpha h/2) \le 1$. Since $\sin^2(\alpha h/2)$ is ≤ 1 the requirement is

$$s^2 = \frac{c^2 k^2}{h^2} \le 1. \tag{14.45}$$

Thus, the stability condition is

$$s = \frac{ck}{h} \le 1. \tag{14.46}$$

Example:

A string of unit length is subjected to the initial displacement $u(x,0) = \sin\pi x$ and zero initial velocity. The time evolution of u is governed by the wave equation

$$u_{tt}(x,t) = u_{xx}(x,t), \quad 0 \le x \le 1, \quad 0 \le t \le 0.5 \tag{14.47}$$

TABLE 14.1
The numerical solution of the wave Eq. (14.48) is obtained from the difference Eq. (14.28) and the exact solution. For each value of t, the first and second rows give the numerical and exact solutions, respectively.

t	$x_0 = 0.0$	$x_4 = 0.2$	$x_8 = 0.4$	$x_{12} = 0.6$	$x_{16} = 0.8$	$x_{20} = 1.0$
0.00	$----$	$----$	$----$	$----$	$----$	$----$
	0.000000	0.587785	0.951057	0.951057	0.587785	0.000000
0.10	$----$	0.559017	0.904508	0.904508	0.559017	$----$
	0.000000	0.559017	0.904508	0.904508	0.559017	0.000000
0.20	$----$	0.475528	0.769421	0.769421	0.475528	$----$
	0.000000	0.475528	0.769421	0.769421	0.475528	0.000000
0.30	$----$	0.345492	0.559017	0.559017	0.345492	$----$
	0.000000	0.345492	0.559017	0.559017	0.345492	0.000000
0.40	$----$	0.181636	0.293893	0.293893	0.181636	$----$
	0.000000	0.181636	0.293893	0.293893	0.181636	0.000000
0.50	$----$	0.005077	0.008215	0.008215	0.005077	$----$
	0.000000	0.000000	0.000000	0.000000	0.000000	0.000000

subjected to the initial and boundary conditions

$$
\begin{aligned}
u(x,0) &= \sin \pi x, & 0 \le x \le 1 & \qquad (14.48\text{a})\\
u_t(x,0) &= g(x) = 0, & 0 \le x \le 1 & \qquad (14.48\text{b})\\
u(0,t) &= v_1 = 0, & 0 \le t \le 0.5 & \qquad (14.48\text{c})\\
u(1,t) &= v_2 = 0, & 0 \le t \le 0.5. & \qquad (14.48\text{d})
\end{aligned}
$$

Determine the solution $u(x,t)$.

In Eq. (14.47) $c^2 = 1$, that is $c = 1$. Choose $h = 0.05$, $k = 0.05$ which give $s = ck/h = 1$. The exact solution for the given conditions is

$$u_e(x,t) = \sin \pi x \cos \pi t. \qquad (14.49)$$

The numerical and the exact solutions for a few values of x for $t = 0, 0.05, 0.1, \ldots, 0.5$ are given in Table 14.1. The difference between u_e and numerically computed u for $x = 0.5$ is plotted in Fig. 14.4.

14.5.5 An Implicit Method

The finite-difference formula (14.28) for Eq. (14.22) is an explicit method since the unknown u at a grid point is determined using the known u at some other grid points. Though the method is simple, a disadvantage of this method is that it is stable only if $0 < s \le 1$. This means, $ck/h \le 1$, that is, $k \le h/c$. The time step k must be sufficiently small. This difficulty is avoided in implicit methods. An *implicit finite-difference formula* is one in which more than one unknown values in the $(j+1)$th row are specified in terms of the known values in the rows $j, j-1, \ldots$. In an implicit method the second-order spatial partial derivative at

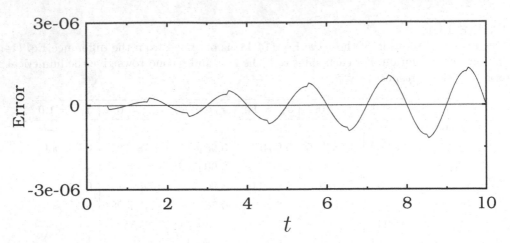

FIGURE 14.4
Error in the numerical solution of Eq. (14.47) for $x = 0.5$ as a function of time.

the grid point (i, j) is taken as the average of the spatial derivative at $(i, j+1)$ and $(i, j-1)$:

$$u_{xx}(x_i, t_j) = \frac{1}{2}[u_{xx}(x_i, t_{j+1}) + u_{xx}(x_i, t_{j-1})] . \tag{14.50}$$

Write

$$u_{xx}(x_i, t_{j+1}) = \frac{1}{h^2}(u_{i+1,j+1} - 2u_{i,j+1} + u_{i-1,j+1}) , \tag{14.51a}$$

$$u_{xx}(x_i, t_{j+1}) = \frac{1}{h^2}(u_{i+1,j+1} - 2u_{i,j+1} + u_{i-1,j+1}) , \tag{14.51b}$$

$$u_{tt}(x_i, t_j) = \frac{1}{k^2}(u_{i,j+1} - 2u_{i,j} + u_{i,j-1}) . \tag{14.51c}$$

At (i, j) the wave equation is

$$u_{tt}(x_i, t_j) = \frac{1}{2}c^2[u_{xx}(x_i, t_{j+1}) + u_{xx}(x_i, t_{j-1})] . \tag{14.52}$$

Substitute (14.51) in (14.52) and obtain

$$u_{i-1,j+1} + du_{i,j+1} + u_{i+1,j+1} = -2eu_{i,j} - u_{i+1,j-1} - du_{i,j-1} - u_{i-1,j-1},$$
$$i = 1, 2, \ldots, n-1 \tag{14.53}$$

where $d = -2(1 + s^2)/s^2$ and $e = 2/s^2$. If u for j and $j - 1$ are assumed to be known then
the right-side of the above equation contains only known terms. Then, Eq. (14.53) simply
generates a system of linear equations for the $n - 1$ unknowns $u_{i,j+1}$, $i = 1, 2, \ldots, n-1$
for $(j + 1)$th time step. Interestingly Eq. (14.53) gives a tridiagonal system. Let us write
Eq. (14.53) explicitly for $i = 1, 2, \ldots, n-1$:

$$u_{0,j+1} + du_{1,j+1} + u_{2,j+1} = b_{1,j} ,$$
$$u_{1,j+1} + du_{2,j+1} + u_{3,j+1} = b_{2,j} ,$$
$$\vdots \tag{14.54}$$
$$u_{n-3,j+1} + du_{n-2,j+1} + u_{n-1,j+1} = b_{n-2,j} ,$$
$$u_{n-2,j+1} + du_{n-1,j+1} + u_{n,j+1} = b_{n-1,j} ,$$

where

$$b_{i,j} = -2eu_{i,j} - u_{i+1,j-1} - du_{i,j-1} - u_{i-1,j-1},$$
$$i = 1, 2, \ldots, n-1, \quad j = 1, 2, \ldots, n-1. \tag{14.55}$$

In each subequation of Eqs. (14.54) more than one unknown appears and hence the method is an implicit one. In matrix form Eqs. (14.54) can be written as

$$
\begin{bmatrix}
d & 1 & & & & & \\
1 & d & 1 & & & & \\
 & 1 & d & 1 & & & \\
 & & & \ddots & & & \\
 & & & 1 & d & 1 & \\
 & & & & 1 & d & 1 \\
 & & & & & 1 & d
\end{bmatrix}
\begin{bmatrix}
u_{1,j+1} \\
u_{2,j+1} \\
u_{3,j+1} \\
\vdots \\
u_{n-3,j+1} \\
u_{n-2,j+1} \\
u_{n-1,j+1}
\end{bmatrix}
$$

$$
=
\begin{bmatrix}
b_{1,j} - u_{0,j+1} \\
b_{2,j} \\
\vdots \\
b_{n-2,j} \\
b_{n-1,j} + u_{n,j+1}
\end{bmatrix}. \tag{14.56}
$$

Since $u_{0,j+1}$ and $u_{n,j+1}$ in Eq. (14.55) are known from the boundary conditions they are brought to right-side of Eq. (14.56). The above tridiagonal system can be solved easily using the procedure discussed in Section 4.8. For each value of j, namely, $j = 1, 2, \ldots, n-1$ the system (14.56) is solved. That is, using $u_{i,0}$ and $u_{i,1}$ the system (14.56) gives $u_{i,2}$. Then, $u_{i,1}$ and $u_{i,2}$ are used to calculate $u_{i,3}$ and so on.

14.5.6 Stability Analysis of the Implicit Method

Substituting $u_{i,j} = e^{i\alpha i h}\xi^j$ and $\xi = e^{\beta k}$ in the implicit formula (14.53) gives

$$\xi^2 \left(\cos\alpha h + \frac{d}{2} \right) + e\xi + \left(\frac{d}{2} + \cos\alpha h \right) = 0 \tag{14.57}$$

which can be rewritten as

$$\xi^2 - \frac{2}{1 + 2s^2 \sin^2(\alpha h/2)}\xi + 1 = 0. \tag{14.58}$$

Its roots are

$$\xi_{\pm} = a \pm \sqrt{a^2 - 1}, \quad a = \frac{1}{1 + 2s^2 \sin^2(\alpha h/2)}. \tag{14.59}$$

$|\xi_{\pm}| < 1$ if $|a| < 1$. This is satisfied without any restriction on h and k values. That is, the implicit method is *unconditionally stable*.

Example:

Solve the wave Eq. (14.47) subjected to the conditions (14.48) by the implicit method. Choose $h = 0.05$ and $k = 0.05$. The numerical solution is displayed in Table 14.2.

TABLE 14.2
The numerical solution of the wave Eq. (14.47) is calculated by the implicit method and the exact solution. Here, $h = k = 0.05$. For each value of t, the first and second rows give the numerical and exact solutions, respectively.

t	$x_0 = 0.0$	$x_4 = 0.2$	$x_8 = 0.4$	$x_{12} = 0.6$	$x_{16} = 0.8$	$x_{20} = 1.0$
0.00	$- - --$	$- - --$	$- - --$	$- - --$	$- - --$	$- - --$
	0.000000	0.587785	0.951057	0.951056	0.587785	0.000000
0.10	$- - --$	0.559191	0.904790	0.904790	0.559191	$- - --$
	0.000000	0.559017	0.904508	0.904508	0.559017	0.000000
0.20	$- - --$	0.476521	0.771027	0.771026	0.476521	$- - --$
	0.000000	0.475528	0.769421	0.769421	0.475528	0.000000
0.30	$- - --$	0.347769	0.562702	0.562702	0.347769	$- - --$
	0.000000	0.345492	0.559017	0.559017	0.345492	0.000000
0.40	$- - --$	0.185387	0.299962	0.299962	0.185387	$- - --$
	0.000000	0.181636	0.293893	0.293893	0.181636	0.000000
0.50	$- - --$	0.005077	0.008215	0.008215	0.005077	$- - --$
	0.000000	0.000000	0.000000	0.000000	0.000000	0.000000

14.5.7 Damped Wave Equation

Consider the damped wave equation

$$u_{tt} = u_{xx} - 2cu_t, \quad 0 \le x \le \infty, \ t \ge 0 \tag{14.60}$$

subjected to the initial and boundary conditions

$$u(x,0) = e^{-x}, \tag{14.61a}$$

$$u_t(x,0) = g(x) = \left(-1 - \sqrt{2}\right)e^{-x}, \tag{14.61b}$$

$$u(0,t) = e^{(-1-\sqrt{2})t}. \tag{14.61c}$$

The exact solution corresponding to the given conditions is

$$u(x,t) = e^{-x}\,e^{(-1-\sqrt{2})t}. \tag{14.62}$$

Replacing the derivatives in (14.60) by the central-difference approximations yields

$$\begin{aligned}
u_{i,j+1} &= \frac{1}{1+ck}\big[-u_{i,j-1} + 2\left(1 - s^2\right)u_{i,j} \\
&\quad + s^2\left(u_{i+1,j} + u_{i-1,j}\right) + cku_{i,j-1}\big],
\end{aligned} \tag{14.63}$$

where $s = k/h$. $u_{i,0}$ is given by the initial condition $u(x,0)$. To calculate $u_{i,1}$ use the initial condition (14.61b). The relation

$$u_t(x,0) = \frac{1}{2k}\left(u_{i,1} - u_{i,-1}\right) = g_i \tag{14.64}$$

gives

$$u_{i,-1} = u_{i,1} - 2kg_i. \tag{14.65}$$

Use the above in Eq. (14.63) for $j = 0$ and obtain

$$u_{i,1} = (1 - ck)kg_i + \frac{1}{2}\left[2\left(1 - s^2\right)u_{i,0} + s^2\left(u_{i+1,0} + u_{i-1,0}\right)\right].$$ (14.66)

For stability investigation substitute

$$u_{i,j} = e^{i\alpha ih}\, e^{i\beta jk} = e^{i\alpha ih}\xi^j$$ (14.67)

in Eq. (14.63) and obtain

$$(1 + ck)\xi^2 - 2\xi\left[1 - 2s^2\sin^2(\alpha h/2)\right] + 1 - ck = 0.$$ (14.68)

Its roots are

$$\xi_{\pm} = \frac{a \pm \sqrt{a^2 - 1 + c^2 k^2}}{1 + ck},$$ (14.69a)

where

$$a = 1 - 2s^2\sin^2(\alpha h/2).$$ (14.69b)

For stability $|\xi_{\pm}| < 1$. If $a < -1$ then $a^2 > 1$, $c^2 k^2 > 0$ and $a^2 - 1 + c^2 k^2 > 0$. So ξ_{-} is < -1. Similarly, if $a > 1$ then ξ_{+} is > 1. The method is thus unstable for $|a| > 1$. Next, consider $-1 < a < 1$. Now, $a^2 < 1$ and, say, $a^2 - 1 + c^2 k^2 > 0$. In this case

$$\sqrt{a^2 - 1 + c^2 k^2} < ck.$$ (14.70)

Write

$$\xi_{\pm} = \frac{[(-1 < a < 1) \pm (< ck)]}{1 + ck},$$ (14.71)

Since the absolute value of the numerator in (14.71) is $< 1 + ck$, $|\xi_{\pm}| < 1$. When $a^2 - 1 + c^2 k^2 < 0$ then

$$\xi_{\pm} = \frac{[(-1 < a < 1) + i(< ck)]}{1 + ck}$$ (14.72)

and

$$|\xi_{\pm}| = \left[\frac{(0 < a^2 < 1) + (< c^2 k^2)}{1 + c^2 k^2 + 2ck}\right]^{1/2}.$$ (14.73)

That is, $|\xi_{\pm}| < 1$. Thus, the stability condition is $|a| \le 1$, that is $s \le 1$ or $k \le h$.

14.6 Parabolic Equations

Some examples of phenomena described by parabolic type differential equations are heat flow, diffusion of particles, transport-reaction in industrial chemicals and materials engineering processes. Specifically, heat flow in a material slab, diffusion of neutrons through atomic piles and time evolution of probability distribution in certain stochastic dynamics are governed by parabolic equations.

In this section, consider the heat equation as an example of parabolic type equations and describe an explicit method and an implicit method for solving it.

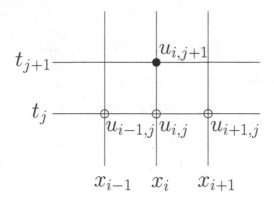

FIGURE 14.5
The grid points (marked by open circles) are involved in the numerical computation of u at the grid point $(i, j + 1)$ (marked by a solid circle) for the heat equation.

14.6.1 An Explicit Method for Heat Equation

Consider the one-dimensional heat conduction equation

$$u_t = c^2 u_{xx}, \quad 0 \leq x \leq a, \ 0 \leq t \leq b. \tag{14.74}$$

Equation (14.74) models the temperature distribution in an insulated rod with both the ends kept at constant temperatures say at g_1 and g_2 and the initial temperature ($t = 0$) distribution along the rod being $f(x)$. Divide the region $R = \{(x, t) : 0 \leq x \leq a, \ 0 \leq t \leq b\}$ into $n \times m$ rectangles with sides $\Delta x = h$ and $\Delta t = k$ as done earlier for the wave equation (see Fig. 14.2). The x_i and t_j are given by Eq. (14.24).

To obtain a finite-difference formula express the derivatives u_t and u_{xx} in terms of central-differences and forward-differences, respectively, as

$$u_t = \frac{1}{k} \left(u_{i,j+1} - u_{i,j} \right) + O(k), \tag{14.75a}$$

$$u_{xx} = \frac{1}{h^2} \left(u_{i-1,j} - 2u_{i,j} + u_{i+1,j} \right) + O\left(h^2\right). \tag{14.75b}$$

Substituting (14.75), after dropping $O(k)$ and $O(h^2)$, in (14.74) and rearranging the terms result in

$$u_{i,j+1} = su_{i-1,j} + (1 - 2s)u_{i,j} + su_{i+1,j}, \tag{14.76}$$

where $s = c^2 k / h^2$. The above formula proposed by Schmidt gives the value of $u_{i,j+1}$ in terms of $u_{i-1,j}$, $u_{i,j}$ and $u_{i+1,j}$ and is shown in Fig. 14.5. Note that for the wave equation, to calculate u at t_{j+1} the values of u at $t = t_{j-1}$ and t_j are required. For the heat equation u values at $t = t_j$ are alone required to calculate it at t_{j+1}.

Next, to determine a stability condition substitute Eq. (14.37) in the difference Eq. (14.76) and obtain

$$e^{i\alpha i h}\xi^{j+1} = (1 - 2s)e^{i\alpha i h}\xi^j + s\,e^{i\alpha i h}\xi^j \left(e^{-i\alpha h} + e^{i\alpha h}\right). \tag{14.77}$$

Simplification of this equation results in

$$\xi = (1 - 2s) + 2s\cos\alpha h. \tag{14.78}$$

TABLE 14.3

The numerical solution of the heat Eq. (14.80) is computed by the explicit formula (14.76) and the exact solution. Here, $h = 0.1$, $k = 0.005$ and $s = 0.5$. For each value of t, the first and second rows give the numerical and exact solutions, respectively.

t	$x_0 = 0.0$	$x_2 = 0.2$	$x_4 = 0.4$	$x_6 = 0.6$	$x_8 = 0.8$	$x_{10} = 1.0$
0.00	$----$	$----$	$----$	$----$	$----$	$----$
	0.000000	0.587785	0.951057	0.951057	0.587785	0.000000
0.10	$----$	0.215449	0.348604	0.348604	0.215449	$----$
	0.000000	0.219072	0.354466	0.354466	0.219072	0.000000
0.20	$----$	0.078972	0.127779	0.127779	0.078972	$----$
	0.000000	0.081650	0.132112	0.132112	0.081650	0.000000
0.30	$----$	0.028947	0.046837	0.046837	0.028947	$----$
	0.000000	0.030432	0.049239	0.049239	0.030432	0.000000
0.40	$----$	0.010610	0.017168	0.017168	0.010610	$----$
	0.000000	0.011342	0.018352	0.018352	0.011342	0.000000
0.50	$----$	0.003889	0.006293	0.006293	0.003889	$----$
	0.000000	0.004227	0.006840	0.006840	0.004227	0.000000

Since $-1 \le \cos \alpha h \le 1$ the condition $|\xi| \le 1$ gives

$$|1 - 2s + 2s| \le 1 \quad \text{or} \quad |1 - 2s - 2s| \le 1. \tag{14.79}$$

The second inequality is satisfied for $s \le 1/2$. Thus, the stability condition for the heat equation is $s \le 1/2$, that is, $c^2 k/h^2 \le 1/2$ or $k \le h^2/(2c^2)$.

Example:

Solve the heat equation

$$u_t = u_{xx} \tag{14.80a}$$

with the initial condition

$$u(x, 0) = f(x) = \sin \pi x \tag{14.80b}$$

and the boundary conditions

$$u(0, t) = v_1 = 0, \quad u(1, t) = v_2 = 0. \tag{14.80c}$$

Here, $c = 1$ and fix $h = 0.1$. The stability condition $k \le h^2/(2c^2)$ gives $k \le 0.005$. Hence, choose $k = 0.005$. The exact solution for the given conditions is

$$u_e(x, t) = \sin \pi x \, e^{-\pi^2 t}. \tag{14.81}$$

The numerical and the exact solutions for a few values of x for $t = 0, 0.05, 0.1, \ldots, 0.5$ are given in Table 14.3. Figure 14.6 shows error versus t for some values of h and k with $s = 0.5$.

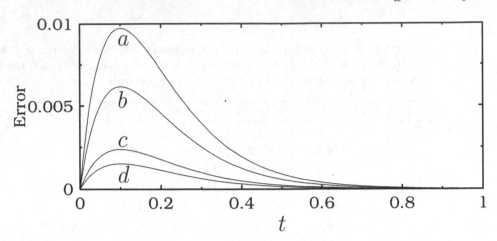

FIGURE 14.6
Error $(u_e - u)$ versus t at $x = 0.5$ for the heat equation. The values of (h, k) are: (a) $(h, k) = (0.125, 0.0078125)$, (b) $(h, k) = (0.1, 0.005)$, (c) $(h, k) = (0.0625, 0.0019531)$ and (d) $(h, k) = (0.05, 0.00125)$.

14.6.2 Crank–Nicholson Implicit Method for Heat Equation

Even though the difference Eq. (14.76) is simple, the method is stable only if $0 < s \le 0.5$. An alternative method was proposed by Crank and Nicholson. In their method, the approximation of the derivatives u_t and u_{xx} at $(x, t + k/2)$ are considered. $u_t(x, t + k/2)$ is given by

$$u_t(x, t + k/2) = \frac{1}{k} \left[u(x, t + k) - u(x, t) \right]$$

or

$$u_t|_{i,j+1/2} = \frac{1}{k} \left(u_{i,j+1} - u_{i,j} \right) . \tag{14.82}$$

In Eq. (14.82) the partial derivative u_t is expressed in terms of central-difference around the half-point. The derivative $u_{xx}(x, t + k/2)$ is approximated as the average of the approximations $u_{xx}(x, t)$ and $u_{xx}(x, t + k)$:

$$
\begin{aligned}
u_{xx}|_{i,j+1/2} = \; & \frac{1}{2h^2} \left(u_{i-1,j+1} - 2u_{i,j+1} + u_{i+1,j+1} \right. \\
& \left. + u_{i-1,j} - 2u_{i,j} + u_{i+1,j} \right) + O\left(h^2 \right) .
\end{aligned} \tag{14.83}
$$

Substitution of (14.82) and (14.83) in the heat Eq. (14.74) gives

$$
\begin{aligned}
-u_{i-1,j+1} + \left(2 + \frac{2}{s} \right) u_{i,j+1} \; - \; & u_{i+1,j+1} \\
= \; & \left(2 - \frac{2}{s} \right) u_{i,j} + u_{i-1,j} + u_{i+1,j} , \\
& i = 1, 2, \ldots, m - 1 \\
& j = 0, 1, \ldots, m - 1 .
\end{aligned} \tag{14.84}
$$

Define

$$b_{1,j} = \left(2 - \frac{2}{s}\right) u_{1,j} + u_{0,j} + u_{2,j} + u_{0,j+1}, \tag{14.85a}$$

$$b_{i,j} = \left(2 - \frac{2}{s}\right) u_{i,j} + u_{i-1,j} + u_{i+1,j+1}, \tag{14.85b}$$

$$b_{n-1,j} = \left(2 - \frac{2}{s}\right) u_{n-1,j} + u_{n-2,j} + u_{n,j} + u_{n,j+1},$$
$$j = 2, 3, \ldots, n-1 \tag{14.85c}$$

$$d = 2 + \frac{2}{s}. \tag{14.85d}$$

Equation (14.84) can be explicitly written as

$$du_{1,j+1} - u_{2,j+1} = b_{1,j},$$
$$-u_{1,j+1} + du_{2,j+1} - u_{3,j+1} = b_{2,j},$$
$$-u_{2,j+1} + du_{3,j+1} - u_{4,j+1} = b_{3,j},$$
$$\vdots \tag{14.86}$$
$$-u_{n-3,j+1} + du_{n-2,j+1} - u_{n-1,j+1} = b_{n-2,j},$$
$$-u_{n-2,j+1} + du_{n-1,j+1} = b_{n-1,j}.$$

Assume that the solution is known at $t = t_j$ and $u_{0,j+1}$ and $u_{n,j+1}$ are also known by the boundary conditions. Then, Eq. (14.86) constitutes a system of linear simultaneous equations for the unknowns $u_{i,j+1}$ which can be solved. In matrix form, Eq. (14.86) is written as

$$
\begin{bmatrix}
d & -1 & & & & & \\
-1 & d & -1 & & & & \\
& -1 & d & -1 & & & \\
& & & \ddots & & & \\
& & & & -1 & d & -1 \\
& & & & & -1 & d
\end{bmatrix}
\begin{bmatrix}
u_{1,j+1} \\
u_{2,j+1} \\
u_{3,j+1} \\
\vdots \\
u_{n-2,j+1} \\
u_{n-1,j+1}
\end{bmatrix}
=
\begin{bmatrix}
b_{1,j} \\
b_{2,j} \\
b_{3,j} \\
\vdots \\
b_{n-2,j} \\
b_{n-1,j}
\end{bmatrix}. \tag{14.87}
$$

Since the system (14.87) is in symmetric tridiagonal form it can be solved easily (refer Section 4.8).

Example:

Solve the heat Eq. (14.80) by the Crank–Nicholson implicit method.

Choose $h = 0.1$ and $k = 0.01$. The numerical solution is given in Table 14.4. Figure 14.7 shows error versus t for four set of values of h and k.

14.7   Elliptic Equations

Elliptic partial differential equations are describing the cases where the dynamical variables do not evolve with time or attain a stationary state. They are often found in the study of steady-state heat conduction and diffusion problems. For example, the heat conduction in

TABLE 14.4
The numerically computed solution of the heat Eq. (14.80) with $h = 0.1$ and $k = 0.01$ using the Crank–Nicholson implicit method and the exact solution. For each value of t, the first and second rows give the numerical and exact solutions, respectively.

t	$x_0 = 0.0$	$x_2 = 0.2$	$x_4 = 0.4$	$x_6 = 0.6$	$x_8 = 0.8$	$x_{10} = 1.0$
0.00	– – – –	– – – –	– – – –	– – – –	– – – –	– – – –
	0.000000	0.587785	0.951057	0.951057	0.587785	0.000000
0.10	– – – –	0.220679	0.357066	0.357066	0.220679	– – – –
	0.000000	0.219072	0.354466	0.354466	0.219072	0.000000
0.20	– – – –	0.082852	0.134057	0.134057	0.082852	– – – –
	0.000000	0.081650	0.132112	0.132112	0.081650	0.000000
0.30	– – – –	0.031106	0.050331	0.050331	0.031106	– – – –
	0.000000	0.030432	0.049239	0.049239	0.030432	0.000000
0.40	– – – –	0.011679	0.018896	0.018896	0.011679	– – – –
	0.000000	0.011342	0.018352	0.018352	0.011342	0.000000
0.50	– – – –	0.004385	0.007094	0.007094	0.004385	– – – –
	0.000000	0.004227	0.006840	0.006840	0.004227	0.000000

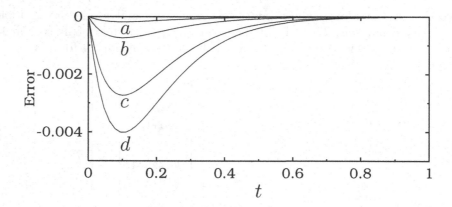

FIGURE 14.7
Error $(u_e - u)$ versus t at $x = 0.5$ for the heat equation solved by the Crank–Nicholson implicit method. The values of (h, k) are: (a) $(h, k) = (0.025, 0.000625)$, (b) $(h, k) = (0.05, 0.0025)$, (c) $(h, k) = (0.1, 0.01)$ and (d) $(h, k) = (0.125, 0.015625)$.

a solid is modelled by Laplace equation. It also arises in the characteristic of electrostatic potentials at points of free space and the velocity potential of an irrotational incompressible fluid flow.

14.7.1　Laplace Equation with Dirichlet Conditions

The heat equation in two-dimension is

$$u_t = \mu\left(u_{xx} + u_{yy}\right),$$

(14.88)

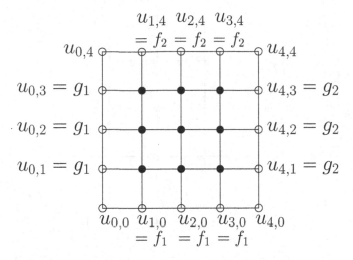

FIGURE 14.8

A 5×5 grid region for Laplace equation. The values of u at the grid points marked by open circles are known from the boundary conditions and those at the grid points marked by solid circles are to be determined.

where μ is the thermal diffusivity and $u(x, y, t)$ is the temperature at (x, y, t). In the case of steady-state transfer of heat $u_t = 0$ giving $u_{xx} + u_{yy} = 0$. This equation is called the *Laplace equation.* The equation $u_{xx} + u_{yy} = f(x, y)$ is termed as the *Poisson equation.*

Consider the two-dimensional Laplace equation

$$u_{xx} + u_{yy} = 0, \quad 0 \le x \le a, \ 0 \le y \le b. \tag{14.89a}$$

Assume that the $x - y$ plane with $0 \le x \le a$ and $0 \le y \le b$ is divided into $(n+1) \times (m+1)$ grid points and $x_i = ih$, $i = 0, 1, \ldots, n$ and $y_j = jk$, $j = 0, 1, \ldots, m$. $h = k$ gives $b/a = m/n$. The boundary conditions are

$$
\begin{aligned}
u(x, 0) &= f_1(x), \quad u(x, b) = f_2(x), & (14.89b)\\
u(0, y) &= g_1(y), \quad u(a, y) = g_2(y). & (14.89c)
\end{aligned}
$$

Figure 14.8 shows the 5×5 grid points of a square region. The values of u on the four sides of the square are given by the boundary conditions (14.89b) and (14.89c). The values of u at the grid points inside the square are unknown and are to be determined.

Replace the second-order partial derivatives by their central-difference approximations. The result is

$$\frac{1}{h^2} (u_{i+1,j} - 2u_{i,j} + u_{i-1,j})$$

$$+ \frac{1}{k^2} (u_{i,j+1} - 2u_{i,j} + u_{i,j-1}) + O\left(h^2 + k^2\right) = 0. \tag{14.90}$$

For the equidistant grid ($h = k$) Eq. (14.90) becomes ($m = n$)

$$u_{i+1,j} + u_{i-1,j} + u_{i,j+1} + u_{i,j-1} - 4u_{i,j} = 0, \tag{14.91}$$

where $i = 1, 2, \ldots, n - 1$ and $j = 1, 2, \ldots, m - 1$. Equation (14.91) gives

$$u_{i,j} = \frac{1}{4} [u_{i+1,j} + u_{i-1,j} + u_{i,j+1} + u_{i,j-1}]. \tag{14.92}$$

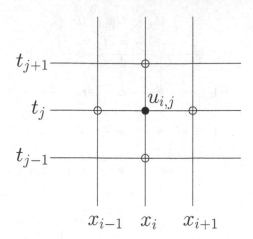

FIGURE 14.9
Representation of the formula (14.92) of the Laplace equation. The solution at the solid circle is given in terms of the solution at the points marked by open circles.

The beauty of Eq. (14.92) is that the solution at the grid point (i, j) is the average of the solution at the grid points adjacent to it. This is depicted in Fig. 14.9.

For a 5×5 grid points ($n = m = 4$) Eq. (14.91) generates a system of 9 linear equations for the solutions at the 9 interior points of the square of Fig. 14.8. These equations are explicitly written as

$$i = 1, \ j = 1: \quad u_{2,1} + u_{0,1} + u_{1,2} + u_{1,0} - 4u_{1,1} = 0, \tag{14.93a}$$

$$i = 2, \ j = 1: \quad u_{3,1} + u_{1,1} + u_{2,2} + u_{2,0} - 4u_{2,1} = 0, \tag{14.93b}$$

$$i = 3, \ j = 1: \quad u_{4,1} + u_{2,1} + u_{3,2} + u_{3,0} - 4u_{3,1} = 0, \tag{14.93c}$$

$$i = 1, \ j = 2: \quad u_{2,2} + u_{0,2} + u_{1,3} + u_{1,1} - 4u_{1,2} = 0, \tag{14.93d}$$

$$i = 2, \ j = 2: \quad u_{3,2} + u_{1,2} + u_{2,3} + u_{2,1} - 4u_{2,2} = 0, \tag{14.93e}$$

$$i = 3, \ j = 2: \quad u_{4,2} + u_{2,2} + u_{3,3} + u_{3,1} - 4u_{3,2} = 0, \tag{14.93f}$$

$$i = 1, \ j = 3: \quad u_{2,3} + u_{0,3} + u_{1,4} + u_{1,2} - 4u_{1,3} = 0, \tag{14.93g}$$

$$i = 2, \ j = 3: \quad u_{3,3} + u_{1,3} + u_{2,4} + u_{2,2} - 4u_{2,3} = 0, \tag{14.93h}$$

$$i = 3, \ j = 3: \quad u_{4,3} + u_{2,3} + u_{3,4} + u_{3,2} - 4u_{3,3} = 0. \tag{14.93i}$$

In the above set of equations, the values of u at the four corners of the square grid in Fig. 14.8 namely, $u_{0,0}$, $u_{4,0}$, $u_{0,4}$ and $u_{4,4}$ are not involved.

Bringing the u's given by the boundary conditions to the right-side and writing Eq. (14.93) in the matrix form $AX = B$ give

$$
\begin{bmatrix}
-4 & 1 & 0 & 1 & 0 & 0 & 0 & 0 & 0 \\
1 & -4 & 1 & 0 & 1 & 0 & 0 & 0 & 0 \\
0 & 1 & -4 & 0 & 0 & 1 & 0 & 0 & 0 \\
1 & 0 & 0 & -4 & 1 & 0 & 1 & 0 & 0 \\
0 & 1 & 0 & 1 & -4 & 1 & 0 & 1 & 0 \\
0 & 0 & 1 & 0 & 1 & -4 & 0 & 0 & 1 \\
0 & 0 & 0 & 1 & 0 & 0 & -4 & 1 & 0 \\
0 & 0 & 0 & 0 & 1 & 0 & 1 & -4 & 1 \\
0 & 0 & 0 & 0 & 0 & 1 & 0 & 1 & -4
\end{bmatrix}
\begin{bmatrix}
u_{1,1} \\
u_{2,1} \\
u_{3,1} \\
u_{1,2} \\
u_{2,2} \\
u_{3,2} \\
u_{1,3} \\
u_{2,3} \\
u_{3,3}
\end{bmatrix}
$$

$$= \begin{bmatrix} -u_{0,1} - u_{1,0} \\ -u_{2,0} \\ -u_{3,0} - u_{4,1} \\ -u_{0,2} \\ 0 \\ -u_{4,2} \\ -u_{0,3} - u_{1,4} \\ -u_{2,4} \\ -u_{3,4} - u_{4,3} \end{bmatrix} . \tag{14.94}$$

For the wave Eq. (14.22) and the heat Eq. (14.74), a system of equations is obtained for the unknown solution by the implicit method. The obtained system is simply a tridiagonal in form. But for the Laplace equation, the coefficient matrix in (14.94) is not in tridiagonal form. However, Eq. (14.94) can be solved by Gauss or Gauss–Jordan method.

Example:

Compute the numerical solution of the Laplace equation

$$u_{xx} + u_{yy} = 0, \quad 0 \le x \le 1, \ 0 \le y \le 1 \tag{14.95a}$$

subjected to the boundary conditions

$$\begin{aligned} u(x,0) &= f_1(x) = \sin \pi x, & \text{(14.95b)} \\ u(x,1) &= f_2(x) = \sin \pi x \, e^{-\pi}, & \text{(14.95c)} \\ u(0,y) &= g_1(y) = 0, & \text{(14.95d)} \\ u(1,y) &= g_2(y) = 0. & \text{(14.95e)} \end{aligned}$$

Choose $h = k = 0.25$, $n = m = 4$. That is, the $x - y$ plane with $x \in [0,1]$ and $y \in [0,1]$ is divided into 5×5 grids. Gauss–Jordan method is used to solve the system of Eq. (14.94). The numerical solution is given along with the exact solution in Table 14.5. The exact solution of Eq. (14.95) is given by

$$u(x,y) = \sin \pi x \, e^{-\pi y}. \tag{14.96}$$

14.7.2 Laplace Equation with Derivative Boundary Conditions

In certain applications of heat flow the edge is insulated so that there is no heat flux through the edge. Suppose, $y = y_m = b$ is insulated. That is, top edge in Fig. 14.8 is fixed. On this edge the boundary condition

$$\frac{\partial}{\partial y} u(x, y_m) = u_y(x, b) = 0 \tag{14.97}$$

must be satisfied. Then, for the grid point (x_i, y_m) Eq. (14.91) becomes

$$u_{i+1,m} + u_{i-1,m} + u_{i,m+1} + u_{i,m-1} - 4u_{i,m} = 0. \tag{14.98}$$

The grid point $(i, m+1)$ lies outside the region R and so $u_{i,m+1}$ is unknown. However, the approximation

$$u_y(x_i, y_m) \approx \frac{1}{2k}(u_{i,m+1} - u_{i,m-1}) = 0 \tag{14.99}$$

TABLE 14.5

The numerical and the exact solutions of the Laplace Eq. (14.95). The numerical solution is obtained by solving (14.94). For each y the first row gives the numerical solution while the second row gives the exact solution.

y	$x = 0.0$	$x = 0.25$	$x = 0.50$	$x = 0.75$	$x = 1.0$
0.00	$- - - -$	$- - - -$	$- - - -$	$- - - -$	$- - - -$
	0.000000	0.707107	1.000000	0.707107	0.000000
0.25	$- - - -$	0.334334	0.472819	0.334334	$- - - -$
	0.000000	0.322397	0.455938	0.322397	0.000000
0.50	$- - - -$	0.157409	0.222610	0.157409	$- - - -$
	0.000000	0.146993	0.207880	0.146993	0.000000
0.75	$- - - -$	0.072692	0.102802	0.072692	$- - - -$
	0.000000	0.067020	0.094780	0.067020	0.000000
1.00	$- - - -$	$- - - -$	$- - - -$	$- - - -$	$- - - -$
	0.000000	0.030557	0.043214	0.030557	0.000000

gives

$$u_{i,m+1} = u_{i,m-1} + O\left(k^2\right). \tag{14.100}$$

Replace $u_{i,m+1}$ by $u_{i,m-1}$ in (14.98) and obtain

$$u_{i+1,m} + u_{i-1,m} + 2u_{i,m-1} - 4u_{i,m} = 0. \tag{14.101}$$

For convenience rewrite the above equation as

$$2u_{i,m-1} + u_{i-1,m} + u_{i+1,m} - 4u_{i,m} = 0, \quad (\textbf{top} - \textbf{edge}). \tag{14.102}$$

For the bottom-edge, the boundary condition is $u_y(x_i, 0) = 0$ and the Laplace difference equation for the grid point (x_i, y_0) is

$$u_{i+1,0} + u_{i-1,0} + u_{i,1} + u_{i,-1} - 4u_{i,0} = 0. \tag{14.103}$$

Approximating

$$u_y\left(x_i, y_0\right) \approx \frac{1}{2k}\left(u_{i,1} - u_{i,-1}\right) = 0 \tag{14.104}$$

gives $u_{i,-1} = u_{i,1}$. Then, Eq. (14.103) takes the form

$$2u_{i,1} + u_{i-1,0} + u_{i+1,0} - 4u_{i,0} = 0, \quad (\textbf{bottom} - \textbf{edge}). \tag{14.105}$$

In a similar way for the left- and right-edges for the boundary conditions $u_x(0, y_j) = 0$ and $u_x(a, y_j) = 0$ the relevant equations are, respectively,

$$2u_{1,j} + u_{0,j-1} + u_{0,j+1} - 4u_{0,j} = 0, \quad (\textbf{left} - \textbf{edge}) \tag{14.106}$$

$$2u_{n-1,j} + u_{n,j-1} + u_{n,j+1} - 4u_{n,j} = 0, \quad (\textbf{right} - \textbf{edge}). \tag{14.107}$$

When the Neumann boundary condition $\partial u(x, y)/\partial N$ is used on certain edges then the value of u on these edges are unknown and are to be determined by the appropriate difference Eqs. (14.103), (14.104), (14.106), (14.107) while the values of u on the other edges are given by the known boundary conditions and the values of u at the interior points are governed by Eq. (14.91).

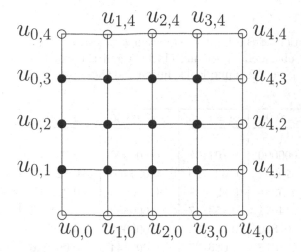

FIGURE 14.10
A five by five grid for the Laplace equation $u_{xx} + u_{yy} = 0$ in the region $0 \le x \le 1$, $0 \le y \le 1$ with a derivative boundary condition at the left-edge. Solution is known at the grid points marked by open circles. The solution is to be found at the grid points marked by the solid circles. There are twelve grid points at which the solution needs to be computed.

Example:

Solve the Laplace equation $u_{xx} + u_{yy} = 0$ in the region $0 \le x \le 1$, $0 \le y \le 1$ subjected to the boundary conditions

$$u(x,0) = \cos \pi x, \tag{14.108a}$$
$$u(x,1) = \cos \pi x \, \mathrm{e}^{-\pi}, \tag{14.108b}$$
$$u_x(0,y) = 0, \tag{14.108c}$$
$$u(1,y) = -\mathrm{e}^{-\pi y}. \tag{14.108d}$$

The grid points at which solution is known and, where the solution is to be determined are depicted in Fig. 14.10. Choose

$$h = k = 0.25 \text{ so that } n = m = 4. \tag{14.109}$$

The derivative boundary condition $u_x(0,y)$ means the solution at the left-edge of the grid plane is unknown. The solution at the right-edge is given by the condition $u(1,y) = -\mathrm{e}^{-\pi y}$. The solution at the bottom- and top-edges are given by $u(x,0) = \cos \pi x$ and $u(x,1) = \cos \pi x \, \mathrm{e}^{-\pi}$, respectively. At the grid points on the left-edge Eq. (14.106) yields

$$-4u_{0,1} + 2u_{1,1} + u_{0,2} = -u_{0,0}, \tag{14.110a}$$
$$u_{0,1} - 4u_{0,2} + 2u_{1,2} + u_{0,3} = 0, \tag{14.110b}$$
$$u_{0,2} - 4u_{0,3} + 2u_{1,3} = -u_{0,4}. \tag{14.110c}$$

For the 9 interior grid points the Laplace difference Eq. (14.91) gives 9 equations. With the above three equations there are, totally, 12 equations. These equations are linear system of equations. This system can be written in matrix form.

TABLE 14.6

The numerical solution of the Laplace equation $u_{xx} + u_{yy} = 0$ subjected to the boundary conditions (14.108) obtained by solving (14.111) and the exact solution $\cos(\pi x)\,e^{-\pi y}$. For each y, the first row gives the numerical solution and the second row gives the exact solution.

y	$x = 0.0$	$x = 0.25$	$x = 0.50$	$x = 0.75$	$x = 1.0$
0.00	$-$ $-$ $--$	$-$ $-$ $--$	$-$ $-$ $--$	$-$ $-$ $--$	$-$ $-$ $--$
	1.000000	0.707107	0.000000	-0.707107	1.000000
0.25	0.470232	0.333129	0.001460	-0.328172	$-$ $-$ $--$
	0.455938	0.322397	0.000000	-0.322397	-0.455938
0.50	0.214671	0.153717	0.000882	-0.151104	$-$ $-$ $--$
	0.207880	0.146993	0.000000	-0.146993	-0.207880
0.75	0.081018	0.066187	-0.000544	-0.069246	$-$ $-$ $--$
	0.094780	0.067020	0.000000	-0.067020	-0.094780
1.00	$-$ $-$ $--$	$-$ $-$ $--$	$-$ $-$ $--$	$-$ $-$ $--$	$-$ $-$ $--$
	0.043214	0.030557	0.000000	-0.030557	-0.043214

In matrix form the 12 system of equations are written as

$$
\begin{bmatrix}
-4 & 2 & 0 & 0 & 1 & 0 & 0 & 0 & 0 & 0 & 0 & 0 \\
1 & -4 & 1 & 0 & 0 & 1 & 0 & 0 & 0 & 0 & 0 & 0 \\
0 & 1 & -4 & 1 & 0 & 0 & 1 & 0 & 0 & 0 & 0 & 0 \\
0 & 0 & 1 & -4 & 0 & 0 & 0 & 1 & 0 & 0 & 0 & 0 \\
1 & 0 & 0 & 0 & -4 & 2 & 0 & 0 & 1 & 0 & 0 & 0 \\
0 & 1 & 0 & 0 & 1 & -4 & 1 & 0 & 0 & 1 & 0 & 0 \\
0 & 0 & 1 & 0 & 0 & 1 & -4 & 1 & 0 & 0 & 1 & 0 \\
0 & 0 & 0 & 1 & 0 & 0 & 1 & -4 & 0 & 0 & 0 & 1 \\
0 & 0 & 0 & 0 & 1 & 0 & 0 & 0 & -4 & 1 & 0 & 0 \\
0 & 0 & 0 & 0 & 0 & 1 & 0 & 0 & 1 & -4 & 1 & 0 \\
0 & 0 & 0 & 0 & 0 & 0 & 1 & 0 & 0 & 1 & -4 & 1 \\
0 & 0 & 0 & 0 & 0 & 0 & 0 & 1 & 0 & 0 & 1 & -4
\end{bmatrix}
$$

$$
\times
\begin{bmatrix}
u_{0,1} \\ u_{1,1} \\ u_{2,1} \\ u_{3,1} \\ u_{0,2} \\ u_{1,2} \\ u_{2,2} \\ u_{3,2} \\ u_{0,3} \\ u_{1,3} \\ u_{2,3} \\ u_{3,3}
\end{bmatrix}
=
\begin{bmatrix}
-u_{0,0} \\ -u_{1,0} \\ -u_{2,0} \\ -u_{3,0} - u_{4,1} \\ 0 \\ 0 \\ 0 \\ -u_{4,2} \\ -u_{0,4} \\ -u_{1,4} \\ -u_{2,4} \\ -u_{3,4} - u_{4,3}
\end{bmatrix}
.
\qquad (14.111)
$$

The system (14.111) is solved by the Gauss–Jordan method. The numerical and the exact solutions $(u(x,y) = \cos \pi x\, e^{-\pi y})$ are given in Table 14.6.

14.7.3 Iterative Methods for Laplace Equation

The method used for computing the solution of the Laplace equation essentially involved solving a system of linear equations. For a $(n+1) \times (m+1)$ grid points (with the solution specified by the boundary conditions at four edges) a system of $(n-1) \times (m-1)$ equations is to be solved for the solution at the interior grid points. For a 5×5 grid points 9 equations to be solved. If 10×10 grid points are used then 64 equations are involved. Thus, it is important to develop methods that will reduce the amount of storage. An example is iterative methods.

Let us consider the Laplace difference Eq. (14.91) with the boundary conditions (14.89b) and (14.89c). Rewrite Eq. (14.91) in the form

$$u_{i,j} = u_{i,j} + s_{i,j}, \tag{14.112a}$$

where the residual term $s_{i,j}$ is given by

$$s_{i,j} = \frac{1}{4} \left(u_{i+1,j} + u_{i-1,j} + u_{i,j+1} + u_{i,j-1} - 4u_{i,j} \right) \tag{14.112b}$$

and $i = 1, 2, \ldots, n-1$, $j = 1, 2, \ldots, m-1$.

The initial guess may be chosen as the average of the $2(n-1) + 2(m-1)$ boundary values on the edges of the region R leaving the corner grid points. Iteration of (14.112) is continued until $|s_{i,j}| < \delta$, where δ is a small preassumed positive constant. Denote u_e and u_n as the exact and numerical solutions, respectively. One cannot realize $|u_e - u_n| \to 0$ as $\delta \to 0$. This is because the formula (14.91) is not exact and has an error $O(h^2)$.

$s_{i,j}$ in Eqs. (14.112) is to change the value of $u_{i,j}$ for one iteration. The speed of convergence can be increased by adding a larger change than this. That is, write

$$u_{i,j} = u_{i,j} + \omega s_{i,j}, \quad 1 \le \omega < 2. \tag{14.113}$$

This is called *successive over-relaxation* (SOR). In the SOR method, the optimal value of ω obtained from the study of eigenvalues of iteration matrices for linear systems is often used and is given by [3]

$$\omega = \frac{4}{2 + \left\{ 4 - [\cos(\pi/(n-1)) + \cos(\pi/(m-1))]^2 \right\}^{1/2}}. \tag{14.114}$$

If the derivative boundary condition is applied on some edges then in addition to Eq. (14.113) (which are for interior grid points) appropriate following equations have to be used.

Top-edge:

$$u_{i,m} = u_{i,m} + \frac{\omega}{4} \left[2u_{i,m-1} + u_{i-1,m} + u_{i+1,m} - 4u_{i,m} \right]. \tag{14.115a}$$

Bottom-edge:

$$u_{i,0} = u_{i,0} + \frac{\omega}{4} \left[2u_{i,1} + u_{i-1,0} + u_{i+1,0} - 4u_{i,0} \right]. \tag{14.115b}$$

Left-edge:

$$u_{0,j} = u_{0,j} + \frac{\omega}{4} \left[2u_{1,j} + u_{0,j-1} + u_{0,j+1} - 4u_{0,j} \right]. \tag{14.115c}$$

TABLE 14.7

The numerical solution of the Laplace Eq. (14.95) obtained by Gauss–Seidel iteration of Eq. (14.113) with $h = k = 0.25$. The solution with $|s_{i,j}| < \delta = 0.001$ is realized at the end of 8th iteration. For each y the first row gives the numerical solution and the second row gives the exact solution.

y	$x = 0.0$	$x = 0.25$	$x = 0.50$	$x = 0.75$	$x = 1.0$
0.00	$--\,--$	$--\,--$	$--\,--$	$--\,--$	$--\,--$
	0.000000	0.707107	1.000000	0.707107	0.000000
0.25	$--\,--$	0.334783	0.473230	0.334522	$--\,--$
	0.000000	0.322397	0.455938	0.322397	0.000000
0.50	$--\,--$	0.157820	0.222986	0.157581	$--\,--$
	0.000000	0.146993	0.207880	0.146993	0.000000
0.75	$--\,--$	0.072880	0.102974	0.072771	$--\,--$
	0.000000	0.067020	0.094780	0.067020	0.000000
1.00	$--\,--$	$--\,--$	$--\,--$	$--\,--$	$--\,--$
	0.000000	0.030557	0.043214	0.030557	0.000000

Right-edge:

$$u_{n,j} = u_{n,j} + \frac{\omega}{4}\left[2u_{n-1,j} + u_{n,j-1} + u_{n,j+1} - 4u_{n,j}\right]. \tag{14.115d}$$

Example:

Solve the Laplace Eq. (14.95) by the iteration method.

Iterate Eq. (14.113) by Gauss–Seidel method, where the updated values of u are used as they become available. Choose $h = k = 0.25$, the tolerance $\delta = 0.001$ and ω given by Eq. (14.114). At the end of 8th iteration solution with $|s_{i,j}| < \delta$ is realized. Table 14.7 gives the numerical solution.

14.7.4   Poisson Equation

The methods discussed for Laplace equation can be extended to the Poisson equation

$$u_{xx} + u_{yy} = g(x, y) \tag{14.116}$$

and the Helmholtz's equation

$$u_{xx} + u_{yy} + f(x, y)u = g(x, y). \tag{14.117}$$

TABLE 14.8

The numerical solution of the Poisson equation $u_{xx} + u_{yy} = x - (y/2)$ subjected to the boundary conditions (14.118). For each y the numerical solution is given in the first row while the exact solution is given in the second row.

y	$x = 0.0$	$x = 0.25$	$x = 0.50$	$x = 0.75$	$x = 1.0$
0.00	$- - - -$	$- - - -$	$- - - -$	$- - - -$	$- - - -$
	0.000000	0.709711	1.020833	0.777419	0.166667
0.25	$- - - -$	0.335636	0.492351	0.403344	$- - - -$
	-0.001302	0.323699	0.475469	0.391407	0.165365
0.50	$- - - -$	0.149596	0.233026	0.217305	$- - - -$
	-0.010417	0.139181	0.218296	0.206889	0.156250
0.75	$- - - -$	0.040140	0.088479	0.107848	$- - - -$
	-0.035156	0.034468	0.080457	0.102176	0.131510
1.00	$- - - -$	$- - - -$	$- - - -$	$- - - -$	$- - - -$
	-0.083333	-0.050172	-0.019286	0.017536	0.083333

Example:

Solve the Poisson Eq. (14.116) with $g(x, y) = x - y/2$ and $0 \le x \le 1, 0 \le y \le 1$ subjected to the boundary conditions

$$u(x, 0) = f_1 = \sin \pi x + \frac{x^3}{6}, \tag{14.118a}$$

$$u(x, 1) = f_2 = \sin \pi x \, e^{-\pi} + \frac{x^3}{6} - \frac{1}{12}, \tag{14.118b}$$

$$u(0, y) = g_1 = -\frac{y^3}{12}, \tag{14.118c}$$

$$u(1, y) = g_2 = \frac{1}{6} - \frac{y^3}{12}. \tag{14.118d}$$

Its exact solution is

$$u(x, y) = \sin \pi x \, e^{-\pi y} + \frac{x^3}{6} - \frac{y^3}{12}. \tag{14.119}$$

The Poisson difference equation (refer Eq. (14.91)) for the solution is

$$u_{i+1,j} + u_{i-1,j} + u_{i,j+1} + u_{i,j-1} - 4u_{i,j} = h^2 g_{i,j}, \tag{14.120}$$

where $i = 1, 2, \ldots, n - 1$, $j = 1, 2, \ldots, m - 1$ and $g_{i,j} = x_i - y_j/2$. The obtained numerical solution is given in Table 14.8.

14.8   First-Order Equation

In this chapter, so far certain second-order linear partial differential equations are considered. In all these equations the starting and end values of x are fixed as, say, a and b and the solution $u(x,t)$ is explicitly specified at $x = a$ and $x = b$, that is $u(a,t)$ and $u(b,t)$ are given. Now, consider a linear first-order equation with $x \in [-\infty, \infty]$. However, assume that $u(x,t) \to \alpha$ as $x \to -\infty$ and $u(x,t) \to \beta$ as $x \to +\infty$. α and β may be zero.

Consider the Fisher equation

$$u_t + cu_x = 0, \tag{14.121}$$

where c is a constant. The exact solution of (14.121) is

$$u(x,t) = \text{sech}(x - ct). \tag{14.122}$$

The solution u has the property $u \to 0$ as $x \to \pm\infty$. To find a numerical solution approximate u_t and u_x as

$$u_t = \frac{1}{2k} \left(u_{i,j+1} - u_{i,j-1} \right), \tag{14.123a}$$

$$u_x = \frac{1}{2h} \left(u_{i+1,j} - u_{i-1,j} \right). \tag{14.123b}$$

Then, Eq. (14.121) gives

$$u_{i,j+1} = u_{i,j-1} - \frac{ck}{h} \left(u_{i+1,j} - u_{i-1,j} \right). \tag{14.124}$$

To determine $u_{i,2}$ the solution at $j = 0$ and 1 are needed. For $u_{i,1}$ approximate

$$u_t = \frac{1}{k} \left(u_{i,j+1} - u_{i,j} \right) \tag{14.125}$$

and substitute (14.123b) and (14.125) in (14.121). This gives for $j = 0$

$$u_{i,1} = u_{i,0} - \frac{ck}{2h} \left(u_{i+1,0} - u_{i-1,0} \right). \tag{14.126}$$

$u_{i,0}$ is given by the initial condition so choose $u_{x,0} = \text{sech}x$. $u_{i,1}$ can be computed from (14.126). Then, Eq. (14.124) determines $u_{i,2}$, $u_{i,3}$ and so on. For stability of Eq. (14.124) the condition is $k \leq h/c$.

Because the initial solution is localized about the origin and the exact solution is a traveling wave moving in the forward direction choose $x \in [-20, 40]$. Fix $h = 0.1$, $k = 0.01$ and $c = 1$. Figure 14.11 shows the numerical solution at $t = 5$ and 15. The exact solution is also shown for comparison.

14.9 Time-Dependent Schrödinger Equation

In Section 13.4, the problem of numerically solving the time-independent Schrödinger equation by the finite-difference method is discussed. This is a boundary-value problem. The present section deals with a method of solving the time-dependent Schrödinger equation. Here again finite-difference scheme is employed. Specifically, the focus is on capturing of time development of certain initial wave profile in a few physically important systems. This is an initial-value problem.

The time-dependent Schrödinger equation of a quantum mechanical particle of mass m in a potential $V(x)$ is given by

$$i\hbar \frac{\partial \psi}{\partial t} = -\frac{\hbar^2}{2m} \frac{\partial^2 \psi}{\partial x^2} + V(x)\psi, \tag{14.127}$$

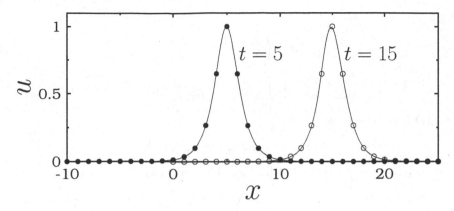

FIGURE 14.11
The numerical solutions (continuous curves) of Eq. (14.121) at $t = 5$ and 15. The exact solutions at selective values of x are represented by solid circles and open circles for $t = 5$ and 15, respectively.

where $\hbar$ is the Planck's constant and ψ is function of x and t. In quantum mechanics, the operator $i\hbar\partial/\partial t$ is the energy operator. For time-independent solutions $i\hbar\partial/\partial t$ is replaced by E. In this case, Eq. (14.127) becomes Eq. (13.28). In the following, the finite-difference method described by Chen [4] is used to solve Eq. (14.127). The final finite-difference formula derived by Chen is equivalent to the one obtained by Goldberg et al [5]. As Eq. (14.127) is complex its solution is also complex.

For convenience set the values of $\hbar$ and m as unity and rewrite Eq. (14.127) as

$$i\frac{\partial\psi}{\partial t} = -\frac{1}{2}\frac{\partial^2\psi}{\partial x^2} + V(x)\psi\,. \tag{14.128}$$

Discretize x in the interval $[a, b]$ with grid points $x_0, x_1, \ldots, x_N$ with $x_{i+1} - x_i = h$ and the time interval $[0, T]$ with grid points $0, t_1, t_2, \ldots, t_m$ with $t_{j+1} - t_j = k$. Equation (14.127) is subjected to the following initial and boundary conditions

$$\psi(x, 0) = \phi(x), \quad \psi(\pm\infty, t) = 0\,. \tag{14.129}$$

Denote $\psi_{i,j}$ as the solution $\psi(x, t)$ at the grid point (i, j). ψ and their partial derivatives $\partial\psi/\partial t$ and $\partial^2\psi/\partial x^2$ are approximated as

$$\psi(x, t) = \psi(x_i, t_j) = \frac{1}{2}(\psi_{i,j} + \psi_{i,j-1})\,, \tag{14.130a}$$

$$\frac{\partial}{\partial t}\psi(x, t) = \frac{1}{k}(\psi_{i,j} - \psi_{i,j-1})\,, \tag{14.130b}$$

$$\frac{\partial^2}{\partial x^2}\psi(x, t) = \frac{1}{2h^2}[(\psi_{i+1,j} - 2\psi_{i,j} + \psi_{i-1,j})$$
$$+ (\psi_{i+1,j-1} - 2\psi_{i,j-1} + \psi_{i-1,j-1})]\,. \tag{14.130c}$$

Replacing ψ, $\partial\psi/\partial t$ and $\partial^2\psi/\partial x^2$ by Eqs. (14.130) in Eq. (14.128) gives

$$\frac{1}{2h^2}\psi_{i+1,j} + \left(\frac{2i}{k} - \frac{1}{h^2} - V_i\right)\psi_{i,j} + \frac{1}{2h^2}\psi_{i-1,j}$$
$$- \left(\frac{2i}{k} + \frac{1}{h^2} + V_i\right)\psi_{i,j-1} + \frac{1}{2h^2}(\psi_{i+1,j-1} + \psi_{i-1,j-1}) = 0\,. \tag{14.131}$$

Defining

$$a = c = \frac{1}{2h^2}, \quad d = \left(\frac{2i}{k} - \frac{1}{h^2} - V_i\right), \tag{14.132a}$$

$$b_i = \left(\frac{2i}{k} + \frac{1}{h^2} + V_i\right)\psi_{i,j-1} - \frac{1}{2h^2}\left(\psi_{i+1,j-1} + \psi_{i-1,j-1}\right) \tag{14.132b}$$

Eq. (14.131) is rewritten as

$$a\psi_{i-1,j} + d\psi_{i,j} + c\psi_{i+1,j} = b_i. \tag{14.133}$$

With $\psi_{0,j} = 0$ and $\psi_{N,j} = 0$ Eq. (14.133) can be written in the following matrix form

$$\begin{pmatrix} 1 & 0 & 0 & & & & \\ a & d & c & & & & \\ 0 & a & d & c & & & \\ & & & \vdots & & & \\ & & & a & d & c & 0 \\ & & & & a & d & c \\ & & & & 0 & 0 & 1 \end{pmatrix}\begin{pmatrix} \psi_{0,j} \\ \psi_{1,j} \\ \psi_{2,j} \\ \vdots \\ \psi_{N-2,j} \\ \psi_{N-1,j} \\ \psi_{N,j} \end{pmatrix} = \begin{pmatrix} 0 \\ b_1 \\ b_2 \\ \vdots \\ b_{N-2} \\ b_{N-1} \\ 0 \end{pmatrix}. \tag{14.134}$$

Since $\psi_{0,j}$ and $\psi_{N,j}$ are set to zero according to the boundary condition given in Eq. (14.129) the system (14.134) becomes

$$\begin{pmatrix} d & c; & & & & \\ a & d & c & & & \\ & a & d & c & & \\ & & & \vdots & & \\ & & a & d & c & \\ & & & a & d & c \\ & & & & a & d \end{pmatrix}\begin{pmatrix} \psi_{1,j} \\ \psi_{2,j} \\ \psi_{3,j} \\ \vdots \\ \psi_{N-3,j} \\ \psi_{N-2,j} \\ \psi_{N-1,j} \end{pmatrix} = \begin{pmatrix} b_1 \\ b_2 \\ b_3 \\ \vdots \\ b_{N-3} \\ b_{N-2} \\ b_{N-1} \end{pmatrix} \tag{14.135}$$

which is of the form $A\psi = B$. An interesting feature of the matrix A is that it is a tridiagonal form and can be solved in a simple manner using the procedure discussed in Section 4.8. Equation (14.133) is the finite-difference formula for the solution of the time-dependent Schrödinger Eq. (14.128).

Now, consider a few quantum mechanical systems and find the time-evolution of initial wave function.

Example 1: Free Particle

For a free particle $V(x) = 0$ and a Gaussian wave packet of it at $t = 0$ is

$$\psi(x,0) = \left(\frac{1}{\sqrt{2\pi}\,\sigma}\right)^{1/2} e^{-(x-x_0)^2/(4\sigma^2)}\, e^{ik_0 x}. \tag{14.136}$$

It is a plane wave with wave number k_0 modulated by a Gaussian profile. Exact $\psi(x,t)$ is given by

$$\psi_e(x,t) = \left(\frac{\sigma}{\sqrt{2\pi}\,\alpha^2}\right)^{1/2} e^{-[\sigma^2 k_0^2 - (\beta^2/(4\alpha^2))]}, \tag{14.137a}$$

where

$$\alpha^2 = \sigma^2 + \frac{i\hbar}{2m}t, \quad \beta = -i\left(x - x_0 - 2i\sigma^2 k_0\right). \tag{14.137b}$$

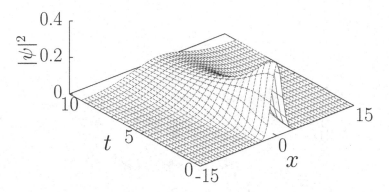

FIGURE 14.12
The numerically computed $|\psi(x,t)|^2$ of a free particle. The initial Gaussian wave packet decays with time.

As $t \to \infty$, $\psi(x,t) \to 0$. To compute numerically $\psi(x,t)$ fix $x_0 = 0$, $k_0 = 0$, $\sigma = 1$, $\hbar = m = 1$, $h = 0.1$, $k = 0.1$ and $x \in [-50, 50]$. $\psi(x,t)$ is complex even if $\psi(x,0)$ is a pure real. Therefore, in the program $\psi(x,t)$ is declared as a complex variable. Real and imaginary parts of ψ can be obtained as

$$\psi_{\mathrm{R}} = [\psi + \mathrm{conjg}(\psi)]/2\,, \tag{14.138a}$$

$$\psi_{\mathrm{I}} = -\mathrm{i}[\psi - \mathrm{conjg}(\psi)]/2\,, \tag{14.138b}$$

where $\mathrm{conjg}(\psi)$ denotes complex conjugate of ψ, that is ψ^*. $|\psi|^2 = \psi^*\psi$ is obtained through

$$|\psi|^2 = \psi \times \mathrm{conjg}(\psi)\,. \tag{14.139}$$

Figure 14.12 shows $|\psi(x,t)|^2$. The initial wave packet spreads in space as time increases and finally flatten.

Example 2: Harmonic Oscillator

The potential of the linear harmonic oscillator is $V(x) = x^2/2$. The eigenfunctions of the time-independent Schrödinger equation are given by Eq. (13.42), where n is the quantum number which takes values $0, 1, \ldots$ and the energy eigenvalues are $E_n = (n + \frac{1}{2})\hbar$. The exact time evolution of the stationary state eigenfunction $\psi_n(x)$ is given by

$$\psi_n(x,t) = \psi_n(x)\,\mathrm{e}^{-\mathrm{i}E_n t/\hbar}\,. \tag{14.140}$$

$\psi_1(x,t)$ is

$$\psi_1(x,t) = 2\left(\frac{1}{2\sqrt{\pi}}\right)^{1/2} x\mathrm{e}^{-x^2/2}\mathrm{e}^{-\mathrm{i}3t/2}\,, \tag{14.141}$$

where $\hbar$ is set as 1. $\psi_1(x,t) \to 0$ as $x \to \pm\infty$ and is periodic in time with period $T = 4\pi/3$. The Schrödinger Eq. (14.128) is solved for $V(x) = x^2/2$ with $\psi_1(x,0)$ as the initial wave function. Fix $h = 0.1$, $k = 0.1$ and $x \in [-5, 5]$. Figure 14.13 depicts the numerically computed ψ_{R} (real part of ψ_1), ψ_{I} (imaginary part of ψ_1) and $|\psi_1|^2$ for $t < 10$. Note that at $t = 0$, $\psi_{\mathrm{I}} = 0$. However, as time increases ψ_{I} evolves. ψ_{R} and ψ_{I} evolve periodically in time while $|\psi(x,t)|^2$ is independent of time.

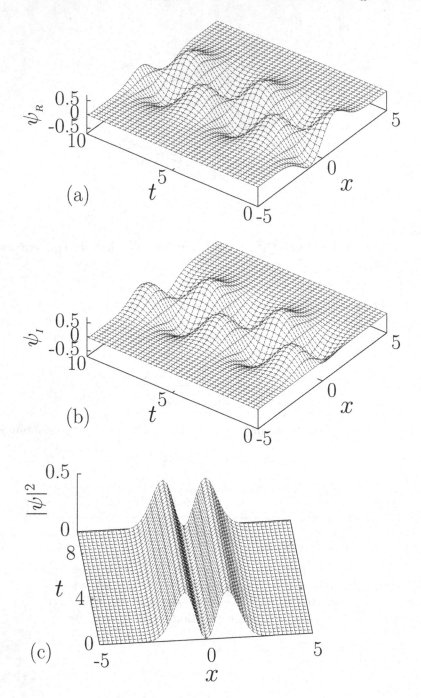

FIGURE 14.13
The numerically computed time evolution of real and imaginary parts of wave function of the harmonic oscillator with $n = 1$. $|\psi(x,t)|^2$ is also plotted. ψ_R and ψ_I evolve periodically in time while $|\psi(x,t)|^2$ is independent of time. The period of the wave function is $4\pi/3 \approx 8.38$.

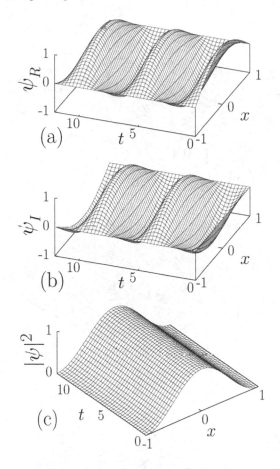

FIGURE 14.14

The numerically computed time evolution of real and imaginary parts of the wave function of a particle in a box with $n = 1$. $|\psi(x,t)|^2$ is also plotted. ψ_R and ψ_I evolve periodically in time while $|\psi(x,t)|^2$ is independent of time. The period of the wave function is $16/\pi \approx 5.093$.

Example 3: Particle in a Box

Consider a particle confined to a box potential $V(x) = 0$ for $|x| < a$ and ∞ for $|x| > a$. Its stationary state energy eigenvalues and eigenfunctions are given by

$$E_n = \frac{\hbar^2\pi^2 n^2}{8ma^2}, \quad n = 1, 2, \ldots \tag{14.142}$$

and

$$\psi_n(x) = \begin{cases} \frac{1}{\sqrt{a}} \sin k_n x, & k_n = \frac{n\pi}{2a}, \quad n = 2, 4, \ldots \\[2mm] \frac{1}{\sqrt{a}} \cos k_n x, & k_n = \frac{n\pi}{2a}, \quad n = 1, 3, \ldots. \end{cases} \tag{14.143}$$

The time evolution of $\psi(x,0)$ is $\psi(x,t) = \psi(x,0)e^{-iE_n t/\hbar}$. In the numerical study $a = 1$, $h = 0.01$, $k = 0.025$, $n = 1$. Figure 14.14 shows the numerical result. The initial wave function is a half-cosine wave. The period of the wave function is $16/\pi \approx 5.093$ which can be clearly noticed in this figure. For $n = 4$, the initial wave is a sine wave. Figure 14.15

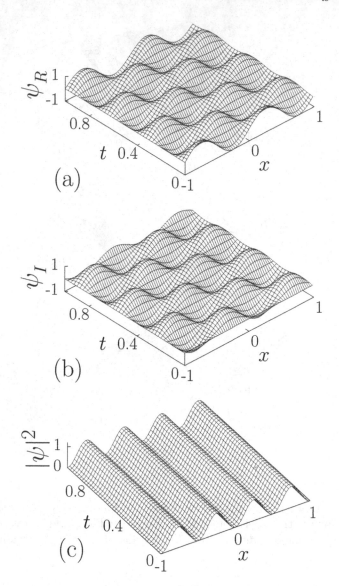

FIGURE 14.15
The numerically computed time evolution of real and imaginary parts of the wave function
of a particle in a box with $n = 4$. $|\psi(x,t)|^2$ is also plotted. ψ_R and ψ_I evolve periodically in
time while $|\psi(x,t)|^2$ is independent of time. The period of the wave function is $1/\pi \approx 0.318$.

presents the numerically computed ψ_R, ψ_I and $|\psi|^2$. The period of $\psi(x,t)$ is $1/\pi \approx 0.318$.

Example 4: Particle in the Presence of a Step Potential

The harmonic oscillator and the particle in a box potential possess stationary state eigenfunctions. These two systems are simple physical systems with bound states with $\psi(x,t) \to 0$
as $x \to \pm\infty$. Next, consider an example of a system, where the wave function incident on
a step potential and undergoes reflection and refraction. Let the potential of the system is

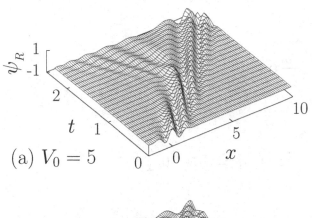

(a) $V_0 = 5$

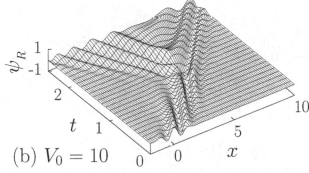

(b) $V_0 = 10$

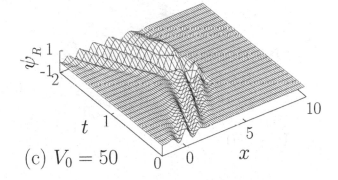

(c) $V_0 = 50$

FIGURE 14.16
The numerically computed time evolution of real part of the wave function $\psi(x,t)$ of a particle in the presence of a step potential with $\psi(x,0)$ given by Eq. (14.144).

$V(x) = 0$ for $x < x_0$ and V_0 for $x > x_0$. Choose the initial wave function as

$$\psi(x,0) = \left(\frac{1}{\sqrt{2\pi}}\right)^{1/2} e^{-x^2} e^{ik_0 x} . \tag{14.144}$$

In the numerical simulation $x_0 = 5$, $k_0 = 5$, $h = 0.05$, $k = 0.05$, $x \in [-10, 20]$. Figure 14.16 shows the ψ_R and $|\psi|^2$ for three values of V_0.

For $V_0 = 5$ and $V_0 = 10$ the incident wave is partly reflected and partly refracted. The amplitude of the reflected wave is relatively large for $V_0 = 10$ when compared to the case of

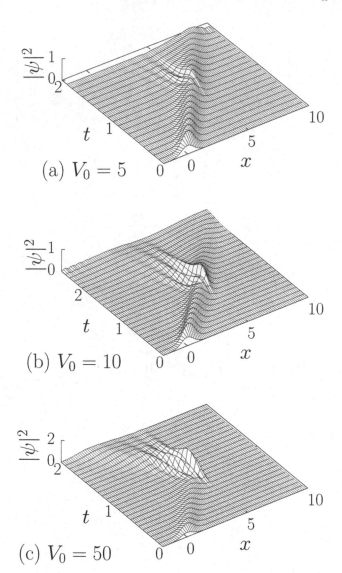

FIGURE 14.17

The numerically computed $|\psi(x,t)|^2$ of a particle in the presence of a step potential with $\psi(x,0)$ given by Eq. (14.144).

$V_0 = 5$. As V_0 increases the amplitude of the reflected wave increases while the amplitude of the refracted wave decreases. For sufficiently large V_0 the wave is completely reflected. This can be clearly seen in the Figs. 14.16 and 14.17 for $V_0 = 50$.

Example 5: Tunnelling

An interesting phenomenon which can occur in certain quantum systems and not observable in classical systems is the tunnelling. If a quantum mechanical particle is impinging on a barrier with an energy relatively lower than the height of the barrier potential, it will not necessarily be reflected by the barrier, but there is always a finite nonzero probability that

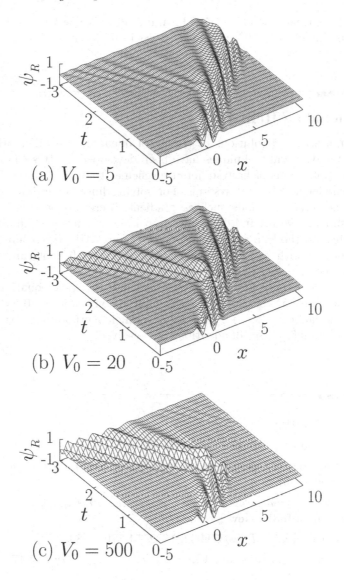

FIGURE 14.18
The numerically computed time evolution of real part of the wave function $\psi = e^{-x^2 + i5x}$ in the presence of a rectangular barrier of height V_0 and width 0.1 at $x = 5$. The effect of V_0 can be clearly seen.

it may cross the barrier and continue its forward motion in the classically forbidden region. This is called *tunnel effect* and can be simulated numerically by solving the time-dependent Schrödinger equation by the finite-difference method.

As an example consider a particle in the presence of a rectangular barrier of height V_0 and width 0.1. The potential is $V(x) = V_0$ for $x \in [5, 5.1]$ and 0 elsewhere. Choose the initial wave function as $e^{-x^2 + i5x}$ and $h = 0.05$, $k = 0.05$, $x \in [-10, 20]$. Figure 14.18 shows the time evolution of the real part of ψ for $V_0 = 5$, 20 and 500. For small values of V_0, ψ_R is nonzero in the forbidden region $x > 5.1$. The amplitude of the wave in this region decreases with increase in V_0. For $V_0 = 500$ in Fig. 14.18 ψ_R is almost negligible in magnitude.

14.10   Concluding Remarks

Numerical computations of solutions of variety of linear partial differential equations are well established and several techniques have been developed. In this chapter, presentation is restricted to applications of finite-difference schemes to a few standard partial differential equations modelling physical systems. For solving linear equations apart from finite-difference schemes other methods such as gradient discretization, finite-volume element, multi-grid, meshfree, spectral and domain composition, etc. are well known. For a brief introduction to these methods, a reader may refer to the ref. [1]. Algorithms for linear partial differential equations with variable coefficients, singular coefficients and stochastic terms have been proposed and tested.

It is also possible to solve numerically the heat and the wave equations subjected to a pure initial-value problem. For example, for the initial condition $u(x,0) = f(x)$, $-\infty < x < \infty$ the explicit method can be used. The use of implicit method is practically not possible since it generates an infinite set of equations to be solved.

14.11   Bibliography

[1]   M. Kumar and G. Mishra, *Appl. Math.* 2:1327, 2011.

[2]   M. Pal, *Numerical Analysis for Scientists and Engineers: Theory and C Programs*. Narosa, New Delhi, 2007.

[3]   J.H. Mathews, *Numerical Methods for Mathematics Science and Engineering*. Prentice-Hall of India, New Delhi, 2005.

[4]   R.L.W. Chen, *Am. J. Phys.* 50:902, 1982; 51:570, 1983.

[5]   A. Goldberg, H.M. Schey and J.L. Schwartz, *Am. J. Phys.* 35:177, 1967.

14.12   Problems

14.1 Solve the wave equation

$$u_{tt} = 4u_{xx}, \quad 0 \le x \le 1, \ 0 \le t \le 1$$

subjected to the conditions

$$
\begin{aligned}
u(x,0) &= \sin \pi x, \quad 0 \le x \le 1 \\
u_t(x,0) &= 0, \quad 0 \le x \le 1 \\
u(0,t) &= 0, \quad u(1,t) = 0 \quad 0 \le t \le 1
\end{aligned}
$$

with $h = 0.1$ and $k = 0.05$. Compare the numerical solution with the exact solution $u_e(x,t) = \sin \pi x \cos 2\pi t$ for $x = 0.5$.

14.2 The exact solution of the wave equation

$$u_{tt} = 4u_{xx}, \quad 0 \le x \le 1, \ 0 \le t \le 1$$

is $u_e(x,0) = \sin \pi x \cos 2\pi t + \sin 2\pi x \cos 4\pi t$. Solve the equation with the conditions

$$
\begin{aligned}
u(x,0) &= \sin \pi x + \sin 2\pi x, \quad 0 \le x \le 1 \\
u_t(x,0) &= 0, \quad 0 \le x \le 1 \\
u(0,t) &= 0, \ u(1,t) = 0, \quad 0 \le t \le 1
\end{aligned}
$$

with (a) $h = 0.1$, $k = 0.05$ and (b) $h = 0.2$, $k = 0.1$. Compare the numerical solution with the exact solution for $x = 0.2$.

14.3 Solve the wave equation $u_{tt} = u_{xx}$, with the conditions

$$
\begin{aligned}
u(x,0) &= (3/4) \sin \pi x - (1/4) \sin 3\pi x, \quad 0 \le x \le 1 \\
u_t(x,0) &= 0, \quad 0 \le x \le 1 \\
u(0,t) &= 0, \ u(1,t) = 0, \quad 0 \le t \le 1.
\end{aligned}
$$

Choose $h = 0.1$ and $k = 0.05$. Construct its exact solution and compare the numerical solution with the exact solution for $x = 0.5$.

14.4 Solve the damped wave equation

$$u_{tt} = u_{xx} - 2cu_t, \quad 0 \le x \le \infty, \ t \ge 0$$

subjected to the initial and boundary conditions

$$
\begin{aligned}
u(x,0) &= e^{-x}, \quad u_t(x,0) = g(x) = \left(-1 - \sqrt{2}\right) e^{-x}, \\
u(0,t) &= e^{(-1-\sqrt{2})t}.
\end{aligned}
$$

Compare the numerical solution with the exact solution

$$u(x,t) = e^{-x} e^{(-1-\sqrt{2})t}.$$

for $t = 5$ and 10 with $h = 0.1$, $k = 0.05$ and $x \in [0, 30]$.

14.5 Obtain the formula for solving the heat equation with the approximation

$$u_t = \frac{1}{2k}(u_{i,j+1} - u_{i,j-1}), \quad u_{xx} = \frac{1}{h^2}(u_{i+1,j} - 2u_{i,j} + u_{i-1,j}).$$

Applying the stability analysis shows that the formula is unconditionally unstable.

14.6 Obtain the numerical solution of the heat equation $u_t = u_{xx}$ for $0 \le x \le 1$, $0 \le t \le 1$ with the boundary conditions

$$u(x,0) = \sin 2\pi x, \quad u(0,t) = \sin 2\pi t, \quad u(1,t) = \sin 2\pi t e^{-2\pi}.$$

Fix $h = 0.1$ and $k = 0.005$. Compare the numerical solution with the exact solution

$$u_e(x,t) = \sin 2\pi x \, e^{-2\pi y} + \sin 2\pi t \, e^{-2\pi x}.$$

14.7 The exact solution of the heat equation $u_t = u_{xx}$ subjected to the boundary conditions $u(x,0) = 0$, $u(0,t) = 0$, $u(1,t) = \sinh\pi \sin\pi t$ is

$$u(x,t) = \sin\pi x \sinh\pi t + \sinh\pi x \sin\pi t.$$

Solve the equation by the Crank–Nicholson method with $h = 0.1$ and $k = 0.01$. Compare the numerical solution with the exact analytical solution.

14.8 Derive an explicit and implicit difference formulas for the parabolic equation $u_t - u_{xx} = g(x)$.

14.9 Construct an iterative formula similar to Eq. (14.112) for the heat equation $u_t = c^2 u_{xx}$. Then for $h = 0.1$, $k = 0.01$, $u(x,0) = \sin\pi x$, $u(0,t) = \sin\pi t$, $u(1,t) = \sin\pi t e^{-\pi}$ find the numerical solution using Gauss–Seidel method.

14.10 Find the numerical solution of the equation $u_{xx} + u_{yy} = -u$, $0 \le x \le \pi$, $0 \le y \le \pi$ with the boundary values $u(x,0) = \cos x$, $u(x,\pi) = \cos x$, $u(0,y) = 1 + \sin y$, $u(\pi,y) = -1 + \sin y$. Compare the numerical solution with the exact solution $u(x,y) = \cos x + \sin y$.

14.11 Consider the Laplace equation $u_{xx} + u_{yy} = 0$, $0 \le x \le 1$, $0 \le y \le 1$, $u(x,0) = \cos(\pi x/2)$, $u(x,1) = \cos(\pi x/2)e^{-\pi/2}$, $u(0,y) = e^{-\pi y/2}$, $u(1,y) = 0$. With $h = k = 0.25$ obtain the numerical solution by the explicit method and the Gauss–Seidel iteration method. What is the exact solution?

14.12 Consider the Laplace equation $u_{xx} + u_{yy} = 0$, $0 \le x \le 1$, $0 \le y \le 1$ with the boundary condition $u(x,0) = \cos\pi x$, $u(x,1) = \cos\pi x e^{-\pi}$, $u_x(1,y) = 0$, $u(1,y) = -e^{-\pi y}$. Show that its exact solution is $u(x,y) = \cos\pi x e^{-\pi y}$. Divide the region $0 \le x \le 1$, $0 \le y \le 1$ into 5×5 grid points and construct the system of equations of the form $Au = B$ for the unknown solution at the grid points. Develop a computer program in C$_{++}$ or Python for solving the Laplace equation by Gauss–Jordan method and then using the program find the numerical solution.

14.13 Construct the exact solution of the Laplace equation $u_{xx} + u_{yy} = 0$, $0 \le x \le 1$, $0 \le y \le 1$ with the boundary conditions $u(0,y) = \cos\pi y$, $u(1,y) = \cos\pi y e^{-\pi}$, $u(x,0) = e^{-\pi x}$, $u_y(x,1) = 0$. For a 5×5 grid points explicitly write down the system of equations for the unknown solutions. Develop a computer program in C$_{++}$ or Python for solving them by Gauss–Jordan method. Using the program obtain the numerical solution.

14.14 Obtain the exact solution of the Laplace equation $u_{xx} + u_{yy} = 0$, $0 \le x \le 1$, $0 \le y \le 1$ with the boundary conditions $u(0,y) = \cos\pi y$, $u(1,y) = \cos\pi y e^{-\pi}$, $u(x,1) = -e^{-\pi x}$, $u_y(x,0) = 0$. For a 5×5 grid points explicitly write down the system of equations for the unknown solution. Develop a computer program in C$_{++}$ or Python for solving them by Gauss–Jordan method. Using the program obtain the numerical solution.

14.15 Develop a nine-point difference formula for the Laplace equation.

14.16 Show that the stability condition for the formula (14.124) is $k \le h/c$.

14.17 Instead of using Eq. (14.126) for $u_{i,1}$ let $u_{i,1} = u_{i,0}$ and then solve Eq. (14.121). Compare the obtained solution at $t = 10$ with the one obtained using Eq. (14.126) and also with the exact solution.

14.18 The normalized stationary state eigenfunction of a particle in a box potential is $\psi(x,0) = \sqrt{15/16}\,(x^2 - a^2)$ for $|x| < a$ and 0 elsewhere. Study the time evolution of ψ.

14.19 Consider the wave function $\psi(x,0) = \frac{1}{2}(\cos\frac{\pi}{2}x + \sin\pi x)$ of a particle in a box potential. It is a sum of the stationary state eigenfunctions ψ_1 and ψ_2. Compute numerically $\psi(x,t)$ and determine its periodicity.

14.20 Tunnelling of a wave function can also be observed with the truncated Gaussian potential $V(x) = V_0 e^{-\beta^2 x^2}$ for $|x| < a$ and 0 elsewhere. Study the time evolution of the initial Gaussian wave packet. Write a note on the effect of V_0 and β on the amplitudes of the reflected and transmitted waves.

15

Nonlinear Partial Differential Equations

15.1 Introduction

The previous chapter dealt with linear partial differential equations (LPDEs). The present chapter is concerned with nonlinear partial differential equations (NLPDEs). Certain class of NLPDEs like Korteweg–de Vries (KdV) equation, sine–Gordon (sG) equation and nonlinear Schrödinger (NLS) equation possess a variety of fascinating localized solutions called *solitary waves* [1–3]. In certain nonlinear wave systems, these waves undergo elastic collisions. Their shapes and speeds are preserved after interactions with other solitary waves. Such wave solutions are named as *solitons* by Zabusky and Kruskal [4]. Zabusky and Kruskal numerically solved the KdV equation with a periodic boundary condition and applied a finite-difference (FD) scheme. The initial cosine wave evolved into a number of solitary waves with different heights and speeds. Zabusky and Kruskal observed elastic collisions of these waves. Their scheme is able to predict accurately the solution during short time interval only and in the long-time limit unphysical oscillations are developed in the numerical solution leading to blowup of the solution. Since then, a great deal of interest has been focused on developing stable numerical methods for NLPDEs. Methods have been developed to simulate and analysis the novel phenomena like elastic collision of localized waves, vortex formation, rogue waves, spiral waves, chimeras, spatio-temporal patterns, quenching dynamics, hierarchical decomposition of an MRI image, etc.

This chapter restricts to NLPDEs possessing solitary waves and solitary wave-like solutions. Various numerical approaches such as FD schemes, spectral method, symplectic and multi-symplectic theories have been developed. In the FD schemes, the partial derivatives are approximated by FD formulas. In the spectral method spatial derivatives are computed in Fourier space and time derivatives are approximated by FD formulas. Since the solitonic equations are conservative equations a scheme developed for them should preserve the energy associated with them. Moreover, the integrable solitonic equations admit infinite number of conservation laws or integrals of motion. They need to be preserved to a desired accuracy in order to generate numerical solutions with desired accuracy. For this purpose, for Hamiltonian PDEs symplectic and multi-symplectic methods were developed [5–8]. For a review on the numerical methods of NLPDEs one may refer to the refs. [9-13].

The symplectic and multi-symplectic methods start with the Hamiltonian of the given equation. In the symplectic approach the spatial derivatives are approximated by FD formulas and then the equation of motion is written in the Hamiltonian form by introducing additional variables. The result is a system of first-order equations. Applying a discretization to this system and then eliminating additional variables introduced lead to a symplectic scheme. In the multi-symplectic method without discretizing the spatial derivatives the given equation is expressed in Hamiltonian form. In this case, for example, for the equation with one time variable and one space variable the conservation law has a component along the time direction and another component along the spatial direction. That is, the conservation law is in multi-symplectic form.

DOI: 10.1201/9781032649931-15

In this chapter, first the Hamiltonian formulation of PDEs and the notion of conservation of symplecticity are introduced. Next, the multi-symplectic integration with the Euler box scheme is presented. Then, the multi-symplectic schemes for the KdV and sG equations are derived and analyzed for their applicability and efficacy by comparing the numerical results with the exact results.

15.2 Hamilton's Equation of Motion

In classical mechanics, there is a class of systems whose evolution equations of dynamical variables can be expressed as

$$\dot{\mathbf{p}} = -\nabla_{\mathbf{q}}H(\mathbf{q},\mathbf{p}), \tag{15.1a}$$

$$\dot{\mathbf{q}} = \nabla_{\mathbf{p}}H(\mathbf{q},\mathbf{p}), \tag{15.1b}$$

where H is called *Hamiltonian* (total energy). In Eq. (15.1) there is a skew(anti)-symmetry between $\mathbf{q}$ and $\mathbf{p}$. Assume that $H = P(\mathbf{p}) + V(\mathbf{q})$. H is a constant of motion (does not change with time) because

$$\frac{\mathrm{d}H}{\mathrm{d}t} = (\nabla_{\mathbf{p}}H)\,\dot{\mathbf{p}} + (\nabla_{\mathbf{q}}H)\,\dot{\mathbf{q}} = \dot{\mathbf{q}}\dot{\mathbf{p}} - \dot{\mathbf{p}}\dot{\mathbf{q}} = 0. \tag{15.2}$$

Further, for $\omega = \mathrm{d}\mathbf{p} \wedge \mathrm{d}\mathbf{q}$ (called *symplectic manifold*), where $\wedge$ represents wedge product and $\mathrm{d}\mathbf{p}$ denotes the vector of differentials,

$$\begin{aligned}
\frac{\mathrm{d}\omega}{\mathrm{d}t} &= \mathrm{d}\mathbf{p}_t \wedge \mathrm{d}\mathbf{q} + \mathrm{d}\mathbf{p} \wedge \mathrm{d}\mathbf{q}_t \\
&= \mathrm{d}\dot{\mathbf{p}} \wedge \mathrm{d}\mathbf{q} + \mathrm{d}\mathbf{p} \wedge \mathrm{d}\dot{\mathbf{q}}_t \\
&= -(\nabla_{\mathbf{qq}}H)\,\mathrm{d}\mathbf{q} \wedge \mathrm{d}\mathbf{q} + \mathrm{d}\mathbf{p} \wedge (\nabla_{\mathbf{pp}}H)\,\mathrm{d}\mathbf{p} \\
&= 0. \tag{15.3}
\end{aligned}$$

In the $(\mathbf{q},\mathbf{p})$ phase space of (15.1) $\mathbf{q}$ and $\mathbf{p}$ evolve in such a way that $\mathrm{d}\omega/\mathrm{d}t = 0$. The time evolution of (15.1) is then said to be *symplectomorphism*. Note that $\mathrm{d}H/\mathrm{d}t = 0$ is conservation of energy while $\mathrm{d}\omega/\mathrm{d}t = 0$ is conservation of symplecticity.

Let

$$H = H(z), \quad z = \begin{pmatrix} p \\ q \end{pmatrix} \tag{15.4}$$

then

$$\dot{p} = -\frac{\partial H}{\partial q}, \quad \dot{q} = \frac{\partial H}{\partial p} \tag{15.5}$$

can be rewritten as

$$J\dot{z} = \nabla_z H(z), \quad J = \begin{pmatrix} 0 & -1 \\ 1 & 0 \end{pmatrix}, \quad z = \begin{pmatrix} p \\ q \end{pmatrix}, \tag{15.6}$$

where J is skew-symmetric. Equation (15.6) can be generalized to z with n components. This Hamiltonian system possesses two conservation laws (refer Chapter 14)

$$\frac{\mathrm{d}H}{\mathrm{d}t} = 0, \quad \frac{\mathrm{d}\omega}{\mathrm{d}t} = 0. \tag{15.7}$$

Let us consider (1+1)-dimensional (one space variable and one-time variable) continuous system described by $H = \displaystyle\int_{-\infty}^{\infty} \mathcal{H}\, dx$, where $\mathcal{H}$ is the Hamiltonian density. Suppose that the system is

$$u_{tt} - u_{xx} + V'(u) = 0\,, \tag{15.8}$$

where $V(u)$ is a smooth function. Define

$$\frac{\delta \mathcal{H}}{\delta \phi} = \frac{\partial \mathcal{H}}{\partial \phi} - \frac{\partial}{\partial x}\left(\frac{\partial}{\partial \phi_x}\right) + \frac{\partial^2}{\partial x^2}\left(\frac{\partial}{\partial \phi_{xx}}\right) + \cdots\,. \tag{15.9}$$

Equation (15.8) can be rewritten as a system of first-order equations as

$$u_t = v\,, \quad v_t = u_{xx} - V'(u)\,. \tag{15.10}$$

One can rewrite the above equation as

$$u_t = \frac{\delta \mathcal{H}}{\delta v} = v\,, \quad v_t = -\frac{\delta \mathcal{H}}{\delta u} = u_{xx} - V'(u)\,, \tag{15.11a}$$

where

$$\mathcal{H} = \frac{1}{2}v^2 + \frac{1}{2}u_x^2 + V(u)\,. \tag{15.11b}$$

In matrix form (15.11) takes the form

$$\begin{pmatrix} 0 & -1 \\ 1 & 0 \end{pmatrix} \frac{\partial}{\partial t}\begin{pmatrix} u \\ v \end{pmatrix} = \begin{pmatrix} \delta H/\delta u \\ \delta H/\delta v \end{pmatrix}\,. \tag{15.12}$$

Equation (15.11a) has a second-order spatial derivative. Write Eq. (15.8) with variables containing at most first derivative in x and t. A choice is

$$-v_t - p_x = V'(u)\,, \quad u_t = v\,, \quad 0 = p + w\,, \quad u_x = w\,. \tag{15.13}$$

That is,

$$\begin{pmatrix} 0 & -1 & 0 & 0 \\ 1 & 0 & 0 & 0 \\ 0 & 0 & 0 & 0 \\ 0 & 0 & 0 & 0 \end{pmatrix}\begin{pmatrix} u_t \\ v_t \\ w_t \\ p_t \end{pmatrix} + \begin{pmatrix} 0 & 0 & 0 & -1 \\ 0 & 0 & 0 & 0 \\ 0 & 0 & 0 & 0 \\ 1 & 0 & 0 & 0 \end{pmatrix}\begin{pmatrix} u_x \\ v_x \\ w_x \\ p_x \end{pmatrix} = \begin{pmatrix} V'(u) \\ v \\ p + w \\ w \end{pmatrix}\,. \tag{15.14}$$

Defining

$$M = \begin{pmatrix} 0 & -1 & 0 & 0 \\ 1 & 0 & 0 & 0 \\ 0 & 0 & 0 & 0 \\ 0 & 0 & 0 & 0 \end{pmatrix}\,, \quad K = \begin{pmatrix} 0 & 0 & 0 & -1 \\ 0 & 0 & 0 & 0 \\ 0 & 0 & 0 & 0 \\ 1 & 0 & 0 & 0 \end{pmatrix}\,, \tag{15.15a}$$

$$z = \begin{pmatrix} u \\ v \\ w \\ p \end{pmatrix}\,, \quad S(z) = \frac{1}{2}v^2 + \frac{1}{2}w^2 + pw + V(u) \tag{15.15b}$$

Eq. (15.14) becomes

$$M z_t + K z_x = \nabla_z S(z)\,. \tag{15.16}$$

In the above form each facet of the equation is arranged neatly. The time derivatives occur in Mz_t and the space derivatives occur in Kz_x. Equation (15.16) is the generalization of Eq. (15.6).

Compare Eqs. (15.12) and (15.16). Equation (15.12) is a Hamiltonian formulation of the system (15.8) on a single symplectic form since in Eq. (15.12) there is only one pre-symplectic operator $\begin{pmatrix} 0 & -1 \\ 1 & 0 \end{pmatrix}$. The same system is in the multi-symplectic form in Eq. (15.16) as there are two pre-symplectic operators M and K. Not only the system (15.8), a large class of PDEs can be written in the form of Eq. (15.16) with M and K being skew-symmetric $n \times n$ matrices with $n \geq 3$ [14]. Examples include KdV, sG, Boussinesq and NLS equations.

15.3   Conservation of Symplecticity, Energy and Momentum

The symplectic form for the system (15.13) is $\omega = \mathrm{d}v \wedge \mathrm{d}u$. Differentiation of ω with respect to t gives $\partial \omega / \partial t = \mathrm{d}v_t \wedge \mathrm{d}u + \mathrm{d}v \wedge \mathrm{d}u_t$. Expanding this one obtains

$$
\begin{aligned}
\frac{\partial \omega}{\partial t} &= (\mathrm{d}u_{xx} - V'(u)\mathrm{d}u) \wedge \mathrm{d}u + \mathrm{d}v \wedge \mathrm{d}v \\
&= \mathrm{d}u_{xx} \wedge \mathrm{d}u \\
&= \frac{\partial}{\partial x}(\mathrm{d}u_x \wedge \mathrm{d}u) .
\end{aligned} \tag{15.17}
$$

Defining $w = u_x$ and $\kappa = \mathrm{d}u \wedge \mathrm{d}w$ (called *flux of symplecticity*) the above equation becomes

$$
\frac{\partial \omega}{\partial t} + \frac{\partial \kappa}{\partial x} = 0 . \tag{15.18}
$$

Equation (15.18) is the conservation law of symplecticity and is a generalization of $\mathrm{d}\omega/\mathrm{d}t = 0$ of Eq. (15.6).

15.3.1   Conservation of Multi-Symplectic Structure

For the multi-symplectic form (15.16) of the system (15.8) the multi-symplectic conservation law is given by Eq. (15.18) with

$$
\begin{aligned}
\omega &= \mathrm{d}z \wedge M \mathrm{d}z \\
&= \mathrm{d}\begin{pmatrix} u \\ v \\ w \\ p \end{pmatrix} \wedge \begin{pmatrix} 0 & -1 & 0 & 0 \\ 1 & 0 & 0 & 0 \\ 0 & 0 & 0 & 0 \\ 0 & 0 & 0 & 0 \end{pmatrix} \mathrm{d}\begin{pmatrix} u \\ v \\ w \\ p \end{pmatrix} \\
&= 2\mathrm{d}v \wedge \mathrm{d}u , \\
\kappa &= \mathrm{d}z \wedge K \mathrm{d}z \\
&= \mathrm{d}\begin{pmatrix} u \\ v \\ w \\ p \end{pmatrix} \wedge \begin{pmatrix} 0 & 0 & 0 & -1 \\ 0 & 0 & 0 & 0 \\ 0 & 0 & 0 & 0 \\ 1 & 0 & 0 & 0 \end{pmatrix} \mathrm{d}\begin{pmatrix} u \\ v \\ w \\ p \end{pmatrix} \\
&= 2\mathrm{d}p \wedge \mathrm{d}u .
\end{aligned}
$$

$$\tag{15.19}$$
$$\tag{15.20}$$

Let us verify Eq. (15.18). Substitute $\omega = dz \wedge Mdz$ and $\kappa = dz \wedge Kdz$ in Eq. (15.18) and obtain

$$
\begin{aligned}
\omega_t + \kappa_x &= dz_t \wedge Mdz + dz \wedge Mdz_t + dz_x \wedge Kdz + dz \wedge Kdz_x \\
&= -Mdz_t \wedge dz + dz \wedge Mdz_t - Kdz_x \wedge dz + dz \wedge Kdz_x \\
&= -(Mdz_t + Kdz_x) \wedge dz + dz \wedge (Mdz_t + Kdz_x) \\
&= -S_{zz}dz \wedge dz + dz \wedge S_{zz}dz \\
&= 0,
\end{aligned}
\tag{15.21}
$$

where the Leibniz's rule, skew-symmetry of M and K and the fact that S_{zz} is symmetric are used. Equation (15.18) implies that at each point (x, t), the multi-symplectic structure is conserved. ω and κ define the symplectic structures associated with the time and space directions, respectively. One may wish to know whether energy and momentum are conserved in the multi-symplectic form.

15.3.2 Conservation Law of Energy

To obtain a conservation law of energy take the inner product of (15.16) with z_t:

$$
\langle z_t, Mz_t + Kz_x \rangle = \frac{\partial S}{\partial t} .
\tag{15.22}
$$

Because $\langle z_t, Mz_t \rangle = 0$ the above equation becomes

$$
\frac{\partial S}{\partial t} - \langle z_t, Kz_x \rangle = 0
\tag{15.23}
$$

or

$$
\begin{aligned}
0 &= \frac{\partial S}{\partial t} - \frac{1}{2}\langle z_t, Kz_x \rangle - \frac{1}{2}\langle z_t, Kz_x \rangle \\
&= \frac{\partial S}{\partial t} - \frac{1}{2}\langle Kz_x, z_t \rangle - \frac{1}{2}\langle Kz_x, z_t \rangle \\
&= \frac{\partial S}{\partial t} - \frac{1}{2}\langle Kz_x, z_t \rangle - \frac{1}{2}\langle Kz_{xt}, z \rangle - \frac{1}{2}\langle Kz_x, z_t \rangle + \frac{1}{2}\langle Kz_{xt}, z \rangle \\
&= \frac{\partial S}{\partial t} - \frac{1}{2}\frac{\partial}{\partial t}\langle Kz_x, z \rangle + \frac{1}{2}\langle Kz_t, z_x \rangle + \frac{1}{2}\langle Kz_{xt}, z \rangle \\
&= \frac{\partial S}{\partial t} - \frac{1}{2}\frac{\partial}{\partial t}\langle Kz_x, z \rangle + \frac{1}{2}\frac{\partial}{\partial x}\langle Kz_t, z \rangle .
\end{aligned}
\tag{15.24}
$$

This equation can be rewritten as

$$
E_t + F_x = 0 ,
\tag{15.25a}
$$

where

$$
E(z) = S(z) - \frac{1}{2}\langle Kz_x, z \rangle ,
\tag{15.25b}
$$

$$
F(z) = \frac{1}{2}\langle Kz_t, z \rangle .
\tag{15.25c}
$$

Equation (15.25) is the local conservation law of energy. E is called *energy density* while F is the *energy flux*.

15.3.3 Conservation Law of Momentum

To derive the momentum conservation law let us take the inner product of (15.16) with z_x:

$$\langle z_x, M z_t + K z_x \rangle = \langle z_x, \nabla_z S \rangle. \tag{15.26}$$

Because $\langle z_x, K z_x \rangle = 0$ the above equation becomes

$$\langle z_x, M z_t \rangle - \frac{\partial S}{\partial x} = 0. \tag{15.27}$$

That is,

$$
\begin{aligned}
0 &= \frac{\partial S}{\partial x} - \frac{1}{2}\langle z_x, M z_t \rangle - \frac{1}{2}\langle z_x, M z_t \rangle \\
&= \frac{\partial S}{\partial x} - \frac{1}{2}\langle M z_t, z_x \rangle - \frac{1}{2}\langle M z_t, z_x \rangle - \frac{1}{2}\langle M z_{xt}, z \rangle + \frac{1}{2}\langle M z_{xt}, z \rangle \\
&= \frac{\partial}{\partial x}\left(S - \frac{1}{2}\langle M z_t, z \rangle \right) + \frac{1}{2}\frac{\partial}{\partial t}\langle M z_x, z \rangle \\
&= \frac{\partial G}{\partial x} + \frac{\partial I}{\partial t},
\end{aligned}
\tag{15.28a}
$$

where

$$G = S - \frac{1}{2}\langle M z_t, z \rangle, \quad I = \frac{1}{2}\langle M z_x, z \rangle. \tag{15.28b}$$

Equation (15.28) is the local conservation law of momentum.

15.3.4 Splitting of the Matrices M and K

Let us split the matrices M and K such that [15]

$$M = M_+ + M_-, \quad K = K_+ + K_- \tag{15.29a}$$

with

$$M_+^{\mathrm{T}} = -M_-, \quad K_+^{\mathrm{T}} = -K_-. \tag{15.29b}$$

In this case, one can easily verify the following:

$$dz \wedge M_+ dz = dz \wedge M_- dz, \quad dz \wedge K_+ dz = dz \wedge K_- dz, \tag{15.30a}$$

$$\omega = dz \wedge M_+ dz, \quad \kappa = dz \wedge K_+ dz, \tag{15.30b}$$

$$E(z) = S(z) - \langle K_+ z_x, z \rangle, \quad F(z) = \langle K_+ z_t, z \rangle, \tag{15.30c}$$

$$G(z) = S(z) - \langle M_+ z_t, z \rangle, \quad I(z) = \langle M_+ z_x, z \rangle. \tag{15.30d}$$

The multi-symplectic PDE (15.16) takes the form

$$M_+ \delta_t z + M_- \delta_t z + K_+ \delta_x z + K_- \delta_x z = \nabla_z S(z). \tag{15.31}$$

15.4 Multi-Symplectic Integrator

A multi-symplectic integrator is a numerical scheme that preserves a discrete version of conservation of symplecticity. Bridges and Reich [7] applied a symplectic integrator to the independent variables x and t and obtained a multi-symplectic integrator.

For the purpose of numerically solving a given PDE with two independent variables x and t consider $x_{\min} \leq x \leq x_{\max}$, $t_0 \leq t \leq t_{\text{end}}$. Discretize the space and time variables with number of grid points as

$$x_j = x_{\min} + jh, \quad j = 0, 1, \ldots, N_x \qquad (15.32\text{a})$$

$$t_n = t_0 + nk, \quad n = 0, 1, \ldots, N_t \qquad (15.32\text{b})$$

where

$$h = (x_{\max} - x_{\min})/N_x, \quad k = (t_{\text{end}} - t_0)/N_t. \qquad (15.32\text{c})$$

h and k are step sizes along space and time directions, respectively. We denote the numerical approximation of z at (x_j, t_n) as z_j^n. The forward- and backward-difference formulas for z_t are given by

$$\delta_t^+ z_j^n = \frac{1}{k}\left(z_j^{n+1} - z_j^n\right), \quad \delta_t^- z_j^n = \frac{1}{k}\left(z_j^n - z_j^{n-1}\right). \qquad (15.33)$$

Similarly, for z_x

$$\delta_x^+ z_j^n = \frac{1}{h}\left(z_{j+1}^n - z_j^n\right), \quad \delta_x^- z_j^n = \frac{1}{h}\left(z_j^n - z_{j-1}^n\right). \qquad (15.34)$$

Let us discretize Eq. (15.31). Apply the two-point forward-difference formula for $\delta_t z$ in the first term and the two-point backward-difference formula for $\delta_t z$ in the second term. Use similar formulas for the spatial derivative in the third and fourth terms. The result is

$$M_+ \delta_t^+ z_j^n + M_- \delta_t^- z_j^n + K_+ \delta_x^+ z_j^n + K_- \delta_x^- z_j^n = \nabla_z S(z_j^n). \qquad (15.35)$$

This method is referred as the *Euler box scheme*.

Example:

Show that the scheme (15.35) satisfies a discrete multi-symplectic conservation law.

Start with the variational equation of (15.35) given by

$$M_+ \delta_t^+ \mathrm{d}z_j^n + M_- \delta_t^- \mathrm{d}z_j^n + K_+ \delta_x^+ \mathrm{d}z_j^n + K_- \delta_x^- \mathrm{d}z_j^n = \nabla_{zz} S(z_i^n) \mathrm{d}z_j^n. \qquad (15.36)$$

The wedge product of $\mathrm{d}z_j^n$ with (15.36) gives, using the result $\mathrm{d}z_j^n \wedge S_{zz} \mathrm{d}z_j^n = 0$,

$$
\begin{aligned}
0 &= \mathrm{d}z_j^n \wedge M_+ \delta_t^+ \mathrm{d}z_j^n + \mathrm{d}z_j^n \wedge M_- \delta_t^- \mathrm{d}z_j^n \\
&\quad + \mathrm{d}z_j^n \wedge K_+ \delta_x^+ \mathrm{d}z_j^n + \mathrm{d}z_j^n \wedge K_- \delta_x^- \mathrm{d}z_j^n \\
&= \mathrm{d}z_j^n \wedge M_+ \delta_t^+ \mathrm{d}z_j^n + \delta_t^- \mathrm{d}z_j^n \wedge M_+ \mathrm{d}z_j^n \\
&\quad + \mathrm{d}z_j^n \wedge K_+ \delta_x^+ \mathrm{d}z_j^n + \delta_x^- \mathrm{d}z_j^n \wedge K_+ \mathrm{d}z_j^n \\
&= \delta_t^+ \left(\mathrm{d}z_j^{n-1} \wedge M_+ \mathrm{d}z_j^n\right) + \delta_x^- \left(\mathrm{d}z_{j-1}^n \wedge K_+ \mathrm{d}z_j^n\right) \\
&= \delta_t^+ \omega_j^n + \delta_x^- \kappa_j^n,
\end{aligned} \qquad (15.37\text{a})
$$

where

$$\omega_j^n = \mathrm{d}z_j^{n-1} \wedge M_+ \mathrm{d}z_j^n, \quad \kappa_j^n = \mathrm{d}z_{j-1}^n \wedge K_+ \mathrm{d}z_j^n. \qquad (15.37\text{b})$$

Through the backward-error analysis, it has been shown that [15] if a multi-symplectic integration scheme is applied to a Hamiltonian PDE, the resulting modified equation is again Hamiltonian. The next two sections apply the multi-symplectic scheme to two ubiquitous NLPDEs.

15.5   Korteweg–de Vries Equation

The KdV equation is

$$u_t + \mu u u_x + \delta u_{xxx} = 0. \tag{15.38}$$

It is a simple nonlinear dispersive wave equation illustrating the combined effect of dispersion and nonlinearity. It describes one-dimensional water wave propagation in a shallow channel. The KdV equation also plays an important role in the plasma physics (ion acoustic waves) and in solid state physics. The KdV and family of KdV equations model internal surface wave in the Andaman Sea between Thailand and Sumatra, acoustic waves in liquids with gas bubbles (bubbly liquids), Great Red Spot (GRS) and certain features in the Jovian atmosphere, generation and interaction of ion-acoustic waves in a cold plasma, acoustic waves on a crystal lattice and so on.

It represents the starting point to investigate soliton behaviour, that is, unscattered (or elastic) interaction of solitary waves (localized, single-hump and bell-shaped waves) solutions [1–3]. The term uu_x in Eq. (15.38) is the nonlinear term.

15.5.1   A Multi-Symplectic Scheme

An equivalent system of first-order equations of the KdV equation is

$$\frac{1}{2}u_t + w_x = 0, \tag{15.39a}$$

$$-\frac{1}{2}\phi_t - \delta v_x = -w + \frac{1}{2}\mu u^2, \tag{15.39b}$$

$$\delta u_x = v, \tag{15.39c}$$

$$-\phi_x = -u. \tag{15.39d}$$

The KdV Eq. (15.39) written in the form of (15.16) has

$$M = \begin{pmatrix} 0 & 1/2 & 0 & 0 \\ -1/2 & 0 & 0 & 0 \\ 0 & 0 & 0 & 0 \\ 0 & 0 & 0 & 0 \end{pmatrix}, \quad K = \begin{pmatrix} 0 & 0 & 0 & 1 \\ 0 & 0 & -\delta & 0 \\ 0 & \delta & 0 & 0 \\ -1 & 0 & 0 & 0 \end{pmatrix}, \tag{15.40a}$$

and

$$z = \begin{pmatrix} \phi \\ u \\ v \\ w \end{pmatrix}, \tag{15.40b}$$

$$S(z) = \frac{1}{2}v^2 - uw + \frac{1}{6}\mu u^3. \tag{15.40c}$$

The multi-symplectic conservation law

$$\frac{\partial \omega}{\partial t} + \frac{\partial \kappa}{\partial x} = \partial_t [dz \wedge M dz] + \partial_x [dz \wedge K dz] = 0 \tag{15.41}$$

becomes

$$\partial_t (d\phi \wedge du) + 2\partial_x (d\phi \wedge dw + \delta dv \wedge du) = 0. \tag{15.42}$$

Splitting M and K as

$$M_+ = \begin{pmatrix} 0 & 1/2 & 0 & 0 \\ 0 & 0 & 0 & 0 \\ 0 & 0 & 0 & 0 \\ 0 & 0 & 0 & 0 \end{pmatrix}, \quad M_- = \begin{pmatrix} 0 & 0 & 0 & 0 \\ -1/2 & 0 & 0 & 0 \\ 0 & 0 & 0 & 0 \\ 0 & 0 & 0 & 0 \end{pmatrix}, \qquad (15.43a)$$

$$K_+ = \begin{pmatrix} 0 & 0 & 0 & 1 \\ 0 & 0 & -\delta & 0 \\ 0 & 0 & 0 & 0 \\ 0 & 0 & 0 & 0 \end{pmatrix}, \quad K_- = \begin{pmatrix} 0 & 0 & 0 & 0 \\ 0 & 0 & 0 & 0 \\ 0 & \delta & 0 & 0 \\ -1 & 0 & 0 & 0 \end{pmatrix} \qquad (15.43b)$$

and applying the Euler box scheme (15.35) give

$$\frac{1}{2k}\left(u_j^{n+1} - u_j^n\right) + \frac{1}{n}\left(w_{j+1}^n - w_j^n\right) = 0, \qquad (15.44a)$$

$$-\frac{1}{2k}\left(\phi_j^n - \phi_j^{n-1}\right) - \frac{\delta}{h}\left(v_{j+1}^n - v_j^n\right) = -w_j^n + \frac{\mu}{2}\left(u_j^n\right)^2, \qquad (15.44b)$$

$$\frac{\delta}{h}\left(u_j^n - u_{j-1}^n\right) = v_j^n, \qquad (15.44c)$$

$$-\frac{1}{h}\left(\phi_j^n - \phi_{j-1}^n\right) = -hu_j^n. \qquad (15.44d)$$

Using Eq. (15.44b) for w_j^n and w_{j+1}^n in Eq. (15.44a) and then eliminating v's and ϕ's using Eqs. (15.44c) and (15.44d), respectively, lead to the multi-symplectic FD scheme

$$\frac{1}{2}\left(\frac{u_j^{n+1} - u_j^n}{k} + \frac{u_{j+1}^n - u_{j+1}^{n-1}}{k}\right) + \mu\left(\frac{u_{j+1}^n + u_j^n}{2}\right)\left(\frac{u_{j+1}^n - u_j^n}{h}\right)$$

$$+\delta\left(\frac{u_{j+2}^n - 3u_{j+1}^n + 3u_j^n - u_{j-1}^n}{h^3}\right) = 0. \qquad (15.45)$$

Comparing this equation with the KdV Eq. (15.38) observe that at the grid point (x_j, t_n), the last term in (15.45) is an FD formula for u_{xxx}, $(u_{j+1}^n - u_j^n)/h$ is the forward-difference formula for u_x and $(u_{j+1}^n + u_j^n)/2$ is the approximation of u. The first term in (15.45) is the discrete version of u_t. It is the average of u_t at (j, n) and $(j+1, n)$. u_t at (j, n)th grid point is approximated by the forward-difference formula while that at $(j+1, n)$th grid point is approximated by the backward-difference formula.

Equation (15.45) can be rewritten as

$$u_j^{n+1} = u_j^n + u_{j+1}^{n-1} - u_{j+1}^n - \frac{\mu k}{h}\left(u_{j+1}^n + u_j^n\right)\left(u_{j+1}^n - u_j^n\right)$$

$$-\frac{2\delta k}{h^3}\left(u_{j+2}^n - 3u_{j+1}^n + 3u_j^n - u_{j-1}^n\right). \qquad (15.46)$$

To compute u_j^{n+1}, the values of u_{j+1}^{n-1}, u_{j-1}^n, u_j^n, u_{j+1}^n and u_{j+2}^n are to be used. This is depicted in Fig. 15.1. Call (15.46) as a five-point formula. In Eq. (15.46) $n = 1, 2, \ldots, N_t$ and $j = 1, 2, \ldots, N_x - 2$ (refer Eq. (15.32)).

15.5.2 Numerical Implementation of the Scheme

When $n = 1$ the formula (15.46) takes the form

$$u_j^2 = u_j^1 + u_{j+1}^0 - u_{j+1}^1 - \frac{\mu k}{h}\left(u_{j+1}^1 + u_j^1\right)\left(u_{j+1}^1 - u_j^1\right)$$

$$-\frac{2\delta k}{h^3}\left(u_{j+2}^1 - 3u_{j+1}^1 + 3u_j^1 - u_{j-1}^1\right). \qquad (15.47)$$

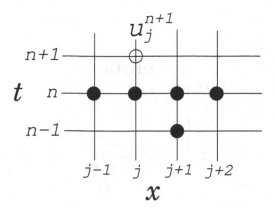

FIGURE 15.1

The grid points involved (marked by solid circles) to compute the numerical solution of KdV equation at the grid point $(j, n+1)$ (marked by an open circle).

To find u_j^2 the values of u at $t = t_0$ and $t = t_0 + k$ are needed. $u(x, t_0)$ is given by the initial condition. $u(x, t_0 + k)$, that is, u_j^1 is need to be known. To compute u_j^1 use the forward-difference formula for u_t:

$$u_t = \frac{1}{k} \left(u_j^{n+1} - u_j^n \right). \tag{15.48}$$

Then, the KdV Eq. (15.38) gives

$$u_j^{n+1} = u_j^n - \frac{\mu k}{2h} \left(u_{j+1}^n + u_j^n \right) \left(u_{j+1}^n - u_j^n \right) - \frac{\delta k}{h^3} \left(u_{j+2}^n - 3u_{j+1}^n + 3u_j^n - u_{j-1}^n \right). \tag{15.49}$$

When $n = 0$ the above equation becomes

$$u_j^1 = u_j^0 - \frac{\mu k}{2h} \left(u_{j+1}^0 + u_j^0 \right) \left(u_{j+1}^0 - u_j^0 \right) - \frac{\delta k}{h^3} \left(u_{j+2}^0 - 3u_{j+1}^0 + 3u_j^0 - u_{j-1}^0 \right). \tag{15.50}$$

To determine u_j^2 (Eq. (15.47)) first generate u_j^0 from the given initial condition and using it compute u_j^1 from (15.50) with

$$u_0^1 = u_0^0, \quad u_{N_x-1}^1 = u_{N_x-1}^0, \quad u_{N_x}^1 = u_{N_x}^0. \tag{15.51}$$

For u_j^{n+1} with $n > 1$ assume that

$$u_0^{n+1} = u_0^n, \quad u_{N_x-1}^{n+1} = u_{N_x-1}^n, \quad u_{N_x}^{n+1} = u_{N_x}^n. \tag{15.52}$$

15.5.3   Stability Analysis

The stability condition for the five-point scheme can be obtained. Substitution of $u = e^{ij\alpha h} \xi^n$ (as done for LPDEs in the previous chapter) in Eq. (15.46) with $(u_{j+1}^n + u_j^n)/2 = u$ gives

$$\xi^2 - a\xi - e^{-i\alpha h} = 0, \tag{15.53a}$$

where

$$a = 1 - e^{i\alpha h} + \frac{\mu k u}{h} \left(1 - e^{i\alpha h} \right) - \frac{2\delta k}{h^3} \left(3 + e^{i2\alpha h} - 3e^{i\alpha h} - e^{-i\alpha h} \right). \tag{15.53b}$$

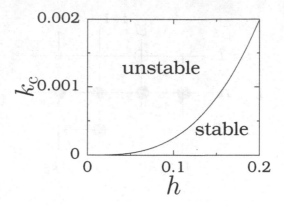

FIGURE 15.2
Plot of k_c versus h for the KdV equation corresponding to the two-solitary wave solution.

The roots of (15.53a) are

$$\xi_\pm = \frac{1}{2}\left[a \pm \sqrt{a^2 + 4e^{-i\alpha h}}\right].$$ (15.54)

Then, the stability condition, $|\text{Re}\xi_\pm| \leq 1$, gives [16]

$$\frac{k}{h}\left[\mu|u_{\max}| + 4\frac{\delta}{h^2}\right] \leq 1,$$ (15.55)

where $|u_{\max}|$ is the maximum of u. That is,

$$k \leq k_c = \frac{h}{\mu|u_{\max}| + 4(\delta/h^2)}.$$ (15.56)

15.5.4 Interaction of Solitary Waves

The KdV equation admits N-solitary wave solutions and the analytical expressions for them are well-known in the literature [1–3]. Solve the KdV equation to simulate a two-solitary wave solution. For this consider the initial-value problem, that is, there is no boundary condition.

For $\mu = 6$ and $\delta = 1$ a two-solitary wave solution is given by

$$u(x,t) = \frac{1}{2}\left(k_2^2 - k_1^2\right)\frac{k_2^2\cosh^2\eta_1 + k_1^2\sinh^2\eta_2}{k_2\cosh\eta_1\cosh\eta_2 - k_1\sinh\eta_1\sinh\eta_2},$$ (15.57a)

where

$$\eta_1 = k_1 x - k_1^3 t + \eta_{10}, \quad \eta_2 = k_2 x - k_2^3 t + \eta_{20}.$$ (15.57b)

In (15.57) k_1, k_2, η_{10} and η_{20} are constants and k_1, $k_2 \neq 0$. Choose $k_1 = 0.2$, $k_2 = \sqrt{3}\,k_1$, $\eta_{10} = \eta_{20} = 0$, $x \in [-150, 150]$ with $h = 0.2$ and the initial value of time t_0 as -900. The maximum value of u, $u_{\max} = 0.06$. Figure 15.2 shows the plot of k_c versus h. The scheme will be stable for the values of h and k in the region marked as stable. Next, verify the numerical solution and the stability condition.

For the choice $k = 0.001$ the stability determining quantity $\Delta = |\text{Re}\xi_\pm| = 0.5018 < 1$. Figure 15.3a shows the initial wave solution ($t = -900$). There are two well-separated solitary waves of different amplitudes. Figures 15.3b-d depict the numerical and exact solutions

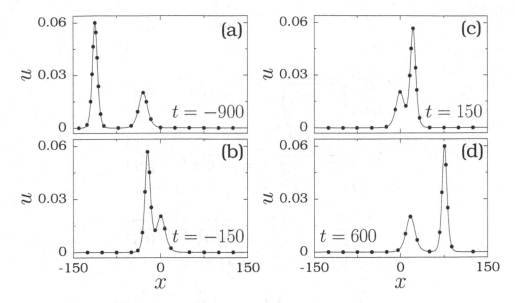

FIGURE 15.3
The numerically computed two-solitary wave solution (continuous line) at a few values of
t illustrating elastic collision. The exact analytical solution is marked by solid circles at
selected values of x. Here, $h = 0.2$ and $k = 0.001$.

for three values of t showing the interaction of the two waves. Numerical solution coincides
with the exact solution for large t also. Let us explain Fig. 15.3. As the initial wave profile
evolves as per the KdV equation, after a sufficient time the waves overlap and interact. The
taller wave catches up the shorter wave. This is shown in Fig. 15.3b. At $t = 0$ both the
waves merge together. However, the two waves reappear for $t > 0$ (Figs. 15.3c and 15.3d).
The taller wave separates from the shorter one, after overtaking it, and asymptotically (as
$t \to \infty$) the two waves regain the initial shape and hence their speeds. The interaction
between the two solitary waves is elastic. Such solitary waves are called *solitons* [1–3].

   What is the effect of k? For $h = 0.2$ and $k = 0.0025 > k_c$ the stability determining
quantity $\Delta = 1.2545$. The method is thus unstable. Figure 15.4 shows the numerical solution.
The two-solitary wave pattern of the solution is destroyed. The amplitude of the numerical
solution grows with time.

15.5.5   Efficiency of the Five-Point Scheme

An important criterion of judging a numerical scheme is to check its ability to preserving
constants of motion or conservation laws. The integrable solitonic equations such as KdV
and sG equations possess infinite number of conservation laws. The numerical compuation
of first few conservation laws can be used to analyze the efficiency of a numerical scheme.

Example:

The first three conservation laws of the KdV equation are [1]

$$I_1(t) \;=\; \int_{-\infty}^{\infty} u \, \mathrm{d}x \,, \quad I_2(t) = \frac{1}{2} \int_{-\infty}^{\infty} u^2 \, \mathrm{d}x \qquad (15.58a)$$

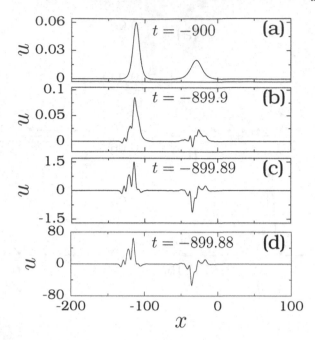

FIGURE 15.4

The numerically computed two-solitary wave solution.

and

$$I_3(t) = \int_{-\infty}^{\infty} \left(\frac{1}{2} \delta u_x^2 - \frac{1}{6} \mu u^3 \right) dx. \tag{15.58b}$$

I_1, I_2 and I_3 are momentum, energy and Hamiltonian, respectively, of the wave solution of the KdV equation. The discrete versions of them are

$$I_1 = h \sum_{j=1}^{N_x} u_j^n, \quad I_2 = \frac{h}{2} \sum_{j=1}^{N_x} \left(u_j^n \right)^2, \tag{15.59a}$$

$$I_3 = h \sum_{j=1}^{N_x} \left[\frac{\delta}{8h^2} \left(u_{j+1}^n - u_{j-1}^n \right)^2 - \frac{\mu}{6} \left(u_j^n \right)^3 \right]. \tag{15.59b}$$

Apart from I_1, I_2 and I_3 consider the maximum error in u, $u_{me}(t)$, given by

$$u_{me}(t) = \text{max.} |u_{num}(x,t) - u_{exact}(x,t)|. \tag{15.60}$$

Compute the percentage of relative errors in the above-mentioned four quantities over a long time. The result is displayed in Fig. 15.5. The errors in all these quantities are $< 0.2\%$. The errors in I_2, I_3 and u_{me} are very small far before and far after the interaction of the solitary waves ($t = 0$ is the time at which two waves merge together and form a single solitary wave).

Can one use the five-point formula to study the evolution of an arbitrary wave profile? To answer this in the next subsection solve the KdV equation with periodic boundary condition.

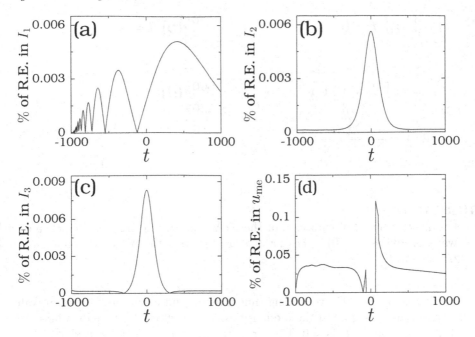

FIGURE 15.5
Variations of the percentage of relative errors in the constants of motion I_1, I_2 and I_3 and the percentage of maximum error in u.

15.5.6 Korteweg–de Vries Equation with Periodic Boundary Condition

The KdV Eq. (15.38) was first numerically solved by Zabusky and Kruskal (ZK) [4] using the FD formula

$$
\begin{aligned}
u_j^{n+1} = {} & u_j^{n-1} - \frac{\mu k}{3}\left(u_{j+1}^n + u_j^n + u_{j-1}^n\right)\left(u_{j+1}^n - u_{j-1}^n\right) \\
& - \frac{\delta k}{h^3}\left(u_{j+2}^n - 2u_{j+1}^n + 2u_{j-1}^n - u_{j-2}^n\right),
\end{aligned} \tag{15.61}
$$

where u_t and u_x are approximated by two-point central-difference formulas while u_{xxx} is approximated by a four-point central-difference formula. The following example shows that the above scheme is numerically unstable while the multi-symplectic scheme (15.46) is stable over a long time.

Example:

Fix the parameters in the KdV equation as $\mu = 1$ and $\delta = 0.0036$ and choose

$$
u(x,0) = \cos \pi x, \quad 0 \le x \le 2, \quad h = 0.01, \quad k = 0.00001, \quad u(x+2,t) = u(x,t). \tag{15.62}
$$

Figure 15.6 shows the outcomes of the ZK scheme. The initial wave profile evolves into four solitary waves. This is clearly evident in Fig. 15.6b for $t = 4$. As t increases the scheme becomes unstable and $u(x,t)$ grows with time and wiggles are generated. These are clearly seen in Figs. 15.6c and d for $t = 12.65$ and 12.6574, respectively.

The result of the multi-symplectic five-point scheme (15.46) is presented in Fig. 15.7. At $t = 4$ observe four solitary waves with distinct amplitudes. In Figs. 15.7c and d for large values of t the solution is bounded and there are four solitary waves. Compute the

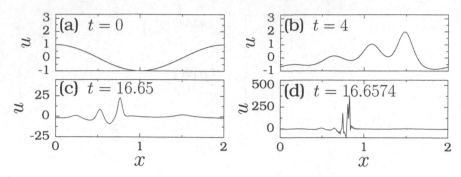

FIGURE 15.6
The numerically computed solutions of the KdV equation using the ZK scheme with the initial wave profile as $u(x,0) = \cos \pi x$, $0 \le x \le 2$ with the periodic boundary condition $u(x+2,t) = u(x,t)$.

conserved quantities $I_1 - I_3$ over a long time interval and calculate their mean values. The percentage of relative errors in these quantities compared to their mean values are plotted in Fig. 15.8. The error in I_1 is ≈ 0. The errors in I_2 and I_3 are $< 0.5\%$ and 5%, respectively. Figure 15.9 shows the numerical solution for $u(x,0) = 2x \sin \pi x$. The initial wave evolves into a five-solitary wave solution. Different number of solitary waves can be generated with different initial conditions.

15.6   Sine-Gordon Equation

Another interesting NLPDE exhibiting soliton solutions is the sG equation

$$u_{tt} - u_{xx} + \sin u = 0. \tag{15.63}$$

The sG equation arises in classical mechanics and electronics. It models the dynamics of a one-dimensional chain of identical pendula connected by a torsion bar and propagation

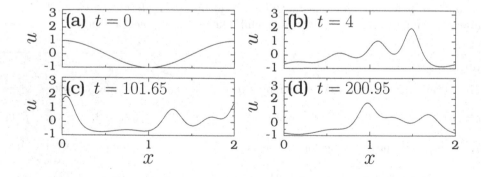

FIGURE 15.7
The numerically computed solutions of the KdV equation using the multi-symplectic scheme (15.46) with the initial wave profile as $u(x,0) = \cos \pi x$, $0 \le x \le 2$ with the periodic boundary condition $u(x+2,t) = u(x,t)$.

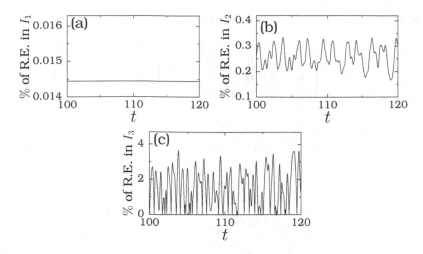

FIGURE 15.8
Variations of the percentage of relative errors in the constants of motion I_1, I_2 and I_3
computed using the numerical solution generated by the multi-symplectic five-point scheme
for the KdV equation subjected to the periodic boundary condition $u(x + 2, t) = u(x, t)$.

of transverse electromagnetic waves on a superconducting strip-line transmission system.
The sG equation is used to model propagation of dislocations in crystals, waves along
lipid membranes, wave propagation along the arrays of Josephson junctions, strain waves,
periodicity of episodic tremor and slow slip events and redistribution and migration of
stresses in the lithosphere.

15.6.1 Multi-Symplectic Four-Point Scheme

In order to express the sG equation in the multi-symplectic form (15.16) write it as a system
of first-order equation by introducing additional variables. An example is

$$-v_t - p_x = \sin u, \tag{15.64a}$$
$$u_t = v, \tag{15.64b}$$
$$p + w = 0, \tag{15.64c}$$
$$u_x = w. \tag{15.64d}$$

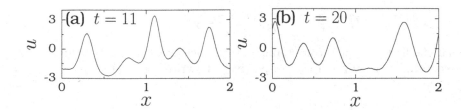

FIGURE 15.9
The numerically computed solution of the KdV equation using the multi-symplectic scheme
(15.46) with the initial wave profile as $u(x, 0) = 2x \sin \pi x$, $0 \le x \le 2$ with the periodic
boundary condition $u(x + 2, t) = u(x, t)$.

The above equation is in the multi-symplectic form (15.16) with

$$M = \begin{pmatrix} 0 & -1 & 0 & 0 \\ 1 & 0 & 0 & 0 \\ 0 & 0 & 0 & 0 \\ 0 & 0 & 0 & 0 \end{pmatrix}, \quad K = \begin{pmatrix} 0 & 0 & 0 & -1 \\ 0 & 0 & 0 & 0 \\ 0 & 0 & 0 & 0 \\ 1 & 0 & 0 & 0 \end{pmatrix}, \tag{15.65a}$$

and

$$z = \begin{pmatrix} u \\ v \\ w \\ p \end{pmatrix}, \tag{15.65b}$$

$$S(z) = \frac{1}{2}v^2 + \frac{1}{2}w^2 + pw + 1 - \cos u. \tag{15.65c}$$

M and K can be splitted into M_+, M_-, K_+ and K_- with $M_+^{\mathrm{T}} = -M_-$ and $K_+^{\mathrm{T}} = -K_-$ as

$$M_+ = \begin{pmatrix} 0 & -1 & 0 & 0 \\ 0 & 0 & 0 & 0 \\ 0 & 0 & 0 & 0 \\ 0 & 0 & 0 & 0 \end{pmatrix}, \quad M_- = \begin{pmatrix} 0 & 0 & 0 & 0 \\ 1 & 0 & 0 & 0 \\ 0 & 0 & 0 & 0 \\ 0 & 0 & 0 & 0 \end{pmatrix}, \tag{15.66a}$$

$$K_+ = \begin{pmatrix} 0 & 0 & 0 & -1 \\ 0 & 0 & 0 & 0 \\ 0 & 0 & 0 & 0 \\ 0 & 0 & 0 & 0 \end{pmatrix}, \quad K_- = \begin{pmatrix} 0 & 0 & 0 & 0 \\ 0 & 0 & 0 & 0 \\ 0 & 0 & 0 & 0 \\ 1 & 0 & 0 & 0 \end{pmatrix}. \tag{15.66b}$$

The Euler box scheme (15.35) gives

$$-\frac{1}{k}\left(v_j^{n+1} - v_j^n\right) - \frac{1}{h}\left(w_{j+1}^n - w_j^n\right) = \sin u_j^n, \tag{15.67a}$$

$$\frac{1}{k}\left(u_j^n - u_j^{n-1}\right) = v_j^n, \tag{15.67b}$$

$$0 = p_j^n + w_j^n, \tag{15.67c}$$

$$\frac{1}{h}\left(u_j^n - u_{j-1}^n\right) = w_j^n. \tag{15.67d}$$

Elimination of v_j^n, v_j^{n+1}, w_j^n and w_j^{n+1} in Eq. (15.67a) using the subequations leads to the explicit five-point scheme

$$\frac{1}{k^2}\left(u_j^{n+1} - 2u_j^n + u_j^{n-1}\right) - \frac{1}{h^2}\left(u_{j+1}^n - 2u_j^n + u_{j-1}^n\right) + \sin u_j^n = 0. \tag{15.68}$$

Note that the first and the second terms in the above equation are the three-point central-difference formulas for u_{tt} and u_{xx}, respectively. Rewrite Eq. (15.68) as

$$u_j^{n+1} = 2u_j^n - u_j^{n-1} + \frac{k^2}{h^2}\left(u_{j+1}^n - 2u_j^n + u_{j-1}^n\right) - k^2 \sin u_j^n. \tag{15.69}$$

The solution at the grid points $(j, n-1)$, $(j-1, n)$, (j, n) and $(j+1, n)$ are used to compute the solution at the grid point $(j, n+1)$ and is depicted in Fig. 15.10.

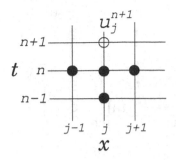

FIGURE 15.10
The grid points involved (marked by solid circles) to compute the numerical solution of the sG equation at the grid point $(j, n+1)$ (marked by a open circle).

15.6.2 Numerical Implementation of the Scheme

In order to use the formula (15.69) the solution at $n = 0$ and $n = 1$ are needed. The solution at $n = 0$, that is, at $t = 0$, is given by the initial condition $u(x, 0)$. The sG equation is second-order in time derivative. Therefore, to solve it the initial condition $u(x, t_0) = f(x)$ and $u_t(x, t_0) = g(x)$ must be specified, where t_0 is the initial time. To calculate $u_{i,1}$, one can make use of the initial condition on u_t.

Using the central-difference formula for the first-derivative the initial condition $u_t(x, t_0) = g(x)$ is written as

$$u_t(x, t_0) = \frac{1}{2k} \left(u_j^{n+1} - u_j^{n-1} \right) |_{n=0} = g(x_j) = g_j. \tag{15.70}$$

That is,

$$u_j^{-1} = u_j^1 - 2kg_j. \tag{15.71}$$

Equation (15.69) for $n = 0$ is

$$u_j^1 = 2u_j^0 - u_j^{-1} - \frac{k^2}{h^2} \left(u_{j+1}^0 - 2u_j^0 - u_{j-1}^0 \right) - k^2 \sin u_j^0. \tag{15.72}$$

Replacing u_j^{-1} by Eq. (15.71) gives

$$u_j^1 = u_j^0 + kg_j - \frac{k^2}{2h^2} \left(u_{j+1}^0 - 2u_j^0 - u_{j-1}^0 \right) - \frac{k^2}{2} \sin u_j^0. \tag{15.73}$$

Equations (15.69) and (15.73) constitute the FD method for solving the sG equation subjected to the initial conditions $u(x, t_0) = f(x)$ and $u_t(x, t_0) = g(x)$.

15.6.3 Stability Condition

To obtain the stability condition on k and h substitute $u_j^n = e^{i\alpha j h} \xi^n$ in (15.69). When the initial disturbance is small then the term $k^2 \sin u_j^n$ is very small and hence it can be neglected. Then, obtain

$$\xi^2 + 2\xi \left(-1 + \frac{k^2}{h^2} - \frac{k^2}{h^2} \cos \alpha h \right) + 1 = 0. \tag{15.74}$$

The condition for $|\text{Re}\xi_{\pm}| \le 1$ is $2 - (k^2/h^2)(1 - \cos \alpha h) \ge 0$. That is, $1 - (k^2/h^2) \sin^2(\alpha h/2) \ge 0$. Thus, the required condition for stability is $k \le h$.

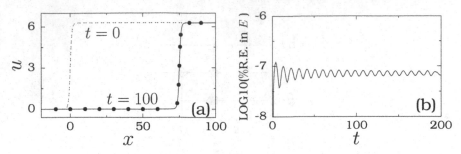

FIGURE 15.11

The numerically predicted kink solution of the sG equation at $t = 100$ (continuous curve). The solid circles represent the analytical solution at some selected values of x.

15.6.4 Numerical Simulation of Different Soliton Solutions

The sG equation admits a variety of exact analytical solution. For example, seeking the travelling wave solution $u(x,t) = u(x - ct) = u(\xi)$ the sG equation becomes

$$\left(1 - c^2\right) u_{\xi\xi} - \sin u = 0 . \tag{15.75}$$

Integrating (15.75) in the standard way one can obtain an elliptic function solution. Choosing the integration constant suitably so that $u \to 0 \pmod{2\pi}$ and $u_\xi \to 0$ as $\xi \to \pm\infty$, the travelling wave solution can be obtained. Its form is [1]

$$u(x,t) = 4\tan^{-1}\left[\exp\left(\frac{\pm(x - ct - x_0)}{\sqrt{1 - c^2}}\right)\right] . \tag{15.76}$$

Fix $x_{\min} = -10$, $x_{\max} = 100$, $h = 0.01$, $k = h/2$, $c = 0.75$ and $x_0 = 0$. The initial wave solution is chosen as the solution (15.76) with '+' sign and $t = 0$. Figure 15.11a shows the initial solution and the numerical solution at $t = 100$. The solution represents a monotonic change in the value of u from 0 to 2π as x increases from $-\infty$ to ∞. Even though it is not a true pulse, the changes in its amplitude occur in a localized region and hence the solution is called a *kink* solution. (The solution (15.76) with the '−' sign corresponds to a monotonic decrease of u from the value 2π at $x = -\infty$ to 0 at $x = \infty$ and is known as an *antikink*). The kink moves in the forward direction with the velocity c.

Example:

Check the accuracy of the multi-symplectic scheme for the sG equation.

Numerically compute the energy of the system given by

$$E = \frac{1}{2} \int_{-\infty}^{\infty} \left(u_t^2 + u_x^2 + 2(1 - \cos u)\right) dx \tag{15.77a}$$

with

$$u_t = \frac{1}{2k}\left(u_j^{n+1} - u_j^{n-1}\right), \quad u_x = \frac{1}{2h}\left(u_{j+1}^n - u_{j-1}^n\right) . \tag{15.77b}$$

The exact energy of the kink solution is 12.09482. In Fig. 15.11b the percentage of relative error in E exhibits a damped oscillation. The error is $< 10^{-6}$.

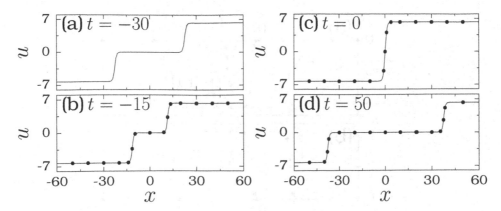

FIGURE 15.12

(a) The two-kink solution chosen as the initial wave profile of the sG equation. (b-d) The numerically computed two-kink solution of the sG equation. The continuous curve is the numerical solution. The solid circles represent the analytical solution at some selected values of x.

Next, consider the two-kink interaction. A kink of velocity $+c$ colliding at the origin with a kink of velocity $-c$ is given by [1]

$$u(x,t) = 4\tan^{-1}\left[\frac{c\sinh(x/\sqrt{1-c^2})}{\cosh(ct/\sqrt{1-c^2})}\right]. \qquad (15.78)$$

Fix $c = 0.75$, $x_{\min} = -60$, $x_{\max} = 60$ and the initial time as -30. For $t \ll 0$ the solution represents a well separated two kinks (see Fig. 15.12a). In Figs. 15.12b-d both numerical and exact solutions are plotted for a few values of t. As t increases the left(right)-side kink moves towards right(left)-side. The two kinks interact with each other. At $t = 0$, the two kinks appear as a single kink since the centre of them coincide. As t increases further the two kinks re-emerge. The kink which is confined to the range $[-2\pi, 0]$ for $t < 0$ becomes the kink with $u \in [0, 2\pi]$.

Figure 15.13 presents the numerical simulation of kink-antikink interaction. The exact solution of the kink-antikink is [1]

$$u(x,t) = 4\tan^{-1}\left[\frac{\sinh(ct/\sqrt{1-c^2})}{c\cosh(x/\sqrt{1-c^2})}\right]. \qquad (15.79)$$

Figure 15.14 shows the change in the energy with time.

Another class of a localized solution of the sG equation is the so-called *breather* solution [1]. This kind of solution can be obtained by the method of separation of variables. The breather solution is given by

$$u(x,t) = 4\tan^{-1}\left[\frac{\beta \sin\omega t}{\omega \cosh\beta(x - x_0)}\right], \qquad (15.80)$$

where $\beta = \sqrt{1 - \omega^2}$ and ω and x_0 are constants. The initial conditions associated with the breather soliton (15.80) with $x_0 = 0$ are

$$u(x,0) = 0, \quad u_t(x,0) = 4\beta\mathrm{sech}x. \qquad (15.81)$$

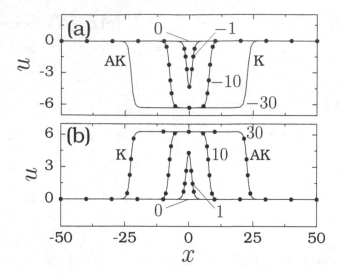

FIGURE 15.13
The numerically computed kink-antikink interaction. The continuous curve is the numerical solution. The solid circles represent the analytical solution at some selected values of x. The wave profile at $t = -30$ is the initial condition with $c = 0.75$.

Fix $\beta = 0.75$ and $\omega = \sqrt{1 - \beta^2}$. The numerical solution is depicted in Fig. 15.15 for various values of t. In Fig. 15.15a as t increases the flat initial solution evolves into a bell shape wave pattern. The amplitude of the wave increases with time and reaches a maximum at $t = \pi/(2\omega)$. When t increases further the amplitude begins to decrease and at $t = \pi/\omega$ the wave flattens. For $t \in [2\pi/\omega, 3\pi/(2\omega)]$ the amplitude increases but $u(x,t) < 0$ (Fig. 15.15c). From $t = 3\pi/(2\omega)$ the amplitude decreases and $u(x,t) = 0$ at $t = 2\pi/\omega$. The above type of evolution repeats in the interval $2\pi/\omega$ to $4\pi/\omega$ and so on. This wave solution is different from the other types of waves such as solitary waves and kinks. The centre of the solitary waves of the KdV equation and the kinks of the sG equation propagates in space, that is the centre moves along the direction of propagation, and the amplitude remains the same.

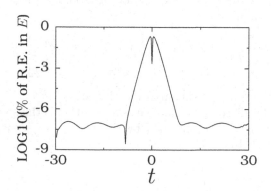

FIGURE 15.14
Variation of the error in the energy of the numerically computed kink-antikink solution of the sG equation.

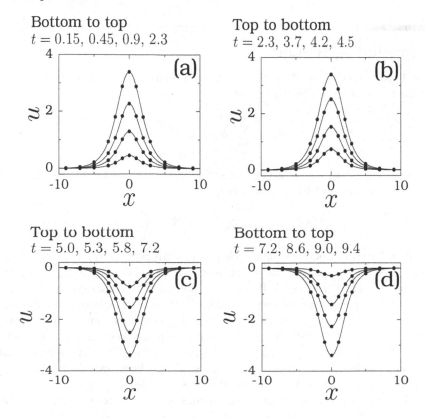

FIGURE 15.15
The numerically computed breather solutions of the sG equation at various values of t.

In contrast, the centre of the wave in Fig. 15.16 remains unchanged while the amplitude oscillates with time. Because the wave pattern resembles the breathing of our heart it is called a *breather*. In Fig. 15.16, the percentage of relative error in E oscillates with t and it is less than 0.1%.

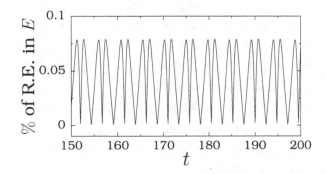

FIGURE 15.16
Variation of the error in the energy of the numerically computed breather solution of the sG equation.

15.7 Concluding Remarks

One can extend the concept of multi-symplectic scheme to a higher space dimension. For L-space dimensions Eq. (15.16) takes the form

$$Mz_t + \sum_{l=1}^{L} K_l z_{x_l} = \nabla_z S(z).$$ (15.82)

This chapter is restricted to the Euler box scheme. An application of the implicit midpoint scheme to (15.16) gives the Preissman scheme

$$M\delta_t^+ z_{j+\frac{1}{2}}^n + K\delta_x^+ z_j^{n+\frac{1}{2}} = \nabla_z S\left(x_{j+\frac{1}{2}}^{n+\frac{1}{2}}\right).$$ (15.83)

For a detailed discussion on Preissman scheme one can refer to [7,17].

Various conservation properties of the methods preserving a multi-symplectic conservation law have been investigated. It has been shown that Gauss–Legendre Runge–Kutta methods preserve energy and momentum laws for linear equations [18]. Multi-symplectic formulation for fluid dynamics using inverse map [19], conservation of wave action [20], nonlinear Schrödinger equation with Runge–Kutta Nystrom methods [21], phase space structure preservation [22], Maxwell's equation [23] and wave equations [24] have been reported. Backward error analysis for multi-symplectic methods [15] has been introduced. Multi-symplectic theory is extended for damped Hamiltonian PDEs [25] also.

Detailed numerical investigations of the nonlinear equations considered in this chapter in the presence of linear and nonlinear dampings, external periodic driving force and weak noise can be treated as project works and may give rise to interesting results.

15.8 Bibliography

[1]   M. Lakshmanan and S. Rajasekar, *Nonlinear Dynamics: Integrability, Chaos and Patterns*. Springer, Berlin, 2003.

[2]   M.J. Ablowitz, P.A. Clarkson, *Solitons, Nonlinear Evolution Equations and Inverse Scattering*. Cambridge University Press, Cambridge, 1992.

[3]   J. Hoppe, *Lectures on Integrable Systems*. Springer, Berlin, 1992.

[4]   N.J. Zabusky and M.D. Kruskal, *Phys. Rev. Lett.* 15:240, 1965.

[5]   R.I. McLachlan, *Numer. Meth.* 6:465, 1994.

[6]   J.E. Marsden, G.P. Patrick and S. Shkoller, *Commun. Math. Phys.* 199:351, 1998.

[7]   T.J. Bridges and S. Reich, *Phys. Lett. A* 284:184, 2001.

[8]   T.J. Bridges and S. Reich, *J. Phys. A: Math. Gen.* 39:5287, 2006.

[9]   E. Tadmor, *Bull. Am. Math. Soc.* 47:507, 2012.

[10] X. Feng, R. Glowinski and M. Neilan, *SIAM Review* 55:205, 2013.

[11] S. Mashayekhi, *Numerical Methods for Nonlinear PDEs in Finance*, Ph.D. Thesis, Univesity of Copenhagen, 2015.

[12] M. Neilan, A. Salgado and W. Zhang, *Acta Numerica* 26:137, 2017.

[13] M. Ratas, A. Salupere and J. Majak, *Math. Model. Anal.* 26:147, 2021.

[14] T.J. Bridges, *Math. Proc. Camb. Phil. Soc.* 121:147, 1997.

[15] B. Moore and S. Reich, *Numer. Math.* 95:625, 2003.

[16] W.H. Ping, W.Y. Shun and H.Y. Ying, *Chin. Phys. Lett.* 25:2335, 2008.

[17] B.E. Moore and S. Reich, *Future Generation Computer Systems* 19:395, 2003.

[18] S. Reich, *J. Comput. Phys.* 157:473, 2000.

[19] C.J. Cotter, D.D. Holm and P.E. Hydon, *Proc. Roy. Soc. A* 463:2671, 2007.

[20] J. Frank, *J. Phys. A: Math. Gen.* 39:5479, 2006.

[21] J. Hong, X. Liu and C. Li, *J. Comput. Phys.* 226:1968, 2007.

[22] A.L. Islas and C.M. Schober, *J. Comput. Phys.* 197:585, 2004.

[23] L. Kong, J. Hong and J. Zhang, *J. Comput. Phys.* 229:4259, 2010.

[24] H. Liu and Z. Zhang, *IMA J. Numer. Anal.* 26:252, 2006.

[25] B.E. Moore, L. Norena and C.M. Schober, *J. Comput. Phys.* 231:214, 2013.

15.9   Problems

15.1 For the system (15.13) with

$$M_+ = \begin{pmatrix} 0 & -1 & 0 & 0 \\ 0 & 0 & 0 & 0 \\ 0 & 0 & 0 & 0 \\ 0 & 0 & 0 & 0 \end{pmatrix} \text{ and } K_+ = \begin{pmatrix} 0 & 0 & 0 & -1 \\ 0 & 0 & 0 & 0 \\ 0 & 0 & 0 & 0 \\ 0 & 0 & 0 & 0 \end{pmatrix}$$

show that $\omega = du \wedge dv$ and $\kappa = du \wedge dp$. Also, verify the energy conservation law (15.18).

15.2 Obtain a semi-discrete equation of $M z_t + K z_x = \nabla_z S(z)$ by applying a symplectic Euler discretization in space. Then, taking inner product of it with z_t^n show that the discretization in space yields the conservation law of energy of the form $\partial_t E_j + \delta_x^+ F_j = 0$ and find E_j and F_j.

15.3 Write the perturbed sG equation

$$u_{tt} - u_{xx} + \sin u = -(a u_t + b \cos \omega t + c \cos kx)$$

in the symplectic form $M z_t + K z_x = \nabla_z S(z) + F(z, x, t)$.

15.4 An exact travelling wave solution of the fifth-order KdV equation

$$u_t + u u_x + u_{xxx} - u_{xxxxx} = 0$$

is [D. Kaya and K. Al-Khaled, *Phys. Lett. A* 363:433, 2007]

$$u = -\frac{72}{169} + \frac{105}{169} \operatorname{sech}^4 \left[\frac{1}{2\sqrt{13}} \left(x + \frac{36}{169} t \right) \right].$$

Express the fifth-order KdV equation as a system of first-order equation in the form of $M z_t + K z_x = \nabla_z S(z)$ and then obtain a multi-symplectic scheme.

15.5 The nonlinear Schrödinger equation $i\psi_t + \psi_{xx} + 2|\psi|^2\psi = 0$ admits the solitary wave solution

$$\begin{aligned}\psi &= 2\beta\exp\left[-2i\left\{\alpha x + 2(\alpha^2 - \beta^2)t + \alpha\delta_1\right\}\right]\\ &\quad \times\mathrm{sech}\left[2\beta(x + 4\alpha t + \delta_2)\right],\end{aligned}$$

where α, β, δ_1 and δ_2 are constants. Letting $\psi = p + iq$ and then introducing two new variables $v = q_x$ and $w = p_x$ express the nonlinear Schrödinger equation in the multi-symplectic form $Mz_t + Kz_x = \nabla_z S(z)$. Determine E, F, G and I in Eqs. (15.25) and (15.28). Develop a multi-symplectic scheme and study its applicability by simulating the above given solitary wave.

15.6 Show that the Boussinesq equation $u_{tt} - u_{xx} - (u^2)_{xx} - u_{xxxx} = 0$ can be written in the multi-symplectic form $Mz_t + Kz_x = \nabla_z S(z)$ [H. Wei-Peng and D. Zi-Chen, *Appl. Math. Mech.* 29:927, 2008]. Also, prove that the conservation law of symplecticity, local energy conservation law and local momentum conservation law are

$$\partial_t(du \wedge dw) + \partial_x(dv \wedge du + dw \wedge dp) = 0, \quad \partial_t e + \partial_x f = 0,$$
$$\frac{1}{2}\partial_t(u\partial_x w - w\partial_x u) + \partial_x\left[S(z) - \frac{1}{2}u\partial_t w - w\partial_t u\right] = 0,$$

respectively, where $v = u_x$, $p = w_x$, $e = S(z) - \frac{1}{2}\langle z_x, Kz\rangle$ and $f = \frac{1}{2}\langle z, Kz_t\rangle$.

15.7 For the Ito type coupled KdV equation [Y. Chen, S. Song and H. Zhu, Appl. Math. Comp. **218**, 5552 (2012)]

$$u_t - 6uu_x - 2vv_x - \gamma u_{xxx} = 0, v_t - 2uv_x = 0$$

with $z = [\phi, \psi, u, v, w, p, q]^\mathrm{T}$ establish that the conservation laws are $\omega_t + \kappa_x = 0$, $E_t + F_x = 0$ and $I_t + G_x = 0$, where

$$\omega = \frac{1}{2}(d\phi \wedge du + d\psi \wedge dv), \quad \kappa = dp \wedge d\phi + dq \wedge d\psi - dw \wedge du,$$
$$E = -u^3 + uv^2 - \frac{1}{2}u_x^2, \quad F = -p\phi_t - q\psi_t + wu_t,$$
$$G = S(z) - \frac{1}{2}(u\phi_t + v\psi_t), \quad I = \frac{1}{2}\left(u^2 + v^2\right).$$

15.8 Develop a multi-symplectic FD scheme for the modified KdV equation

$$u_t + \alpha u^2 u_x + \delta^2 u_{xxx} = 0.$$

Study its applicability by numerically simulating the two-soliton solution

$$u(x,t) = \frac{2\left[k_1\cosh\left(\eta_2 + A_{12}/2\right) + k_2\cosh\left(\eta_1 + A_{21}/2\right)\right]}{\alpha_1\left[\cosh s_1 + \alpha_2\cosh s_2 + \alpha_3\right]},$$

where

$$A_{12} = \log\frac{\alpha_1^2}{4k_2^2}, \quad A_{21} = \log\frac{\alpha_1^2}{4k_1^2},$$
$$\alpha_1 = \frac{k_1 - k_2}{k_1 + k_2}, \quad \alpha_2 = \left(\frac{k_1 + k_2}{k_1 - k_2}\right)^2, \quad \alpha_3 = \frac{4k_1k_2}{(k_1 + k_2)^2},$$

and

$$s_1 = \left(\eta_1 + \eta_2 + \frac{R_4}{2}\right), \quad s_2 = \left(\eta_1 - \eta_2 + \frac{R_1}{2} - \frac{R_2}{2}\right),$$

$$R_1 = \log\frac{1}{4k_1^2}, \quad R_2 = \log\frac{1}{4k_2^2}, \quad R_4 = \log\frac{\alpha_1^4}{16k_1^2k_2^2},$$

$$\eta_1 = k_1x - k_1^3t + \eta_{10}, \quad \eta_2 = k_2x - k_2^3t + \eta_{20}.$$

15.9 The Burgers equation modelling fluid turbulence in a channel, unidirectional sound waves in a gas and shock waves in real fluids is given by $u_t + uu_x - u_{xx} = 0$. A shock wave solution of it is [P. Sachdev, *Nonlinear Diffusive Waves*. Cambridge University Press, Cambridge, 1987; W. Malfliet, *Am. J. Phys.* 60:650, 1992]

$$u(x,t) = c\left\{1 - \tanh\left[\frac{c}{2}(x - ct)\right]\right\},$$

where c is the velocity of the wave. Develop a multi-symplectic scheme and using it study the accuracy of the method for the initial wave profile corresponding to the above given shock wave solution.

15.10 The Fisher equation $u_t - u(1 - u) - u_{xx} = 0$ is used to model a one-dimensional habitat, neutron population in a nuclear reaction, chemical kinetics and so on [E. Infeld and G. Rowlands, *Nonlinear Waves, Solitons and Chaos*. Cambridge University Press, Cambridge, 1990; P. Gray and S. Scott, *Chemical Oscillations and Instabilities*. Clarendon, Oxford, 1990]. The shock wave solution of it is given by

$$u(x,t) = \frac{1}{4}\left\{1 - \tanh\left[\frac{1}{2\sqrt{6}}\left(x - \frac{5}{\sqrt{6}}t\right)\right]\right\}^2.$$

Construct a multi-symplectic scheme and analyze its accuracy for the initial wave profile corresponding to the above given shock wave solution.

16

Fractional Order Ordinary Differential Equations

16.1 Introduction

In 1965 L'Hopital asked a question about differentiation of order 1/2 to Leibniz. The reply was 'This is an apparent paradox from which, one day, useful consequences will be drawn' [1]. In the following centuries, considerable progress has been made on fractional calculus of derivatives and integrals. Fractional derivatives are nonlocal operators and appropriate for the description of the presence of hereditary and memory properties in physical processes and in certain materials [2]. In Chapters 8 and 10, the fractional order derivatives and integrals, respectively, are introduced and methods to evaluate them numerically are presented.

In the past few decades, fractional calculus has found notable applications in physics and engineering, particularly, in signal processing and control engineering [3,4], electromagnetism [5,6], viscoelasticity [7,8], biology [9], electrochemistry [10], statistical mechanics [11], diffusion process [12-19], relaxation oscillation [20], fluid flow [21,22], heat conduction [23] and nanotechnology [24]. One may realize fractional order multiploes in electromagnetism [5]. Fractional order models have been proposed for the dynamics of premotor neurons in the vestibulo-ocular reflex system [25]. Electrical circuits may have *fractance* which corresponds to a circuit element with fractional order impedance. Fractional order capacitor is also realized. Certain fractional-order circuits are analyzed [26-34]. For a review on the applications of fractional calculus to science and engineering refer to the ref. [35].

Consider the fractional differential equations of the form

$$D^{\alpha}x(t) = f(x(t), t), \quad \alpha > 0. \tag{16.1}$$

It is desired to find the solution of (16.1) for, say, $t \in [0, T]$. *How many initial conditions are to be specified for the initial-value problem of (16.1)?* The number of initial conditions needed is $m = [\alpha] + 1$ where $[\alpha]$ is the integer part of α. For $0 < \alpha < 1$ the value of m is 1, that is, one initial condition is sufficient. For $\alpha > 0$, the initial conditions are

$$x^{(k)}(0) = x_0^{(k)}, \quad k = 0, 1, \ldots, m - 1 \tag{16.2}$$

and are assumed to be given. Exact solution is not known for most of the fractional order differential equations. Some of the methods for finding approximate solutions are Adomian decomposition, homotopy, variational and perturbation analysis and generalized differential transform. Approximate analytical solutions are found for certain linear and nonlinear fractional differential equations, for example see the refs. [36-42]. Several numerical methods are developed and analyzed [43-71] for fractional order ordinary differential equations. Note that as the fractional integral is nonlocal, the numerical integration of factional order differential equations is expensive taking into the account of the memory needed and computational effect.

DOI: 10.1201/9781032649931-16

The present chapter is devoted to the numerically solving the fractional order ordinary differential equations. Particularly, the following four methods are considered:

1.   Backward-Difference methods.

2.   Fractional Euler method.

3.   Adams–Bashforth–Moulton method.

4.   A Two-Step Adams–Bashforth method.

First, fractional order differential operators are introduced. Next, the above four methods are presented. For each method, the formula for the numerical integration of differential equation of the form (16.1) is developed. The methods are applied to one or two linear differential equations for which exact solutions are known. The accuracies of the methods are discussed by comparing the numerical solution with the exact solution. Then, applicability of the fractional Euler method to certain nonlinear differential equations is demonstrated by verifying the qualitative behaviour of numerical solution with the results of equilibrium point analysis. The occurrence of period-doubling route to a complicated dynamics called *chaotic behaviour* in the ubiquitous Duffing oscillator is illustrated by solving the system by the fractional Euler method.

16.2   Fractional Order Differential Operators

In this section, certain definitions of fractional derivatives are reviewed.

1. Grünwald–Letnikov Operator

The Grünwald and Letnikov (GL) fractional derivative of order α ($\alpha > 0$) is defined as [51,72-75]

$$D_a^\alpha f(t) = \lim_{h \to 0} \frac{\Delta_h^\alpha f(t)}{h^\alpha} = \lim_{h \to 0} \frac{1}{h^\alpha} \sum_{k=0}^{n} (-1)^k \begin{pmatrix} \alpha \\ k \end{pmatrix} f(t - kh) , \qquad (16.3)$$

where $a \le t \le b$, $h = (t - a)/n$ and $\Delta_h^\alpha f(t)$ is a fractional formulation of a backward-difference for an arbitrary function $f(t)$. In the above equation

$$\begin{pmatrix} \alpha \\ k \end{pmatrix} = \frac{\alpha!}{k!(\alpha - k)!} = \frac{\Gamma(\alpha + 1)}{\Gamma(k + 1)\Gamma(\alpha - k + 1)} \quad \text{and} \quad \begin{pmatrix} \alpha \\ 0 \end{pmatrix} = 1, \qquad (16.4)$$

where $\Gamma(\gamma)$ is the gamma function with argument γ and is given by

$$\Gamma(\gamma) = \int_0^\infty z^{\gamma - 1} e^{-z} \mathrm{d}z . \qquad (16.5)$$

D_a^α is the left-GL derivative. The right-GL derivative of $f(t)$ is

$$D_b^\alpha f(t) = \lim_{h \to 0} \frac{1}{h^\alpha} \sum_{k=0}^{n} (-1)^k \begin{pmatrix} \alpha \\ k \end{pmatrix} f(t + kh) , \qquad (16.6)$$

where $h = (b - t)/n$.

An important result of the GL derivative is

$$
\begin{aligned}
D_a^\alpha f(t) &= \sum_{k=0}^{m-1} \frac{f^{(k)}(a)}{\Gamma(k+1-\alpha)} (t-a)^{k-a} \\
&+ \frac{1}{\Gamma(m-\alpha)} \int_a^t (t-\tau)^{m-1-\alpha} f^{(m)}(\tau) d\tau,
\end{aligned}
\tag{16.7}
$$

where m is the integer with $m-1 < \alpha \le m$. $m = [\alpha] + 1$ with $[\alpha]$ being the integer part of α. That is, m is greater than α but the smallest integer. For $0 < \alpha < 1$, the value of m is 1. For simplicity, the left-fractional derivative is called *fractional derivative*.

For $D^\alpha f(t_j)$ if the limit $h \to 0$ is not performed then the result is the finite GL operator

$$
D^\alpha f(t_j) = \frac{1}{h^\alpha} \sum_{k=0}^{j} (-1)^k \binom{\alpha}{k} f(t_j - kh), \quad j = 0, 1, \ldots, n.
\tag{16.8}
$$

2. Riemann–Liouville Operator

The Riemann–Liouville (RL) integral operator of order $\alpha > 0$ is given by

$$
I^\alpha f(t) = D^{-\alpha} f(t) = \frac{1}{\Gamma(\alpha)} \int_0^t (t-\tau)^{\alpha-1} f(\tau) d\tau,
\tag{16.9}
$$

where the integral is pointwise defined on $[0, \infty]$. The RL fractional derivative operator D^α, $\alpha > 0$ is defined as [76-78]

$$
\begin{aligned}
D^\alpha f(t) &= \frac{d^m}{dt^m} I^{m-\alpha} f(t) \\
&= \frac{1}{\Gamma(m-\alpha)} \frac{d^m}{dt^m} \int_0^t (t-\tau)^{m-\alpha-1} f(\tau) d\tau,
\end{aligned}
\tag{16.10}
$$

where m is the integer with $m-1 < \alpha \le m$. One can write

$$
D^\alpha I^\alpha f(t) = f(t)
\tag{16.11}
$$

and

$$
I^\alpha D^\alpha f(t) = f(t) - \sum_{k=0}^{m-1} f^{(k)}(0^+) \frac{t^k}{k!}, \quad t > 0.
\tag{16.12}
$$

3. Caputo Fractional Derivatives

For a function $f(t)$, the Caputo type fractional derivative of order $\alpha > 0$ is defined as [76-82]

$$
D^\alpha f(t) = \frac{1}{\Gamma(m-\alpha)} \int_0^t (t-\tau)^{m-\alpha-1} f^{(m)}(\tau) d\tau,
\tag{16.13}
$$

where again m is the integer with $m-1 < \alpha \le m$. This operator was introduced by Michele Caputo in his study on the theory of viscoelasticity [79]. $D^\alpha f(t)$ can also be defined by

$$
D^\alpha f(t) = I^{m-\alpha} f^{(m)}(t),
\tag{16.14}
$$

where $f^{(m)}$ is the ordinary mth derivative of f and I is the RL integral operator of order $\beta = m - \alpha > 0$.

The Caputo–Fabrizio fractional derivative of f is

$$D^{\alpha} f(t) = \frac{M(\alpha)}{1 - \alpha} \int_0^t f'(\tau) \exp\left[-\frac{\alpha(t - \tau)}{1 - \alpha}\right] d\tau, \qquad (16.15)$$

where $M(\alpha)$ is the normalization function with $M(0) = M(1) = 1$ [83]. Certain other types of fractional derivatives are Hadamard [84], Riesz [85], Sonin–Letnikov [85] and Miller–Ross [84].

For a suitable smooth $f(t)$, $t \in [a, b]$, $m - 1 < \alpha < m$ with m being a positive integer $_{RL}D^{\alpha} f(t) = {}_{GL}D^{\alpha} f(t)$ and

$$_{RL}D^{\alpha} f(t) = {}_{C}D^{\alpha} f(t) + \sum_{k=0}^{m-1} \frac{f^{(k)}(0) t^{k-\alpha}}{\Gamma(k + 1 - \alpha)}, \qquad (16.16)$$

where $_{C}D^{\alpha}$ is the Caputo fractional derivative operator.

16.3 Backward-Difference Methods

This section set up the numerical algorithms [51] for the equations of the form given in (16.1) treating D^{α} as the GL derivative. The formulas are obtained for $x(0) = 0$ and $x(0) \neq 0$.

16.3.1 The Algorithms

Consider the fractional differential equations of the form

$$D^{\alpha} x(t) = f(x(t), t), \quad x(0) = 0, \quad 0 < \alpha < 1, \qquad (16.17)$$

where D^{α} is the GL differential operator. The case $x(0) \neq 0$ will be considered later. It is desired to find the solution $x(t)$ for $t \in [0, T]$. The interval $[0, T]$ is discretized with step size h with $n + 1$ mesh points given by $t_m = mh$, $m = 0, 1, \ldots, n$, where $h = T/n$. The solution $x(t)$ at $t = t_m$ is denoted as $x(t_m)$ or simply x_m and $f(x_m, t_m)$ is denoted as f_m. In the limit of $h \to 0$, $D^{\alpha} x(t)$ is given by Eq. (16.3) with f replaced by x. Relaxing the limit $h \to 0$ and for finite value of h, $D^{\alpha} x$ is written as

$$D^{\alpha} x(t_m) = \frac{1}{h^{\alpha}} \sum_{k=0}^{m} \omega_k x(t_m - kh), \quad m = 0, 1, \ldots, n, \qquad (16.18a)$$

where

$$\omega_k = (-1)^k \binom{\alpha}{k} = \left(1 - \frac{1 + \alpha}{k}\right) \omega_{k-1}, \quad \omega_0 = 1. \qquad (16.18b)$$

For $\alpha = 0.5$, ω_k values are $\omega_0 = 1$, $\omega_1 = -0.5$, $\omega_2 = -0.125$, $\omega_3 = -0.0625$, $\ldots$, $\omega_{100} = -0.00028$, $\ldots$, $\omega_{1000} = -0.0000089$, $\ldots$. Substitution of Eq. (16.18a) in Eq. (16.17) gives

$$\frac{1}{h^{\alpha}} \sum_{k=0}^{m} \omega_k x_{m-k} = f_m. \qquad (16.19)$$

Separating the case $k = 0$ out of the summation, the above equation can be rewritten as

$$\frac{1}{h^\alpha}x_m + \frac{1}{h^\alpha}\sum_{k=1}^{m}\omega_k x_{m-k} = f_m\,.\tag{16.20}$$

That is,

$$x_m = h^\alpha f_m - \sum_{k=1}^{m}\omega_k x_{m-k},\quad m = 1,2,\ldots,n\tag{16.21}$$

with $x_0 = 0$. This is called *backward-difference algorithm* [51]. If the dependent variable x occurs explicitly in $f(x(t),t)$ then the first term in the right-side of Eq. (16.21) also contains x_m. For f containing the linear term $-cx$

$$x_m = h^\alpha(-cx_m) + h^\alpha f(t_m) - \sum_{k=1}^{m}\omega_k x_{m-k},\quad m = 1,2,\ldots,n.\tag{16.22}$$

x_m is then written as

$$x_m = \frac{1}{1+ch^\alpha}\left[h^\alpha f(t_m) - \sum_{k=1}^{m}\omega_k x_{m-k}\right],\quad m = 1,2,\ldots,n.\tag{16.23}$$

For the case of f being nonlinear in x, each subequation in Eq. (16.21) becomes nonlinear in x_m. They can be solved by applying the fixed point method. With $x_0 = 0$, the solution $x_1, x_2, \ldots, x_n$ can be computed successively from Eq. (16.21). This formula is a first-order numerical method.

Next, in each step introduce an additional starting weight ω_m' given by [51]

$$\omega_m' = \frac{1}{m^\alpha\Gamma(m-\alpha)} - \sum_{k=0}^{m}\omega_k\,.\tag{16.24}$$

This leads to the modification of the formula (16.21) as

$$x_m = h^\alpha f_m - \sum_{k=1}^{m}\omega_k x_{m-k} - \omega_m' x_0,\quad m = 1,2,\ldots,n,\tag{16.25}$$

where $x_0 = x(0)$. The error in this formula is $O(h)$. For $x(0) \neq 0$, the algorithm (16.25) can be used. For $x(0) = 0$, this algorithm becomes the one given by (16.21).

16.3.2   Numerical Examples

Let us apply the backward-difference method to a few systems. The first system is

$$D^\alpha x(t) = f(x,t) = t\,.\tag{16.26}$$

It is easy to obtain the exact analytical solution of the above equation. Fractional integration of Eq. (16.26) is written as $I^\alpha D^\alpha x(t) = I^\alpha t$. This gives

$$x(t) - \sum_{k=0}^{m'-1}x^{(k)}(0^+)\frac{t^k}{k!} = \frac{1}{\Gamma(\alpha)}\int_0^t (t-\tau)^{\alpha-1}\tau d\tau\,,\tag{16.27}$$

TABLE 16.1

x_n, x_e, $|x_e - x_n|$ of Eq.(16.26) at certain values of t computed using the backward-difference algorithm (Eq. (16.21)) with $h = 0.001$ and $x(0) = 0$.

t	x_n	x_e	$|x_e - x_n|$	t	x_n	x_e	$|x_e - x_n|$
0.1	0.02388	0.02379	0.00009	0.6	0.34983	0.34962	0.00021
0.2	0.06741	0.06728	0.00013	0.7	0.44080	0.44057	0.00023
0.3	0.12376	0.12361	0.00015	0.8	0.53852	0.53827	0.00025
0.4	0.19048	0.19031	0.00017	0.9	0.64255	0.64228	0.00027
0.5	0.26616	0.26596	0.00020	1.0	0.75253	0.75225	0.00028

where $m' = [\alpha] + 1$. For $0 < \alpha < 1$, the value of m' is 1 and integrating the integral in the above equation by parts once results in

$$x(t) - x(0) = \frac{1}{\Gamma(\alpha + 2)} t^{\alpha + 1}. \tag{16.28}$$

The exact solution of a linear fractional order differential equation with constant coefficients can be determined by applying the Laplace transform (see Problem 4 at the end of the present chapter). For $\alpha = 0.5$, the exact solution after substituting $\Gamma(2.5) = 1.5 \times 0.5 \times \Gamma(0.5)$ and $\Gamma(0.5) = \sqrt{\pi}$ is

$$x(t) = x(0) + \frac{1}{\Gamma(2.5)} t^{3/2} = x(0) + \frac{4}{3\sqrt{\pi}} t^{3/2}. \tag{16.29}$$

First choose $x(0) = 0$ for which the formula (16.21) is desired. Assume that the solution is to be determined for $t \in [0, 1]$.

For the given equation f is independent of x. Hence, in (16.21) f_m is simply $f(t_m)$ and the right-side of it not contain x_m. For $m = 1, 2, 3$ the formula (16.21) gives

$$x_1 = h^\alpha f(t_1) - \omega_1 x_0, \tag{16.30a}$$
$$x_2 = h^\alpha f(t_2) - \omega_1 x_1 - \omega_2 x_0, \tag{16.30b}$$
$$x_3 = h^\alpha f(t_3) - \omega_1 x_2 - \omega_2 x_1 - \omega_3 x_0. \tag{16.30c}$$

The solution is computed with $h = 0.001$. Table 16.1 presents the numerically computed solution x_n, the exact solution x_e and the absolute difference between them, $|x_e - x_n|$, at $t = 0.1, 0.2, \ldots, 1.0$.

Next, consider Eq. (16.26) with $x(0) = x_0 = 1$. The relevant backward-difference formula is (16.25). For each mesh point or node t_m the quantity ω'_m to be calculated that involves $\Gamma(m - \alpha)$. To compute it for $0 < \alpha < 1$ define $g(1) = \Gamma(1 - \alpha) = \Gamma(\bar{\alpha})$ with

$$\Gamma(\bar{\alpha}) = \int_0^\infty z^{\bar{\alpha} - 1} e^{-z} dz. \tag{16.31}$$

Then, to find $\Gamma(m - \alpha)$, consider the relations

$$g(2) = \Gamma(2 - \alpha) = (1 - \alpha)\Gamma(1 - \alpha) = (1 - \alpha)g(1), \tag{16.32a}$$
$$g(3) = \Gamma(3 - \alpha) = (2 - \alpha)(1 - \alpha)\Gamma(1 - \alpha) = (2 - \alpha)g(2). \tag{16.32b}$$

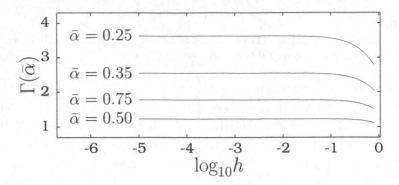

FIGURE 16.1
The numerically computed $\Gamma(\bar{\alpha})$ versus $\log_{10} h$ for four values of $\bar{\alpha}$.

In general, with $g(1) = \Gamma(1 - \alpha)$,

$$g(m) = \Gamma(m - \alpha) = (m - \alpha)g(m - 1), \quad m = 2, 3, \dots . \tag{16.33}$$

In this way, $\Gamma(m - \alpha)$ can be computed. For $\alpha = 0.5$, the exact value of $\Gamma(\bar{\alpha}) = \Gamma(0.5) = \sqrt{\pi} \approx 1.77245$. For other values of α in the interval $[0, 1]$ the values of the gamma function need to be evaluated numerically. *Can one apply the composite quadrature formulas to evaluate the gamma function given by Eq. (16.31)?* Here $0 < \alpha < 1$ for which $0 < \bar{\alpha} < 1$ and $\bar{\alpha} - 1$ is negative. The integrand in Eq. (16.31) has a singularity at $z = 0$. It is easy to overcome this difficulty. Let us perform integration by parts once and obtain

$$\Gamma(\bar{\alpha}) = \frac{1}{\bar{\alpha}} \int_0^\infty z^{\bar{\alpha}} e^{-z} dz \tag{16.34}$$

which has no singularity. The integrand $z^{\bar{\alpha}} e^{-z}$ rapidly decays to zero. So, the integral in Eq. (16.34) can be evaluated numerically, for example, applying the composite trapezoidal rule. With the application of this rule, Eq. (16.34) takes the form

$$\Gamma(\bar{\alpha}) \approx \frac{h}{\bar{\alpha}} \sum_{k=1}^\infty z_k^{\bar{\alpha}} e^{-z_k}, \tag{16.35}$$

where h is the step size in z and $z_k = kh$, $k = 1, 2, \dots$. In practice, the summation can be stopped, for example, when $z_k > 10$ and $z_k^{\bar{\alpha}} e^{-z_k} < 10^{-6}$.

Using Eq. (16.35), the values of the gamma function are computed for a few values of $\bar{\alpha}$ for a set of values of h in the interval $\left[10^{-1}, 10^{-7}\right]$. Figure 16.1 depicts the variation of $\Gamma(\bar{\alpha})$ with $\log_{10} h$. For each value of $\bar{\alpha}$, $\Gamma(\bar{\alpha})$ approaches a constant value with h. The values of $\Gamma(\bar{\alpha})$ obtained are

$$\Gamma(0.25) = 3.62560, \quad \Gamma(0.35) = 2.54614, \quad \Gamma(0.5) = 1.77245, \quad \Gamma(0.75) = 1.22541. \tag{16.36}$$

For further clarity, in Fig. 16.2 $\log_{10} E$ is plotted as a function of $\log_{10} h$ for $\bar{\alpha} = 0.5$ where E is the absolute difference between the exact value of $\Gamma(0.5) = \sqrt{\pi}$ and the numerically computed value of $\Gamma(0.5)$. For $h = 10^{-4}$ the error is $< 10^{-5}$. So, this value of h can be chosen for the numerical computation of the gamma function. However, before using the numerically computed value of the gamma function it is necessary to ensure the desired accuracy in it by suitably choosing the value of h and the tolerance in $z_k^{\bar{\alpha}} e^{-z_k}$.

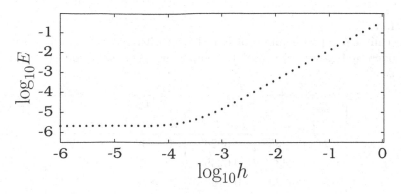

FIGURE 16.2
Variation of the error E in the numerical computation of $\Gamma(\bar{\alpha})$ with $\log_{10} h$ for $\bar{\alpha} = 0.5$.

In Table 16.2, $|x_e - x_n|$ for three values of h are presented for selective values of t along with the exact solution. The error is found to decrease with decrease in the value of h.

In Eq. (16.26) the function $f(x(t), t)$ is independent of $x(t)$. Now, deal with the case of f being a linear function of $x(t)$. A simple system of this form is

$$D^\alpha x(t) = f(x(t), t) = -x(t). \quad x(0) = x_0, \ 0 < \alpha < 1. \tag{16.37}$$

Its exact solution is [56] (also see Problem 4 at the end of this chapter)

$$x(t) = E_\alpha(-t^\alpha) = x_0 \sum_{k=0}^{\infty} \frac{(-t^\alpha)^k}{\Gamma(\alpha k + 1)}, \tag{16.38}$$

where $E_\alpha(z)$ is the Mittag–Leffler function of order α. Rewrite the algorithm (16.25) as

$$x_m = -h^\alpha x_m + h^\alpha f(t_m) - \sum_{k=1}^{m} \omega_k x_{m-k} - \omega'_m x_0. \tag{16.39}$$

TABLE 16.2
Comparison of $|x_e - x_n|$ for three values of h at ten values of t for Eq. (16.26) with $x(0) = 1$.

t	x_e	$|x_e - x_n|$ for $h = 10^{-3}$	$h = 10^{-4}$	$h = 10^{-5}$
0.1	1.02379	0.05350	0.01684	0.00532
0.2	1.06728	0.03764	0.01190	0.00376
0.3	1.12361	0.03065	0.00971	0.00307
0.4	1.19031	0.02648	0.00840	0.00266
0.5	1.26596	0.02364	0.00751	0.00238
0.6	1.34962	0.02154	0.00685	0.00217
0.7	1.44057	0.01990	0.00634	0.00201
0.8	1.53827	0.01859	0.00593	0.00188
0.9	1.64228	0.01749	0.00559	0.00177
1.0	1.75225	0.01656	0.00530	0.00168

TABLE 16.3
The numerically computed solution of Eq. (16.37) with $\alpha = 0.5$ for three values of h. The algorithm given by Eq. (16.40) is used to compute x_n. The exact solutions x_e are also given for comparison.

t	x_e	x_n for $h = 10^{-3}$	$h = 10^{-4}$	$h = 10^{-5}$
0.1	0.72358	0.69200	0.71360	0.72041
0.2	0.64379	0.62550	0.63798	0.64194
0.3	0.59202	0.57908	0.58790	0.59071
0.4	0.55361	0.54362	0.55042	0.55259
0.5	0.52316	0.51506	0.52057	0.52233
0.6	0.49803	0.49124	0.49586	0.49733
0.7	0.47670	0.47089	0.47484	0.47611
0.8	0.45825	0.45317	0.45662	0.45773
0.9	0.44202	0.43753	0.44058	0.44156
1.0	0.42758	0.42357	0.42630	0.42717

That is,

$$x_m = \frac{1}{1+h^\alpha}\left[h^\alpha f_m - \sum_{k=1}^{m}\omega_k x_{m-k} - \omega'_m x_0\right], \quad m = 1, 2, \ldots, n, \tag{16.40}$$

where $f_m = f(t_m)$. For Eq. (16.37) the term $h^\alpha f_m = 0$. For $h = 10^{-3}$, 10^{-4} and 10^{-5} and for $\alpha = 0.5$ with $x_0 = 1$ the numerically computed solutions x_n are given in Table 16.3. The exact solution is also given for comparison. x_n approaches the exact solution with decreasing the value of h.

16.4 Fractional Euler Method

This section develops a generalization of the classical Euler method applicable for fractional order differential equations. This method has been proposed by Odibat and Momani [56].

16.4.1 Basic Definitions

The RL fractional integral operator I^α of order $\alpha > 0$ for a function $f(t)$ with $t > 0$ is given by Eq. (16.9). For the properties of I^α one may refer to the refs. [76,86,87]. Two notable properties of I^α and D^α are given by Eqs. (16.11) and (16.12). The RL fractional order derivative appears as the left-inverse of the RL fractional integral. The Caputo fractional derivative given by Eq. (16.13) is a modified version of the RL fractional derivative given by Eq. (16.10). From Eq. (16.12), for $0 < \alpha < 1$,

$$I^\alpha D^\alpha f(t) = f(t) - f(0^+). \tag{16.41}$$

By definition

$$I^\alpha D^\alpha f(t) = \frac{1}{\Gamma(\alpha)} \int_0^t (t-\tau)^{\alpha-1} D^\alpha f(\tau) d\tau. \tag{16.42}$$

According to the intermediate-value theorem for integrals (refer Section 1.6),

$$\int_0^t (t-\tau)^{\alpha-1} D^\alpha f(\tau) d\tau = D^\alpha f(\xi) \int_0^t (t-\tau)^{\alpha-1} d\tau \doteq D^\alpha f(\xi) \frac{t^\alpha}{\alpha}. \tag{16.43}$$

Then,

$$I^\alpha D^\alpha f(t) = \frac{1}{\alpha \Gamma(\alpha)} D^\alpha f(\xi) t^\alpha = \frac{1}{\Gamma(\alpha+1)} D^\alpha f. \tag{16.44}$$

Comparison of Eqs. (16.41) and (16.44) gives

$$f(t) = f(0^+) + \frac{1}{\Gamma(\alpha)} D^\alpha f(\xi) \cdot t^\alpha. \tag{16.45}$$

Denoting $D^{n\alpha} = D^\alpha D^\alpha \cdots D^\alpha$ (n times) the following two relations can be easily proved [56]:

$$I^{n\alpha} D^{n\alpha} f(t) - I^{(n+1)\alpha} D^{(n+1)\alpha} f(t) = \frac{t^{n\alpha}}{\Gamma(n\alpha+1)} D^{n\alpha} f(0^+), \tag{16.46}$$

and

$$f(t) - \sum_{k=0}^{n} \frac{t^{k\alpha}}{\Gamma(k\alpha+1)} D^{k\alpha} f(0^+) + \frac{D^{(n+1)\alpha} f(\xi)}{\Gamma[(n+1)\alpha+1]} t^{(n+1)\alpha}, \quad 0 < \xi < t, \tag{16.47}$$

The $f(t)$ given by Eq. (16.47) is the generalized Taylor's formula.

The modified trapezoidal rule for the numerical computation of $I^\alpha f(t)$ in the interval $t \in [0, T]$ with step size $h = T/n$ and the nodes $t_m = mh$, $m = 0, 1, \ldots, n$ is (refer Eq. (10.85))

$$
\begin{aligned}
(I^\alpha f(t))(T) ={}& \frac{h^\alpha}{\Gamma(2+\alpha)} \Bigg\{ \left[(n-1)^{\alpha+1} - (n-\alpha-1)n^\alpha \right] f(0) \\
&+ \sum_{m=1}^{n-1} \Big[(n-m+1)^{\alpha+1} - 2(n-m)^{\alpha+1} \\
&+ (n-m-1)^{\alpha+1} \Big] f(t_m) + f(T) \Bigg\}.
\end{aligned} \tag{16.48}
$$

16.4.2   Fractional Euler Formula

The initial-value problem is

$$D^\alpha x(t) = f(x(t), t), \quad x(0) = x_0, \ 0 < \alpha \le 1, \ t > 0. \tag{16.49}$$

The solution is to be determined for $t \in [0, T]$ at the set of nodes $t_m = mh$, $m = 0, 1, \ldots, n$, $h = T/n$. Assume that $x(t)$, $D^\alpha x(t)$ and $D^{2\alpha} x(t)$ are continuous in the interval $t \in [0, T]$.

Expansion of $x(t)$ as the generalized Taylor's formula (16.47) about $t = t_0 = 0$ (with $n = 1$) is

$$x(t) = x(t_0) + \frac{t^\alpha}{\Gamma(\alpha + 1)} \left(D^\alpha x(t)\right)(t_0) + \frac{t^{2\alpha}}{\Gamma(2\alpha + 1)} \left(D^{2\alpha} x(t)\right)(\xi_1), \qquad (16.50)$$

where ξ_1 is a value that exists for each value of t. Suppose $(D^\alpha x(t)) t_0 = f(x(t_0), t_0)$. Then, at $t = t_1 = h$, Eq. (16.50) gives

$$x(t_1) = x(t_0) + \frac{h^\alpha}{\Gamma(\alpha + 1)} f(x(t_0), t_0) + \frac{h^{2\alpha}}{\Gamma(2\alpha + 1)} \left(D^{2\alpha} x(t_1)\right)(\xi_1). \qquad (16.51)$$

For small h, the last term in Eq. (16.51) can be neglected so that

$$x(t_1) = x(t_0) + \frac{h^\alpha}{\Gamma(\alpha + 1)} f(x(t_0), t_0). \qquad (16.52)$$

In general, from t_m to t_{m+1},

$$x_{m+1} = x_m + \frac{h^\alpha}{\Gamma(\alpha + 1)} f(x_m, t_m), \qquad (16.53a)$$

$$t_{m+1} = t_m + h, \quad m = 0, 1, \ldots, n - 1. \qquad (16.53b)$$

When $\alpha = 1$, the above formula reduces to the well-known classical Euler formula.

Let us proceed to develop an algorithm making use of the modified trapezoidal rule and the formula (16.53) [56]. Write $x(t)$ as

$$x(t) = I^\alpha f(x(t), t) + x(0). \qquad (16.54)$$

At $t = t_1$,

$$x(t_1) = (I^\alpha f(x(t), t))(t_1) + x(0). \qquad (16.55)$$

The modified trapezoidal rule (16.48) for the interval $[t_0 = 0, t_1 = h]$ becomes $(n = 1)$

$$(I^\alpha f(x(t), t))(t_1) = \frac{\alpha h^\alpha}{\Gamma(\alpha + 2)} f(x(t_0), t_0) + \frac{h^\alpha}{\Gamma(\alpha + 2)} f(x(t_1), t_1). \qquad (16.56)$$

Then,

$$x(t_1) = x(0) + \frac{\alpha h^\alpha}{\Gamma(\alpha + 2)} f(x(t_0), t_0) + \frac{h^\alpha}{\Gamma(\alpha + 2)} f(x(t_1), t_1). \qquad (16.57)$$

Note that the right-side of Eq. (16.57) also contains $x(t_1)$. Therefore, approximate $x(t_1)$ in the right-side of the above equation by Eq. (16.52). This gives

$$\begin{aligned}
x(t_1) &= x(0) + \frac{\alpha h^\alpha}{\Gamma(\alpha + 2)} f(x(t_0), t_0) \\
&\quad + \frac{h^\alpha}{\Gamma(\alpha + 2)} f\left(x(t_0) + \frac{h^\alpha}{\Gamma(\alpha + 1)} f(x(t_0), t_0), t_1\right).
\end{aligned} \qquad (16.58)$$

Next, generalize the Eq. (16.58). For the interval $[t_0, t_{m+1}]$, $m = 0, 1, 2, \ldots, n - 1$

$$\begin{aligned}
x(t_{m+1}) &= x(0) + \frac{h^\alpha}{\Gamma(\alpha + 2)} \left[m^{\alpha+1} - (m - \alpha)(m + 1)^\alpha\right] f(x(t_0), t_0) \\
&\quad + \frac{h^\alpha}{\Gamma(\alpha + 2)} \sum_{k=1}^{m} \left[(m - k + 2)^{\alpha+1} - 2(m - k + 1)^{\alpha+1}\right. \\
&\quad \left. + (m - k)^{\alpha+1}\right] f(x(t_k), t_k) \\
&\quad + \frac{h^\alpha}{\Gamma(\alpha + 2)} f(x(t_{m+1}), t_{m+1}).
\end{aligned} \qquad (16.59)$$

Define

$$a_{k,m+1} = \begin{cases} m^{\alpha+1} - (m-\alpha)(m+1)^{\alpha}, & \text{if } k = 0 \\ (m-k+2)^{\alpha+1} - 2(m-k+1)^{\alpha+1} \\ \quad + (m-k)^{\alpha+1}, & \text{if } 1 \le k \le m \\ 1, & \text{if } k = m+1. \end{cases} \tag{16.60}$$

Then, the second term in the right-side of (16.60) can be brought to the third term with the summation running index starting from 0. Thus,

$$x(t_{m+1}) = x(0) + \frac{h^{\alpha}}{\Gamma(\alpha+2)} \sum_{k=0}^{m} a_{k,m+1} f(x(t_k), t_k)$$
$$+ \frac{h^{\alpha}}{\Gamma(\alpha+2)} f(x(t_{m+1}), t_{m+1}). \tag{16.61}$$

In Eq. (16.61) the unknown quantity $x(t_{m+1})$ occurs on both sides. If f is linear in x then from Eq. (16.61) an expression for $x(t_{m+1})$ can be written. In general, for nonlinear function one cannot solve Eq. (16.61) directly for $x(t_{m+1})$. In this case, a preliminary approximation can be sought. Suppose, approximate $x(t_{m+1})$ by the Euler formula (16.53a). Then,

$$x(t_{m+1}) = x(0) + \frac{h^{\alpha}}{\Gamma(\alpha+2)} \sum_{k=0}^{m} a_{k,m+1} f(x(t_k), t_k)$$
$$+ \frac{h^{\alpha}}{\Gamma(\alpha+2)} f\left(x(t_m) + \frac{h^{\alpha}}{\Gamma(\alpha+1)} f(x(t_m), t_m), t_{m+1}\right). \tag{16.62}$$

This formula is known as the *fractional Euler method* or the *fractional Euler algorithm*. Instead of approximating $x(t_{m+1})$ in the right-side of Eq. (16.61) by the Euler formula (16.53a) an iterative process can be used. This is considered in the next section.

16.4.3   Examples

Let us apply the fractional Euler method to two equations, one with order α, $0 < \alpha < 1$ and the other with order greater than α.

1. Example 1

First, apply the algorithm to the system (16.37) where f is a linear function of x. $\Gamma(\alpha+2)$ is calculated by numerically evaluating the integral in Eq. (16.5) by the composite trapezoidal rule with step size 0.0001. Table 16.4 shows the numerically computed solution x_n and the exact solution (16.38) for $\alpha = 0.5$ and 0.75. The step size used is $h = 0.001$. x_n is close to the exact solution x_e. Accuracy can be improved by decreasing the value of h.

2. Example 2

Consider the linear and inhomogeneous equation

$$D^2 x + D^{1.5} x + x = 1 + t, \quad x(0) = x'(0) = 1. \tag{16.63}$$

A. Exact Solution

The exact solution of Eq. (16.63) can be obtained by applying the Laplace transform

TABLE 16.4

The numerical solution of Eq. (16.37) computed for $\alpha = 0.5$ and 0.75 with step size $h = 0.001$ and $x(0) = 1$ by employing the fractional Euler algorithm.

t	$\alpha = 0.5$		$\alpha = 0.75$	
	x_n	x_e	x_n	x_e
0.1	0.72399	0.72358	0.82826	0.82825
0.2	0.64412	0.64379	0.73259	0.73259
0.3	0.59229	0.59202	0.66034	0.66034
0.4	0.55384	0.55361	0.60212	0.60212
0.5	0.52336	0.52316	0.55360	0.55360
0.6	0.49821	0.49803	0.51228	0.51229
0.7	0.47687	0.47670	0.47655	0.47656
0.8	0.45840	0.45825	0.44529	0.44529
0.9	0.44216	0.44202	0.41768	0.41768
1.0	0.42771	0.42758	0.39310	0.39311

[88]. The Laplace transform of Caputo fractional derivatives of order $\alpha > 0$ with $m = [\alpha] + 1$ is

$$
\begin{aligned}
L\left[D^\alpha f(t)\right] &= L\left[I^{m-\alpha} f^{(m)}(t)\right] \\
&= \frac{L\left[f^{(m)}(t)\right]}{s^{m-\alpha}} \\
&= \frac{s^m F(s) - s^{m-1} f(0) - \cdots - f^{(m-1)}(0)}{s^{m-\alpha}},
\end{aligned}
\tag{16.64}
$$

where $L[f(t)] = F(s)$. Now, taking the Laplace transform on both sides of Eq. (16.63) gives

$$
s^2 F(s) - sx(0) - x'(0) + \frac{1}{s^{1/2}}\left[s^2 F(s) - sx(0) - x'(0)\right] + F(s) = \frac{1}{s} + \frac{1}{s^2}.
\tag{16.65}
$$

Simplifying this equation results in

$$
\left(s^2 + s^{3/2} + 1\right) F(s) = \left(s^2 + s^{3/2} + 1\right)\left(\frac{1}{s} + \frac{1}{s^2}\right).
\tag{16.66}
$$

That is,

$$
F(s) = \frac{1}{s} + \frac{1}{s^2}.
\tag{16.67}
$$

Taking the inverse Laplace transform gives the solution as $x(t) = 1 + t$.

B. Equivalent System of Equations

To numerically solve Eq. (16.63) write it in the form of system of equations with each equation of the form of $D^\beta x = f(x,t)$, $0 < \beta < 1$. Equation (16.63) is rewritten as

$$
\begin{aligned}
D^{1/2} x_1 &= x_2, & \text{(16.68a)} \\
D^{1/2} x_2 &= x_3, & \text{(16.68b)} \\
D^{1/2} x_3 &= x_4, & \text{(16.68c)} \\
D^{1/2} x_4 &= -x_4 - x_1 + 1 + t. & \text{(16.68d)}
\end{aligned}
$$

TABLE 16.5
The variation of $|x_e - x_n|$ with the step size h for Eq. (16.63). The numerical solution is computed by the fractional Euler method.

t	$h = 10^{-2}$	$h = 10^{-3}$	$h = 10^{-4}$	$h = 10^{-5}$
		$|x_e - x_n|$ for		
0.1	0.00666	0.00079	0.00008	0.00001
0.2	0.00705	0.00079	0.00008	0.00001
0.3	0.00704	0.00077	0.00008	0.00001
0.4	0.00684	0.00073	0.00008	0.00001
0.5	0.00652	0.00069	0.00007	0.00001
0.6	0.00610	0.00064	0.00007	0.00001
0.7	0.00560	0.00058	0.00006	0.00001
0.8	0.00502	0.00051	0.00005	0.00001
0.9	0.00438	0.00044	0.00005	0.00001
1.0	0.00369	0.00036	0.00004	0.00001

It is easy to find the initial conditions $x_i(0)$, $i = 1, 2, 3, 4$ corresponding to the exact analytical solution $x_1(t) = 1 + t$. From this solution,

$$x_1(0) = 1, \tag{16.69a}$$
$$x_3(0) = Dx_1(t)|_{t=0}$$
$$= 1. \tag{16.69b}$$

Using the definition of $D^\alpha f(t)$, $0 < \alpha < 1$

$$x_2 = D^\alpha(1+t)$$
$$= \frac{1}{\Gamma(1-\alpha)} \int_0^t (t-\tau)^{-\alpha} d\tau$$
$$= \frac{1}{\Gamma(2-\alpha)} t^{1-\alpha}, \tag{16.69c}$$
$$x_4 = D^{1+\alpha}(1+t)$$
$$= D^\alpha 1$$
$$= 0. \tag{16.69d}$$

Thus, $x_2(0) = 0$ and $x_4(0) = 0$.

C. Numerical Solution

With the initial conditions

$$x_1(0) = 1, \quad x_2(0) = 0, \quad x_3(0) = 1, \quad x_4(0) = 0 \tag{16.70}$$

apply the fractional Euler method to each of the subequations of Eqs. (16.68) and compute the solutions $x_i(t)$, $i = 1, 2, 3, 4$ for $t \in [0, 1]$ with a fixed h. Table 16.5 presents the values of $|x_e - x_n|$ for four values of h.

A fractional order differential equation with the higher order of the derivative greater than 1 can be numerically solved by equivalently rewriting it as a system of equations with each subequation with appropriate order in the range $0 < \beta < 1$. Each subequation need not be of the same fractional order.

16.5   Adams–Bashforth–Moulton Method

Diethelm, Ford and Freed [48] developed an algorithm to solve fractional differential equations. Their method is a generalization of the Adams–Bashforth–Moulton method of ordinary differential equations with integer derivatives and is applicable to a wide class of fractional differential equations. This method is applied to certain linear and nonlinear systems [61]. Error analysis on this method is performed and reported in [49,69].

16.5.1   Review of the Classical Method

For the equation $Dx(t) = f(x(t), t)$ with $x(0) = x_0$ divide the time interval $[0, T]$ into n nodes with step size $h = T/n$. The nodes are $t_m = mh$, $m = 0, 1, \ldots, n$. A unique solution is assumed to exist in the interval $[0, T]$. From the known $x(t_m)$ an approximation of $x(t_{m+1})$ is written as

$$x(t_{m+1}) = x(t_m) + \int_{t_m}^{t_{m+1}} f(x(\tau), \tau)\mathrm{d}\tau. \tag{16.71}$$

Use the trapezoidal rule for the integral in the above equation. Then,

$$x(t_{m+1}) = x(t_m) + \frac{h}{2}\left[f(x(t_m), t_m) + f(x(t_{m+1}), t_{m+1})\right]. \tag{16.72}$$

The unknown quantity $x(t_{m+1})$ occurs on both sides of this equation. If f is linear in x then from (16.72) an expression for $x(t_{m+1})$ can be written. In general, for a nonlinear function Eq. (16.72) cannot be solved directly for $x(t_{m+1})$. In this case, an iterative scheme is useful with an approximation. This approximation is usually termed as *predictor* and denoted as x_{m+1}^{p}. For determining the predictor use the rectangle rule (refer the subsection 10.2.3). The rectangle rule is

$$\int_{t_m}^{t_{m+1}} f(x(\tau), \tau)\mathrm{d}\tau = (t_{m+1} - t_m)f(x(t_m), t_m) = hf(x(t_m), t_m). \tag{16.73}$$

Then, Eq. (16.71) gives (with $x^{\mathrm{p}}(t_{m+1})$ for $x(t_{m+1})$)

$$x^{\mathrm{p}}(t_{m+1}) = x(t_m) + hf(x(t_m), t_m). \tag{16.74}$$

This is known as the *Euler formula* or *one-step Adams–Bashforth method* [90].

With the predictor $x^{\mathrm{p}}(t_{m+1})$, Eq. (16.72) becomes

$$x(t_{m+1}) = x(t_m) + \frac{h}{2}\left[f(x(t_m), t_m) + f(x^{\mathrm{p}}(t_{m+1}), t_{m+1})\right]. \tag{16.75}$$

This is the *one-step Adams–Bashforth–Moulton method*. The order of convergence of this method is 2.

16.5.2 Construction of the Algorithm

The equation of our interest is

$$D^\alpha x(t) = f(x(t), t), \ \alpha > 0 \tag{16.76}$$

subjected to the initial conditions $x^{(k)}(0) = x_0^{(k)}$, $k = 0, 1, \ldots, [\alpha]$ where $[\alpha]$ is the integer part of α and $x^{(k)}$ is the ordinary kth derivative of x. D^α is chosen as the Caputo type. Consideration of the integration of Eq. (16.76) according to Eq. (10.79) gives

$$x(t) = \sum_{k=0}^{[\alpha]} x_0^{(k)} \frac{t^k}{k!} + \frac{1}{\Gamma(\alpha)} \int_0^t (t - \tau)^{\alpha-1} f(x(\tau), \tau) d\tau. \tag{16.77}$$

This equation is a *second-order Volterra equation*.

A. Corrector Formula

For discretized t, Eq. (16.77) takes the form

$$x(t_{m+1}) = \sum_{k=0}^{[\alpha]} \frac{t_{m+1}^k}{k!} x_0^{(k)} + \int_0^{t_{m+1}} (t_{m+1} - \tau)^{\alpha-1} f(x(\tau), \tau) d\tau. \tag{16.78}$$

To evaluate the integral in the above equation, the product trapezoidal rule can be applied. This gives

$$\int_0^{t_{m+1}} (t_{m+1} - \tau)^{\alpha-1} f(x(\tau), \tau) d\tau$$
$$\approx \int_0^{t_{m+1}} (t_{m+1} - \tau)^{\alpha-1} \tilde{f}_{m+1}(x(\tau), \tau) d\tau, \tag{16.79}$$

where $\tilde{f}_{m+1}$ is the piecewise linear interpolant for f. Then,

$$\int_0^{t_{m+1}} (t_{m+1} - \tau)^{\alpha-1} \tilde{f}_{m+1} d\tau$$
$$= \frac{h^\alpha}{\alpha(\alpha+1)} \sum_{k=0}^{m+1} a_{k,m+1} f(x(t_k), t_k), \tag{16.80a}$$

where

$$a_{k,m+1} = \begin{cases} m^{\alpha+1} - (m - \alpha)(m + 1)^\alpha, & \text{if } k = 0 \\ (m - k + 2)^{\alpha+1} - 2(m - k + 1)^{\alpha+1} + (m - k)^{\alpha+1}, & \text{if } 1 \le k \le m \\ 1, & \text{if } k = m + 1. \end{cases} \tag{16.80b}$$

Thus, the *corrector formula* is

$$x(t_{m+1}) = \sum_{k=0}^{[\alpha]-1} \frac{t_{m+1}^k}{k!} x_0^{(k)} + \frac{h^\alpha}{\Gamma(\alpha+2)} \sum_{k=0}^m a_{k,m+1} f(x(t_k), t_k)$$
$$+ \frac{h^\alpha}{\Gamma(\alpha+2)} f(x^P(t_{m+1}), t_{m+1}), \tag{16.81}$$

where $\alpha > 0$. Compare this formula with the fractional Euler formula (16.62) in the previous section for the case $0 < \alpha < 1$.

B. Predictor Formula

Next, procced to find an approximation for $x^{\mathrm{p}}(t_{m+1})$. Now, approximate the integral in Eq. (16.77) by the product rectangle rule

$$\int_0^{t_{m+1}} (t_{m+1} - \tau)^{\alpha-1} f(x(\tau), \tau) \mathrm{d}\tau \approx \frac{h^\alpha}{\alpha} \sum_{k=0}^m b_{k,m+1} f(x(t_k), t_k), \qquad (16.82a)$$

where

$$b_{k,m+1} = (m+1-k)^\alpha - (m-k)^\alpha. \qquad (16.82b)$$

Then, the predicted value $x^{\mathrm{p}}(t_{m+1})$ is given by the fractional Adams–Bashforth method as

$$x^{\mathrm{p}}(t_{m+1}) = \sum_{k=0}^{[\alpha]} \frac{t_{m+1}^k}{k!} x_0^{(k)} + \frac{h^\alpha}{\Gamma(\alpha+1)} \sum_{k=0}^m b_{k,m+1} f(x(t_k), t_k). \qquad (16.83)$$

The combined algorithm given by Eqs. (16.81) and (16.83) is called the *fractional Adams–Bashforth–Moulton (FABM) method*.

A detailed error analysis has shown that the error behaves as [49]

$$\max|X(t_k) - x(t_k)| = O(h^p), \qquad (16.84)$$

where $k = 0, 1, 2, \ldots, n$, $p = \min(2, 1+\alpha)$ and X and x denote the exact solution and numerically predicted solution, respectively.

16.5.3   Example

Let us apply the FABM method to Eq. (16.37). In developing a program to implement the ABM method, $\Gamma(\alpha+1)$ can be computed by evaluating the integral in Eq. (16.5) numerically by the composite trapezoidal rule with step size $h = 0.0001$. Then, $\Gamma(\alpha+2)$ can be calculated using $\Gamma(\alpha+2) = (\alpha+1)\Gamma(\alpha+1)$. For $0 < \alpha < 1$, $[\alpha] = 0$ and the first term in the right-side of Eqs. (16.81) and (16.83) become simply $x_0^{(0)} = x(0) = 1$.

The numerical solution is computed for $t \in [0, 1]$ for $h = 10^{-2}$, 10^{-3} and 10^{-4} for $\alpha = 0.5$. The result is summarized in Table 16.6. The numerically computed solution is in very good agreement with the exact solution for even $h = 10^{-2}$. For a value of t, the numerical solutions computed for the above-mentioned three values of h are all almost the same. $|x_{\mathrm{e}} - x_{\mathrm{n}}|$ is computed for $\alpha = 0.5$, 0.75 and 1 with $h = 0.01$. Table 16.7 shows the result. The quantity $|x_{\mathrm{e}} - x_{\mathrm{n}}|$ is < 0.00003 for $\alpha = 0.5$ while it is < 0.00002 for $\alpha = 0.75$ and 1.

Compare the fractional Euler formula (16.62) and the FABM formula given by Eqs. (16.81) and (16.83). In the fractional Euler method $x(t_{m+1})$ in the right-side of Eq. (16.61) is approximated simply by

$$x(t_m) + (h^\alpha/\Gamma(\alpha+1)) f(x(t_m), t_m), \qquad (16.85)$$

that is, approximated using the value of f at only one node. But in the FABM method, from Eq. (16.83) note that for the solution at t_{m+1} the values of f at the nodes t_0 to t_m are used. FABM involves more calculations than the fractional Euler method. Comparison of the errors involved in these two methods for Eq. (16.37) is left as an exercise to the readers.

TABLE 16.6
The numerical solutions x_n of the equation $D^\alpha x = -x$, $x(0) = 1$ for $\alpha = 0.5$ computed by the Adams–Bashforth–Moulton method with the step sizes $h = 10^{-2}$, 10^{-3} and 10^{-4}.

t	x_e	x_n for $h = 0.01$	x_n for $h = 0.001$	x_n for $h = 0.0001$
0.1	0.72358	0.72354	0.72358	0.72358
0.2	0.64379	0.64380	0.64379	0.64379
0.3	0.59202	0.59205	0.59202	0.59202
0.4	0.55361	0.55364	0.55361	0.55361
0.5	0.52316	0.52319	0.52316	0.52316
0.6	0.49803	0.49806	0.49803	0.49802
0.7	0.47670	0.47674	0.47670	0.47670
0.8	0.45825	0.45828	0.45825	0.45825
0.9	0.44202	0.44205	0.44202	0.44202
1.0	0.42758	0.42761	0.42758	0.42758

16.6 A Two-Step Adams–Bashforth Method

In [63] Gnitchogna and Atangana proposed an algorithm for fractional order partial and ordinary differential equations and is presented in this section.

16.6.1 Derivation of the Algorithm

Consider the general fractional partial differential equation of the form

$$D_t^\alpha u(x,t) = A_L u(x,t) + A_N u(x,t),\qquad(16.86)$$

where A_L and A_N are a linear operator and a nonlinear operator, respectively, and D_t^α is the Caputo type fractional partial derivative. Taking the Laplace transform on both sides

TABLE 16.7
The variations of $|x_e - x_n|$ for the equation $D^\alpha x = -x$, $x(0) = 1$ for three values of α. x_n is computed by the Adams–Bashforth–Moulton method with the step size $h = 10^{-2}$.

t	$|x_e - x_n|$ for $\alpha = 0.5$	$\alpha = 0.75$	$\alpha = 1.0$	t	$|x_e - x_n|$ for $\alpha = 0.5$	$\alpha = 0.75$	$\alpha = 1.0$
0.1	0.00004	0.00002	0.00000	0.6	0.00003	0.00001	0.00001
0.2	0.00001	0.00001	0.00000	0.7	0.00004	0.00001	0.00000
0.3	0.00003	0.00000	0.00000	0.8	0.00003	0.00002	0.00000
0.4	0.00003	0.00001	0.00000	0.9	0.00003	0.00002	0.00001
0.5	0.00003	0.00001	0.00001	1.0	0.00003	0.00001	0.00001

of the above equation with respect to the space variable x gives

$$D_t^\alpha u(t) = L\left[A_L u(x,t) + A_N u(x,t)\right] = F(u,t), \tag{16.87}$$

where $u(t) = u(s,t)$. Application of the Caputo fractional integral operator on (16.87) gives

$$u(t) = u(t_0) + \frac{1}{\Gamma(\alpha)} \int_0^t (t-\tau)^{\alpha-1} F(u,\tau) d\tau. \tag{16.88}$$

Discretization of time as t_0, t_1, ... with step size h leads to the equation

$$u_m = u(t_m) = u_0 + \frac{1}{\Gamma(\alpha)} \int_0^{t_m} (t_m-\tau)^{\alpha-1} F(u,\tau) d\tau. \tag{16.89}$$

Then,

$$\begin{aligned}
u_{m+1} - u_m &= \frac{1}{\Gamma(\alpha)} \int_0^{t_{m+1}} (t_{m+1}-\tau)^{\alpha-1} F(u,\tau) d\tau \\
&\quad - \frac{1}{\Gamma(\alpha)} \int_0^{t_m} (t_m-\tau)^{\alpha-1} F(u,\tau) d\tau.
\end{aligned} \tag{16.90}$$

Since t is discretized

$$\begin{aligned}
u_{m+1} - u_m &= \frac{1}{\Gamma(\alpha)} \sum_{k=0}^{m} \int_{t_k}^{t_{k+1}} (t_{m+1}-\tau)^{\alpha-1} F d\tau \\
&\quad - \frac{1}{\Gamma(\alpha)} \sum_{k=0}^{m-1} \int_{t_k}^{t_{k+1}} (t_m-\tau)^{\alpha-1} F d\tau \\
&= \frac{1}{\Gamma(\alpha)} (I_1 - I_2).
\end{aligned} \tag{16.91}$$

To evaluate the integrals I_1 and I_2, replace F by the Lagrange polynomial approximation

$$F(u,t) \approx P(t) = \frac{t - t_{m-1}}{h} F_m + \frac{t - t_m}{h} F_{m-1}, \tag{16.92}$$

where $F_m = F(u,t_m)$. The first integral I_1 in Eq. (16.91) then becomes

$$\begin{aligned}
I_1 &= \sum_{k=0}^{n} \left[\frac{F_m}{h} \int_{t_k}^{t_{k+1}} (t_{m+1}-t)^{\alpha-1}(t - t_{m-1}) dt \right. \\
&\quad \left. - \frac{F_{m-1}}{h} \int_{t_k}^{t_{k+1}} (t_{m+1}-t)^{\alpha-1}(t - t_m) dt \right].
\end{aligned} \tag{16.93}$$

Carrying out the integration, gives

$$\begin{aligned}
\int_{t_k}^{t_{k+1}} &(t_{m+1}-t)^{\alpha-1}(t - t_{m-1}) dt \\
&= \frac{1}{\alpha+1} \left[(t_{m+1} - t_{k+1})^{\alpha+1} - (t_{m+1} - t_k)^{\alpha+1} \right] \\
&\quad - \frac{2h}{\alpha} \left[(t_{m+1} - t_{k+1})^{\alpha} - (t_{m+1} - t_k)^{\alpha} \right]
\end{aligned} \tag{16.94}$$

and

$$\int_{t_k}^{t_{k+1}} (t_{m+1} - t)^{\alpha-1}(t - t_m)dt$$

$$= \frac{1}{\alpha+1}\left[(t_{m+1} - t_{k+1})^{\alpha+1} - (t_{m+1} - t_k)^{\alpha+1}\right]$$

$$-\frac{h}{\alpha}\left[(t_{m+1} - t_{k+1})^{\alpha} - (t_{m+1} - t_k)^{\alpha}\right]. \tag{16.95}$$

Using Eqs. (16.94) and (16.95) in Eq. (16.93), after simplification results in

$$I_1 = h^{\alpha}\left[\frac{2(m+1)^{\alpha}}{\alpha} - \frac{(m+1)^{\alpha+1}}{\alpha+1}\right]F_m$$

$$-h^{\alpha}\left[\frac{(m+1)^{\alpha}}{\alpha} - \frac{(m+1)^{\alpha+1}}{\alpha+1}\right]F_{m-1}. \tag{16.96}$$

Similarly, I_2 is worked out as

$$I_2 = h^{\alpha}\left[\frac{m^{\alpha}}{\alpha} - \frac{m^{\alpha+1}}{\alpha+1}\right]F_m + h^{\alpha}\frac{m^{\alpha+1}}{\alpha+1}F_{m-1}. \tag{16.97}$$

Using the obtained expressions of I_1 and I_2 in Eq. (16.91) leads to

$$u_{m+1} - u_m = \frac{h^{\alpha}}{\Gamma(\alpha)}\left[\left(\frac{2(m+1)^{\alpha} - m^{\alpha}}{\alpha} + \frac{m^{\alpha+1} - (m+1)^{\alpha+1}}{\alpha+1}\right)F_m\right.$$

$$\left.-\left(\frac{(m+1)^{\alpha}}{\alpha} + \frac{m^{\alpha+1} - (m+1)^{\alpha+1}}{\alpha+1}\right)F_{m-1}\right]. \tag{16.98}$$

Taking inverse Laplace transform on the above equation gives the solution $u(x,t)$. This algorithm is called a *two-step Laplace Adams–Bashforth algorithm* for the partial differential Eq. (16.86) [63].

For the equation $D^{\alpha}x(t) = f(x(t), t)$ the left-side of Eq. (16.98) becomes $x(t_{m+1}) - x(t_m)$ and $F_m = f(x(t_m), t)$. In this case, the algorithm is termed as a *two-step Adam–Bashforth algorithm*. The error in the algorithm is shown to be [63]

$$|R_m^{\alpha}| \le \frac{h^{\alpha+2}}{8\Gamma(\alpha+1)}\max\left\{F^{(2)}(u, \xi)\right\}((m+1)^{\alpha} + m^{\alpha}) < +\infty, \tag{16.99}$$

where $\xi \in (0, t_{m+1})$. To find $x(t_{m+1})$ using Eq. (16.98) the solution $x(t_m)$ and $x(t_{m-1})$ are required. That is, to find $x_2(= x(t_2))$ the solutions x_1 and x_0 are needed. x_0 is the initial condition. To find x_1, the fractional Euler formula

$$x_1 = x_0 + \frac{h^{\alpha}}{\Gamma(\alpha)}F(x_0, t) \tag{16.100}$$

can be used. Thus, knowing x_0 and x_1 the solutions at $t_2, t_3, \ldots, t_{m+1}$ can be computed employing (16.98).

Treating (16.98) as a predictor, in [68] the following an implicit method with the corrector has been proposed:

$$x(t_{m+1}) = x(t_m) + \frac{h^{\alpha}}{\Gamma(\alpha)}\left[\frac{(m+1)^{\alpha}}{\alpha} + \frac{m^{\alpha+1} - (m+1)^{\alpha+1}}{\alpha+1}\right]F_{m+1}$$

$$+\frac{h^{\alpha}}{\Gamma(\alpha)}\left[-\frac{m^{\alpha}}{\alpha} + \frac{(m+1)^{\alpha+1} - m^{\alpha}}{\alpha+1}\right]F_m. \tag{16.101}$$

TABLE 16.8

The numerical solution x_n of $D^\alpha x(t) = t$, $x(0) = 0$ computed employing the two-step Adam–Bashforth method with $h = 0.001$ for $\alpha = 0.5$. The exact solution x_e and the error in the numerical solution $|x_e - x_n|$ are also given.

t	x_n	x_e	$|x_e - x_n|$	t	x_n	x_e	$|x_e - x_n|$
0.1	1.02376	1.02379	0.00003	0.6	1.34959	1.34962	0.00003
0.2	1.06726	1.06728	0.00002	0.7	1.44054	1.44057	0.00003
0.3	1.12358	1.12361	0.00003	0.8	1.53825	1.53827	0.00002
0.4	1.19028	1.19031	0.00003	0.9	1.64226	1.64228	0.00002
0.5	1.26594	1.26596	0.00002	1.0	1.75223	1.75225	0.00002

16.6.2 Numerical Examples

Consider the system (16.26) with $x(0) = 1$, $\alpha = 0.5$ and $h = 0.001$. It is easy to develop a C++ or Python code for solving a fractional differential equation using the formula (16.98). Table 16.8 presents the x_n, x_e and $|x_e - x_n|$ for ten values of t. The numerical solution is in close agreement with the analytical solution.

The second example is the system (16.37). The numerical solutions of this system for $\alpha = 0.5$ and 0.75 along with the exact solutions are given in Table 16.9. The step size used is $h = 0.001$.

TABLE 16.9

The numerical solution x_n of $D^\alpha x(t) = -x(t)$, $x(0) = 1$ computed using the two-step Adams–Bashforth algorithm with $h = 0.001$ for $\alpha = 0.5$ and 0.75. x_e denotes the exact solution.

t	$\alpha = 0.5$ x_n	x_e	$\alpha = 0.75$ x_n	x_e
0.1	0.73338	0.72358	0.82912	0.82825
0.2	0.64945	0.64379	0.73252	0.73259
0.3	0.59444	0.59202	0.65916	0.66034
0.4	0.55332	0.55361	0.59971	0.60212
0.5	0.52052	0.52316	0.54989	0.55360
0.6	0.49331	0.49803	0.50722	0.51229
0.7	0.47013	0.47670	0.47013	0.47656
0.8	0.45000	0.45825	0.43750	0.44529
0.9	0.43224	0.44202	0.40853	0.41768
1.0	0.41639	0.42758	0.38263	0.39311

16.7 Nonlinear Fractional Differential Equations

So far, the applicability of the numerical methods for fractional differential equations is discussed by considering certain linear fractional differential equations for which exact solutions are known. It is important to analyze the applicability of these methods to nonlinear systems. The backward-difference method is difficult to apply for nonlinear systems. The other methods are suitable for nonlinear systems also. In this section, the behaviours of certain nonlinear systems are explored by applying the fractional Euler method. For most of the nonlinear systems, exact analytical solutions are not known, however, useful qualitative behaviours can be extracted applying theoretical approaches. For autonomous systems, that is equations not having explicit occurrence of the independent variable, time t, equilibrium point analysis is useful. In this section, linear stability analysis is performed for two nonlinear systems and the predictions are verified numerically. Then, the periodically driven fractional Duffing oscillator is numerically solved and the occurrence of interesting period-doubling route to chaotic dynamics is presented. To start with, let us review the linear stability analysis [90].

16.7.1 Linear Stability Analysis

Consider a system of two-coupled differential equations, each of order α $(0 < \alpha < 1)$, of the form

$$D^\alpha x_1 = f_1(x_1, x_2), \quad D^\alpha x_2 = f_2(x_1, x_2), \tag{16.102}$$

where x_1 and x_2 are functions of time t. The system (16.102) is an autonomous system. This system can be rewritten as

$$D^\alpha X = F(X), \quad X = (x_1, x_2)^{\mathrm{T}} \in R_n, \quad F = (f_1, f_2)^{\mathrm{T}}. \tag{16.103}$$

Any allowed solution of $F(X) = 0$ is called a *fixed point* or an *equilibrium point* of the system and is denoted as $X^* = (x_1^*, x_2^*)$. The equilibrium points of (16.102) are thus obtained by setting $f_1(x_1, x_2) = 0$ and $f_2(x_1, x_2) = 0$.

An equilibrium point X^* is said to be *stable* if the neighbouring trajectories of it approach it in the limit of $t \to \infty$. If the neighouring trajectories of X^* deviate or move away from it in the long-time limit then X^* is called as an *unstable* equilibrium point. If the above two are not happening then X^* is *neutrally* stable.

The stability condition for the fractional order system is determined in [90]. To find the stability of X^* let us slightly disturb it and write

$$x_1 = x_1^* + \xi_1(t), \quad x_2 = x_2^* + \xi_2(t). \tag{16.104}$$

Write the Taylor series expansion of f_1 and f_2 about (x_1^*, x_2^*) as

$$f_i(x_1^* + \xi_1, x_2^* + \xi_2) \approx f_i(x_1^*, x_2^*) + \frac{\partial f_i}{\partial x_1}\bigg|_{x_1^*, x_2^*} \cdot \xi_1 + \frac{\partial f_i}{\partial x_2}\bigg|_{x_1^*, x_2^*} \cdot \xi_2$$
$$+ \text{ higher-order terms in } (\xi_1, \xi_2), \tag{16.105}$$

where $i = 1, 2$. Substitution of Eqs. (16.104) and (16.105) in Eqs. (16.102) gives

$$D^\alpha \xi_i(t) = \frac{\partial f_i}{\partial x_1}\bigg|_{x_1^*, x_2^*} \cdot \xi_1 + \frac{\partial f_i}{\partial x_2}\bigg|_{x_1^*, x_2^*} \cdot \xi_2, \quad i = 1, 2 \tag{16.106}$$

where the higher-order terms in (ξ_1, ξ_2) are neglected. This system can be rewritten as

$$D^\alpha \xi = M\xi, \quad \xi = \begin{pmatrix} \xi_1 \\ \xi_2 \end{pmatrix}, \quad M = \begin{pmatrix} a & b \\ c & d \end{pmatrix} \tag{16.107a}$$

and

$$a = \left.\frac{\partial f_1}{\partial x_1}\right|_{x_1^*, x_2^*}, \quad b = \left.\frac{\partial f_1}{\partial x_2}\right|_{x_1^*, x_2^*}, \quad c = \left.\frac{\partial f_2}{\partial x_1}\right|_{x_1^*, x_2^*}, \quad d = \left.\frac{\partial f_2}{\partial x_2}\right|_{x_1^*, x_2^*}. \tag{16.107b}$$

For the systems of the form (16.102) with $\alpha = 1$, a detailed stability analysis predicts that (x_1^*, x_2^*) becomes *asymptotically stable if the real part of both the eigenvalues of the matrix M are negative* [91]. If λ_1 and λ_2 are the eigenvalues of M, B is the eigenvectors of M and $C = \begin{pmatrix} \lambda_1 & 0 \\ 0 & \lambda_2 \end{pmatrix}$ then $MB = BC$ and $M = BCB^{-1}$. Equation (16.107a) becomes,

$$D^\alpha \xi = M\xi = \left(BCB^{-1}\right)\xi, \quad D^\alpha\left(B^{-1}\xi\right) = C\left(B^{-1}\xi\right). \tag{16.108}$$

Defining $\eta = B^{-1}\xi = \begin{pmatrix} \eta_1 \\ \eta_2 \end{pmatrix}$, the second subequation in Eqs. (16.108) takes the form $D^\alpha \eta = C\eta$. That is,

$$D^\alpha \eta_1 = \lambda_1 \eta_1, \quad D^\alpha \eta_2 = \lambda_2 \eta_2. \tag{16.109}$$

This is a linear system and its solution is [92] (also see Problem 4 at the end of the present chapter)

$$\eta_i(t) = \sum_{n=1}^{\infty} \frac{\lambda_i^n t^{n\alpha}}{\Gamma(n\alpha + 1)} \eta_i(0) = E_\alpha\left(\lambda_i t^\alpha\right)\eta_i(0), \quad i = 1, 2 \tag{16.110}$$

where $E_\alpha(\lambda, t)$ is the Mittag–Leffler function. The asymptotic stability of the system (16.109) (that is, both $|\eta_1| \to 0$ and $|\eta_2| \to 0$ as $t \to \infty$) has been analyzed by Matignon [92]. If all the eigenvalues of M satisfy the condition

$$|\arg(\lambda_i)| > \alpha\pi/2, \quad i = 1, 2 \tag{16.111}$$

then the solution $\eta_i(t) \to 0$ as $t \to \infty$. For a detailed proof of this see the ref. [93]. Write an eigenvalue as $\lambda = \lambda_R + i\lambda_I$. Then, $\arg(\lambda = \lambda_R + i\lambda_I)$ is given by

$$\arg\left(\lambda_R + i\lambda_I\right) = \begin{cases} \tan^{-1}(\lambda_I/\lambda_R), & \lambda_R > 0, \\ \tan^{-1}(\lambda_I/\lambda_R) + \pi, & \lambda_R < 0, \ \lambda_I \geq 0, \\ \tan^{-1}(\lambda_I/\lambda_R) - \pi, & \lambda_R < 0, \ \lambda_I < 0, \\ \pi/2, & \lambda_R = 0, \ \lambda_I > 0, \\ -\pi/2, & \lambda_R = 0, \ \lambda_I < 0, \\ \text{undefined}, & \lambda_R = 0, \ \lambda_I = 0. \end{cases} \tag{16.112}$$

From (16.111) and (16.112) note that, as pointed out in [91], *if the real part of both λ_1 and λ_2 are negative then (x_1^*, x_2^*) becomes stable.*

In general, for $D^\alpha X = F(X)$ with X and F having n components, an equilibrium point of this system is asymptotically stable if

$$|\arg(\lambda_i)| > \alpha\pi/2, \quad i = 1, 2, \ldots, n. \tag{16.113}$$

For the system $D^\alpha x = f(x)$ the stability condition for an equilibrium point is $f'(x^*) < 0$.

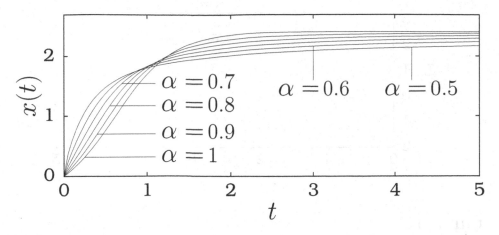

FIGURE 16.3
The numerically computed solution of the nonlinear system (16.114) by the fractional Euler method for six values of α with $x(0) = 0$. All the trajectories approach the equilibrium point $x_+^* = 2.41421$.

16.7.2   Stability Analysis of a System of the Form $D^\alpha x = f(x)$

Consider the nonlinear system

$$D^\alpha x = f(x) = 1 + 2x - x^2, \quad 0 < \alpha \le 1. \tag{16.114}$$

For $\alpha = 1$, it is an exactly solvable system with the solution (with $x(0) = 0$)

$$x(t) = 1 + \sqrt{2}\tanh\left[\sqrt{2}\,t + \frac{1}{2}\ln\left(\frac{\sqrt{2}-1}{\sqrt{2}+1}\right)\right]. \tag{16.115}$$

The exact analytical solution is difficult to find for the system (16.114). Therefore, numerically solve it by the fractional Euler method and check whether the qualitative behaviour of the numerical solution is in accordance with the linear stability analysis of the equilibrium points of the system.

The equilibrium points of (16.114) are obtained by letting $D^\alpha x = 0$. This gives two equilibrium points $x_\pm^* = 1 \pm \sqrt{2} = -0.41421$ and 2.41421. Here, for an equilibrium point to be stable the condition is $f'(x^*) < 0$. For x_+, $f'(x_+^*) = -2\sqrt{2} < 0$. x_+^* is thus stable. The trajectories started in the neighbourhood of it asymptotically approach x_+^*. For x_-^*, $f'(x_-^*) = 2\sqrt{2} > 0$. x_-^* is an unstable equilibrium point. In the neighbourhood of it the trajectories diverge. Figure 16.3 shows the numerically computed solution for $\alpha = 1$, 0.9, 0.8, 0.7, 0.6 and 0.5 with $x(0) = 0$ and $h = 0.001$. For each value of α, the solution $x(t)$ approaches to $x_+^* = 2.41421$. The rate of convergence of $x(t)$ depends on α and is clearly seen in Fig. 16.3.

In Fig. 16.4, $x(t)$ for $\alpha = 0.7$ is plotted for five different initial values of x. For the first trajectory from the bottom $x(0) = -0.42$. This initial condition is left to $x_-^* = -0.41421$ which is the unstable equilibrium point. Consequently, $x(t)$ starting with $x(0) = -0.42$ moves away from it. For the second, third and fourth trajectories from bottom in Fig. 16.4 the values of $x(0)$ are right to x_-^* and left to x_+^*. The corresponding trajectories move away from x_-^* and attracted to the stable equilibrium point x_+^*. The top most trajectory is corresponding to an initial condition right to x_+^* and is also approaching x_+^*. The qualitative

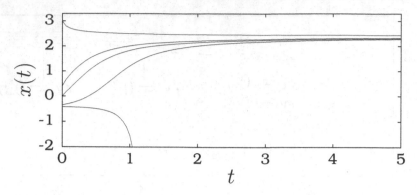

FIGURE 16.4

The variation of $x(t)$ of the system (16.114) for $\alpha = 0.7$ and for different values of $x(0)$. For details see the text.

behaviour of the numerical solution of the system (16.114) computed by the fractional Euler method is in agreement with the linear stability analysis of the equilibrium points of the system.

16.7.3   A Fractional Order Duffing Oscillator

A prototype nonlinear oscillator is the Duffing oscillator given by [91]

$$D^2x + dDx - x + x^3 = f \sin \omega t. \tag{16.116}$$

First, consider this system with $f = 0$. Suppose, a fractional order form of (16.116) is

$$D^{2\alpha}x + dD^\alpha x - x + x^3 = 0, \quad 0 < \alpha < 1. \tag{16.117}$$

This equation can be rewritten as

$$D^\alpha x \;=\; y = f_1(x,y), \tag{16.118a}$$
$$D^\alpha y \;=\; -dy + x - x^3 = f_2(x,y). \tag{16.118b}$$

This system has three equilibrium points $(x^*, y^*) = (0,0), (\pm 1, 0)$. The stability determining eigenvalues are the eigenvalues of the matrix

$$
\begin{aligned}
M \;&=\; \left(\begin{array}{cc} \partial f_1/\partial x & \partial f_1/\partial y \\ \partial f_2/\partial x & \partial f_2/\partial y \end{array} \right) \Bigg|_{x^*, y^*} \\
&=\; \left(\begin{array}{cc} 0 & 1 \\ 1 - 3x^{*2} & -d \end{array} \right).
\end{aligned} \tag{16.119}
$$

For $(x^*, y^*) = (0,0)$, the stability determining eigenvalues are

$$\lambda_\pm = \frac{1}{2}\left[-d \pm \sqrt{d^2 + 4}\,\right]. \tag{16.120}$$

That is, $\lambda_- < 0$ while $\lambda_+ > 0$. The equilibrium point $(0,0)$ is unstable.

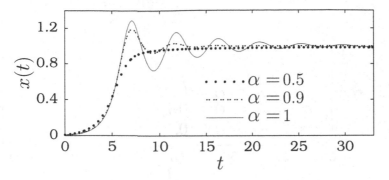

FIGURE 16.5
$x(t)$ with t for the system (16.118) for three values of α. Here, $h = 0.001$, $x(0) = 0.01$, $y(0) = 0$ and $d = 0.3$.

For the other two equilibrium points $(\pm 1, 0)$ the matrix $M = \begin{pmatrix} 0 & 1 \\ -2 & -d \end{pmatrix}$ and the eigenvalues are

$$\lambda_\pm = \frac{1}{2}\left[-d \pm \sqrt{d^2 - 8}\right]. \tag{16.121}$$

For $d^2 < 8$, $\lambda_\pm$ are complex conjugate with negative real part. The equilibrium points are thus stable. In the neighbourhood of each of these equilibrium points the trajectories spiral around it and reach it in the limit of $t \to \infty$. The corresponding equilibrium point is called a *stable focus*. For $d^2 > 8$, the eigenvalues λ_+ are real negative and distinct and the equilibrium points are said to be *stable nodes*. In this case, the trajectories in the neighbourhood of them approach them along parabolic paths in the $x - y$ phase space.

Equations (16.118) are solved by the fractional Euler method for $d = 0.3$ with step size $h = 0.001$ and with the initial conditions $x(0) = 0.01$ and $y(0) = 0$. The chosen $(x(0), y(0))$ is close to the unstable equilibrium point $(0, 0)$. Numerical solution is obtained for $\alpha = 1$, 0.7 and 0.5. Figure 16.5 shows the numerical solutions. The numerical solutions $x(t)$ diverge out of $x^* = 0$ and approach the stable focus point $x^* = 1$ by exhibiting damped oscillation as predicted by the linear stability analysis of the equilibrium points.

Next, choose $d = 3$ for which $d^2 > 8$. The numerically computed solutions for $\alpha = 1$, 0.9, 0.7 and 0.5 with $x(0) = 0.01$ and $y(0) = 0$ are presented in Fig. 16.6. As predicted by the linear stability analysis the solutions $x(t)$ approach $x^* = 1$ along parabolic paths. The effect of α on the rate of convergence of the solution to x^* is clearly seen. As the value of α decreases from 1 the rate of convergence of the solution to x^* also decreases as expected. The results in the Figs. 16.5 and 16.6 clearly indicate the very good agreement of the qualitative behaviour of the numerical solution computed by the fractional Euler method with the outcome of the linear stability analysis.

16.7.4 A Fractional Order Damped Duffing Oscillator

Suppose the damping term Dx in Eq. (16.116) is fractional order so that the equation is

$$D^2 x + dD^\alpha x - x + x^3 = f \sin \omega t, \quad \alpha = 3/2. \tag{16.122}$$

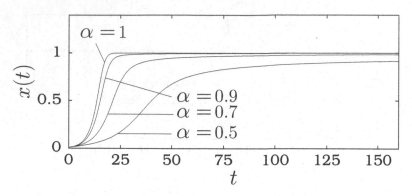

FIGURE 16.6
$x(t)$ with t for the system (16.117) for four values of α. Here, $h = 0.001$, $x(0) = 0.01$, $y(0) = 0$ and $d = 3$.

To numerically solve it, rewrite it in the form of (16.1) as

$$D^{1/2}x_1 = x_2, \tag{16.123a}$$
$$D^{1/2}x_2 = x_3, \tag{16.123b}$$
$$D^{1/2}x_3 = x_4, \tag{16.123c}$$
$$D^{1/2}x_4 = -dx_4 + x_1 - x_1^3 + f\sin\omega t. \tag{16.123d}$$

Case 1: $f = 0$.

For $f = 0$, this equation is an autonomous system and linear stability analysis gives useful information of the dynamics of the system. A detailed analysis shows that there are three equilibrium points $(x_1, x_2, x_3, x_4) = (0, 0, 0, 0)$ and $(x_1, x_2, x_3, x_4) = (\pm1, 0, 0, 0)$. The equilibrium point $(0, 0, 0, 0)$ is always unstable. $(\pm1, 0, 0, 0)$ are stable foci for $d^2 < 8$ and stable node for $d^2 > 8$. Equations (16.123) are solved by the fractional Euler method with step size $h = 0.001$ and for $t \in [0, 50]$. The initial conditions are chosen near the origin which is the unstable equilibrium point. In Fig. 16.7 two solutions for $\alpha = 1$ and 0.5 are plotted in the $x_1 - x_3$, that is $(x_1 - \dot{x}_1)$ phase space.

For the initial condition $(-0.1, 0, 0, 0)$ chosen left to the origin the trajectories spiral about the left equilibrium point $(-1, 0, 0, 0)$ while for the initial condition $(0.1, 0, 0, 0)$ right to the origin the trajectories approach the right equilibrium point $(1, 0, 0, 0)$. This result is in accordance with the stability nature of the equilibrium points of the system.

Case 2: $f \neq 0$.

Finally, consider the system (16.123) with $d = 0.5$, $\omega = 1$ and treat f as the control parameter. For a range of values of f numerical solutions are computed with the initial condition $(1.2, 0, 0, 0)$ and with the step size in the fractional Euler method as $h = 0.005$. For $f \neq 0$, as $f\sin\omega t$ is an explicit time-dependent external forcing term, linear stability analysis cannot be performed and the solution never ends up on any of the equilibrium points determined for $f = 0$. The integer order system (16.116) exhibits a variety of regular and complex dynamics when the parameter f or ω is varied from a small value. One of the interesting dynamics is the period doubling route to chaotic dynamics. This nonlinear dynamics is displayed by the fractional order Duffing oscillator also. Let us illustrate this dynamics.

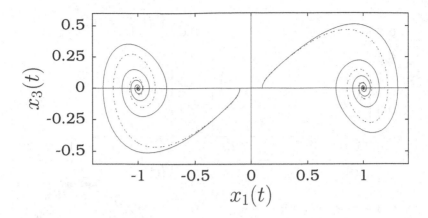

FIGURE 16.7
The numerically computed solutions of the system (16.123) for $d = 0.5$ and for $\alpha = 1$ and
0.5. The continuous and dashed trajectories represent the solution for $\alpha = 1$ and $\alpha = 0.5$,
respectively. Here, $(x_1(0), x_2(0), x_3(0), x_4(0)) = (\pm 0.1, 0, 0, 0)$.

In Figs. 16.5-16.7, a trajectory started from an initial condition approaches a stable
equilibrium point X^* and once it reaches X^* it remains there. This is the case for various
initial conditions in the neighbourhood of a stable equilibrium point. As a stable equilibrium
point attracts nearby trajectories, it is an *attractor* and is a point attractor. The part
of a trajectory between the initial state and the stable equilibrium point in Figs. 16.5-
16.7 is termed as *transient*. In studying the long-time behaviour of a dynamical system
the transient motion has to be discarded. In numerically solving Eq. (16.122), starting
with an initial condition, the solution is computed with the step size $h = 0.005$ and the
solution corresponding to $t < 600$ is left as a transient. Then, the nature of the solution
for $t \in [600, 1000]$ is plotted in $x_1 - x_3$ phase space and the periodicity of the solution is
determined.

A. Period Doubling and Chaotic Dynamics

Figure 16.8 presents the plot of x_1 versus x_3 for some selected values of f where $d = 0.5$
and $\omega = 1$. For $f = 0.252$ the orbit in Fig. 16.8a is closed and is a periodic orbit with period
$T = 2\pi/\omega$. When $f = 0.275$ (see Fig. 16.8b) the orbit is also a closed one but the period
of it is $2T$, that is, twice the period of the orbit corresponding to $f = 0.252$. A period-$4T$
solution is realized for $f = 0.282$ (Fig. 16.8c).

The orbit shown in Fig. 16.8d is clearly different from the orbits shown in Figs. 16.8a-
c. This is a nonperiodic orbit. An important characteristic property of this solution is
that it is highly sensitive to any small change in the initial condition. Essentially, two
solutions started from two nearby initial conditions diverge exponentially until they become
completely uncorrelated and long time future prediction is highly inaccurate. The distance
between two such trajectories neither decays to zero (as is the case for the periodic motions
shown in Figs. 16.8a-c) nor diverges continuously but varies irregularly with time. The
readers may verify this for the solution shown for $f = 0.288$. Such a solution is called a
chaotic solution.

For the f values in Figs. 16.8a-d the orbit existing for $x_1 > 0$ is shown. For these values
of f there is another orbit confined to $x_1 < 0$ and is not shown in Fig. 16.8. In Fig. 16.8e the
orbit coexisting with $x_1 < 0$ is shown. As f is increased from 0.288 the two coexisting orbits

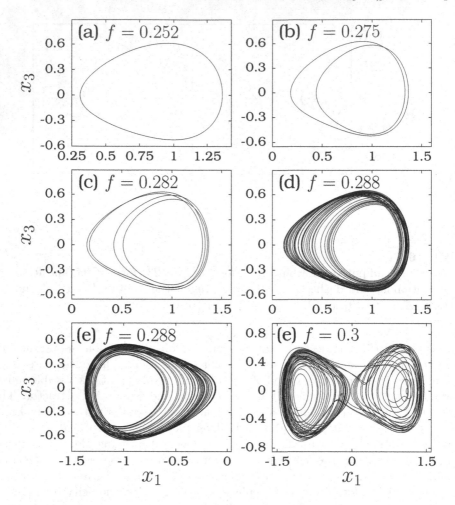

FIGURE 16.8
The numerically computed solution of the system (16.123) for $d = 0.5$, $\omega = 1$ and for a few values of f.

expand and at a value of f merge together and form a single chaotic orbit. An example of such a chaotic orbit is presented in Fig. 16.8e for $f = 0.3$.

B. Bifurcation Diagram

In order to understand what is happening when the value of f is varied from 0 the so-called *bifurcation diagram* is useful. It can be constructed as described below.

For the system (16.123) for a value of f collect the values of $X(t) = (x_1(t), x_2(t), x_3(t), x_4(t))$ at integer multiples of $t = T = 2\pi/\omega$. Leave first, say, 200 values of X as a transient and consider next 100 values of X. These values of X are called *Poincaré points*. Do this for f values in the interval $f \in [f_{\min}, f_{\max}]$ with step size Δf. For a period-T orbit one value of X repeats. Similarly, for a period-mT orbits m values of X repeat. For the nonperiodic solution, like a chaotic solution, the X values are nonrepeating but bounded. Plot x_1 versus f. This plot will describe the qualitative change in the behaviour of the solution of the system when the parameter f is varied.

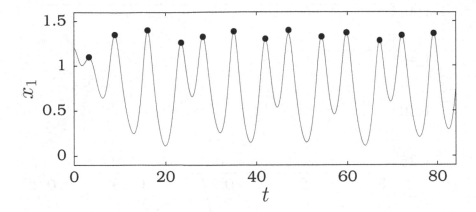

FIGURE 16.9
$x(t)$ versus t for the system (16.123) for $f = 0.288$.

There is an alternative approach to draw a bifurcation diagram. $X(t)$ oscillates with time. The local maximum values of $X(t)$ can be used for bifurcation diagram. Consider the part of the solution for $f = 0.288$ shown in Fig. 16.9. The solid circles in this figure represent the local maxima of $x_1(t)$. Collect these values of $x_1(t)$ after a sufficient transient. *How does one collect such values of $x_1(t)$?* When $x_3(t - h)$ (that is, $\dot{x}_1(t - h) > 0$) and $x_3(t) < 0$ find the value of $t = t'$ at which $x_3(t') = 0$. This t' value can be determined using the interpolation formula

$$t' - t - h - \frac{hx_3(t - h)}{x_3(t) - x_3(t - h)}. \tag{16.124}$$

Then, using the values of $x_1(t-h)$, $x_1(t)$, $t-h$, t and t' the value of $x_1(t')$ can be determined again using the interpolation method. $x_1(t')$ is given by

$$x_1(t') = x_1^{\max}(t') = x_1(t - h) + (t' - t + h)\frac{x_1(t) - x_1(t - h)}{h}. \tag{16.125}$$

For the system (16.123) 200 values of f in the interval $f \in [0.25, 0.29]$ with $\Delta f = 0.0002$ are chosen. For each value of f after leaving a sufficient transition 100 values of $x_1^{\max}$ are collected. Figure 16.10 presents the plot of $x_1^{\max}$ versus f. For $f < 0.26$ one value of $x_1^{\max}$ is repeating and this correponds to a period-T solution. For $0.26 < f < 0.2776$ two values of $x_1^{\max}$ are repeating. The solution is thus a periodic with period-$2T$. As f is further varied, period-$4T$, $8T$, ... solutions are developed. At $f \approx 0.2840$ this period doubling phenomenon accumulates and a chaotic solution is developed. In this case, $x_1^{\max}$ is nonperiodic. In Fig. 16.10 the region of f where such solutions occur can be clearly seen.

The numerical study of the nonlinear systems considered in this section clearly indicates the applicability of the fractional Euler method to nonlinear systems.

16.8 Concluding Remarks

Four methods for numerically solving initial-value problem of fractional order ordinary differential equations are described and their applicability and efficacy are illustrated by

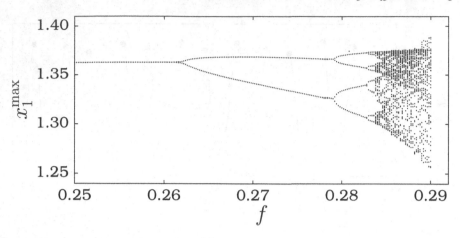

FIGURE 16.10
Bifurcation diagram of the system (16.123) for $d = 0.5$ and $\omega = 1$.

considering certain linear and nonlinear systems. Among the four methods considered the fractional Euler method is a better choice. Its applicability is demonstrated by considering single-order and multi-order systems. Numerical predictions are found to be in close agreement with the exact analytical solutions and the expected results in the cases of unknown analytical solutions. Compared to integer order equations numerical computation of solutions of even very simple fractional order equations involves enormous calculations in increase in time due to the memory effect.

Methods have been proposed in the mathematics literature to solve time-delayed, distributed order and stiff fractional order differential equations. Methods for fractional order systems subjected to boundary conditions are not considered in this book. For some methods see the refs. [96-100].

16.9   Bibliography

[1]   M. Rehman and R.A. Khan, *Appl. Math. Model* 60:2630, 2011.

[2]   Y.A. Rossikhin and M.V. Shitikova, *Appl. Mech. Rev.* 63, 2010, 52 pages.

[3]   F.B.M. Duarte and J.A.T. Machado, *Nonl. Dyn.* 29:342, 2002.

[4]   O.P. Agrawal, *Nonl. Dyn.* 38:323, 2004.

[5]   N. Engheta, *IEEE Trans. Ant. Prop.* 44:554, 1996.

[6]   J.F. Gomez-Aguilar, R.F. Escobar-Jimenez, M.G. Lopez-Lopez and V.M. Alvarado-Martínez, *J. Electro. Waves and Appl.* 30:1937, 2016.

[7]   C. Lederman, J.M. Roquejoffre and N. Wolanski, *Ann. Mat. Pure Appl.* 183:173, 2004.

[8]   F.C. Meral, T.J. Royston and R.L. Magin, Commun. *Nonl. Sci. Num. Simul.* 15:939, 2010.

[9]   R.L. Magin, *Comput. Math. Appl.* 59:1586, 2010.

[10] K.B. Oldham, *Adv. Eng. Soft.* 41:9, 2010.

[11] F. Mainardi, *Fractional Calculus: Some Basic Problems in Continuum and Statistical Mechanics.* Springer, Vienna, 1997.

[12] D.A. Benson, S.W. Wheatcraft and M.M. Meerschaert, *Water Resource Research* 36:1403, 2000.

[13] Y. Pachepsky, D. Benson and W. Rawls, *Soil Sci. Soc. Am. J.* 4:1234, 2000.

[14] L. Zhou and H.M. Selim, *Soil Sci. Soc. Am. J.* 67:1079, 2003.

[15] M.M. Meerschaert and C. Tadjeran, *J. Comp. Appl. Math.* 172:65, 2004.

[16] S. Shen and F. Liu, *ANZIAM J.* 46E:C871, 2005.

[17] G. Huang, Q. Huang and H. Zhan, *J. Contam. Hydr.* 85:53, 2006.

[18] V.G. Ychuk, B. Datsko and V. Meleshko, *J. Comput. Appl. Math.* 220:215, 2008.

[19] E. Sousa, *J. Comp. Phys.* 228:4038, 2009.

[20] F. Mainardi, *Chaos, Solitons & Fractals* 7:1461, 1996.

[21] D.N. Slinn and J.J. Riley, *J. Comput. Phys.* 144:550, 1998.

[22] A. Atangana, *Proceedings of the International Conference on Algebra and Applied Analysis* 6, 2012, 20 pages.

[23] X.J. Yang and D. Baleanu, *Thermal Sci.* 17:625, 2013.

[24] D. Baleanu, B. Guvenc, J.A. Tenreino-Machado, *New Trends in Nanotechnology and Fractional Calculus Applications.* Springer, New York, 2010.

[25] T.J. Anastasio, *Biol. Cybernet.* 72:69, 1994.

[26] A. Le Mehaute and G. Crepy, *Solid State Ionics* 9-10:17, 1983.

[27] Y. Jiang, B. Zhang, X. Shu and Z. Wei, *J. Adv. Res.* 25:217, 2020.

[28] T. Kaczorek and K. Rogowski, *Fractional Linear Systems and Electrical Circuits.* Springer, London, 2007.

[29] M. Sivarama Krishna, S. Das, K. Biswas and B. Goswami, *IEEE Trans. Elect. Dev.* 58:4067, 2011.

[30] A.G. Radwan, *IEEE J. Emer. Sel. Top. Circ. Sys.* 3:2156, 2013.

[31] A. Alsaedi, J.J. Nieto and V. Venktesh, *Adv. Mech. Eng.* 7:1, 2015.

[32] J.I. Hidalgo-Reyes, J.F. Gomez-Aguilar, R.F. Escobar-Jimenez, V.M. Alvarado-Martinez, M.G. Lopez-Lopez, *Int. J. Circuit Theory and Appl.* 47:1225, 2019.

[33] M.S. Semary, M. Fouda and A. Radwan, *J. Adv. Res.* 18:147, 2019.

[34] S. Kapoulea, G. Tsirimokou, C. Psychalinos and A.S. Elwakil, *Circ. Syst. Signal Process*, 2019. https://doi.org/10.1007/s00034-019-01252-5.

[35] L. Debnath, *IJMMS* 54:3413, 2003.

[36] J.H. He, *Comput. Meth. Appl. Mech. Eng.* 167:57, 1998.

[37] N. Shawagfeh, *Appl. Math. Comput.* 131:517, 2002.

[38] Z. Odibat and S. Momani, *Int. J. Nonlin. Sci. Numer. Simulat.* 7:15, 2006.

[39] S. Momani and Z. Odibat, *Phys. Lett. A* 355:271, 2006.

[40] S. Momani and Z. Odibat, *Appl. Math. Comput.* 177:488, 2006.

[41] Z. Odibat and S. Momani, *Appl. Math. Comput.* 181:767, 2006.

[42]   S. Momani, *Chaos, Solitons & Fractals* 28:930, 2006.

[43]   L. Blank, *Numerical Analysis Report 287*, Manchester Centre for Computational Mathematics, Manchester, 1996.

[44]   K. Diethelm, *Elec. Transact. Numer. Anal.* 5:1, 1997.

[45]   K. Diethelm and G. Walz, *Numer. Algor.* 16:231, 1997.

[46]   K. Diethelm and A.D. Freed, *On the solutions of nonlinear fractional differential equations used in the modeling of viscoplasticity*. In Scientific Computing in Chemical Engineering II − Computational Fluid Dynamics, Reaction Engineering and Molecular Properties. F. Keil, W. Mackens, H. Voβ and J. Werther (Eds.). Springer, Heidelberg, 1999. pages 217-224.

[47]   N.J. Ford and A.C. Simpson, Numer. Algor. **26**, 333 (2001).

[48]   K. Diethelm, N.J. Ford and A.D. Freed, *Nonl. Dyn.* 29:3, 2002.

[49]   K. Diethelm, N.J. Ford and A.D. Freed, *Numer. Algor.* 36:31, 2004.

[50]   A.M.A. El-Sayed, A.E.M. El-Mesiry and H.A.A. El-Saka, *Comput. Appl. Math.* 23:33, 2004.

[51]   M. Weilbeer, *Efficient Numerical Methods for Fractional Differential Equations and Their Analytical Background*. Verlag, Nicht, Ermittelbar, 2005.

[52]   S. Momani, *Math. Comput. Simul.* 70:110, 2005.

[53]   K. Diethelm, N.J. Ford, A.D. Freed and Y. Luchko, *Comput. Meth. Appl. Mech. Eng.* 194:743, 2005.

[54]   K. Diethelm, J.M. Ford, N.J. Ford and M. Weilbeer, *J. Comput. Appl. Math.* 186:482, 2006.

[55]   S. Momani and Z. Odibat, *Chaos, Solitons & Fractals* 31:1248, 2007.

[56]   Z.M. Odibat and S. Momani, *J. Appl. Math. & Inform.* 26:15, 2008.

[57]   C. Li and Y. Wang, *Comput. Math. Appl.* 57:1672, 2009.

[58]   J. Peinado, J. Ibanez, E. Arias and V. Hernandez, *Comput. Math. Appl.* 60:3032, 2010.

[59]   S. Esmaeili, M. Shamsi and Y. Luchko, *Comput. Math. Appl.* 62:918, 2011.

[60]   J. Cao and C. Xu, *J. Comput. Phys.* 238:154, 2013.

[61]   H.M. Baskonus and H. Bulut, *Open Math.* 13:547, 2015.

[62]   R.B. Albadarneh, M. Zerqat and I.M. Batiha, *Int. J. Pure Appl. Math.* 106:859, 2016.

[63]   R. Gnitchogna and A. Atangana, *Numer. Meth. Part. Diff. Eqs.* 34:1739, 2017.

[64]   H.F. Ahmed, *J. Egypt. Math. Soc.* 26:38, 2018.

[65]   H. Singh and H.M. Srivastava, *Front. Phys.* 8, 2020, article number 120.

[66]   A. Daraghmeh, N. Qatanani and A. Saadeh, *Appl. Math.* 11:1100, 2020.

[67]   H. Wang, F. Wu and D. Lei, *AIMS Math.* 6:5596, 2021.

[68]   N.A. Zabidi, Z.A. Majid, A. Kilicman and Z.B. Ibrahim, *Adv. Cont. Disc. Models* 26, 2022; https://doi.org/10.1186/s13662-022-03697-6.

[69]   C.W.H. Green and T. Yan, *Foundations* 2:839, 2022.

[70]   M. Harker, *Fractional Differential Equations: Numerical Methods for Applications*. Springer, New York, 2022.

[71]   R. Garappa, *Mathematics* 6:16, 2018.

[72]   A.K. Grünwald, *Z. Angrew. Math. Phys.* 12:441, 1867.

[73]   A.V. Letnikov, *Mat. Sh.* 3:1, 1868.

[74]   I. Podlubny, *Fractional Differential Equations: An Introduction to Fractional Derivatives, Fractional Differential Equations, to Methods of Their Solution and Some of Their Applications.* Elsevier, Amsterdam, 1998.

[75]   M.D. Ortigueira and F. Coito, *Int. J. Theory Appl.* 7:459, 2004.

[76]   K.B. Oldham and J. Spanier, *The Fractional Calculus Theory and Applications of Differentiation and Integration to Arbitrary Order.* Elsevier, Amsterdam, 1974.

[77]   S.G. Samko, A.A. Kilbas and O.I. Marichev, *Fractional Integrals and Derivatives.* Gordon and Breach Science Publishers, Lausanne, 1993.

[78]   K. Diethelm, *The Analysis of Fractional Differential Equations: An Application-Oriented Exposition Using Differential Operators of Caputo Type*, Springer, Berlin, 2010.

[79]   M. Caputo, *Geophys. J. Int.* 13:529, 1967.

[80]   C. Li, D. Qian and Y.Q. Chen, *Disc. Dyn. Nat. Soc.* 2011, article ID 562494; https://doi.org/10.1155/2011/562494.

[81]   R. Hilfer, *Threshold introduction to fractional derivatives.* In Anomalous Transport: Foundations and Applications. R. Kloges et al. (Eds.). Wiley-VCH, Weinheim, 2008.

[82]   R. Herrmann, *Fractional Calculus: An Introduction for Physicists.* World Scientific, Singapore, 2014.

[83]   M. Caputo and M. Fabrizio, *Prog. Fract. Differ. Appl.* 1:1, 2015.

[84]   D. Oliveira, E. Capelas, T. Machado and J. António, *Math. Probl. Eng.* 2014; doi:10.1155/2014/238459.

[85]   Aslan Ismail, *Math. Meth. Appl. Sci.* 38:27, 2015.

[86]   K.S. Miller and B. Ross, *An Introduction to the Fractional Calculus and Fractional Differential Equations.* John Wiley and Sons, New York, 1993.

[87]   Y. Luchko and R. Gorneflo, *The Initial Value Problem for Some Fractional Differential Equations with the Caputo Derivative.* Fachbreich Mathematic and Informatik, Freic Universitat Berlin, 1998.

[88]   S. Kazem, *Int. J. Nonl. Sci.* 16:3, 2013.

[89]   E. Hairer, S.P. Norsett and G. Wanner, *Solving Ordinary Differential Equations I: Nonstiff Problems.* Springer, Berlin, 1993.

[90]   E. Ahmed, A.M.A. El-Sayed and H.A.A. El-Saka, *J. Math. Anal. Appl.* 325:542, 2007.

[91]   M. Lakshmanan and S. Rajasekar, *Nonlinear Dynamics: Integrability, Chaos and Patterns.* Springer, Berlin, 2002.

[92]   D. Matignon, *Comput. Eng. Syst. Appl.* 2:963, 1996.

[93]   D. Qian, C. Li, R.P. Agarwal and P.J.Y. Wong, *Math. Comput. Model* 52:862, 2010.

[94]   Y.Y.Y. Noupoue, Y. Tandogdu and M. Awadalla, *Adv. Difference Eqs.* 2019-108, 2019, 13 pages.

[95]   I. Grigorenko and E. Grogorenko, *Phys. Rev. Lett.* 91:034101, 2003.

[96]   N.J. Ford and M. Luisa Morgado, *Fract. Cal. Appl. Anal.* 14:554, 2011.

[97]   M.V. Rehman and R.A. Khan, *Appl. Math. Model.* 36:894, 2012.

[98]   Q.M. Al-Mdallah and M.I. Syam, *Commun. Nonl. Sci. Numer. Simulat.* 17:2299, 2012.

[99]   W.K. Zahra and S.M. Elkholy, *Numer. Algor.* 59:373, 2012.

[100] A. Jajarmi and D. Baleanu, *Front. Phys.* 8:220, 2020.

16.10   Problems

16.1 Write the Grünwald–Letnikov fractional integral of order α.

16.2 Prove that $\omega_k = (-1)^k \begin{pmatrix} \alpha \\ k \end{pmatrix} = \left(1 - \dfrac{1+\alpha}{k}\right)\omega_{k-1}$ with $\omega_0 = 1$.

16.3 Represent the formula (16.21) in matrix form for all the nodes.

16.4 Define the Laplace transform of $D^\alpha x(t)$ for $0 < \alpha < 1$ as $[sF(s) - x(0)]/s^{1-\alpha}$ where $F(s)$ is the Laplace transform of $x(t)$. Then, find the solutions of the equations $D^\alpha x(t) + ax(t) = 0$, $x(0) = x_0$ and $D^\alpha x(t) = t$, $x(0) = x_0$, $0 < \alpha < 1$.

16.5 Applying the backward-difference method and using a calculator find the solution of $D^\alpha x(t) = e^{-t}$ at $t = 0.001$ and 0.002 with $x(0) = 0$ and $\alpha = 0.5$ and 0.75.

16.6 Using a calculator compute the solution of $D^\alpha x(t) = x(t) + t$, $x(0) = 1$ and $\alpha = 0.5$ at $t = 0.001$ and 0.002 applying the backward-difference method.

16.7 Prove the following relations [56]:

(i) $\left(I^{n\alpha}D^{n\alpha}f\right)(t) - \left(I^{(n+1)\alpha}D^{(n+1)\alpha}f\right)(t)$

$$= \frac{t^{n\alpha}}{\Gamma(n\alpha + 1)}\left(D^{n\alpha}f\right)(0^+).$$

(ii) $f(t) = \displaystyle\sum_{k=0}^{n} \frac{t^{k\alpha}}{\Gamma(k\alpha + 1)}\left(D^{k\alpha}f\right)(0^+)$

$$+ \frac{\left(D^{(n+1)\alpha}f\right)(\xi)}{\Gamma[(n+1)\alpha + 1]}t^{(n+1)\alpha}, \quad 0 \le \xi \le t.$$

16.8 Find the exact solution of $D^\alpha x(t) = t^\beta$, $\beta > 0$, $x(0) = 0$, $0 < \alpha < 1$ applying the Laplace transform technique. Then, numerically solve this equation by the fractional Euler method for $\alpha = 0.5$, $\beta = 3$ and for $t \in [0,1]$ and compare the numerical solution with the exact solution.

16.9 For the equation $D^{0.5}x + x + 1 = 0$, $x(0) = 0$ by the fractional Euler method compute $x(0.0001)$ and $x(0.0002)$. Do the calculations using a calculator.

16.10 Using a calculator and applying the fractional Euler method find the solution of $D^{0.5}x = x + t$, $x(0) = 1$ at $t = 0.001$ and 0.002.

16.11 Applying the ABM method find the numerical solution of the equation $D^{0.5}x = t$, $x(0) = 1$ at $t = 0.01$ and 0.02. Use a calculator for the calculations.

16.12 For the equation $D^{0.25}x = -x$, $x(0) = 1$ compute $x(0.01)$ and $x(0.02)$ by the ABM method and using a calculator.

16.13 Develop a program to solve Eq. (16.37) by the ABM method and verify the results presented in Table 16.6.

16.14 Using a calculator and applying the fractional Euler method find the solution of $D^{0.5}x = t$, $x(0) = 1$ at $t = 0.001$ and 0.002.

16.15 For the equation $D^{0.5}x = -x$, $x(0) = 1$ compute $x(0.001)$ and $x(0.002)$ by the two-step Adams–Bashforth method and using a calculator.

16.16 The fractional order logistic differential equation is $D^\alpha N = rN(1 - (N/k))$, $0 < \alpha < 1$, $0 < r \le 1$ and $k > 0$ [94]. Show that the equilibrium point $N^* = 0$ is always unstable and $N^* = k$ is always stable. For $\alpha = 0.25$, 0.5 and 0.75 numerically solve this equation with different values of $N(0)$ by the Euler method and verify that $N(t) \to N^*$ in the long-time limit.

16.17 Consider the stability determining eigenvalues of the equilibrium points $(x^*, y^*) = (\pm 1, 0)$ of the system (16.118). Vary the value of d from 0 to 5 in step size 0.1 and compute the values of $|\arg(\lambda_+)|$ and $|\arg(\lambda_-)|$ using Eq. (16.112). Verify that both $|\arg(\lambda_\pm)|$ are $> \pi/2$ there by they are $> \alpha\pi/2$ for $0 < \alpha < 1$ and so $(\pm 1, 0)$ are stable equilibrium points.

16.18 The fractional order Lotka–Volterra predator-prey model equation is [90]

$$D^\alpha x_1 = x_1(4 - ax_1 - bx_2), \quad D^\alpha x_2 = x_2(-d + cx_1),$$

where r, a, b, c, $d > 0$. Verify that the equilibrium points of the system are $(x_1^*, x_2^*) = (0,0)$, $(r/a, 0)$ and $(d/c, (rc - ad)/bc)$. Show that the equilibrium point $(0,0)$ is always unstable and $(r/a, 0)$ is stable for $0 < r < ad/c$. For a value of α and $a = b = c = d = 1$ solve the system by the fractional Euler method for several initial conditions and plot the trajectories in the $x_1 - x_2$ phase space for $r = 0.5$ and 1.5. Verify the predictions of linear stability analysis.

16.19 For the system $D^\alpha x = \mu - x^2$, $D^\alpha y = -y$ with $\mu > 0$ determine all the equilibrium points and identify their stability. Verify the stability nature of the equilibrium points by numerically solving the equations for various initial conditions by the fractional Euler method.

16.20 The fractional order Lorenz equations are [95]

$$D^\alpha x = \sigma(y - x), \quad D^\beta y = rx - y - xz^\rho, \quad D^\gamma z = xy - bz,$$

where $r > 0$, $\sigma > 0$, $\rho > 0$ and $b > 0$ are parameters. Choose $\rho = 1$, $\sigma = 10$, $b = 8/3$, $\alpha = \beta = \gamma$. One of the equilibrium points of this equations is $(x^*, y^*, z^*) = (0,0,0)$. This equilibrium point is stable for $r < 1$ and unstable for $r > 1$ as per the linear stability analysis. Numerically, solve these equations by the fractional Euler method and draw the trajectories in the neighbourhood of the equilibrium point $(0,0,0)$ for $r = 0.5$ and $r = 1.5$ and verify the predictions of the linear stability analysis.

17

Fractional Order Partial Differential Equations

17.1   Introduction

Certain numerical methods for fractional order ordinary differential equations are described in the previous chapter. The present chapter is devoted to the fractional order partial differential equations. Consider the linear partial differential equations of the form

$$a(x,t)\frac{\partial^m}{\partial t^m}u(x,t) + b(x,t)\frac{\partial^n}{\partial x^n}u(x,t) = f(x,t), \tag{17.1}$$

where x and t are space and time variables, respectively, and m, $n = 0, 1, 2, \ldots$. Classifications of the equations of the form (17.1) for integer values of m and n are well-known in the field of partial differential equations. For some details see Chapter 14. In the equations of the form (17.1) the time fractional order derivatives are often treated as the Caputo type or the Riemann–Liouville sense or the Riez sense [1,2]. For Eqs. (17.1) with $f = 0$, a and b are constants the classification of them according to the noninteger values of m and n in the interval $[0,2]$ is reported by Changpin Li and An Chen [2].

An excellent review on numerical methods of solving the fractional order linear partial differential equations with the order in the interval $[0,2]$ is presented in the refs. [2,3]. A great deal of interest has been focused on diffusion equations and variants of diffusion equations. Various finite-difference schemes for time-fractional diffusion, diffusion-wave, advection-diffusion and diffusion-convection like equations are proposed and analyzed [4-22]. Space-fractional diffusion and advection-dispersion equations with special emphasize on finite-difference schemes are dealt [23-34]. Numerical methods for equations with both space and time derivatives being fractional order have been reported [35-43]. Apart from finite-difference schemes, other approaches, namely, Galerkin methods [44-49], space spectral time Adams–Bashforth Moulton method [50], McCormack method [51], multi-step method [52], kernel scheme [53], radial point interpolation method [54], semi-discrete finite element method [55], Dufort-Frankel scheme [56], Crank-Nicolson method [57-58], Fourier method [59] and higher-order approximation methods [60-66] are developed and tested. Interest has been focused on fractional order nonlinear partial differential equations in the refs. [67-72].

In this present chapter, certain fractional order linear partial differential equations with a time variable and one space variable, that is, equations with 1+1 dimension are considered. Particularly, the following equations are taken for analysis:

1.   Time-fractional order diffusion equation

2.   Time-fractional order advection-diffusion equation

3.   Time-fractional order wave equation

4.   Time-fractional order damped wave equation

5.   Time-fractional order Fisher equation

6.   Space-fractional order diffusion equation

DOI: 10.1201/9781032649931-17

For the time-fractional diffusion equation without an external force or the source term a finite-difference scheme is developed and the stability condition for the scheme is obtained. The scheme is verified by considering the system with the exact analytical solution. The system in the presence of an external force is examined and approach of the numerical solution towards an equilibrium solution is tested. Then, the advection-diffusion equation is taken for analysis. Next, the time-fractional order wave equation, the damped wave equation and the Fisher equation are discussed. For the space-fractional order diffusion equation with appropriate initial and boundary conditions, a finite-difference scheme is arrived. Applicability of the scheme is studied by considering a decaying solution and a nonzero equilibrium solution.

17.2 Time-Fractional Diffusion Equation

The mean-square displacement of the anomalous diffusion is given by

$$\langle x^2(t) \rangle \sim \frac{2K_\alpha}{\Gamma(1+\alpha)} t^\alpha, \quad t \to \infty. \tag{17.2}$$

In Eq. (17.2) α is the anomalous diffusion exponent and K_α is the diffusion coefficient. When $\alpha = 1$ the diffusion is called *Brownian* and $0 < \alpha < 1$ corresponds to subdiffusion. Define $u(x,t)$ as the probability density of determining a particle at x at time t. The equation used to model the diffusion process with $\alpha = 1$ is

$$\frac{\partial}{\partial t} u(x,t) = D \frac{\partial^2}{\partial x^2} u(x,t), \quad D = K_1. \tag{17.3}$$

In the case of anomalous diffusion with $0 < \alpha < 1$ the relevant model is the fractional order diffusion equation [73-76]

$$\frac{\partial^\alpha}{\partial t^\alpha} u(x,t) = K_\alpha \frac{\partial^2}{\partial x^2} u(x,t), \quad t \geq 0. \tag{17.4}$$

This equation can be rewritten as

$$\frac{\partial u}{\partial t} = K_\alpha D_t^{1-\alpha} \frac{\partial^2 u}{\partial x^2}, \quad u = u(x,t), \tag{17.5}$$

where D_t^γ denotes γ-order time derivative. An explicit finite-difference algorithm for Eq. (17.5) (or Eq. (17.4)) has been developed and its applicability was investigated by Yuste and Acedo [4]. Their method is followed in this section for Eq. (17.5).

Fractional order diffusion-like equations are employed in the model of information diffusion in a social network [77], diffusion of second messenger signalling molecule inosital-1,4,5-trisphosphate (IP$_3$) in spiny dendrites of Purkinje neurons (in neuroscience) [78] and diffusion process due to high velocity contrasts in heterogeneous porous media [79]. In financial mathematics, in the formalism of the risk-neutral approach, the exponential option pricing models are governed by the fractional order diffusion equations [81]. In electrodynamics, averaging of charge density in a special manner gives fractional order time-derivatives in Maxwell equations.

17.2.1 A Finite-Difference Scheme

Discretize the space and time variables as

$$x_j = j\Delta x, \quad t_m = m\Delta t, \tag{17.6}$$

where Δx and Δt are the step sizes of the variables x and t, respectively. Denote $u(x_j, t_m)$ as $u_j^{(m)}$. The finite-difference scheme for numerically solving the diffusion equation $\partial u/\partial t = D\partial^2 u/\partial x^2$ is

$$\frac{1}{\Delta t}\left(u_j^{(m+1)} - u_j^{(m)}\right) = \frac{D}{(\Delta x)^2}\left(u_{j-1}^{(m)} - 2u_j^{(m)} + u_{j+1}^{(m)}\right) + T(x,t), \tag{17.7}$$

where the truncation term is $T(x,t)$. For Eq. (17.5), an equation similar to Eq. (17.7) after neglecting the truncation term is

$$\frac{1}{\Delta t}\left(u_j^{(m+1)} - u_j^{(m)}\right) = K_\alpha D_t^{1-\alpha}\frac{1}{(\Delta x)^2}\left(u_{j-1}^{(m)} - 2u_j^{(m)} + u_{j+1}^{(m)}\right). \tag{17.8}$$

Treat $D_t^{1-\alpha}$ in Eq. (17.8) as the Riemann–Liouville fractional order derivative operator. The Grünwald–Letnikov (GL) form of $D_t^{1-\alpha}f(t)$ is [82-86] (refer Eq. (16.3))

$$D_t^{1-\alpha}f(t) = \lim_{\Delta t\to 0}\frac{1}{(\Delta t)^{1-\alpha}}\sum_{k=0}^{m}\omega_k^{(1-\alpha)}f(t - k\Delta t), \tag{17.9a}$$

where m is the integer part of $t/\Delta t$ and

$$\omega_k^{(1-\alpha)} = (-1)^k \binom{1-\alpha}{k}. \tag{17.9b}$$

$\omega_k^{(\gamma)}$ can be computed recursively as

$$\omega_0^{(\gamma)} = 1, \quad \omega_k^{(\gamma)} = \left(1 - \frac{\gamma+1}{k}\right)\omega_{k-1}^{(\gamma)}. \tag{17.9c}$$

Use of the GL form of $D_t^{1-\alpha}$ in Eq. (17.8) leads to the finite-difference scheme [4]

$$u_j^{(m+1)} = u_j^{(m)} + S_\alpha\sum_{k=0}^{m}\omega_k^{(1-\alpha)}\left(u_{j-1}^{(m-k)} - 2u_j^{(m-k)} + u_{j+1}^{(m-k)}\right), \tag{17.10a}$$

where

$$S_\alpha = \frac{K_\alpha(\Delta t)^\alpha}{(\Delta x)^2} \tag{17.10b}$$

and $m = 0, 1, 2, \ldots$. To implement the above finite-difference scheme to Eq. (17.5), for $m = 0$ the values of $u_j^{(0)}$ are needed and are given by the initial condition $u(x,0)$. Notice the major difference in the formulas given by Eqs. (17.3) and (17.5). For Eq. (17.3) $u_j^{(m+1)}$ at $(j, m+1)$ is evaluated using the values of u at the mesh points $(j-1, m)$, (j, m) and $(j+1, m)$. In contrast to this, for the time-fractional order diffusion equation the value of u at $(j, m+1)$ is computed using the values of u at the mesh points (j, k), $(j-1, k)$ and $(j+1, k)$, $k = 0, 1, 2, \ldots, m$. That is, for each value of m, the value of k starts from 0. Though the scheme can be straight-forwardly implemented through a computer program, it is necessary to store all the estimates $u_{j-1}^{(k)}$, $u_j^{(k)}$ and $u_{j+1}^{(k)}$, $k = 0, 1, 2, \ldots, m$ to compute $u_j^{(m+1)}$.

17.2.2   Stability of the Finite-Difference Scheme

Let us perform the von Neumann stability analysis (refer Subsection 14.5.6) and obtain the condition for the stability of the scheme obtained [4].

Assume the solution as $u_j^{(m)} = \zeta_m e^{iqj\Delta x}$ with q as the spatial wave number and ζ_m as the time-dependent term. Substitution of this $u_j^{(m)}$ in Eq. (17.10a) leads to

$$\zeta_{m+1} e^{iqj\Delta x} = \zeta_m e^{iqj\Delta x} + S_\alpha \sum_{k=0}^{m} \omega_k^{(1-\alpha)} \zeta_{m-k} e^{iqj\Delta x} \left[e^{-iq\Delta x} - 2 + e^{iq\Delta x} \right]. \tag{17.11}$$

That is,

$$\zeta_{m+1} = \zeta_m - 4S_\alpha \sin^2 (q\Delta x/2) \sum_{k=0}^{m} \omega_k^{(1-\alpha)} \zeta_{m-k}. \tag{17.12}$$

Let $\zeta_{m+1} = \xi\zeta_m$. From this it is easy to note that $\zeta_{m-k} = \xi^{-k}\zeta_m$. With this, Eq. (17.12) becomes (see Problem 17.2 at the end of the present chapter)

$$\xi = 1 - 4S_\alpha \sin^2 (q\Delta x/2) \sum_{k=0}^{m} \omega_k^{(1-\alpha)} \xi^{-k}. \tag{17.13}$$

As $\zeta_{m+1} = \xi\zeta_m$, the time part ζ grows with time if $|\xi| > 1$ for certain q implying unstable aspect of the numerical solution. For stability substitute the extreme value of ξ as -1 and that of $\sin^2(q\Delta x/2)$ as 1 in Eq. (17.13) and find the condition on S_α. This results in

$$S_\alpha \leq \frac{1}{2\sum_{k=0}^{m}(-1)^k \omega_k^{(1-\alpha)}}. \tag{17.14}$$

In the limit of $m \to \infty$ find the value of the summation term in Eq. (17.14) [4]. As

$$\omega_0^{(1-\alpha)} = 1, \quad \omega_1^{(1-\alpha)} = \alpha - 1, \quad \omega_2^{(1-\alpha)} = (\alpha - 1)\alpha/2, \ldots \tag{17.15}$$

the summation term becomes

$$\sum_{k=0}^{m}(-\xi)^k \omega_k^{(1-\alpha)} = 1 - (\alpha - 1)\xi + \frac{1}{2!}(\alpha - 1)\alpha\xi^2 - \ldots = (1 - \xi)^{1-\alpha}. \tag{17.16}$$

When $\xi = -1$, the summation term is $2^{1-\alpha}$. Then, from Eq. (17.14)

$$S_\alpha = \frac{K_\alpha(\Delta t)^\alpha}{(\Delta x)^2} < \frac{1}{2^{2-\alpha}} \tag{17.17}$$

which gives

$$\Delta t < (\Delta t)_c = \left(\frac{(\Delta x)^2}{2^{2-\alpha}K_\alpha} \right)^{1/\alpha}. \tag{17.18}$$

Table 17.1 presents the values of $(\Delta t)_c$ for some values of α and Δx with $K_\alpha = 1$.

Treating the so-called L1 approximation [87] for $D_t^{1-\alpha}$ an implicit scheme for the diffusion equation has been proposed [5].

TABLE 17.1
$(\Delta t)_c$ values for certain chosen values of α and Δx for $K_\alpha = 1$.

α	Δx	$(\Delta t)_c$	α	Δx	$(\Delta t)_c$
1.00	1/40	0.0003125	0.70	1/25	0.0000280
0.90	1/40	0.0001180	0.60	1/10	0.0000920
0.75	1/25	0.0000590	0.50	1/10	0.0000125

17.2.3   Examples

Now, test the applicability of the scheme (17.10) by considering the system (17.4) with two different types of solutions.

Example 1: Decaying Solution

Consider the system with absorbing boundaries and the initial conditions as

$$u(0,t) = u(1,t) = 0, \quad u(x,0) = x(1-x), \quad 0 \le x \le 1. \tag{17.19}$$

The exact analytical solution for the given $u(x,0)$ can be determined by the variable separable method. The exact solution is

$$u(x,t) = \frac{8}{\pi^3} \sum_{n=0}^{\infty} \frac{1}{(2n+1)^3} \sin[(2n+1)\pi x] E_\alpha \left[-K_\alpha (2n+1)^2 \pi^2 t^\alpha \right], \tag{17.20}$$

where $E_\alpha[.]$ is the Mittag–Leffler function (refer Eq. (16.110)). Figure 17.1a shows the numerically computed solution (represented by solid circles) at $t = 0.5$ for four values of α and $K_\alpha = 1$ along with the exact solution (represented by continuous line). For $\alpha = 1, 0.9$, 0.75 and 0.5 the values of $(\Delta x, \Delta t)$ used are $(1/40, 0.00025)$, $(1/40, 0.0001)$, $(1/25, 0.00005)$ and $(1/10, 0.00001)$, respectively. The values of Δt are $< (\Delta t)_c$ (see Table 17.1). In Fig. 17.1b the numerically computed $u(x,t)$ is shown for a few values of t. $u(x,t) \to 0$ in the long time limit. Very good agreement between the numerical solution and the exact solution is clearly seen.

Example 2: An Equilibrium Solution

The finite-difference scheme (17.10) can be extended to the equation of the form

$$\frac{\partial^\alpha u}{\partial t^\alpha} = K_\alpha \frac{\partial^2 u}{\partial x^2} + f(x,t) \tag{17.21}$$

with $u(0,t) = u(1,t) = 0$ and an appropriate initial condition. Rewrite this equation as

$$\frac{\partial u}{\partial t} = K_\alpha D_t^{1-\alpha} \frac{\partial^2 u}{\partial x^2} + D_t^{1-\alpha} f(x,t). \tag{17.22}$$

Making use of Eqs. (17.9), the finite-difference scheme for the above equation is

$$u_j^{(m+1)} = u_j^{(m)} + S_\alpha \sum_{k=0}^{m} \omega_k^{(1-\alpha)} \left(u_{j-1}^{(m-k)} - 2u_j^{(m-k)} + u_{j+1}^{(m-k)} \right)$$

$$+ (\Delta t)^\alpha \sum_{k=0}^{m} \omega_k^{(1-\alpha)} f_j^{(m-k)}, \quad m = 0, 1, 2, \ldots. \tag{17.23}$$

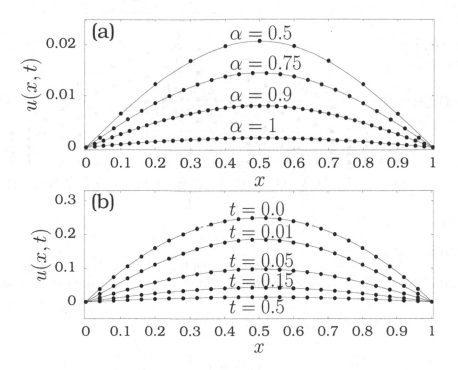

FIGURE 17.1

(a) $u(x,t)$ of the diffusion equation computed using Eqs. (17.10) for four values of α at $t = 0.5$. The solid circles are the numerical solutions and the continuous lines are the exact solutions. (b) $u(x,t)$ versus x for five values of t for $\alpha = 0.75$ and $K_\alpha = 1$. $u(x,t)$ decreases with increase in t.

If $D_t^{1-\alpha} f(x,t) = g(x,t)$ is easy to determine analytically then the last term in Eq. (17.23) can be replaced by $\Delta t g(x,t) = \Delta t g_j^{(m)}$. *What is the advantage of this?*

What is a stable equilibrium solution $u^*(x)$ of Eq. (17.21)? By a stable equilibrium solution, it is meant that an initial wave profile chosen near u^* approach it in the long-time limit. Let us seek such a solution. For simplicity assume that $K_\alpha = 1$. $u^*(x)$ is the solution of

$$u_{xx}^* + S(x) = 0. \tag{17.24}$$

Choose $f(x,t) = S(x) = \sin \pi x$. Then, Eq. (17.24) gives

$$u^*(x) = \frac{1}{\pi^2} \sin \pi x + ax + b, \tag{17.25}$$

where a and b are to be determined by the given boundary conditions. The condition $u(0,t) = 0$ gives $b = 0$ while $u(1,t) = 0$ gives $a = 0$. Therefore,

$$u^*(x) = \frac{1}{\pi^2} \sin \pi x. \tag{17.26}$$

Then, the exact solution of Eq. (17.21) is

$$u(x,t) = \frac{1}{\pi^2} \left[1 - E_\alpha \left(-\pi^2 t^\alpha \right) \right] \sin \pi x. \tag{17.27}$$

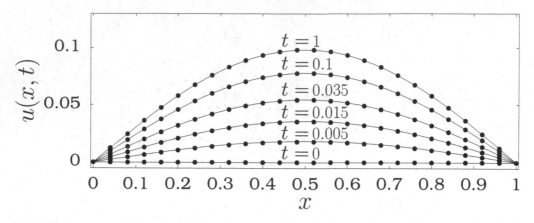

FIGURE 17.2
Plot of $u(x,t)$ for various values of t of the diffusion Eq. (17.21) with $f(x,t) = \sin \pi x$. Here, the solid circles are the numerical solutions and the continuous lines are the exact solution.

With $f(x,t) = \sin \pi x$ the term $D_t^{1-\alpha} \sin \pi x$ is obtained as

$$
\begin{aligned}
D_t^{1-\alpha} \sin \pi x &= \frac{1}{\Gamma(\alpha)} \frac{\partial}{\partial t} \int_0^t (t-\tau)^{\alpha-1} f(x,\tau) d\tau \\
&= \frac{\sin \pi x}{\Gamma(\alpha)} \frac{\partial}{\partial t} \int_0^t (t-\tau)^{\alpha-1} d\tau \\
&= \frac{\sin \pi x}{\Gamma(\alpha)} t^{\alpha-1}.
\end{aligned}
\tag{17.28}
$$

Then, replace the last term in Eq. (17.23) by $\Delta t (\sin \pi x) t^{\alpha-1}/\Gamma(\alpha)$.

Equation (17.22) is numerically solved using the algorithm (17.23) with $K_\alpha = 1$, $\Delta x = 1/25$, $\Delta t = 0.00005$ and $\alpha = 0.75$. The result is shown in Fig. 17.2 for some selected values of t. For sufficiently large values of $\pi^2 t^\alpha$, the term $E_\alpha\left(-\pi^2 t^\alpha\right) \approx 0$ giving $u(x,t) = \sin \pi x/\pi^2$. That is, as t increases $u(x,t)$ approaches the equilibrium solution u^* and is evident in Fig. 17.2. Though the initial solution is 0 the solution $u(x,t)$ evolves to the nonzero stationary solution.

17.3   Time-Fractional Advection-Diffusion Equation

The time-fractional order advection-diffusion equation is [20]

$$
\frac{\partial^\alpha u}{\partial t^\alpha} = K_\alpha \frac{\partial^2 u}{\partial x^2} - V_\alpha \frac{\partial u}{\partial x} + f(x,t)
\tag{17.29}
$$

subjected to the conditions

$$
u(x,0) = g_1(x), \quad u(0,t) = g_2(t), \quad u(L,t) = g_3(t), \quad 0 \le x \le L.
\tag{17.30}
$$

In Eq. (17.29) $0 < \alpha \le 1$, $K_\alpha > 0$ is the diffusion coefficient and V_α is the advection coefficient. In [20] making use of the rth degree interpolation method, some schemes for

Caputo derivative of order $\alpha \in [0,1]$ are proposed and are applied to Eq. (17.29). In the present section, the finite-difference approach discussed in the previous section is extended to Eq. (17.29).

With $D_t^{1-\alpha}$ as the Riemann–Liouville fractional derivative operator Eq. (17.29) can be written as

$$\frac{\partial u}{\partial t} = K_\alpha D_t^{1-\alpha}\frac{\partial^2 u}{\partial x^2} - V_\alpha D_t^{1-\alpha}\frac{\partial u}{\partial x} + D_t^{1-\alpha}f(x,t). \tag{17.31}$$

Discretize x and t as in Eq. (17.6) and denote $u(x_j,t_m)$ as u_j^m. For Eq. (17.31) the finite-difference scheme with

$$\frac{\partial}{\partial x}u_j^{(m)} = \frac{1}{2\Delta x}\left(u_{j+1}^{(m)} - u_{j-1}^{(m)}\right) \tag{17.32}$$

is

$$
\begin{aligned}
\frac{1}{\Delta t}\left(u_j^{(m+1)} - u_j^{(m)}\right) &= \frac{K_\alpha}{(\Delta x)^2}D_t^{1-\alpha}\left(u_{j-1}^{(m)} - 2u_j^{(m)} + u_{j+1}^{(m)}\right) \\
&\quad -\frac{V_\alpha}{2\Delta x}D_t^{1-\alpha}\left(u_{j+1}^{(m)} - u_{j-1}^{(m)}\right) \\
&\quad +D_t^{1-\alpha}f_j^{(m)}, \quad m = 0, 1, 2, \ldots .
\end{aligned} \tag{17.33}
$$

Considering Eqs. (17.9) for $D_t^{1-\alpha}f$ the above equation takes the form

$$
\begin{aligned}
u_j^{(m+1)} &= u_j^{(m)} + S_\alpha \sum_{k=0}^{m}\omega_k^{(1-\alpha)}\left(u_{j-1}^{(m-k)} - 2u_j^{(m-k)} + u_{j+1}^{(m-k)}\right) \\
&\quad -R_\alpha \sum_{k=0}^{m}\omega_k^{(1-\alpha)}\left(u_{j+1}^{(m-k)} - u_{j-1}^{(m-k)}\right) \\
&\quad +(\Delta t)^\alpha \sum_{k=0}^{m}\omega_k^{(1-\alpha)}f_j^{(m-k)}, \quad m = 0, 1, 2, \ldots ,
\end{aligned} \tag{17.34a}
$$

where

$$S_\alpha = \frac{K_\alpha(\Delta t)^\alpha}{(\Delta x)^2}, \quad R_\alpha = \frac{V_\alpha(\Delta t)^\alpha}{2\Delta x}. \tag{17.34b}$$

Performing the stability analysis, similar to the diffusion equation, to Eq. (17.29) with $f(x,t) = 0$ gives the stability condition for the finite-difference scheme (17.34) as (see Problem 17.8 at the end of the present chapter)

$$\frac{(\Delta t)^{2\alpha}}{4(\Delta x)^4}\left(V_\alpha^2(\Delta x)^2 + 16K_\alpha^2\right) \leq \frac{1}{2^{2-2\alpha}}. \tag{17.35}$$

For $V_\alpha \Delta x \ll 1$ the term $V_\alpha^2(\Delta x)^2$ can be neglected so that

$$\Delta t < (\Delta t)_c = \left(\frac{(\Delta x)^2}{2^{2-\alpha}K_\alpha}\right)^{1/\alpha}. \tag{17.36}$$

This condition is same as the condition (17.18) obtained for Eq. (17.4).

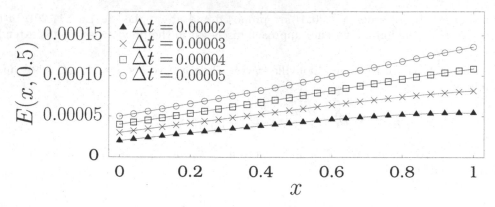

FIGURE 17.3

Plot of $E(x, 0.5) = |u_e(x, 0.5) - u_n(x, 0.5)|$, where u_e and u_n are the exact solution and the numerically computed solution, respectively, of Eq (17.29) for four different values of Δt.

Example:

For simplicity assume that $K_\alpha = V_\alpha = 1$ and the solution as $u(x, t) = te^x$ with $0 \le x \le 1$, $u(x, 0) = 0$, $u(0, t) = t$ and $u(1, t) = te$. Substitution of this solution in Eq. (17.29) with $\partial^\alpha u / \partial t^\alpha = t^{1-\alpha}/\Gamma(2-\alpha)$ gives $f(x, t) = t^{1-\alpha}e^x/\Gamma(2-\alpha)$. For $\Delta x = 1/25$ and $\alpha = 0.75$ the value of $(\Delta t)_c = 0.000058$. Numerical $u(x, t)$ is computed applying the finite-difference scheme (17.34) for four values of Δt and for $t \le 0.5$. Note that $u(x, t)$ is independent of α. The numerical solution and the error E in the numerical solution, that is the absolute difference between the exact and the numerically computed solutions given by

$$E(x, t) = \big|u_e(x, t) - u_n(x, t)\big|, \tag{17.37}$$

where u_e and u_n are the exact solution and the numerically computed solution, respectively, are calculated at $t = 0.5$ for $0 \le x \le 1$. The obtained $E(x, 0.5)$ for chosen values of Δt are presented in Fig. 17.3. Notice that $E < 10^{-3}$ and decreases with decrease in the value of Δt. The advection-diffusion system can also possess an equilibrium solution because of the presence of the external forcing term $f(x, t)$. For an example, see Problem 17.10 at the end of the present chapter).

17.4   Time-Fractional Wave Equation

The time-fractional wave equation is

$$\frac{\partial^\alpha u}{\partial t^\alpha} = \frac{\partial^2 u}{\partial x^2} + f(x, t), \quad 0 \le x \le \pi, \tag{17.38}$$

where $\alpha \in [1, 2]$ and the relevant initial and boundary conditions are

$$u(x, 0) = \varphi(x), \quad u_t(x, 0) = \phi(x), \quad u(0, t) = g_1(t), \quad u(\pi, t) = g_2(t). \tag{17.39}$$

When $\alpha = 1$ it is the classical diffusion equation. The equation with $\alpha = 2$ is the wave equation and it governs sound wave propagation through ideal conducting media. For $\alpha \in [1, 2]$

it is thought to interpolate the diffusion and wave equations. Time fractional wave equation describes the propagation of waves in viscoelastic media, particularly, in viscoelastic solids with a power-law creep and oscillations of smart material cable. It also describes travelling of sound waves in an inhomogeneous medium exhibiting power-law attenuation with frequency [88,89].

Operation of the fractional integral operator $D_t^{1-\alpha}$ with $\alpha \in [1,2]$ on Eq. (17.38) gives

$$\frac{\partial u}{\partial t} = \frac{\partial u}{\partial t}\bigg|_{t=0} + D_t^{1-\alpha}\frac{\partial^2 u}{\partial x^2} + D_t^{1-\alpha}f(x,t). \tag{17.40}$$

Discretize the variables x and t as in Eq. (17.6). Similar to the diffusion equation, the finite-difference scheme for the wave equation is

$$u_j^{(m+1)} = u_j^{(m)} + \Delta t\phi(x) + S_\alpha \sum_{k=0}^{m} \omega_k^{(1-\alpha)}\left(u_{j-1}^{(m-k)} - 2u_j^{(m-k)} + u_{j+1}^{(m-k)}\right)$$

$$+(\Delta t)^\alpha \sum_{k=0}^{m} \omega_k^{(1-\alpha)} f_j^{(m-k)}, \quad S_\alpha = \frac{(\Delta t)^\alpha}{(\Delta x)^2}, \tag{17.41}$$

and $m = 0, 1, 2, \ldots$. The stability condition for the above scheme is same as that of the diffusion Eq. (17.4), however, here $\alpha \in [1,2]$. That is,

$$\Delta t < (\Delta t)_c = \left(\frac{(\Delta x)^2}{2^{2-\alpha}}\right)^{1/\alpha}, \quad \alpha \in [1,2]. \tag{17.42}$$

Example: An Equilibrium Solution

The equilibrium solution u^* is the solution of the equation $\partial^2 u^*/\partial x^2 + f(x) = 0$. For $f(x) - \sin x$, u^* is $\sin x + ax + b$. Assume that $u(0,t) = g_1(t) = 0$ and $u(\pi,t) - g_2(t) = 0$. The first and the second conditions give $b = 0$ and $a = 0$, respectively, and thus $u^* = \sin x$. For $f(x,t) = \sin x$ the last term in Eq. (17.40) can be easily evaluated. As $\alpha \in [1,2]$, write $\alpha = 1 + \beta$, $\beta \in [0,1]$. Then,

$$\begin{aligned} D_t^{1-\alpha}f(x,t) &= {}_{\mathrm{RL}}D_t^{1-1-\beta}f(x,t) \\ &= \frac{\sin x}{\Gamma(\beta)}\int_0^t (t-\tau)^{\beta-1}\mathrm{d}\tau \\ &= \frac{\sin x}{\Gamma(1+\beta)}t^\beta \\ &= \frac{\sin x}{\Gamma(\alpha)}t^{\alpha-1}, \quad \alpha \in [1,2]. \end{aligned} \tag{17.43}$$

The last term in Eq. (17.41) containing $f_j^{(m-k)}$ can be replaced by $\Delta t \sin x\, t^{\alpha-1}/\Gamma(\alpha)$.

Next, determine $\phi(x)$ in Eq. (17.41) which is given by $u_t(x,0)$. As the wave equation is linear its exact solution can be determined, for example, by the variable separable method. Writing $u(x,t) = U(t)\sin x$, the equation for $U(t)$ is

$$\frac{\mathrm{d}^\alpha U}{\mathrm{d}t^\alpha} = -U + 1. \tag{17.44}$$

Its solution is

$$U(t) = 1 + E_\alpha(-t^\alpha), \tag{17.45}$$

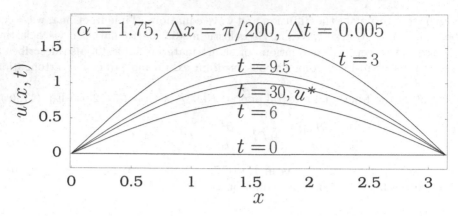

FIGURE 17.4
The numerically computed solution $u(x,t)$ of the wave Eq. (17.38) for $\alpha = 1.75$.

where E_α is the Mittag-Leffler function (refer Eq. (16.110)). Then,

$$
\begin{aligned}
u(x,t) &= (1 + E_\alpha(-t^\alpha)) \sin x \\
&= \sin x (1 + 1 - t^\alpha / \Gamma(\alpha + 1) + t^{2\alpha} / \Gamma(2\alpha + 1) - \ldots). \quad (17.46)
\end{aligned}
$$

This gives

$$
u_t(x,t) = \sin x (-\alpha t^{\alpha - 1} / \Gamma(\alpha + 1) + 2\alpha t^{2\alpha - 1} / \Gamma(2\alpha + 1) - \ldots) \quad (17.47)
$$

and $\phi(x) = u_t(x,0) = 0$.

Fix $\alpha = 1.75$, $\Delta x = \pi/200$ and $\Delta t = 0.005$ whereas $(\Delta t)_c = 0.00786$. The initial wave profile is chosen as $u(x,0) = 0$. Figure 17.4 shows the time evolution of the zero initial wave. The solution appears as oscillating and in this figure at $t = 30$ the numerically computed solution $u(x,t)$ becomes $u^*(x)$. In order to know what is happening, collect the value of, for example, $u(\pi/2, t)$ for a wide range of values of t. The result is the Fig. 17.5. $u(\pi/2, t)$ exhibits damped oscillation. Though the initial solution is zero because of the term $f(x)$ the system displays an equilibrium solution in the long-time limit.

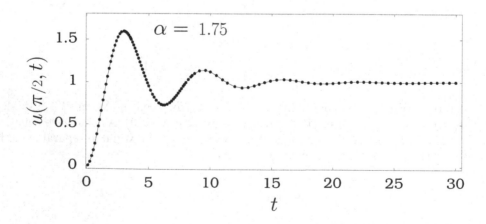

FIGURE 17.5
Time variation of $u(\pi/2, t)$ of the diffusion Eq. (17.38) for $\alpha = 1.75$.

In the scheme given by Eq. (17.41) choose $\phi(x) = \sin x$ and compute the numerical solution and find whether it approaches $u^* = \sin x$ or not. Explain your observation.

17.5 Time-Fractional Damped Wave Equation

The time-fractional damped wave equation in $1 + 1$ dimension is

$$\frac{\partial^2 u}{\partial t^2} + \frac{\partial^\alpha u}{\partial t^\alpha} - \frac{\partial^2 u}{\partial x^2} = f(x, t), \tag{17.48}$$

where $0 < \alpha < 1$, $0 \leq x \leq 1$ and $u(x, 0) = \psi(x)$, $u_t(x, 0) = \phi(x)$ and $u(0, t) = u(1, t) = 0$. Equation (17.48) with the additional term $u(x, t)$ is called *fractional telegraph equation*. For this equation, two finite-difference schemes have been developed [56]. In this section, Eq. (17.48) is considered.

17.5.1 A Finite-Difference Scheme

Rewrite Eq. (17.48) as

$$\frac{\partial u}{\partial t} + D_t^{1-\alpha} \frac{\partial^2 u}{\partial t^2} - D_t^{1-\alpha} \frac{\partial^2 u}{\partial x^2} = D_t^{1-\alpha} f(x, t). \tag{17.49}$$

Define $\Delta x = 1/j_x$, $x_j = j\Delta x$, $j = 0, 1, \ldots, j_x$, increment in t as Δt, $u_j^{(m)} = u(x_j, t_m)$,

$$\frac{\partial}{\partial t} u_j^{(m)} \quad - \quad \frac{1}{2\Delta t} \left(u_j^{(m+1)} - u_j^{(m-1)} \right), \tag{17.50a}$$

$$\frac{\partial^2}{\partial x^2} u_j^{(m)} \quad = \quad \frac{1}{(\Delta x)^2} \left(u_{j+1}^{(m)} - 2u_j^{(m)} + u_{j-1}^{(m)} \right), \tag{17.50b}$$

$$\frac{\partial^2}{\partial t^2} u_j^{(m)} \quad = \quad \frac{1}{(\Delta t)^2} \left(u_j^{(m+1)} - 2u_j^{(m)} + u_j^{(m-1)} \right), \tag{17.50c}$$

$$D_t^{1-\alpha} f_j^{(m)} \quad = \quad \frac{1}{(\Delta t)^{1-\alpha}} \sum_{k=0}^{m} \omega_k^{(1-\alpha)} f_j^{(m-k)}. \tag{17.50d}$$

Substitution of Eqs. (17.50) in Eq. (17.49) gives

$$
\begin{aligned}
u_j^{(m+1)} \quad = \quad & u_j^{(m-1)} - 2(\Delta t)^{\alpha-2} \sum_{k=0}^{m} \omega_k^{(1-\alpha)} \left(u_j^{(m+1-k)} - 2u_j^{(m-k)} + u_j^{(m-1-k)} \right) \\
& +2\frac{(\Delta t)^\alpha}{(\Delta x)^2} \sum_{k=0}^{m} \omega_k^{(1-\alpha)} \left(u_{j+1}^{(m-k)} - 2u_j^{(m-k)} + u_{j-1}^{(m-k)} \right) \\
& +2(\Delta t)^\alpha \sum_{k=0}^{m} \omega_k^{(1-\alpha)} f_j^{(m-k)}, \quad m = 0, 1, 2, \ldots.
\end{aligned}
\tag{17.51}
$$

When $k = 0$ the second term in the right-side of Eq. (17.51) contains the terms $u_j^{(m+1)}$, $u_j^{(m)}$ and $u_j^{(m-1)}$. The quantity $u_j^{(m+1)}$ to be determined occurs in both sides of the equation. Therefore, separate out the terms in this summation for $k = 0$ and so the starting value of

k becomes 1. Then, bringing the $u_j^{(m+1)}$ term $-2(\Delta t)^{\alpha-2}w_0^{(1-\alpha)}u_j^{(m+1)}$ to the right-side of Eq. (17.51) leads to the finite-difference scheme

$$
\begin{aligned}
u_j^{(m+1)} \;=\; \frac{1}{S_1}\Bigg[& u_j^{(m-1)} - S_2\left(-2u_j^{(m)} + u_j^{(m-1)}\right) \\
& -S_3\sum_{k=1}^{m}\omega_k^{(1-\alpha)}\left(u_j^{(m+1-k)} - 2u_j^{(m-k)} + u_j^{(m-1-k)}\right) \\
& +S_4\sum_{k=0}^{m}\omega_k^{(1-\alpha)}\left(u_{j+1}^{(m-k)} - 2u_j^{(m-k)} + u_{j-1}^{(m-k)}\right) \\
& +S_5\sum_{k=0}^{m}\omega_k^{(1-\alpha)}f_j^{(m-k)}\Bigg], \quad m = 0,\,1,\,2,\,\ldots
\end{aligned}
\tag{17.52a}
$$

where

$$
S_1 \;=\; 1 + 2(\Delta t)^{\alpha-2}w_0^{(1-\alpha)}, \quad S_2 = S_1 - 1, \tag{17.52b}
$$

$$
S_3 \;=\; 2(\Delta t)^{\alpha-2}, \quad S_4 = 2\frac{(\Delta t)^{\alpha}}{(\Delta x)^2}, \quad S_5 = 2(\Delta t)^{\alpha}. \tag{17.52c}
$$

From Eqs. (17.52) note that to calculate $u_j^{(2)}$ the values of $u_j^{(-1)}$, $u_j^{(0)}$, $u_j^{(1)}$, $u_{j+1}^{(0)}$, $u_{j+1}^{(1)}$ and $u_{j-1}^{(1)}$ are needed. Except $u_j^{(-1)}$ and $u_j^{(1)}$ the values of all other quantities are given by the initial and boundary conditions. To find $u_j^{(-1)}$ and $u_j^{(1)}$ consider the first three terms in the Taylor series expansion of them about $u_j^{(0)}$. This gives

$$
\begin{aligned}
u_j^{(1)} \;=\; & u_j^{(0)} + \Delta t (u_t)_j^{(0)} + \frac{1}{2}(\Delta t)^2 (u_{tt})_j^{(0)} \\
\;=\; & u_j^{(0)} + \Delta t\phi_j + \frac{1}{2}(\Delta t)^2\left[-\left(\frac{\partial^\alpha u}{\partial t^\alpha}\right)_j^{(0)} + \left(\frac{\partial^2 u}{\partial x^2}\right)_j^{(0)} + f_j^{(0)}\right].
\end{aligned}
\tag{17.53}
$$

As $D_t^\alpha u_j^{(m)} = (1/(\Delta t)^\alpha)\sum_{k=0}^{m}\omega_k^{(\alpha)}u_j^{(m-k)}$ the expression for $D_t^\alpha u_j^{(0)}$ is

$$
D_t^\alpha u_j^{(0)} = \left(\frac{\partial^\alpha u}{\partial t^\alpha}\right)_j^{(0)} = \frac{1}{(\Delta t)^\alpha}\omega_0^{(\alpha)}u_j^{(0)}. \tag{17.54}
$$

Then, with $\omega_0^{(\alpha)} = 1$

$$
\begin{aligned}
u_j^{(1)} \;=\; & u_j^{(0)} + \Delta t\phi_j - \frac{1}{2}(\Delta t)^{2-\alpha}u_j^{(0)} + \frac{(\Delta t)^2}{2(\Delta x)^2}\left(u_{j+1}^{(0)} - 2u_j^{(0)} + u_{j-1}^{(0)}\right) \\
& +\frac{1}{2}(\Delta t)^2 f_j^{(0)}
\end{aligned}
\tag{17.55}
$$

and an expression for $u_j^{(-1)}$ can be obtained from

$$
u_j^{(-1)} = u_j^{(0)} - \Delta t (u_t)_j^{(0)} + \frac{1}{2}(\Delta t)^2 (u_{tt})_j^{(0)}. \tag{17.56}
$$

17.5.2   Example

The exact solution of Eq. (17.48) for

$$
f(x,t) = \left(\frac{t^{1-\alpha}}{\Gamma(2-\alpha)} + \pi^2 t\right)\sin \pi x \tag{17.57}
$$

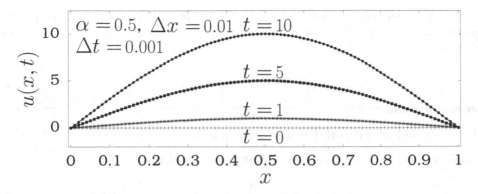

FIGURE 17.6
The numerical solutions of Eq. (17.48) for three values of t computed employing the finite-difference scheme (17.52) for $\alpha = 0.5$. Continuous lines and the solid circles represent the exact and the numerically computed solutions, respectively.

is $u(x,t) = t \sin \pi x$. For the choice $0 \leq x \leq 1$ the conditions are $u(x,0) = 0$, $u(0,t) = 0$, $u(1,t) = 0$ and $u_t(x,0) = \phi(x) = \sin \pi x$. Further, $u_j^{(1)} = u_j^{(0)} + \Delta t \sin \pi x_j$ and $u_j^{(-1)} = u_j^{(0)} - \Delta t \sin \pi x_j$. Choose $\Delta x = 0.01$ and $\Delta t = 0.001$. The numerical solutions computed at $t = 1$, 5 and 10 are plotted in Fig. 17.6. The exact solution is independent of α. This is verified from the numerical solution computed for several values of α. The error E in the computed solution is given by Eq. (17.37). In Fig. 17.7 E at $x = 0.5$ and $t = 10$ for $\alpha = 0.5$ is plotted as a function of the space step size Δx. As $u(x,t)$ is maximum at $x = 0.5$ the error E at this value of x is considered in Fig. 17.7. E decreases with decrease in Δx. For $\Delta x = 0.01$ the error $\overline{E}(0.5, 10)$ is $< 10^{-3}$.

Instead of Eq. (17.50a) the approximation

$$\frac{\partial}{\partial t} u_j^{(m)} = \frac{1}{\Delta t} \left(u_j^{(m+1)} - u_j^{(m)} \right) \tag{17.58}$$

can be considered. Then, the finite-difference scheme is given by Eq. (17.52) with the first term in the right-side of it is replaced by $u_j^{(m)}$ and with S_1 to S_5 without the multiplication

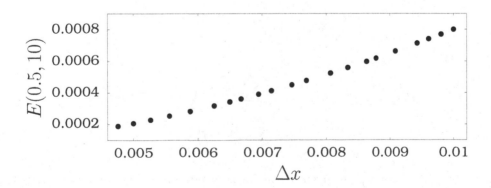

FIGURE 17.7
The variation of the error in the numerical solution at $t = 10$ and $x = 0.5$ for $\alpha = 0.5$, $\Delta t = 0.001$ with Δx for the damped wave Eq. (17.48).

factor 2. The numerical solution obtained with the approximation (17.58) is found to be almost the same as the solution obtained with the approximation (17.50a). The reader may verify this.

17.6   Time-Fractional Fisher Equation

The time-fractional order version of the Fisher Eq. (14.121) is

$$\frac{\partial^{\alpha} u}{\partial t^{\alpha}} + c\frac{\partial u}{\partial x} = f(x,t)\,. \tag{17.59}$$

For simplicity choose $c = 1$ and $\alpha \in [0,1]$. As usual, rewrite this equation as

$$\frac{\partial u}{\partial t} = -cD_t^{1-\alpha}\frac{\partial u}{\partial x} + D_t^{1-\alpha}f(x,t)\,. \tag{17.60}$$

17.6.1   Finite-Differences Schemes

With

$$\frac{\partial u}{\partial t} = \frac{1}{\Delta t}\left(u_j^{(m+1)} - u_j^{(m)}\right)\,, \quad \frac{\partial u}{\partial x} = \frac{1}{2\Delta x}\left(u_{j+1}^{(m)} - u_{j-1}^{(m)}\right)\,, \tag{17.61}$$

the finite-difference scheme for Eq. (17.60) is

$$u_j^{(m+1)} = u_j^{(m)} - S_\alpha \sum_{k=0}^{m} \omega_k^{(1-\alpha)}\left(u_{j+1}^{(m-k)} - u_{j-1}^{(m-k)}\right)$$

$$+ (\Delta t)^\alpha \sum_{k=0}^{m} \omega_k^{(1-\alpha)} f_j^{(m-k)}\,, \quad m = 0,1,\dots \tag{17.62a}$$

where

$$S_\alpha = \frac{(\Delta t)^\alpha}{2\Delta x}\,, \quad \omega_0^{(1-\alpha)} = 1\,, \quad \omega_k^{(1-\alpha)} = \left(1 - \frac{2+\alpha}{k}\right)\omega_{k-1}^{(1-\alpha)}\,. \tag{17.62b}$$

One may choose

$$\frac{\partial u}{\partial t} = \frac{1}{2\Delta t}\left(u_j^{(m+1)} - u_j^{(m-1)}\right)\,. \tag{17.63}$$

In this case

$$u_j^{(m+1)} = u_j^{(m-1)} - S_\alpha' \sum_{k=0}^{m} \omega_k^{(1-\alpha)}\left(u_{j+1}^{(m-k)} - u_{j-1}^{(m-k)}\right)$$

$$+ 2(\Delta t)^\alpha \sum_{k=0}^{m} \omega_k^{(1-\alpha)} f_j^{(m-k)}\,, \quad m = 0,1,\dots \tag{17.64}$$

with $S_\alpha' = (\Delta t)^\alpha/\Delta x$. $u_j^{(0)}$ is given by the initial condition $u(x,0)$. Equation (17.64) gives $u_j^{(2)}$, $u_j^{(3)}$, $\dots$. The values of $u_j^{(1)}$ need to be calculated. For this use

$$\frac{\partial u}{\partial t} = \frac{1}{\Delta t}\left(u_j^{(m+1)} - u_j^{(m)}\right) \tag{17.65}$$

and obtain the approximation

$$u_j^{(1)} = u_j^{(0)} - S_\alpha \omega_0^{(1-\alpha)} \left(u_{j+1}^{(0)} - u_{j-1}^{(0)} \right) + (\Delta t)^\alpha \omega_0^{(1-\alpha)} f_j^{(0)}. \tag{17.66}$$

It is easy to obtain the stability condition for the finite-difference schemes (17.62) and (17.64).

17.6.2 Examples

For illustrative purpose, assume the solution of Eq. (17.59) as $u(x,t) = t^\alpha e^x$ and find $f(x,t)$. As

$$D_t^\alpha u(x,t) = \Gamma(1+\alpha) e^x \tag{17.67}$$

Eq. (17.59) gives

$$f(x,t) = \Gamma(1+\alpha) e^x + t^\alpha e^x. \tag{17.68}$$

$D_t^{1-\alpha} f(x,t)$ in Eq. (17.60) becomes

$$D_t^{1-\alpha} f(x,t) = \frac{\Gamma(1+\alpha)}{\Gamma(\alpha)} t^{\alpha-1} e^x + e^x D_t^{1-\alpha} t^\alpha. \tag{17.69}$$

Then, in Eq. (17.62a) the last term in the right-side takes the form

$$\frac{\Gamma(1+\alpha)}{\Gamma(\alpha)} t_m^{\alpha-1} e^{x_j} \Delta t + (\Delta t)^\alpha e^{x_j} \sum_{k=0}^{m} \omega_k^{(1-\alpha)} t_{m-k}^\alpha. \tag{17.70}$$

The initial and the boundary conditions are

$$u(x,0) = 0, \quad u(x_{\min},t) = t^\alpha e^{x_{\min}}, \quad u(x_{\max},t) = t^\alpha e^{x_{\max}}. \tag{17.71}$$

Fix $x_{\min} = 0$, $x_{\max} = 1$, $\Delta x = 1/25$ and $\Delta t = 0.00005$. The numerically computed solution at $t = 2$ employing the scheme (17.62) for $\alpha = 1$, 0.75 and 0.5 is presented in Fig. 17.8a along with the exact solution. Figure 17.8b shows $u(x,t)$ at $t = 0.5, 1, 1.5$ and 2 for $\alpha = 0.75$.

In Section 14.8, the Fisher Eq. (17.59) with $\alpha = 1$ is considered with $u(x,t) = \text{sech}(x-t)$. Shape preserving and forwardly moving nature of the solution is realized in the numerical solution also. For the Eq. (17.59) with $\alpha \neq 1$ suppose $u(x,0) = \text{sech}(2x)$, $x \in [-10,15]$, $\Delta x = 1/25$ and $\Delta t = 0.00005$. The time evolution of this single hump wave is shown in Fig. 17.9. Because the value of α is < 1, the system is a damped system and hence the wave solution is spreading over space with decreasing amplitude with time.

17.7 Diffusion Equation with Fractional Order Space Derivative

In Section 17.2, diffusion equation with fractional order time derivative is considered. This section is concerned with the diffusion equation with a fractional order space derivative. The system is

$$\frac{\partial u}{\partial t} = \frac{\partial^\alpha u}{\partial x^\alpha} + f(x,t), \tag{17.72}$$

where $u = u(x,t)$, $\alpha \in [1,2]$, $0 \leq x \leq 1$ and subjected to appropriate initial and boundary conditions. Finite-difference schemes for Eq. (17.72) are obtained by treating $D_x^\alpha = \partial^\alpha/\partial x^\alpha$ as the Riemann–Liouville derivative operator [23] and Grünwald–Lutnikov derivative operator [26].

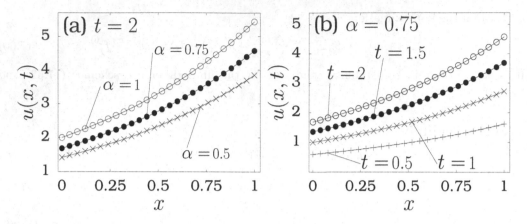

FIGURE 17.8
(a) $u(x,t)$ of the Fisher Eq. (17.59) with $f(x,t) = \Gamma(1+\alpha)\,\mathrm{e}^x + t^\alpha \mathrm{e}^x$ obtained using the scheme (17.62). The symbols and the continuous lines represent the numerically computed and the exact analytical solutions, respectively. (b) $u(x,t)$ at four values of t for $\alpha = 0.75$.

17.7.1 D^α as the Riemann–Liouville Derivative Operator

$D^\alpha_{a,x}F(x)$ is given by

$$D^\alpha_{a,x}F(x) = \frac{1}{\Gamma(m-\alpha)}\frac{\partial^m}{\partial x^m}\int_a^x (x-\tau)^{m-1-\alpha}F(\tau)\mathrm{d}\tau\,, \qquad (17.73)$$

where m is an integer with $m > \alpha > m-1$. For $\alpha \in [1,2]$ the value of m is 2. As

$$_{\mathrm{RL}}D^\alpha_{0,x}F(x) = {}_{\mathrm{C}}D^\alpha_{0,x}F(x) + \sum_{k=0}^{m-1}\frac{F^{(k)}(0)x^{k-\alpha}}{\Gamma(k+1-\alpha)}\,, \qquad (17.74)$$

one can write, for $\alpha \in [1,2]$,

$$_{\mathrm{RL}}D^\alpha_{0,x}F(x) = \frac{F(0)x^{-\alpha}}{\Gamma(1-\alpha)} + \frac{F'(0)x^{1-\alpha}}{\Gamma(2-\alpha)} + {}_{\mathrm{C}}D^\alpha_{0,x}F(x)\,. \qquad (17.75)$$

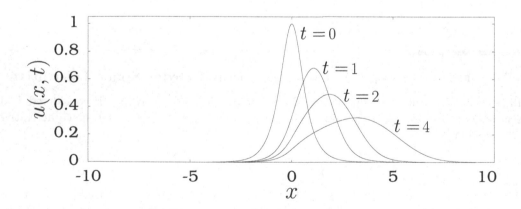

FIGURE 17.9
The time evolution of $u(x,0) = \mathrm{sech}(2x)$ according to Eq. (17.59) for $\alpha = 0.75$.

Let us discretize the derivative at $x = x_j$ as

$$\left[_{\mathrm{RL}}D_{0,x}^{\alpha}F(x)\right]_{x=x_j} = \frac{F(0)x_j^{-\alpha}}{\Gamma(1-\alpha)} + \frac{F'(0)x_j^{1-\alpha}}{\Gamma(2-\alpha)}$$
$$+ \frac{1}{\Gamma(2-\alpha)}\sum_{k=0}^{j-1}\int_{x_j}^{x_{j+1}}\tau^{1-\alpha}F''(x_j-\tau)\mathrm{d}\tau. \qquad (17.76)$$

Approximating $F''(x_j - \tau)$ on each interval $[x_j, x_{j+1}]$ as [2]

$$\frac{1}{(\Delta x)^2}\left[F(x_j - x_{k-1}) - 2F(x_j - x_k) + F(x_j - x_{k+1})\right] \qquad (17.77)$$

Eq. (17.76) becomes [23]

$$\left[_{\mathrm{RL}}D_{0,x}^{\alpha}F(x)\right]_{x=x_j} = \frac{F(0)x_j^{-\alpha}}{\Gamma(1-\alpha)} + \frac{F'(0)x_j^{1-\alpha}}{\Gamma(2-\alpha)} + \frac{(\Delta x)^{-\alpha}}{\Gamma(3-\alpha)}$$
$$\times \sum_{k=0}^{j-1}b_k\left(F_{j-k-1} - 2F_{j-k} + F_{j-k+1}\right), \qquad (17.78a)$$

where

$$b_k = (k+1)^{2-\alpha} - k^{2-\alpha}. \qquad (17.78b)$$

In Eq. (17.78a), for $x_j = 0$, the first two terms in the right-side are singular and are to be determined from the boundary conditions.

Returning to Eq. (17.72), approximate $\partial u_j^{(m)}/\partial t$ as $\left(u_j^{(m+1)} - u_j^{(m)}\right)/\Delta t$ and $D_x^{\alpha}u_j^{(m)}$ by Eq. (17.78a). The result is

$$u_j^{(m+1)} = u_j^{(m)} + \frac{\Delta t}{\Gamma(1-\alpha)}u_0^{(m)}x_j^{-\alpha} + \frac{\Delta t\, x_j^{1-\alpha}}{\Gamma(2-\alpha)}u_x|_{x=0,t=t_m}$$
$$+ \frac{\Delta t(\Delta x)^{-\alpha}}{\Gamma(3-\alpha)}\sum_{k=0}^{j-1}b_k\left(u_{j-k-1}^{(m)} - 2u_{j-k}^{(m)} + u_{j-k+1}^{(m)}\right)$$
$$+ \Delta t f_j^{(m)}. \qquad (17.79)$$

17.7.2 Appropriate Initial and Boundary Conditions

The singular terms, namely the second and the third terms in Eq. (17.79) can be avoided by appropriate boundary conditions [23]. The boundary condition $u_x(0,t) = 0$ removes the singular term, the third term in the right-side of Eq. (17.79). In order to remove the other singular term $\Delta t u_0^{(m)}x_j^{-\alpha}/\Gamma(1-\alpha)$ choose $u(0,t) = 0$. Then, the appropriate initial and boundary conditions are

$$u(x,0) = g(x), \quad u(1,t) = 0, \quad u(0,t) = 0, \quad u_x(0,t) = 0. \qquad (17.80)$$

The last two boundary conditions remove the singular terms in Eq. (17.79). Then, the finite-difference scheme becomes

$$u_j^{(m+1)} = u_j^{(m)} + \frac{\Delta t(\Delta x)^{-\alpha}}{\Gamma(3-\alpha)}\sum_{k=0}^{j-1}b_k\left(u_{j-k-1}^{(m)} - 2u_{j-k}^{(m)} + u_{j-k+1}^{(m)}\right) + \Delta t f_j^{(m)}. \qquad (17.81)$$

The stability condition for the scheme given by Eq. (17.81) is $\Delta t < (\Delta t)_c = (\Delta x)^\alpha$ [23].

In the case of $u(x,t) \neq 0$, the singular term $\Delta t u_0^{(m)} x_j^{-\alpha}/\Gamma(1-\alpha)$ is present. To remove it, consider the modified version of Eq. (17.72) as

$$\frac{\partial u}{\partial t} = D_x^\alpha(u(x,t) - u(0,t)) + f(x,t)\,, \tag{17.82}$$

The first singular term in the fractional derivative given by Eq. (17.78a) with $F = u(x,t) - u(0,t)$ is

$$\frac{F(0)x^{-\alpha}}{\Gamma(1-\alpha)} = [u(0,t) - u(0,t)]\frac{x^{-\alpha}}{\Gamma(1-\alpha)} = 0\,. \tag{17.83}$$

17.7.3   Determination of $u(0,t)$ for $t > 0$

If $u(0,t) \neq 0$ for $t > 0$ and is unknown then an expression for it can be easily obtained [23]. $u(x = 0,t)$ at $t = 0$ is known from the given initial condition. Consider the solution at $t = \Delta t$. From the finite-difference formula note that $u_0^{(\Delta t)}$ is not required for calculating $u_j^{(\Delta t)}$, $j = 1, 2, \ldots, j_x - 1$ but it is needed for calculating the solution at $t = 2\Delta t$.

$u_j^{(\Delta t)}$ for $j = 1, 2, \ldots, j_x - 1$ can be calculated from the finite-difference scheme. Then, an expression for $u_0^{(\Delta t)}$ involving $u_1^{(\Delta t)}$ and $u_2^{(\Delta t)}$ can be arrived. From this expression, one can write the expression for $u_0^{(m+1)}$ as described below. For this purpose, consider the expansion of $u_1^{(\Delta t)}$ near $x = 0$ as

$$u_1^{(\Delta t)} = u_0^{(\Delta t)} + \Delta x (u_x)_0^{(\Delta t)} + (\Delta x)^\alpha \left(\frac{\partial^\alpha u}{\partial x^\alpha}\right)_0^{(\Delta t)}. \tag{17.84}$$

As $u_x(0,t) = 0$

$$u_1^{(\Delta t)} = u_0^{(\Delta t)} + (\Delta x)^\alpha \left(\frac{\partial^\alpha u}{\partial x^\alpha}\right)_0^{(\Delta t)}. \tag{17.85}$$

Similarly,

$$u_2^{(\Delta t)} = u_0^{(\Delta t)} + 2^\alpha (\Delta x)^\alpha \left(\frac{\partial^\alpha u}{\partial x^\alpha}\right)_0^{(\Delta t)}. \tag{17.86}$$

This equation gives

$$(\Delta x)^\alpha \left(\frac{\partial^\alpha u}{\partial x^\alpha}\right)_0^{(\Delta t)} = \frac{1}{2^\alpha}\left(u_2^{(\Delta t)} - u_0^{(\Delta t)}\right). \tag{17.87}$$

Use of this in Eq. (17.85) leads to the result

$$u_0^{(\Delta t)} = \frac{u_1^{(\Delta t)} - 2^{-\alpha}u_2^{(\Delta t)}}{1 - 2^{-\alpha}}. \tag{17.88}$$

In general,

$$u_0^{(t+\Delta t)} = \frac{u_1^{(t+\Delta t)} - 2^{-\alpha}u_2^{(t+\Delta t)}}{1 - 2^{-\alpha}}\,, \tag{17.89}$$

that is,

$$u_0^{(m+1)} = \frac{u_1^{(m+1)} - 2^{-\alpha}u_2^{(m+1)}}{1 - 2^{-\alpha}}. \tag{17.90}$$

17.7.4 An Equivalent Alternative Form of Equation (17.81)

Reorganization of the sum in the second term in the right-side of Eq. (17.81) leads to an equivalent form of this term as [23]

$$\sum_{k=0}^{j-1} b_k \left(u_{j-k-1}^{(m)} - 2u_{j-k}^{(m)} + u_{j-k+1}^{(m)} \right) = \sum_{k=-1}^{j} W_k u_{j-k}^{(m)}, \tag{17.91a}$$

where

$$\begin{aligned}
W_0 &= 2^{2-\alpha} - 3, \quad W_{-1} = 1, \tag{17.91b}\\
W_k &= (k+2)^{2-\alpha} - 3(k+1)^{2-\alpha} + 3k^{2-\alpha} \\
&\quad -(k-1)^{2-\alpha}, \quad 1 \le k \le j-2, \tag{17.91c}\\
W_{j-1} &= -2j^{2-\alpha} + 3(j-1)^{2-\alpha} - (j-2)^{2-\alpha}, \tag{17.91d}\\
W_j &= j^{2-\alpha} - (j-1)^{2-\alpha}. \tag{17.91e}
\end{aligned}$$

Then,

$$u_j^{(m+1)} = u_j^{(m)} + \frac{\Delta t (\Delta x)^{-\alpha}}{\Gamma(3-\alpha)} \sum_{k=-1}^{j} W_k u_{j-k}^{(m)} + \Delta t f_j^{(m)}. \tag{17.92}$$

This formula is for Eq. (17.72) with the conditions given by Eq. (17.80). For Eq. (17.82) with $u(0,t) \ne 0$ the finite-difference scheme is

$$u_j^{(m+1)} = u_j^{(m)} + \frac{\Delta t (\Delta x)^{-\alpha}}{\Gamma(3-\alpha)} \sum_{k=-1}^{j} W_k \left(u_{j-k}^{(m)} - u_0^{(m)} \right) + \Delta t f_j^{(m)}. \tag{17.93}$$

17.7.5 Application of the Scheme (17.81)

Let us study the applicability of the finite-difference scheme (17.81) to the diffusion equation by considering a few specific solutions $u(x,t)$.

Example 1: Decaying Solution

Assume the solution of (17.72) as

$$u(x,t) = 16e^{-t}x^2(1-x)^2, \quad 0 \le x \le 1. \tag{17.94}$$

The initial and the boundary conditions (17.80) for this solution become

$$\begin{aligned}
u(x,0) &= g(x) = 16x^2(1-x)^2, \quad u(1,t) = 0, \quad u(0,t) = 0, \tag{17.95a}\\
u_x(0,t) &= 32e^{-t}(x - 3x^2 + 2x^3)_{x=0} = 0. \tag{17.95b}
\end{aligned}$$

Note that for the assumed solution $u(0,t) = 0$. $f(x,t)$ in Eq. (17.72) for the solution (17.93) is determined as

$$f(x,t) = -16e^{-t}x^2(1-x)^2 + \frac{32e^{-t}}{\Gamma(2-\alpha)} \left[\frac{x^{2-\alpha}}{\Gamma(3-\alpha)} + \frac{12x^{4-\alpha}}{\Gamma(5-\alpha)} - \frac{6x^{3-\alpha}}{\Gamma(4-\alpha)} \right]. \tag{17.96}$$

Discretize the space variable x by dividing the interval $[0,1]$ into j_x equal intervals with grid points $x_j = j\Delta x$, $j = 0, 1, \ldots, j_x$ where $\Delta x = 1/j_x$. For $\Delta x = 1/40 = 0.025$

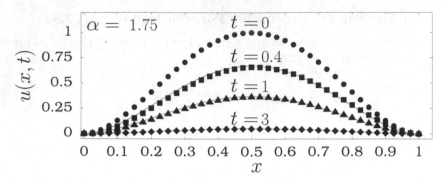

FIGURE 17.10
The numerically computed solution (marked by the symbols) and the exact analytical solution (marked by the continuous curve) of Eq. (17.72) with $f(x,t)$ given by Eq. (17.96). For $t = 0$ the solid circles represent the analytical solution. The scheme (17.81) is used with $\Delta x = 0.025$ and $\Delta t = 0.0001$.

and $\alpha = 1.75$ the value of $(\Delta t)_c$ is 0.00157. Fix the value of Δt as 0.0001. Figure 17.10 presents the numerical solutions (symbols) at $t = 0.4$, 1 and 3 along with the exact solutions (continuous curves) and the solution at $t = 0$. Closeness of the numerical solution with the exact solution is clearly seen. The solution is independent of α. This is verified with the numerical solution computed for several values of α. As expected, the solution decays to 0 in the long-time limit.

The influence of Δx on the error in the numerical solution is studied. For $\alpha = 1.75$ and $\Delta t = 0.0001$ the maximum error in the numerical solution at t is defined as

$$E_{\max}(t) = \max. \left(\left| u_e(x_j, t) - u_n(x_j, t) \right|, \quad j = 1, 2, \ldots, j_x - 1 \right), \tag{17.97}$$

where u_e is the exact solution and u_n is the numerically computed solution. $E_{\max}$ is computed at $t = 0.4$ for a range of values of Δx and the result is shown in Fig. 17.11. As Δx decreases the error decreases almost linearly. Instead of $E_{\max}$, another quantity called *average error*, E_{ave}, defined as [23]

$$E_{\text{ave}}(t) = \left[\frac{\sum_{j=0}^{j_x} \left(u_e(x_j, t) - u_n(x_j, t) \right)^2}{\sum_{j=0}^{j_x} (u_e(x_j, t))^2} \right]^{1/2} \tag{17.98}$$

can also be used to compare the numerical solution with the exact solution.

Example 2: Nonzero Equilibrium Solution

The solution given by Eq. (17.94) decays with time. For a suitable chosen $f(x,t)$ the system can admit a nonzero stationary solution [23]. Assume that $f(x,t)$ is independent of time and is $S(x)$. For an equilibrium solution $\partial u/\partial t = 0$. So, an equilibrium solution u^* of Eq. (17.72) is the solution of the equation $D_x^\alpha u^* + S(x) = 0$ where $f(x,t) = S(x)$, independent of time. For a chosen $S(x)$ integration of this equation gives u^*. For illustrative purpose, let $S(x) = 1 - x^{2-\alpha}$ [23]. Then,

$$u^*(x) = -D_{0,x}^{-\alpha} S(x) + ax + b, \tag{17.99}$$

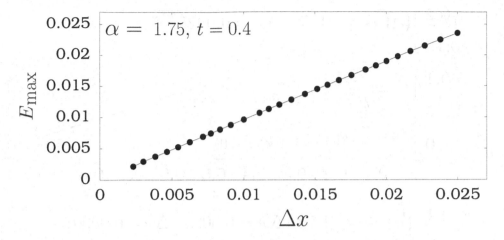

FIGURE 17.11
$E_{\max}$, the maximum error in the numerical solution for $x \in [0,1]$, versus the spatial step size Δx for Eq. (17.72) at $t = 0.4$ and $\alpha = 1.75$.

where a and b are to be determined employing the boundary conditions $u_x(0,t) = u(1,t) = 0$. $D_{0,x}^{-\alpha} S(x)$ is obtained as

$$
\begin{aligned}
D_{0,x}^{-\alpha} S(x) &= \frac{1}{\Gamma(\alpha)} \int_0^x (x-\tau)^{\alpha-1} \left(1 - \tau^{2-\alpha}\right) d\tau \\
&= \frac{1}{\Gamma(\alpha)} \int_0^x (x-\tau)^{\alpha-1} d\tau - \frac{1}{\Gamma(\alpha)} \int_0^x (x-\tau)^{\alpha-1} \tau^{2-\alpha} d\tau \\
&= -\frac{1}{\Gamma(1+\alpha)} (x-\tau)^{\alpha} \Big|_0^x - \frac{1}{\Gamma(\alpha)} L^{-1} \frac{\Gamma(3-\alpha)}{s^{3-\alpha}} \frac{\Gamma(\alpha)}{s^{\alpha}} \\
&= \frac{1}{\Gamma(1+\alpha)} x^{\alpha} - \frac{\Gamma(3-\alpha)}{2} x^2,
\end{aligned}
\tag{17.100}
$$

where L^{-1} is the inverse Laplace transform. Then,

$$
u^*(x) = -\frac{1}{\Gamma(1+\alpha)} x^{\alpha} + \frac{\Gamma(3-\alpha)}{2} x^2 + ax + b.
\tag{17.101}
$$

Applying the boundary condition $u_x(0,t) = 0$ gives $a = 0$. The other condition $u(1,t) = 0$ gives

$$
b = \frac{1}{\Gamma(1+\alpha)} - \frac{\Gamma(3-\alpha)}{2}.
\tag{17.102}
$$

Therefore,

$$
u^*(x) = \frac{1}{\Gamma(1+\alpha)} \left(1 - x^{\alpha}\right) - \frac{\Gamma(3-\alpha)}{2} \left(1 - x^2\right).
\tag{17.103}
$$

For Eq. (17.72) with $f(x,t) = S(x) = 1 - x^{2-\alpha}$ and the initial condition $u(x,0)$ chosen not too far away from the equilibrium solution $u^*(x)$ the solution $u(x,t)$ evolves towards u^*.

In the numerical solution fix $\alpha = 1.75$, $\Delta x = 1/400 = 0.0025$ and $\Delta t = 0.00001$. $u(x,t)$ is computed for two choices of $u(x,0)$. $u_j^{(m+1)}$ for $j = 1, 2, \ldots, j_x - 1$ are computed using the

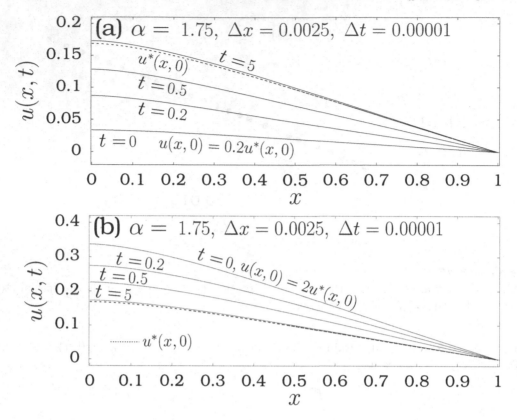

FIGURE 17.12
Plots of $u(x,t)$ of the diffusion Eq. (17.72) for $\alpha = 1.75$ and $f(x,t) = 1 - x^{2-\alpha}$. In the subplot (a) the initial wave profile is $u(x,0) = 0$ while it is $2u^*(x)$ in the subplot (b). The dashed line represents $u^*(x)$.

scheme given by Eq. (17.81) and $u_0^{(m+1)}$ is computed using Eq. (17.90) while $u_{j_x}^{(m+1)} = 0$. Figure 17.12a presents the time evolution of $u(x,0) = 0.2u^*(x)$. Even for the initial solution as zero solution, because $f(x,t)$ is nonzero the solution $u(x,t)$ is found to end up on the equilibrium state u^*. In Fig. 17.12b the initial wave profile is $u(x,0) = 2u^*(x)$. In this case, also $u(x,t)$ becomes u^* for sufficiently large values of t. $u(x,t)$ approaches to u^* with increase in t. $u(0.1,t)$ is computed for $t \in [0,6]$. The result is shown in Fig. 17.13 for both $u(x,0) = 0.2u^*(x)$ and $u(x,0) = 2u^*(x)$. $u(0.1,t)$ asymptotically approaches to $u^*(0.1)$ and in general, $u(x,t) \to u^*(x)$ in the long time limit. The convergence to u^* is a nonlinear function of time. In Fig. 17.14 $u(x,t)$ attained the equilibrium state for, say, $t > 3$. The maximum error $E_{\max}$ in the numerical solution is calculated at $t = 5$ using Eq. (17.97) with $u(x,0) = 2u^*(x)$ for some selected values of Δx. In Fig. 17.14 $E_{\max}$ is found to decrease with decrease in the value of Δx.

17.7.6   A Semi-Implicit Scheme

A semi-implicit scheme is developed in [23]. In this scheme, the terms in the summation in Eq. (17.93) forming a tridiagonal matrix are collected and used in the time advanced part.

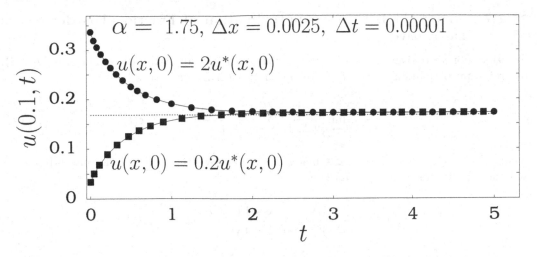

FIGURE 17.13
Time development of $u(0.1,0)$ for Eq. (17.72) with $f(x,t) = 1 - x^{2-\alpha}$. The horizontal dotted line repesents $u^*(0.1)$.

The scheme is given by

$$
u_j^{(m+1)} - \Delta t \left[W_j u_{j-1}^{(m+1)} + W_0 u_j^{(m+1)} + W_{-1} u_{j+1}^{(m+1)} \right]
$$
$$
= u_j^{(m)} + \Delta t \left[\sum_{k=2}^{j} W_k \left(u_{j-k}^{(m)} - u_0^{(m)} \right) - (W_{-1} + W_0 + W_1) u_0^{(m)} + f_j^{(m)} \right] , \quad (17.104)
$$

where W_k's are given by Eqs. (17.91b)-(17.91e).

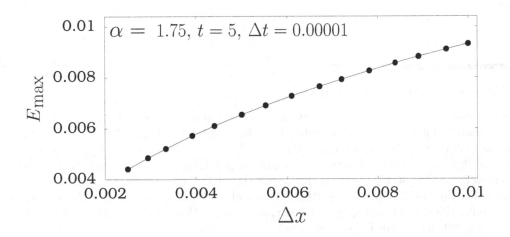

FIGURE 17.14
Variation of $E_{\max}$ given by Eq. (17.97) in the numerical solution at $t = 5$ of the diffusion Eq. (17.72) for the case of $u(x,0) = 2u^*(x)$ with Δx.

17.7.7 Grünwald–Letnikov Approximation of Fractional Order Space Derivative

Fractional order derivatives can be discretized by the GL approximation. The standard GL discrete approximation of $\partial^\alpha u/\partial x^\alpha$ for $\alpha \in [1, 2]$ with $L \le x \le R$ is

$$\frac{\partial^\alpha u}{\partial x^\alpha} = \frac{1}{\Gamma(-\alpha)} \lim_{j_x \to \infty} \frac{1}{(\Delta x)^\alpha} \sum_{k=0}^{j_x} \frac{\Gamma(k - \alpha)}{\Gamma(k + 1)} u(x - k\Delta x, t), \quad (17.105)$$

where $\Delta x = (x - L)/j_x$. With Δt being the increment step size of t the finite-difference scheme for the diffusion Eq. (17.72) is

$$u_j^{(m+1)} = u_j^{(m)} + \frac{\Delta t}{(\Delta x)^\alpha \Gamma(-\alpha)} \sum_{k=0}^{j} \frac{\Gamma(k - \alpha)}{\Gamma(k + 1)} u_{j-k}^{(m)} + \Delta t f_j^{(m)}. \quad (17.106)$$

This scheme is found to be unstable (see Problem 20 at the end of the present chapter) [24].

The shifted Grünwald formula is appropriate for Eq. (17.72). For $\alpha \in [1, 2]$ this formula for $D^\alpha f(x)$ is

$$D^\alpha f(x) = \lim_{j_x \to \infty} \frac{1}{(\Delta x)^\alpha} \sum_{k=0}^{j_x} g_k f(x - (k - 1)\Delta x) + O(\Delta x), \quad (17.107)$$

where $g_k = \Gamma(k - \alpha)/(\Gamma(-\alpha)\Gamma(k + 1))$. Applying this shifted Grünwald formula, the discretized version of Eq. (17.72) is [26]

$$u_j^{(m+1)} = u_j^{(m)} + \beta \sum_{k=0}^{j+1} g_k u_{j-k+1}^{(m)} + \Delta t f_j^{(m)}, \quad (17.108)$$

where $\beta = \Delta t/(\Delta x)^\alpha$. The above explicit Euler scheme is shown to be stable for $\Delta t \le (\Delta x)^\alpha/\alpha$ [26].

17.8 Concluding Remarks

In this chapter, a few fractional order linear partial differential equations with and without the external force $f(x, t)$ are considered. Finite-difference schemes are developed and the stability condition for the schemes are obtained. In the illustrative examples, the time step size is chosen such that the stability condition is satisfied. A reader can choose the time step as out of the stability region and notice the unstable nature of the numerical solution developed by the numerical algorithm. By suitably choosing the external force $f(x, t)$ one can make the system to display nondecaying solution. How to choose $f(x, t)$ for realizing a nonzero stationary solution is illustrated.

For space-fractional partial differential equations one example is considered. Appropriate finite-difference scheme for other linear partial differential equations can be obtained and analyzed. The numerical analysis of fractional order nonlinear partial differential equations is not well established at present and hence is not considered in the present book. For some details one may refer to the refs. [67-72].

17.9 Bibliography

[1] C.P. Li and F.H. Zeng, *Numerical Methods for Fractional Calculus.* Chapman and Hall/CRC Press, Boca Raton, 2015.

[2] C. Li and A. Chen, *Int. J. Comput. Math.* 95:1048, 2018.

[3] C. Li and F. Zeng, *Int. J. Bifur. & Chaos* 22:1230014, 2012.

[4] S.B. Yuste and L. Acedo, *SIAM J. Numer. Anal.* 42:1862, 2005.

[5] T.A.M. Langlands an B.I. Henry, *J. Comput. Phys.* 205:719, 2005.

[6] Z.Z. Sun and X. Wu, *Appl. Numer. Math.* 56:193, 2006.

[7] Y.M. Lin and C.J. Xu, *J. Comput. Phys.* 225:1533, 2007.

[8] E. Sousa, *J. Comput. Phys.* 228:4038, 2009.

[9] M.R. Cui, *J. Comput. Phys.* 228:7792, 2009.

[10] R. Du, W.R. Cao and Z.Z. Sun, *Appl. Math. Model.* 34:2998, 2010.

[11] K. Mustapha and J. Al Mustawa, *Numer. Algor.* 61:525, 2012.

[12] H.M. Nasir, B.L.K. Gunawardana and H.M.N. Abeyrathna, *Int. J. Appl. Phys. Math.* 3:237, 2013.

[13] J.F. Huang, Y.F. Tang, L. Vazquez and J.Y. Yang, *Numer. Algor.* 64:707, 2013.

[14] A.A. Alikhanov, *J. Comput. Phys.* 280:424, 2015.

[15] G.H. Gao, H.W. Sun and Z.Z. Sun, *J. Comput. Phys.* 280:510, 2015.

[16] Y.M. Wang, *BIT Numer. Math.* 55:1187, 2015.

[17] A. Chen and C. Li, *Numer. Funct. Anal. Opt.* 37(1):19, 2016.

[18] X. Hu, F. Liu, I. Turner and V. Anh, *Numer. Algor.* 72:393, 2016.

[19] H. Chen, S. Lu and W. Chen, *J. Comput. Phys.* 315:84, 2016.

[20] H. Li, J. Cao and C. Li, *J. Comput. Appl. Math.* 299:159, 2016.

[21] W. Bu, A. Xiao and W. Zeng, *J. Sci. Comput.* 72:422, 2017.

[22] S. Cheng, N. Du, H. Wang and Z. Yang, *Fract. Frac.* 6:525, 2022.

[23] V.E. Lynch, B.A. Carreras, D. del-Castillo-Negrete, K.M. Ferriera-Mejias and H.R. Hicks, *J. Comput. Phys.* 192:406, 2003.

[24] M.M. Meerschaert and C. Tadjeran, *J. Comput. Appl. Math.* 172:65, 2004.

[25] M. Illic, F. Liu, I. Turner and V. Anh, *Fract. Calc. Appl. Anal.* 8:323, 2005.

[26] M.M. Meerschaert and C. Tadjeran, *Appl. Numer. Math.* 56:80, 2006.

[27] M. Illic, F. Liu, I. Turner and V. Anh, *Fract. Calc. Appl. Anal.* 9:333, 2006.

[28] Q. Yang, F. Liu and I. Turner, *Appl. Math. Model.* 34:200, 2010.

[29] H. Wang and T.S. Basu, *SIAM J. Sci. Comput.* 34:A2444, 2012.

[30] H. Hejazi, T. Moroney and F. Liu, *J. Comput. Appl. Math.* 255:684, 2014.

[31] M.H. Chen and W.H. Deng, *Appl. Math. Model.* 38:3244, 2014.

[32] J.F. Huang, N.M. Nie and Y.F. Tang, *Sci. China Math.* 57:1303, 2014.

[33] W.Y. Tian, H. Zhou and W.H. Deng, *Math. Comput.* 84:1703, 2015.

[34] H.F. Ding and C.P. Li, *Fract. Calc. Appl. Anal.* 20:722, 2017.

[35] R. Metzler and T.F. Nonnenmacher, *Chem. Phys.* 284:67, 2002.

[36] M. Meerschaert, D. Benson, H.P. Scheffler and B. Baeumer, *Phys. Rev. E* 65:1103, 2002.

[37] W.H. Deng, *SIAM J. Numer. Anal.* 47:204, 2008.

[38] S.J. Shen, F. Liu and V.Anh, *Numer. Algor.* 56:383, 2011.

[39] M.H. Chen, W.H. Deng and Y.J. Wu, *Appl. Numer. Math.* 70:22, 2013.

[40] Y. Liu, Y. Du, H. Li, S. He and W. Cao, *Comput. Math. Appl.* 70:573, 2015.

[41] Q. Liu, F.H. Zeng and C.P. Li, *Int. J. Comput. Math.* 92:1439, 2015.

[42] L.B. Feng, P. Zhuang, F. Liu, I. Turner and Y.T. Gu, *Numer. Algor.* 72:749, 2016.

[43] H. Sun, Z.Z. Sun and G.H. Hao, *Appl. Math. Comput.* 281:356, 2016.

[44] W.H. Deng and J.S. Hesthaven, *Math. Model. Numer. Anal.* 47:1845, 2013.

[45] M. Li, D. Xu and M. Luo, *J. Comput. Phys.* 255:471, 2013.

[46] W.P. Bu, Y.F. Tang and J.Y. Yang, *J. Comput. Phys.* 276:26, 2014.

[47] K. Mustapha. B. Abdallah and K.M. Furati, *SIAM J. Numer. Anal.* 52:2512, 2014.

[48] B. Jin, R. Lazarov, Y. Liu and Z. Zhou, *J. Comput. Phys.* 281:825, 2015.

[49] Y. Liu, Y.W. Du, H. Li and J.F. Wang, *J. Appl. Math. Comput.* 47:103, 2015.

[50] A. Sohail, K. Maqbool and R. Ellahi, *Numer. Meth. Part. Diff. Eqs.* 34:19, 2017.

[51] A.R. Haghighi, A. Dadv and H.H. Ghejlo, *Commun. Adv. Comput. Sci. Appl.* 2014. Article ID Cascsa-00024, 11 pages.

[52] J.Y. Yang, J.F. Huang, D.M. Liang and Y.F. Tang, *Appl. Math. Model.* 38:3652, 2014.

[53] D. Baffet and J.S. Hesthaven, *SIAM J. Numer. Anal.* 55:496, 2017.

[54] V.R. Hosslini, E. Shivanian and W. Chen, *J. Comput. Phys.* 312:307, 2016.

[55] R. Lazarov and Z. Zhou, *SIAM J. Numer. Anal.* 51:445, 2013.

[56] M. Modanli, *Adv. Diff. Eqs.* 2018:333, 2018.

[57] C. Cem and M. Duman, *J. Comput. Phys.* 231:1743, 2012.

[58] W.P. Bu, Y.F. Tang, Y.C. Wu and J.Y. Yang, *Appl. Math. Comput.* 257:355, 2015.

[59] C.M. Chen, F. Liu, I. Turner and V. Anh, *J. Comput. Phys.* 227:886, 2007.

[60] Y.G. Jiang and J.T. Ma, *J. Comput. Appl. Math.* 235:3285, 2011.

[61] C.P. Li and H.F. Ding, *Appl. Math. Model.* 38:3802, 2014.

[62] M.H. Chen and W.H. Deng, *SIAM J. Numer. Anal.* 52:1418, 2014.

[63] H.W. Sun and Z.Z. Sun *J. Comput. Phys.* 298:337, 2015.

[64] M.H. Ding and C.P. Li, *Numer. Meth. Part. Diff. Eqs.* 32:213, 2016.

[65] H. Li, J. Cao and C. Li, *J. Comput. Appl. Math.* 299:159, 2016.

[66] Z. Li, Z. Liang and Y. Yan, *J. Sci. Comput.* 71:785, 2017.

[67] C.P. Li, Z.G. Zhao and Y.Q. Chen, *Comput. Math. Appl.* 62:855, 2011.

[68] F.H. Zeng, F. Liu, C.P. Li, K. Burrage, I. Turner and V. Anh, *SIAM J. Numer. Anal.* 52:2599, 2014.

[69] W. Cao, F.H. Zeng, Z. Zheng and G.E. Karniadakis, *SIAM J. Sci. Comput.* 38:A3070, 2016.

[70] C.P. Li, Q. Yi and A. Chen, *J. Comput. Phys.* 316:614, 2016.

[71] A.Q.M. Khaliq, X. Liang and K.M. Furati, *Numer. Algor.* 75:147, 2017.

[72] X. Liang, A.Q.M. Khaliq, H. Bhatt and K.M. Furati, *Numer. Algor.* 76:939, 2017.

[73] V. Balakrishnan, *Physica A* 132:569, 1985.

[74] W. Wyss, *J. Math. Phys.* 27:2782, 1986.

[75] W.R. Schneider and W. Wyss, *J. Math. Phys.* 30:134, 1989.

[76] R. Metzler and J. Klafter, *Phys. Rep.* 339:1, 2009.

[77] P. Sakrajda and M. Sławomir Wiraszka, *Fractional Order Diffusion Model for Social Networks*. Proceedings of International Conference on Fractional Differentiation and its Applications (ICFDA) 2018. Available at SSRN: https://ssrn.com/abstract=3271330 or http://dx.doi.org/10.2139/ssrn.3271330.

[78] Nirupama Bhattacharya, *Fractional Diffusion: Numerical Methods and Applications in Neuroscience*. Ph.D. Thesis, University of California, San Diego, 2014.

[79] D. Benson, S. Wheatcraft and M. Meerschaert, *Water Resour. Res.* 36:1403, 2000.

[80] J.P. Aguilar, J. Korbel and Y. Luchko, *Mathematics* 7:796, 2019.

[81] A.V. Pskhu and S.Sh. Rekhviashvili, *Theor. Math. Phys.* 75:316, 2020.

[82] M. Weilbeer, *Efficient Numerical Methods for Fractional Differential Equations and Their Analytical Background*. Verlag, Nicht, Ermittelbar, 2005.

[83] A.K. Grünwald, *Z. Angrew. Math. Phys.* 12:441, 1867.

[84] A.V. Letnikov, *Mat. Sh.* 3:1, 1868.

[85] I. Podlubny, *Fractional Differential Equations: An Introduction to Fractional Derivatives, Fractional Differential Equations, to Methods of Their Solution and Some of Their Applications*. Elsevier, Amsterdam, 1998.

[86] M.D. Ortigueira and F. Coito, *Int. J. Theory Appl.* 7:459, 2004.

[87] K.B. Oldham and J. Spanier, *The Fractional Calculus*. Academic Press, New York, 1974.

[88] P. Straka, M.M. Meerschaert, R.J. McGough and Y. Zhou, *Fract. Cal. Appl. Anal.* 16:262, 2013.

[89] T. Sandev and Z. Tomovski, *Fractional wave equations*. In: Fractional Equations and Models. Developments in Mathematics, vol 61, 2019. Springer, Cham. https://doi.org/10.1007/978-3-030-29614-8_5.

[90] O.P. Agrawal, *Z. Angew. Math. Mech.* 83(4):265, 2003.

17.10 Problems

17.1 Show that when $\alpha = 1$ Eq. (17.10a) becomes the classical finite-difference scheme (17.7) of Eq. (17.3).

17.2 Substituting $\zeta_{m+1} = \xi\zeta_m$ in Eq. (17.12) arrive at the Eq. (17.13).

17.3 For the diffusion Eq. (17.4) construct the exact analytical solution subjected to the conditions given by (17.19) by the variable separable method.

17.4 Study the time evolution of the initial Gaussian wave function

$$u(x,0) = \frac{1}{\sigma\sqrt{2\pi}}e^{-(x-x_0)^2/(2\sigma^2)}, \quad x \in [-20, 20]$$

according to the diffusion Eq. (17.4) with $x_0 = 0$, $\sigma = 0.5$ and $\alpha = 0.75$.

17.5 Consider Eq. (17.21) with $f(x,t) = \sin \pi x$. Compute the numerical solution $u(x,t)$ with $u(x,0) = 0$, $u(0,t) = u(1,t) = 0$, $\Delta x = 1/25$ and $\Delta t = 0.00005$. Plot $u(0.48,t)$ versus t and verify that as t increases from 0, $u(0.48,t)$ approaches to 0.10112.

17.6 For the diffusion Eq. (17.21) with $K_\alpha = 1$ choose $u(x,t) = t\sin 2\pi x$. For this solution find $f(x,t)$. Compute the numerical solution for $\alpha = 0.75$, $\Delta x = 1/25$ and $\Delta t = 0.00005$. Compare the numerical solution with the exact solution for some selected values of t.

17.7 For the time-fractional diffusion Eq. (17.21) with $f(x,t) = x(1-x)$, $u(0,t) = u(1,t) = 0$ and $u(x,0) = 0$ determine the equilibrium solution $u^*(x)$ and then $u(x,t)$. Numerically solve the Eq. (17.21) with $\Delta x = 0.05$, $\Delta t = 0.0001$ and $\alpha = 0.75$. Compare the numerical solution with the exact solution.

17.8 Perform the stability analysis for the finite-difference scheme (17.34) for the advection-diffusion Eq. (17.29) and arrive the result given by Eq. (17.35).

17.9 The exact solution of Eq. (17.29) for $K_\alpha = V_\alpha = 1$ with $f(x,t) = 2x - 2 + 2t^{2-\alpha}/\Gamma(3-\alpha)$ is $u(x,t) = x^2 + t^2$, $0 \le x \le 1$. Compute the numerical solution $u(x,t)$ with $\Delta x = 1/25$, $\Delta t = 0.00005$ at $t = 0.5$ for several values of α in the interval $[0,1]$. Compare the numerical solution with the exact solution.

17.10 For the advection-diffusion Eq. (17.29) with $K_\alpha = V_\alpha = 1$ and $f(x,t) = 3 - 2x$ determine the equilibrium solution $u^*(x)$. With $u(x,0) = 0$ and $u(x,0) = 1.5u^*(x)$ study the time evolution of the initial wave profile by suitably choosing Δx and Δt for $\alpha = 0.75$. Plot $u(x,t)$ as a function of t for a value of x and state the observation.

17.11 For the wave Eq. (17.38) subjected to the conditions $u(x,0) = u_t(x,0) = 0 = u(0,t) = u(\pi,t) = 0$ and $f(x,t) = \sin x \left[\Gamma(2+\alpha)t + t^{1+\alpha}\right]$ the exact solution is $u(x,t) = t^{1+\alpha}\sin x$. With $\Delta x = \pi/200$ compute the numerical solution at $t = 0.5$ for $\alpha = 1$, 1.25, 1.5, 1.75 and 2. For each value of α find $(\Delta t)_c$ and then choose Δt appropriately. Compare the numerical solution with the exact solution.

17.12 Consider the wave Eq. (17.38) with $f(x,t) = \sin x$ and driven by the external time-periodic force $F\cos(2\pi t)$. With $\Delta x = \pi/200$, $\Delta t = 0.005$ and $\alpha = 1.75$ compute the numerical solution and plot $u(\pi/2,t)$ for $t \in [0,30]$. Verify that after a transient evolution, $u(\pi/2,t)$ exhibits periodic oscillation with period 1. Further, confirm the periodicity of $u(x,t)$ by plotting the solution for $t \in [30, 33]$.

17.13 Obtain the stability condition of the scheme (17.52) for the damped wave Eq. (17.48).

17.14 For the damped wave Eq. (17.48) determine the exact analytical solution for $f(x,t) = \sin x$. Numerically compute the solution for a chosen value of α and compare it with the exact solution.

17.15 For the Fisher equation

$$\frac{\partial^\alpha u}{\partial t^\alpha} + \frac{\partial u}{\partial x} = \frac{t^{1-\alpha}}{\Gamma(2-\alpha)} \sin x + t \cos x, \quad x \in [0, 2\pi], \quad \alpha \in [0, 1]$$

determine the exact analytical solution. Applying the finite-difference scheme (17.62) compute the numerical solutions for $\alpha = 1, 0.75$ and 0.5 at $t = 0.5$. What is the effect of α on $u(x, t)$? Compare the numerical solutions with the exact solutions by plotting the error in the solution with x.

17.16 Show that for the assumed solution (17.94) of the diffusion Eq. (17.72) the $f(x, t)$ is given by Eq. (17.96).

17.17 Determine $f(x, t)$ in the diffusion Eq.(17.72) corresponding to the solution $u(x, t) = e^{-t}x(1 - x)$. Then, numerically compute the solution for $\alpha = 1.75$ and compare it with the exact solution at $t = 0.5$.

17.18 For the space-fractional diffusion Eq. (17.72) determine the stationary solution $u^*(x)$ for $f(x, t) = x(1 - x)$ subjected to the conditions (17.80). Compute the numerical solution for $u(x, 0) = 0$ and $\alpha = 1.75$ and verify that it approaches to u^* in the long-time limit.

17.19 Assume the initial wave solution of the space-fractional diffusion Eq. (17.72) as

$$u(x, 0) = \frac{1}{\sigma\sqrt{2\pi}} x^2 e^{-(x-x_0)^2/(2\sigma^2)}, \quad x \in [-20, 20]$$

with $x_0 = 0$ and $\sigma = 0.5$. Study the time evolution of this initial solution by computing the numerical solution for $\alpha = 1.75$.

17.20 Consider the finite-difference scheme (17.106) of the diffusion Eq. (17.72) obtained using the standard Grünwald approximation for the fractional order space derivative. Assume that the perturbed value of $u_j^{(0)}$ as $\overline{u}_j^{(0)} = u_j^{(0)} + \xi_j^{(0)}$ and it leads the perturbation in $u_j^{(1)}$ as $\overline{u}_j^{(1)} = u_j^{(1)} + \xi_j^{(1)}$. Obtain an expression for $\xi_j^{(1)}$ in term of $\xi_j^{(0)}$ and then $\xi_j^{(m)} = \mu_j^{(m)}\xi_j^{(0)}$ [24]. Show that $\mu_j^{(m)}$ is > 1 and hence the scheme is unstable.

18

Fourier Analysis and Power Spectrum

18.1   Introduction

When a given data set exhibits a periodic behaviour it is desired to construct an appropriate trigonometric function through a curve fitting. In many engineering problems, signal analysis and dynamical systems theory it is often required to know the various frequency components present in the solutions. The frequencies present in a solution can be identified through a spectral analysis. For a sufficiently large data set, one can construct an appropriate Fourier series and Fourier transform.

Fourier analysis is a standard tool for decomposing periodic as well as non-periodic solutions or non-periodic functions into an infinite number of sinusoidal functions such as sine and cosine terms. Fourier series is an appropriate tool for approximating periodic functions while the Fourier transform is the most suitable tool for non-periodic functions.

This chapter is devoted to the numerical construction of Fourier series and Fourier transform. For a complex form of Fourier series expansion of a function, the analytical calculation of the Fourier coefficients is reviewed. The numerical computation of the Fourier coefficients by employing the composite trapezoidal rule is outlined with an example. As this way of calculation of the Fourier coefficients is time-consuming the fast Fourier transform technique is explained. Next, the power spectrum of a function is defined and its significance is illustrated with an example. The power spectrum is determined exactly for a constant and a periodic function. Then, the problem of discrete Fourier transform is introduced and the computation of its frequency components by a simplified fast Fourier transform is described. The features of the numerically computed power spectrum of periodic, quasiperiodic and chaotic solutions are discussed. The power spectrum of a set of uniformly distributed random numbers is presented.

18.2   Fourier Series

Problems involving oscillations or vibrations are common in physics, engineering and chemistry. A vibrating tuning fork, to and fro motion of a wall-clock pendulum and sound waves are some examples of such systems. Mathematically, such oscillatory/vibrating processes are described by appropriate periodic functions, often by sine or cosine functions. Musical instruments give a fundamental tone accompanied by a number of overtones of frequencies which are integral multiples of the fundamental frequency. The combination of the fundamental and overtones is generally a complicated periodic function. If the goal is to construct the complicated function as a sum of terms corresponding to various overtones (harmonics) including fundamental, it will require, in general, an infinite series of terms called *Fourier series*.

DOI: 10.1201/9781032649931-18

The complex form of a Fourier series expansion of a function $f(t)$ with period-T is given by

$$f(t) = \sum_{k=-\infty}^{\infty} C_k e^{ik\omega t}, \quad \omega = 2\pi/T, \tag{18.1}$$

where C_k is the Fourier coefficient corresponding to the wave number k. C_k provides information about the various harmonics contributing to $f(t)$ and their strengths. Let us first outline the calculation of the Fourier coefficients.

For convenience write Eq. (18.1) as

$$f(t) = C_0 + \sum_{k=1}^{\infty} (C_k + C_{-k}) \cos k\omega t + \sum_{k=1}^{\infty} i (C_k - C_{-k}) \sin k\omega t. \tag{18.2}$$

Defining

$$a_k = C_k + C_{-k}, \quad b_k = i(C_k - C_{-k}), \quad \frac{a_0}{2} = C_0 \tag{18.3}$$

Eq. (18.2) becomes

$$f(t) = \frac{a_0}{2} + \sum_{k=1}^{\infty} a_k \cos k\omega t + \sum_{k=1}^{\infty} b_k \sin k\omega t. \tag{18.4}$$

The term $a_0/2$ in the Fourier series allows the possibility that $f(t)$ oscillates about an average value other than zero. Substitution of $\omega = 2\pi/T$ in Eq. (18.4) gives

$$f(t) = \frac{a_0}{2} + \sum_{k=1}^{\infty} a_k \cos \frac{2\pi k}{T} t + \sum_{k=1}^{\infty} b_k \sin \frac{2\pi k}{T} t. \tag{18.5}$$

The equations for a's and b's are obtained as follows.

Integration of Eq. (18.5) with respect to t from 0 to T leads to

$$\int_0^T f(t)\, dt = \int_0^T \left[\frac{a_0}{2} + \sum_{k=1}^{\infty} a_k \cos \frac{2\pi k}{T} t + \sum_{k=1}^{\infty} b_k \sin \frac{2\pi k}{T} t \right] dt$$

$$= \frac{a_0}{2} T. \tag{18.6}$$

That is,

$$a_0 = \frac{2}{T} \int_0^T f(t)\, dt. \tag{18.7a}$$

To determine a_k multiply Eq. (18.5) by $\cos(2\pi k' t/T)$ and integrate from 0 to T:

$$\int_0^T f(t) \cos \frac{2\pi k'}{T} t\, dt = \int_0^T \left\{ \frac{a_0}{2} + \sum_{k=1}^{\infty} a_k \cos \frac{2\pi k}{T} t \right.$$

$$\left. + \sum_{k=1}^{\infty} b_k \sin \frac{2\pi k}{T} t \right\} \cos \frac{2\pi k'}{T} t\, dt$$

$$= \frac{a_{k'}}{2} T. \tag{18.8}$$

Therefore,

$$a_k = \frac{2}{T} \int_0^T f(t) \cos \frac{2\pi k}{T} t \, dt. \tag{18.7b}$$

To determine b_k's multiply Eq. (18.5) by $\sin(2\pi k't/T)$ and integrate from 0 to T. The result is

$$\int_0^T f(t) \sin \frac{2\pi k'}{T} t \, dt = \frac{b_{k'}}{2} T. \tag{18.9}$$

That is,

$$b_k = \frac{2}{T} \int_0^T f(t) \sin \frac{2\pi k}{T} t \, dt. \tag{18.7c}$$

When $T = 2\pi$ with $0 \le t < 2\pi$ the Fourier series is written as

$$f(t) = \frac{a_0}{2} + \sum_{k=1}^{\infty} a_k \cos kt + \sum_{k=1}^{\infty} b_k \sin kt, \tag{18.10a}$$

where

$$a_k = \frac{1}{\pi} \int_0^{2\pi} f(t) \cos kt \, dt, \quad k = 0, 1, 2, \ldots \tag{18.10b}$$

$$b_k = \frac{1}{\pi} \int_0^{2\pi} f(t) \sin kt \, dt, \quad k = 1, 2, \ldots. \tag{18.10c}$$

For $f(t)$ of period-2π the Fourier series (18.1) is rewritten as

$$f(t) \approx 2 \sum_{k=0}^{\infty} |C_k| \cos(\theta_k + kt), \tag{18.11a}$$

where

$$C_k = |C_k| \, e^{i\theta_k} \quad \text{and} \quad C_k = \frac{1}{2\pi} \int_0^{2\pi} f(t) e^{-ikt} \, dt. \tag{18.11b}$$

In Eq. (18.11b), $f(t)$ is expressed as a sum of simple harmonic oscillations. The kth mode is

$$f_k = 2 |C_k| \cos(\theta_k + kt). \tag{18.12}$$

f_k is periodic with period $2\pi/k$ with phase angle θ_k. It has an amplitude $2|C_k|$, frequency $k/2\pi$ and an angular frequency k. $|C_k|$ is a measure of the extent to which a simple harmonic motion of the angular frequency k is present in the total motion.

For a certain periodic forms of $f(t)$ like rectangular, triangular and linear functions the Fourier coefficients can be determined analytically by exactly integrating the integrals in Eqs. (18.10). However, analytic integration is possible only for very limited functions. If $f(t)$ is given as a table of, say, N values or if it is difficult to evaluate the integrals in Eqs. (18.10) analytically for a given analytical form of $f(t)$ it is desirable to evaluate these integrals numerically. The next section discusses the construction of the Fourier coefficients numerically.

18.3 Numerical Calculation of the Fourier Coefficients

Let $f(t)$ is periodic with period-2π and the values of $f(t)$ are known in the interval 0 to 2π at equidistant points $t_m = m\Delta t$, $m = 1, 2, ..., N$ with $\Delta t = 2\pi/N$. Denote $f(t_m)$ as f_m. Application of the composite trapezoidal rule

$$\int_0^{2\pi} f(t)\,\mathrm{d}t \approx \Delta t \sum_{m=1}^{N} f(m\Delta t) \tag{18.13}$$

to the integrals in Eqs. (18.10) results in

$$a_0 = \frac{\Delta t}{\pi} \sum_{m=1}^{N} f_m, \tag{18.14a}$$

$$a_k = \frac{\Delta t}{\pi} \sum_{m=1}^{N} f_m \cos k(m\Delta t), \quad k = 1, 2, \ldots \tag{18.14b}$$

$$b_k = \frac{\Delta t}{\pi} \sum_{m=1}^{N} f_m \sin k(m\Delta t), \quad k = 1, 2, \ldots. \tag{18.14c}$$

The above approximation is first-order in Δt and hence numerically computed coefficients may not coincide exactly with the true values of the coefficients. In practice, if the absolute value of a Fourier coefficient is less than a pre-assumed small number δ (say 10^{-3}) then its value is assumed as zero. Equations (18.14) are used to compute Fourier coefficients of certain periodic functions.

Example:

Compute the Fourier coefficients of the square-wave

$$f(t) = \begin{cases} \alpha, & 0 \leq t \leq \pi \\ -\alpha, & \pi < t < 2\pi. \end{cases} \tag{18.15}$$

The exact Fourier series of the above square-wave is

$$f(t) = \alpha \sum_{k=\text{odd}}^{\infty} \frac{4}{k\pi} \sin kt. \tag{18.16}$$

The first few values of a_k and b_k were computed numerically using Eqs. (18.14) with $N = 2^{14} = 16384$, $\alpha = 1$ are given in Table 18.1. Table 18.2 presents the numerically computed values of a_0, a_1 and b_1 for various values of N. As N increases the values of a_0, a_1 and b_1 (as well as other a's and b's) approach the true values.

The calculation of a_k's and b_k's using Eqs. (18.14) is time-consuming. These coefficients can be calculated very quickly by the fast Fourier transform technique.

18.4 Fourier Transform and Power Spectrum

In order to express a function $f(t)$ in a Fourier series, it has to be periodic with period T, T being finite. For a nonperiodic function defined over infinite interval of t, one needs a

TABLE 18.1

The numerically calculated some of the Fourier coefficients (a_k and b_k) and their exact values (denoted as a_{ke} and b_{ke}, respectively) for the square-wave given by Eq. (18.15), where $\alpha = 1$.

k	a_k	a_{ke}	b_k	b_{ke}
0	0.00000	0.00000		
1	-0.00024	0.00000	1.27324	1.27324
2	0.00000	0.00000	0.00000	0.00000
3	-0.00024	0.00000	0.42441	0.42441
4	0.00000	0.00000	0.00000	0.00000
5	-0.00024	0.00000	0.25465	0.25465
6	0.00000	0.00000	0.00000	0.00000
7	-0.00024	0.00000	0.18189	0.18189
8	0.00000	0.00000	0.00000	0.00000
9	-0.00024	0.00000	0.14147	0.14147
10	0.00000	0.00000	0.00000	0.00000
11	-0.00024	0.00000	0.11575	0.11575
12	0.00000	0.00000	0.00000	0.00000
13	-0.00024	0.00000	0.09794	0.09794
14	0.00000	0.00000	0.00000	0.00000
15	-0.00024	0.00000	0.08488	0.08488
16	0.00000	0.00000	0.00000	0.00000
17	-0.00024	0.00000	0.07490	0.07490
18	0.00000	0.00000	0.00000	0.00000
19	-0.00024	0.00000	0.06701	0.06701
20	0.00000	0.00000	0.00000	0.00000

TABLE 18.2

Variations of the coefficients a_0, a_1 and b_1 with N for the square-wave (Eq. (18.15)) with $\alpha = 1$. The exact values of a_0, a_1 and b_1 are 0, 0 and 1.27324, respectively.

N	a_0	a_1	b_1	N	a_0	a_1	b_1
2^8	0	-0.01562	1.27318	2^{12}	0	-0.00098	1.27324
2^9	0	-0.00781	1.27322	2^{13}	0	-0.00049	1.27324
2^{10}	0	-0.00391	1.27324	2^{14}	0	-0.00024	1.27324
2^{11}	0	-0.00195	1.27324	2^{15}	0	-0.00012	1.27324

different representation called *Fourier integral* or *Fourier transform*. The Fourier integral is thought of as the limiting case of a Fourier series of $f(t)$ defined in the interval $-T < t < T$ as $T \to \infty$. This representation forms the basis for the Fourier transform.

The Fourier transform of a function $f(t)$ is given by

$$\bar{f}(\Omega) = \frac{1}{\sqrt{2\pi}} \int_{-\infty}^{\infty} f(t)e^{-i\Omega t}\, dt\,. \tag{18.17}$$

Knowing $\bar{f}(\Omega)$ the corresponding function is obtained through the inverse transform

$$f(t) = \frac{1}{\sqrt{2\pi}} \int_{-\infty}^{\infty} \bar{f}(\Omega)e^{i\Omega t}\, d\Omega\,. \tag{18.18}$$

Essentially, the Fourier transform decomposes a signal into complex exponential functions of different frequencies. $f(t)$ is thought of as a signal in time domain. In this case $\bar{f}(\Omega)$ denotes the signal in frequency domain. If $\bar{f}(\Omega)$ is large at $\Omega = \Omega'$ then $f(t)$ has a dominant spectral component at $\Omega = \Omega'$. That is, the major part of $f(t)$ is made of frequency Ω'. When $\bar{f}(\Omega) = 0$ or very small for $\Omega = \Omega'$ then $f(t)$ does not contain the frequency Ω'. $\bar{f}(\Omega)$ is generally a complex-valued function. Therefore, one can define a real-valued function by taking the modulus square of $\bar{f}(\Omega)$ and call the real function as power spectrum $P(\Omega)$ of $f(t)$, that is,

$$P(\Omega) = |\bar{f}(\Omega)|^2\,. \tag{18.19}$$

18.4.1   Significance of $\bar{f}(\Omega)$ and Power Spectrum

Suppose that the inverse transform $f(t)$ given by Eq. (18.18) is considered as the superposition of sinusoidal oscillations of all possible frequencies. Then, $\bar{f}(\Omega)$ gives the intensity of $f(t)$ in the interval Ω to $\Omega + d\Omega$ where $d\Omega$ is small. For an oscillatory system the integral

$$\int_{-\infty}^{\infty} |\bar{x}(\Omega)|^2 d\Omega \tag{18.20}$$

is interpreted as the total energy [1] where $x(t)$ is the displacement of the system from the equilibrium position or rest state. For example, the Hamiltonian or the energy of a linear harmonic oscillator is

$$E = H = \frac{1}{2}v^2 + \frac{1}{2}\omega_0^2 x^2\,, \tag{18.21}$$

where $v = \dot{x}$ is the velocity and ω_0^2 is a constant. The equation of motion of the oscillator is

$$\ddot{x} + \omega_0^2 x = 0\,. \tag{18.22}$$

The general solution of Eq. (18.22) is

$$x(t) = a_1 \cos \omega_0 t + b_1 \sin \omega_0 t \tag{18.23}$$

which can be rewritten as

$$x(t) = C_1 e^{i\omega_0 t} + C_{-1} e^{-i\omega_0 t}\,. \tag{18.24}$$

Then,

$$\begin{aligned}
E &= -\frac{1}{2}\omega_0^2 \left[C_1 e^{i\omega_0 t} - C_{-1} e^{-i\omega_0 t} \right]^2 + \frac{1}{2}\omega_0^2 \left[C_1 e^{i\omega_0 t} + C_{-1} e^{-i\omega_0 t} \right]^2 \\
&= 2\omega_0^2 C_1 C_{-1} \\
&= 2\omega_0^2 |C_1|^2\,.
\end{aligned} \tag{18.25}$$

Energy is proportional to the square of the amplitude $|C_1|$. Note that if $x(t)$ is represented by a Fourier series then the result is a series of squares $|C_k|^2$ instead of only one $|C_1|^2$.

The set of quantities $|C_k|^2$ is called a *discrete power spectrum*. Comparison of Eqs. (18.11) and (18.17) indicates that for a periodic function with period-2π (or T) C_k is the Fourier transform of $x(t)$.

Example 1:

Calculate the Fourier transform $\bar{f}(\Omega)$ and the power spectrum of $f(t) = \alpha$, a constant.

Equation (18.17) becomes

$$\bar{f}(\Omega) = \frac{1}{\sqrt{2\pi}} \int_{-\infty}^{\infty} \alpha e^{-i\Omega t} \, dt = \frac{\alpha}{\sqrt{2\pi}} \frac{e^{-i\Omega t}}{-i\Omega} \bigg|_{-\infty}^{\infty} . \tag{18.26}$$

Then,

$$P(\Omega) = |\bar{f}(\Omega)|^2 = \frac{\alpha^2}{2\pi} \frac{e^{-i\Omega t} e^{i\Omega t}}{-i\Omega \times i\Omega} \bigg|_{-\infty}^{\infty} = \frac{\alpha^2}{2\pi\Omega^2} . \tag{18.27}$$

Thus, as Ω increases $P(\Omega)$ decays to zero.

Example 2:

Find the Fourier transform of $f(t) = \cos \omega t$.

The result is

$$\begin{aligned} \bar{f}(\Omega) &= \frac{1}{\sqrt{2\pi}} \text{ Real part of } \int_{-\infty}^{\infty} e^{i\omega t} e^{-i\Omega t} \, dt \\ &= \frac{1}{\sqrt{2\pi}} \text{ R.P. } 2\pi\delta(\omega - \Omega), \end{aligned} \tag{18.28}$$

where $\delta(\omega - \Omega)$ is the Dirac-delta function.

If $f(t)$ is given for a discrete set of values of t then the resultant Fourier transform is called a *discrete Fourier transform* [2]. For $f = \cos \omega t$ from (18.28) note that the discrete Fourier transform $\bar{f}(\Omega)$ has sharp peaks at $\Omega = \omega$ and at integral multiples of Ω. The next section deals with the problem of discrete Fourier transform.

18.5   Discrete Fourier Transform

Let us consider a function $f(t)$ written in time series as $f_m = f(t_m)$, where $t_m = m\Delta t$, $m = 0, 1, \ldots, N-1$ and Δt is the time duration with which the data are collected. Applying the composite trapezoidal rule

$$\int_0^{2\pi} y(t) \, dt \approx \frac{2\pi}{N} \sum_{m=0}^{N-1} y\left(2\pi m/N\right), \tag{18.29}$$

where $y(t)$ is a periodic function with period 2π, to the integral in Eq. (18.11b) gives

$$C_k = \frac{1}{N} \sum_{m=0}^{N-1} f\left(2\pi m/N\right) e^{-i2\pi km/N} . \tag{18.30}$$

Now, represent a periodic orbit or a time series $\{f_1^p, f_2^p, \ldots, f_q^p\}$ with period $q = 2^p$ by the Fourier expansion

$$f_k^p = \sum_j C_j^p \, e^{i2\Omega_j k} , \tag{18.31}$$

where the frequency $\Omega_j = 2\pi j/2^p$, $j = 0, 1, \ldots, 2^{p-1}$. f_k^p is called a *discrete Fourier transform* [2] of the given time series. Note that the complex Fourier transform of $e^{i\omega kt}$ is

$$2\pi\delta(\Omega - \omega) . \tag{18.32}$$

Therefore, the Fourier transform of the above periodic orbit is

$$\bar{f}(\Omega) = 2\pi \sum_j C_j^p \delta(\Omega - \omega_j) . \tag{18.33}$$

Thus, the power spectrum

$$P(\Omega) = |\bar{f}(\Omega)|^2 \tag{18.34}$$

will consist of delta functions at the frequencies ω_j with amplitudes $|C_j^p|^2$. Later, computed power spectra of different types of periodic and nonperiodic time series and their features will be presented.

From Eq. (18.30) notice that the evaluation of a particular C_k needs N additions. Therefore, for the calculation of N numbers of C's totally $O(N^2)$ operations are required. Consequently, a computer program will take an amount of computing time proportional to N^2. For 10^3 sample points, the number of operations required is $O(10^6)$ and is a serious time-consuming computation. That is, the straight forward calculation of discrete Fourier transform is very inefficient. An alternate and efficient method is the fast Fourier transform and is described in the next section.

18.6   Fast Fourier Transform

In order to reduce the number of arithmetic operations in the calculation of C_k, the fast Fourier transform (FFT) algorithm was proposed by Cooley and Tukey [3]. In this method, the number of arithmetic operations is only $O(N \log_2 N)$. For 10^3 sample points the discrete Fourier transform will need $O(10^6)$ arithmetic operations while the FFT requires only $O(10^3 \log_2 10^3)$.

18.6.1   A Simplified Procedure

The FFT uses the advantages of symmetry properties of trigonometric functions at the points of calculation. Essentially, the terms in the summation in Eq. (18.30) are rearranged and the summation is performed in a hierarchical way as outlined in the following.

In Eq. (18.30) separate the odd and even terms as

$$\begin{aligned}
C_k &= \frac{1}{N} \sum_{m=0}^{N/2-1} f_{2m} e^{-i2\pi(2m)k/N} + \frac{1}{N} \sum_{m=0}^{N/2-1} f_{2m+1} e^{-i2\pi(2m+1)k/N} \\
&= u_k + v_k e^{-i2\pi k/N} ,
\end{aligned} \tag{18.35a}$$

where

$$u_k \;=\; \frac{1}{N} \sum_{m=0}^{N/2-1} f_{2m} e^{-i2\pi mk/(N/2)} , \tag{18.35b}$$

$$v_k \;=\; \frac{1}{N} \sum_{m=0}^{N/2-1} f_{2m+1} e^{-i2\pi mk/(N/2)} . \tag{18.35c}$$

In Eq. (18.35) C_k is written into two summations. Each summation has $N/2$ terms. Repeat the above process until there would be only two terms in each summation. This is possible if $N = 2^M$, where M is an integer.

In the case of $2^{M=3} = 8$ data points $f_0,\ f_1,\ \ldots,\ f_7$

$$
\begin{aligned}
C_k \;&=\; \sum_{m=0}^{7} f_m e^{-i2\pi mk/N} \\[4pt]
&=\; \sum_{m=0}^{3} f_{2m} e^{-i2\pi 2mk/N} + \sum_{m=0}^{3} f_{2m+1} e^{-i2\pi (2m+1)k/N} \\[4pt]
&=\; \sum_{m=0}^{1} f_{4m} e^{-i2\pi 4mk/N} + \sum_{m=0}^{1} f_{4m+2} e^{-i2\pi (4m+2)k/N} \\[4pt]
&\quad + \sum_{m=0}^{1} f_{4m+1} e^{-i2\pi (4m+1)k/N} + \sum_{m=0}^{1} f_{4m+3} e^{-i2\pi (4m+3)k/N} .. \tag{18.36}
\end{aligned}
$$

There is another symmetry between the coefficients C_k for $k < N/2$ and $k \geq N/2$, namely,

$$C_k \;=\; u_k + v_k e^{-i2\pi k/N} , \tag{18.37a}$$

$$C_{k+N/2} \;=\; u_k - v_k e^{i2\pi k/N} , \quad k = 0, 1, \ldots, \frac{N}{2} - 1. \tag{18.37b}$$

The above equations suggest that immediately after computing $v_k e^{i2\pi k/N}$ one can compute both the coefficients C_k and $C_{k+N/2}$.

Once the summation in Eq. (18.30) is split then M times (with each summation containing only two terms) the individual data points have to be added in pairs. But the points in each pair at the first level of additions can be very far apart in the time series because of grouping of even and odd terms in each level of splitting. For example, in Eq. (18.36) with $2^M = 8$ data points the first summation contains addition of the data points f_0 and f_4, the second summation contains addition of the data points f_2 and f_6 and so on. To handle this situation Cooley and Tukey suggested indexing the data string with binary numbers. The advantage of this is that pair of the data in each summation is represented by a binary number and its bit reversed order. For 8 data points the indexes representing the data are 0 to 7. In binary representation they are

$$000,\ 001,\ 010,\ 011,\ 100,\ 101,\ 110,\ 111. \tag{18.38}$$

The bit reversed-order of this sequence is

$$000,\ 100,\ 010,\ 110,\ 001,\ 101,\ 011,\ 111, \tag{18.39}$$

that is, in number system the sequence is 0, 4, 2, 6, 1, 5, 3, 7. Now, perform the first level of addition between f_0 and f_4, f_2 and f_6, f_1 and f_5 and f_3 and f_7. After the bit reversal apply Eq. (18.37) repeatedly with the addition of pairs which are 2^{l-1} spaces apart. Here, the index l indicates the level of additions.

The following is the systematic procedure of computing power spectrum for a real set of data by FFT:

(1) Read the $N = 2^M$ data and store them in the variable CR (real part of the data). Set CI's (imaginary part of the data) zero.

(2) Rearrange the data in the bit reversed order.

(3) Perform the addition of pair of points obtained from step (2) at M levels and obtain the Fourier coefficients.

(4) Calculate the power spectral density of the Fourier coefficients.

18.6.2 Advantage of the Fast Fourier Transform

To appreciate the feature of FFT algorithm, consider a case with $N = 2^M$ data points. In the basic discrete Fourier transform equation N complex multiplications are needed to compute one coefficient C_k. Since, there are N coefficients C_k the number of complex multiplications is N^2. In FFT algorithm after the bit reversal the number of levels of additions becomes M and each level consists of $N/2$ additions. Therefore, the total number of operations is $M \cdot (N/2)$. Since $M = \log_2 N$ the above number is $(N/2) \log_2 N$ or $O(N \log_2 N)$. Thus, the total computing time in FFT is proportional to $O(N \log_2 N)$ and is significantly less than the direct implementation of discrete Fourier transform.

18.6.3 Fourier Coefficients from the Fast Fourier Transform

The Fourier coefficients a's and b's can be obtained from the coefficients C's through Eqs. (18.3). Calculation of Fourier series through FFT technique is much faster than the calculation of it using Eqs. (18.14).

Table 18.3 gives the numerically calculated values of the Fourier coefficients for the square-wave given by Eq. (18.15). Since the exact values of a_0, a_k's and even b_k's are zero, only the odd values of b_k's are given. The values of the coefficients a_0, a_1 and b_1 are calculated for a few values of N, the number of data points. The values of the Fourier coefficients converge to the exact values with increase in the number of data N. Table 18.4 presents the result.

Table 18.5 presents the Fourier coefficients for the function

$$f(t) = \begin{cases} t/\pi, & 0 \le t \le \pi \\ \\ 0, & \pi < t < 2\pi. \end{cases} \tag{18.40}$$

The exact Fourier series of (18.40) is given by

$$a_0 = 1/2, \quad a_k = \begin{cases} 0, & k = 2, 4, \ldots \\ \\ -2/\left(k^2\pi^2\right), & k = 1, 3, \ldots \end{cases} \tag{18.41a}$$

$$b_k = -\frac{(-1)^k}{k\pi}, \quad k = 1, 2, \ldots. \tag{18.41b}$$

TABLE 18.3

The numerically calculated Fourier coefficients of the square-wave, Eq. (18.15) (with $\alpha = 1$), through FFT. Here, $a_0 = 0$, $a_k = 0.00024$, $k = 1, 2, \ldots$ and $b_k = 0$ for $k = 2, 4, \ldots$. b_{kn} and b_{ke} denotes the numerically computed and the exact values of b's.

k	b_{kn}	b_{ke}	k	b_{kn}	b_{ke}
1	1.27324	1.27324	27	0.04716	0.04716
3	0.42441	0.42441	29	0.04390	0.04390
5	0.25465	0.25465	31	0.04107	0.04107
7	0.18189	0.18189	33	0.03858	0.03858
9	0.14147	0.14147	35	0.03638	0.03638
11	0.11575	0.11575	37	0.03441	0.03441
13	0.09794	0.09794	39	0.03265	0.03265
15	0.08488	0.08488	41	0.03105	0.03105
17	0.07490	0.07490	43	0.02961	0.02961
19	0.06701	0.06701	45	0.02829	0.02829
21	0.06063	0.06063	47	0.02709	0.02709
23	0.05536	0.05536	49	0.02598	0.02598
25	0.05093	0.05093			

TABLE 18.4

The numerically computed values of the Fourier coefficients a_0, a_1 and b_1 for a few values of N for the square-wave, Eq. (18.15) (with $\alpha = 1$).

N	a_0	a_1	b_1	N	a_0	a_1	b_1
2^8	0.01563	0.0	1.27318	2^{12}	0.00098	0.0	1.27324
2^9	0.00781	0.0	1.27322	2^{13}	0.00049	0.0	1.27324
2^{10}	0.00391	0.0	1.27324	2^{14}	0.00024	0.0	1.27324
2^{11}	0.00195	0.0	1.27324				

18.6.4 Power Spectrum of Certain Functions

In real physical and engineering problems a time series $x(t)$ can be thought of as a component of the solution of the equation of motion of a system or a signal of an electrical system and so on. A simple way to check for periodic and nonperiodic nature of $x(t)$ is to compute its power spectrum.

Let us present the power spectrum computed using FFT for various types of $x(t)$. The spectrum of a constant, that is $x(t) =$ constant, contains a single peak at zero frequency. The power spectrum of a periodic $x(t)$ with frequency ω_1 has Dirac-delta peaks at ω_1 and its various harmonics. This will be the case if the Fourier series of $x(t)$ has an infinite number of nonzero Fourier coefficients a's and b's. If the Fourier series of $x(t)$ has only a finite number of sinusoidal functions with frequencies ω_i, $i = 1, 2, \ldots, n$ then the power spectrum will have peaks only at the frequencies ω_i.

TABLE 18.5
The numerically calculated Fourier coefficients of the function given by Eq. (18.40) through FFT. a_{kn} and b_{kn} are the numerically computed values of the Fourier coefficients. a_{ke} and b_{ke} are their respective exact values.

k	a_{kn}	a_{ke}	b_{kn}	b_{ke}
0	0.49994	0.50000		
1	−0.20258	−0.20264	0.31831	0.31831
2	−0.00006	0.00000	−0.15915	−0.15915
3	−0.02245	−0.02252	0.10610	0.10610
4	−0.00006	0.00000	−0.07958	−0.07958
5	−0.00804	−0.00811	0.06366	0.06366
6	−0.00006	0.00000	−0.05305	−0.05305
7	−0.00407	−0.00414	0.04547	0.04547
8	−0.00006	0.00000	−0.03979	−0.03979
9	−0.00244	−0.00250	0.03537	0.03537
10	−0.00006	0.00000	−0.03183	−0.03183
11	−0.00161	−0.00167	0.02894	0.02894
12	−0.00006	0.00000	−0.02653	−0.02653
13	−0.00114	−0.00120	0.02449	0.02449
14	−0.00006	0.00000	−0.02274	−0.02274
15	−0.00084	−0.00090	0.02122	0.02122

For all the following examples, $2^{14} (= 16384)$ data are used. The data are collected at every integer multiple of the time interval $2\pi/100$ unless otherwise specified.

A. Sinusoidal Time Series

Figure 18.1a presents the power spectrum of $x(t) = \sin 5t$. Since, $x(t)$ is a sinusoidal function with a single frequency $\omega = 5$ the power spectrum shows a peak at $\omega(= \Omega/2\pi) = 5$. The repetition period of $\sin 5t$ is $2\pi/5$. Figure 18.1b depicts the power spectrum corresponding to $x(t) = \sin 5t + 10 \sin 12t$. It has peaks at $\Omega = 5$ and 12. The Fourier coefficients of this $x(t)$ are $b_5 = 1$ and $b_{12} = 10$ and all other coefficients are 0. The amplitudes of the peaks at $\omega = 5$ and 12 in $\log_{10}$ scale are 0 and 2, respectively. The power spectrum gives $\log_{10}(b_5^2) = 0$ and $\log_{10}(b_{12}^2) = 2$, that is, $b_5 = \sqrt{10^0} = 1$ and $b_{12} = \sqrt{10^2} = 10$ and are in fact the exact values of the Fourier coefficients. The point is that from the power spectrum one can obtain the various frequency components present in the time series $x(t)$ and as well as their respective amplitudes.

B. Square-Wave

Next, consider the square-wave given by Eq. (18.15). Its exact Fourier series given by Eq. (18.16) has an infinite number of Fourier components. Odd integer coefficients b_k's are nonzero. Its power spectrum shown in Fig. 18.1c has peaks at the fundamental frequency and odd integral multiples of the fundamental frequency. The period of the square-wave considered here is 2π. The number of components of the power spectrum will double when the period of $x(t)$ is doubled. An example is given in Fig. 18.1d. This spectrum is for the

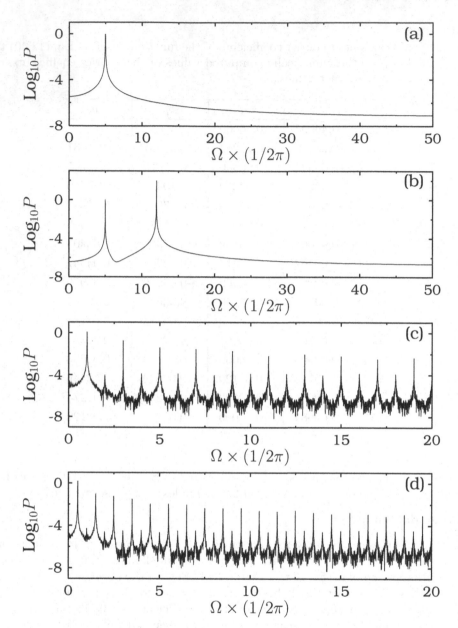

FIGURE 18.1
The power spectra of (a) $x(t) = \sin 5t$, (b) $x(t) = \sin 5t + 10 \sin 12t$, (c) square-wave of period 2π, Eq. (18.15), and (d) square-wave of period 4π, Eq. (18.42).

square-wave with period-4π defined as

$$f(t) = \begin{cases} \alpha, & 0 \le t \le 2\pi \\ -\alpha, & 2\pi < t < 4\pi, \end{cases} \tag{18.42}$$

where $\alpha = 1$.

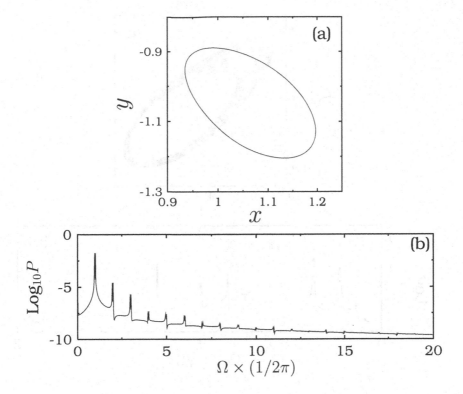

FIGURE 18.2
(a) A period-T solution of Eq. (18.43) and (b) the power spectrum of $x(t)$. The values of the parameters are fixed as $\beta = 1$, $\nu = 0.015$, $a = -1.02$, $b = -0.55$ and $\omega_1 = 1$, $\omega_2 = 0$, $f_1 = 0.1$ and $f_2 = 0$.

C. Periodic Time Series

Consider periodic, quasiperiodic and chaotic solutions of a differential equation. For an illustrative purpose choose the Murali–Lakshmanan–Chua circuit model equation

$$\dot{x} = y - h(x), \tag{18.43a}$$

$$\dot{y} = -\beta(1+\nu)y - \beta x + f_1 \sin \omega_1 t + f_2 \sin \omega_2 t, \tag{18.43b}$$

where

$$h(x) = bx + 0.5(a-b)(|x+1| - |x-1|). \tag{18.43c}$$

Equation (18.43) is solved numerically by the fourth-order Runge–Kutta method with time step, say, $2\pi/100$. Figure 18.2a shows the solution in the $x-y$ plane for $\beta = 1$, $\nu = 0.015$, $a = -1.02$, $b = -0.55$ and $\omega_1 = 1$, $\omega_2 = 0$, $f_1 = 0.1$ and $f_2 = 0$. The solution corresponding to first 1000 drive cycles, that is $1000 \times (2\pi/100)$ time, is left as a transient. The solution after the transient, is periodic with period $T = 2\pi/\omega_1 = 2\pi$. Its power spectrum is shown in Fig. 18.2b. Dominant peaks occur at the frequencies that are integer multiples of $1/2\pi$.

D. Quasiperiodic Orbit

A quasiperiodic orbit is written as

$$x(t) = \phi(\omega_1 t, \omega_2 t, \dots, \omega_n t), \tag{18.44}$$

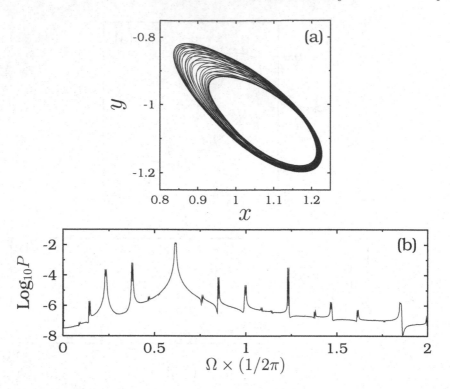

FIGURE 18.3

(a) A quasiperiodic solution of Eq. (18.43) and (b) the power spectrum of the component $x(t)$. The parameters are fixed as $\beta = 1$, $\nu = 0.015$, $a = -1.02$, $b = -0.55$ and $\omega_1 = 0.75$, $\omega_2 = (\sqrt{5} - 1)/2$, $f_1 = 0.01$ and $f_2 = 0.04$.

where ϕ is periodic of period 2π, in each of the $n(> 1)$ arguments. Further, the frequencies $(\omega_1, \omega_2, \dots, \omega_n)$ must have the following properties:

1. They are linearly independent, that is, there does not exist a nonzero set of integers $(l_1, l_2, \dots, l_n)$ such that

$$l_1\omega_1 + l_2\omega_2 + \cdots + l_n\omega_n = 0. \tag{18.45}$$

2. For each i, $\omega_i \neq |l_1\omega_1 + l_2\omega_2 + \cdots + l_n\omega_n|$ for some integers $(l_1, l_2, \dots, l_n)$, that is, there is no rational relation between the ω's.

A quasiperiodic orbit is essentially a sum of periodic orbits each of whose frequency is one of the various sums and differences of the finite set of frequencies $(\omega_1, \omega_2, \dots, \omega_n)$. Such a quasiperiodic motion takes place on a n-dimensional torus T^n. Suppose this torus is an attracting set, that is nearby trajectories get attracted to it, then T^n is said to be a *quasiperiodic attractor* [4].

An example of a system with quasiperiodic solution is the Murali–Lakshmanan–Chua circuit model Eq. (18.43). Fix $\beta = 1$, $\nu = 0.015$, $a = -1.02$, $b = -0.55$ and $\omega_1 = 0.75$, $\omega_2 = (\sqrt{5} - 1)/2$, $f_1 = 0.01$ and $f_2 = 0.04$. Use the initial condition $(x(0), y(0)) = (1, -1)$. The solution after transient in $x - y$ plane is shown in Fig. 18.3a. It is neither a definite periodic motion nor an irregular orbit but a quasiperiodic orbit. The quasiperiodic character of the solution $x(t)$ can be verified from its power spectrum. A quasiperiodic motion

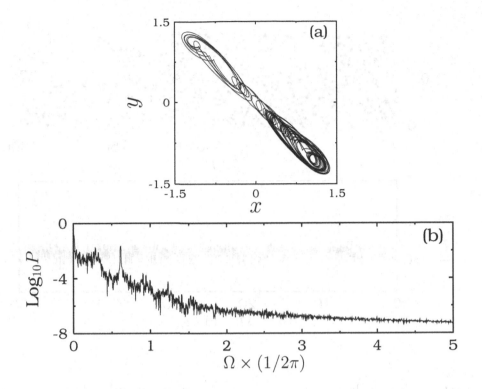

FIGURE 18.4

(a) A chaotic solution of Eq. (18.43) for $\beta = 1$, $\nu = 0.015$, $a = -1.02$, $b = -0.55$ and $\omega_1 = 0.75$, $\omega_2 = (\sqrt{5} - 1)/2$, $f_1 = 0.01$ and $f_2 = 0.08$. (b) The power spectrum of the component $x(t)$ of the solution shown in Fig. 18.4a.

with m rationally independent frequencies $\omega_1, \omega_2, \ldots, \omega_m$ will have Dirac-delta peaks at all linear combinations of the basic frequencies. Figure 18.3b depicts the numerically computed power spectrum using the data collected at $t = (2\pi/100)$. In Fig. 18.3b, there are peaks at $\Omega(\times 1/2\pi) = 0.235, 0.38, 0.615, 0.85, 0.995$ and 1.23. These data give two fundamental frequencies $\Omega_1 = 0.235$ and $\Omega_2 = 0.615$ and these linear combinations $\Omega_2 - \Omega_1 = 0.38$, $\Omega_2 + \Omega_1 = 0.85$, $2\Omega_2 - \Omega_1 = 0.995$ and $2\Omega_2 = 1.23$.

E. Chaotic Solution

Another interesting type of solution of a dynamical system is a chaotic solution. A non-periodic solution of a nonlinear system with highly sensitive to initial condition is called a *chaotic solution* [4]. By sensitive dependence on initial conditions it is meant that the trajectories started from two nearby initial conditions diverge exponentially until they become completely uncorrelated and future prediction becomes inaccurate. More precisely, any small error δ_0 in the specification of the initial state or round-off error in the numerical computation can get amplified exponentially fast in a finite time interval and hence become as large as the system size itself. Figure 18.4a shows the plot of the solution of Eq. (18.43) for $f_1 = 0.01$ and $f_2 = 0.08$. The solution is not a definite periodic and not a quasiperiodic but chaotic. The power spectrum of the chaotic solution $x(t)$ is plotted in Fig. 18.4b. This spectrum is quite different from that of the periodic and quasiperiodic solutions. It has con-

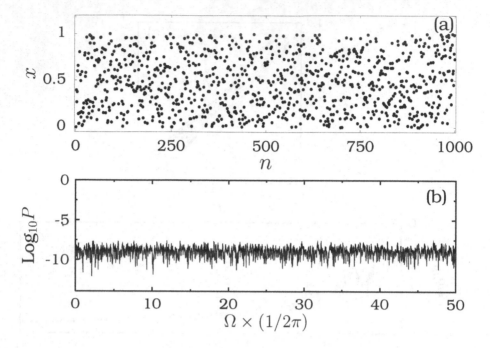

FIGURE 18.5

(a) A set of uniformly distributed random numbers in the interval $[0, 1]$ and (b) its power spectrum.

tinuous and broad-band nature. It is also quite common for spectrum of a chaotic solution to contain spikes indicating the predominant frequencies of the solution.

F. Uniform Random Numbers

Finally, consider a set of random numbers uniformly distributed in the interval $[0,1]$. Figure 18.5a shows the random numbers generated using the Park and Miller method (see Section 19.3.7). Its power spectrum shown in Fig. 18.5b is very noisy, a typical pattern for a highly nonuniform signal. It has no dominant peak which implies that the sequence of numbers considered has no periodicity.

18.7   Concluding Remarks

One of the advantages of Fourier series representation over some others, for example, with Taylor series is that it may represent a discontinuous function also. Secondly, in the Fourier series representation the odd and even functions are conveniently expressed as sine series ($a_k = 0$) and cosine series ($b_k = 0$), respectively.

Fourier transform converts a signal from time domain into frequency domain. In FFT the speed of the computation is greatly enhanced by decreasing the number of computations. From the power spectrum one can obtain the various frequency components present in the time series $x(t)$ and as well as their respective amplitudes. The power spectrum is not an invertible transform since it has no information about the phase. *What is power spectral density and how is it obtained?* There are numerous applications of FFT and power spectrum analysis in everyday life. In recent years, neurofeedback power spectrum has been utilized to diagnose many neuropsychiatric disorders. Changes due to gamma-ray exposure can be detected by a power spectrum analysis. The power spectrum analysis is utilized to seek patterns in the frequency characteristics of heart rate, blood pressure, central venous pressure data, etc.

18.8   Bibliography

[1]   E. Kreyszig, *Advanced Engineering Mathematics*. John Wiley, New York, 1999. 8th edition.

[2]   W.H. Press, S.A. Teukolsky, W.T. Vetterling and B.P. Flannery, *Numerical Recipes in Fortran*. Foundation Books, New Delhi, 1993. Indian edition.

[3]   J.W Cooley and J.W. Tukey, *Mathematics by Computation* 19:297, 1965.

[4]   M. Lakshmanan and S. Rajasekar, *Nonlinear Dynamics: Integrability, Chaos and Patterns*. Springer, Berlin, 2003.

18.9   Problems

18.1 Some of the periodic functions, their mathematical representation and the Fourier series are given below. Using the FFT algorithm compute numerically the Fourier coefficients. Use the data in the interval $[0, 2\pi]$ with $\omega = 1$, $f = 1$ and $T = 2\pi$. Sketch the graphs of the given functions and their corresponding Fourier series. Draw the graph of the function using its Fourier series by considering first 2 terms, 5 terms, 10 terms and 20 terms. Prepare a report on the above. Compute the power spectrum for each function and draw the power spectrum plot.

(a) **Symmetric Saw-Tooth Wave**

A symmetric saw-tooth wave is mathematically represented as

$$F_{\text{sst}}(t) = \begin{cases} \dfrac{4ft}{T}, & (2n-2)\dfrac{\pi}{\omega} \leq t < \left(\dfrac{4n-3}{2}\right)\dfrac{\pi}{\omega} \\[2ex] -\dfrac{4ft}{T} + 2f, & \left(\dfrac{4n-3}{2}\right)\dfrac{\pi}{\omega} \leq t < \left(\dfrac{4n-1}{2}\right)\dfrac{\pi}{\omega} \\[2ex] \dfrac{4ft}{T} - 4f, & \left(\dfrac{4n-1}{2}\right)\dfrac{\pi}{\omega} \leq t < 2n\dfrac{\pi}{\omega}, \end{cases}$$

where $T = 2\pi/\omega$ is the period of the wave and $n = 1, 2, \ldots$. The numerical implementation of the wave is given by

$$
F_{sst}(t) = \begin{cases}
\dfrac{4ft}{T}, & 0 \le t < \dfrac{\pi}{2\omega} \\[2mm]
-\dfrac{4ft}{T} + 2f, & \dfrac{\pi}{2\omega} \le t < \dfrac{3\pi}{2\omega} \\[2mm]
\dfrac{4ft}{T} - 4f, & \dfrac{3\pi}{2\omega} \le t < \dfrac{2\pi}{\omega},
\end{cases}
$$

where t is taken as $\mathrm{mod}(2\pi/\omega)$. Its Fourier series is

$$
F_{sst}(t) = \frac{8f}{\pi^2} \sum_{n=1}^{\infty} \frac{(-1)^{(n+1)}}{(2n-1)^2} \sin(2n-1)\omega t.
$$

(b) Asymmetric Saw-Tooth Wave

The mathematical representation of an asymmetric saw-tooth wave form is

$$
F_{ast}(t) = \begin{cases}
\dfrac{2ft}{T}, & (2n-2)\dfrac{\pi}{\omega} \le t < (2n-1)\dfrac{\pi}{\omega} \\[2mm]
\dfrac{2ft}{T} - 2f, & (2n-1)\dfrac{\pi}{\omega} \le t < 2n\dfrac{\pi}{\omega}, \quad n = 1, 2, \dots .
\end{cases}
$$

The above wave form can be rewritten as

$$
F_{ast}(t) = \begin{cases}
\dfrac{2ft}{T}, & 0 \le t < \dfrac{\pi}{\omega} \\[2mm]
\dfrac{2ft}{T} - 2f, & \dfrac{\pi}{\omega} \le t < \dfrac{2\pi}{\omega},
\end{cases}
$$

where t is taken as $\mathrm{mod}(2\pi/\omega)$. The Fourier series of $F_{ast}(t)$ is given by

$$
F_{ast}(t) = \frac{2f}{\pi} \sum_{n=1}^{\infty} \frac{(-1)^{n+1}}{n} \sin n\omega t.
$$

(c) Modulus of Sine Wave

The modulus of sine wave is given as

$$
F_{msi}(t) = f|\sin(\omega t/2)|.
$$

Its Fourier series is

$$
F_{msi}(t) = \frac{2f}{\pi} - \frac{4f}{\pi} \sum_{n=1}^{\infty} \frac{n}{(4n^2-1)} \cos n\omega t.
$$

(d) Rectified Sine Wave

The mathematical representation of a rectified sine wave is

$$
F_{rsi}(t) = \begin{cases}
f, & (2n-2)\dfrac{\pi}{\omega} \le t < (2n-1)\dfrac{\pi}{\omega} \\[2mm]
0, & (2n-1)\dfrac{\pi}{\omega} \le t < 2n\dfrac{\pi}{\omega}, \quad n = 1, 2, \dots .
\end{cases}
$$

The numerical implementation of the above wave form is

$$F_{rsi}(t) = \begin{cases} f, & 0 \le t < \dfrac{\pi}{\omega} \\[2ex] 0, & \dfrac{\pi}{\omega} \le t < \dfrac{2\pi}{\omega}. \end{cases}$$

The Fourier series of the rectified sine wave is

$$F_{rsi}(t) = \frac{2f}{\pi} - \frac{4f}{\pi} \sum_{n=1}^{\infty} \frac{1}{(4n^2 - 1)} \cos 2n\omega t$$

for $0 \le t < \pi/\omega$ and 0 for $\pi/\omega \le t < 2\pi/\omega$.

18.2 Verify that the power spectrum of $x(t) = 10$ has a peak of magnitude 2 at the zero frequency and 0 at all other frequencies.

18.3 An amplitude-modulated sine wave is given by

$$F(t) = (f + 2g \cos \omega t) \sin \Omega t.$$

Show that its Fourier series is

$$f \sin \Omega t + g \sin(\Omega + \omega)t + g \sin(\Omega - \omega)t.$$

Choose $\Omega = 3$, $\omega = 2$, $f = 1$ and $g = 2$. Compute the Fourier coefficients using the FFT algorithm. Also, calculate its power spectrum and show that it has peaks only at the frequencies 1, 3 and 5 (multiplied by $1/2\pi$).

18.4 Consider the amplitude-modulated wave

$$F(t) = (f + 2g \sin \omega t) \sin \Omega t.$$

Show that its Fourier series is

$$F(t) = f \sin \Omega t + g \cos(\Omega - \omega)t - g \cos(\Omega + \omega)t.$$

Fix $\Omega = 5$, $\omega = 4$, $f = 1$ and $g = 2$. Compute the Fourier coefficients of $F(t)$ using the FFT algorithm. Then, obtain its power spectrum. Verify that it has dominant peaks only at the frequencies 1, 5 and 9 (multiplied by $1/2\pi$). Draw the power spectrum plot.

18.5 Compute the Fourier series and the power spectrum $P(\Omega)$ of

$$f(t) = 10 + 10 \sin t + \cos t + 5 \sin 2t - \sqrt{75} \cos 2t.$$

Verify that the power spectrum has peaks only at the frequencies 0, 1 and 2 (multiplied by $1/2\pi$). Also, verify that the $P(\Omega)$ at these frequencies are all ≈ 100, that is, $\log_{10} P(0) \approx 2$, $\log_{10} P(1) \approx 2$ and $\log_{10} P(2) \approx 2$.

18.6 The motion of a particle in a double-well potential driven by a periodic external force is modelled by the equation

$$\ddot{x} + \alpha \dot{x} - \omega_0^2 x + \beta x^3 = f \sin \omega t.$$

Fix $\alpha = 0.5$, $\omega_0^2 = 1$, $\beta = 1$, $\omega = 1$ and $x(0) = -1$, $\dot{x}(0) = 0$. Using the fourth-order Runge–Kutta integration method numerically solve the equation with time step $(2\pi/\omega)/100$ for $f = 0.33$. Leave $x(t)$ for $t = 0$ to $200 \times (2\pi/\omega)$. Then, collect 2^{12} values of x at t equal to every integer multiples of $(2\pi/\omega)/10$. From its power spectrum identify the frequency and period of the solution. Repeat the above analysis for $f = 0.35$, 0.357 and 0.365. Write a short note on your result and observation.

18.7 The Duffing–van der Pol oscillator equation is given by

$$\ddot{x} - p\left(1 - x^2\right)\dot{x} + \omega_0^2 x + \beta x^3 = f\cos\omega t\,.$$

Choose the parameters in the equation as $p = 0.2$, $\omega_0^2 = 0.011$, $\beta = 1$ and $f = 1$. Obtain the numerical solutions after leaving $x(t)$ for $t = 0$ to $(2\pi/\omega) \times 200$ as a transient for $\omega = 0.92$ and $\omega = 0.94$. Plot x versus $\dot{x}$. Compute the power spectra of the solutions and from them show that the solution for $\omega = 0.92$ is quasiperiodic and the solution for $\omega = 0.94$ is chaotic.

18.8 Consider the Duffing–van der Pol equation in the previous problem with $p = 0.2$, $f = 0$, $\omega_0^2 = 0.011$, $\beta = 1$. Obtain the long-time solution. From the power spectrum identify the period of the solution.

18.9 Consider the quasiperiodically driven pendulum equation

$$\ddot{x} + d\dot{x} + \sin x = K + f\left(\cos\omega_1 t + \cos\omega_2 t\right)$$

with $\omega_1 = (\sqrt{5} - 1)/2$, $\omega_2 = 1$, $d = 3$ and $f = 0.55$. For $K = 1.33$ obtain the numerical solution $x(t)$ after leaving a sufficient transient evolution. Compute the power spectrum of the solution. Verify that the number of peaks, $N(\sigma)$, in the power spectrum exceeding a threshold amplitude σ scales as $N(\sigma) \propto \sigma^{-\alpha}$, $1 < \alpha < 2$.

18.10 The two-dimensional map

$$\begin{aligned} x_{n+1} &= x_n + \Omega - (K/2\pi)\sin 2\pi x_n + b y_n \pmod{1}, \\ y_{n+1} &= b y_n - (K/2\pi)\sin 2\pi x_n \end{aligned}$$

has a quasiperiodic solution for $K = 0.8$. Iterating the map and leaving the first 1000 iterations as a transient collect 2^{14} values of (x_n, y_n). Plot x_n versus y_n. The obtained power spectrum of the quasiperiodic solution shows that the number of peaks $N(\sigma)$ in the power spectrum exceeding a threshold amplitude σ scales as $N(\sigma) \propto \log(1/\sigma)$. (Hint: Compute $\log|P(\Omega)/\max.P(\Omega)|$ and then calculate $N(\sigma)$ for a range of values of σ. Draw a graph between $N(\sigma)$ and $\log(1/\sigma)$. The data will fall roughly on a straight-line).

19

Random Numbers

19.1   Introduction

What are random variables? What are random numbers? An important concept in the theory of probability is the definition of a (real) random variable. It is a set function $X(\omega)$ and attaches a real number x to an outcome ω. The term random variable is commonly used to emphasize that one does not know the specific value this variable will assume. If a random variable X takes on values from a discrete set x_1, x_2, ..., x_n it is called a *discrete random variable*. A random variable X is said to be *continuous* if it takes on any value out of a certain interval $[a, b]$.

There is no proper and satisfactory definition of *randomness*. A sequence of numbers is called a *random*, if it is generated by a random physical process. Physical processes such as radioactivity decay, thermal noise in electronic devices, cosmic ray arrival times, etc., give rise to a sequence of random numbers. Interestingly, it is possible to generate a random number sequence employing simple algorithms which take very little time and memory. Such numbers generated by a deterministic algorithm are called *pseudo-random numbers* because they are predictable and reproducible.

A sequence of random numbers is often characterized by statistical quantities such as mean, variance, moments and probability density function. The *mean* (or expectation value or average or the first moment) of a discrete set of N random numbers, x_i, $i = 1, 2, , \ldots, N$, is given by

$$\bar{x} = \lim_{N \to \infty} \frac{1}{N} \sum_{i=1}^{N} x_i \, . \tag{19.1}$$

Generally, a random sequence is not adequately described by just stating its mean value. It is important to have some idea about how the numbers are dispersed about the mean. One such measure is the *variance*. It is given by

$$\sigma^2 = \lim_{N \to \infty} \frac{1}{N} \sum_{i=1}^{N} (x_i - \bar{x})^2 \, . \tag{19.2}$$

Its positive square-root is known as the *standard deviation* σ. The variance σ^2 is the expected value of the square of the deviation of x_i from its mean value $\bar{x}$. Essentially, it characterizes the spreading of the values x_i around the mean value $\bar{x}$. The unit of mean and standard deviation are dimensional quantities and have the same units as x_i.

Another important quantity associated with a sequence of random numbers is the *kth moment* defined as

$$m_k = \lim_{N \to \infty} \frac{1}{N} \sum_{i=1}^{N} (x_i - \bar{x})^k \, . \tag{19.3}$$

DOI: 10.1201/9781032649931-19

For continuous distributions in the range $[\alpha, \beta]$

$$\bar{x} = \frac{1}{\beta - \alpha} \int_\alpha^\beta x \, dx \qquad (19.4)$$

and

$$\sigma^2 = \frac{1}{\beta - \alpha} \int_\alpha^\beta \left(x - \frac{\alpha + \beta}{2} \right)^2 dx. \qquad (19.5)$$

The plan of this chapter is as follows. First, the importance of study of noise or random numbers is presented. The probability density, mean and variance of a single uniformly distributed random variable in an interval are determined. A simple procedure for generating a set of uniform random numbers is enumerated. Then, the Park and Miller's method of generation of uniform random numbers is described. The uniformity, autocorrelation and periodicity tests for checking the randomness of a random sequence are discussed. Next, the algorithms to generate random numbers with Gaussian distribution by the Box–Muller formula and Fernandez–Criado method are considered. Then, the inversion method, rejection technique and the Metropolis algorithm for generating random numbers with desired distribution are explained with specific examples. Methods of developing Cauchy, Lévy and dichotomous random numbers and their features are presented. Quantities to compare two distributions are also defined.

19.2   Importance of Study of Noise and Random Numbers

Why should one study about noise? Is the study of noise important? What can one do with random numbers? Study of noise is very crucial in laser systems, analog simulations with electronic circuits, computer networks and in communication. Noise in the form of random numbers is used in solving and simulating certain complex physical, biological and engineering problems and processes. They play a key role in the Monte Carlo method (see next chapter) which is a numerical method of solving mathematical problems by means of random sampling. Presence of noise in electrical and mechanical devices and random disturbances by an external environment in dynamical systems alter their normal functions. Noise can be modelled by an appropriate pseudo-random sequence in the mathematical modelling of systems driven by noise. Effect of noise can be studied theoretically without performing actual experiments [1,2].

In poor weather conditions, millimetre waves offer a much greater penetration over the visible spectrum through small dust particles (aerosols), rain and fog. Antenna arrays capable of detecting millimetre waves can be constructed. This design utilizes radiometry which is the science of using passive detection techniques to detect background radiation. In this passive detection system, the signals are inherently noise. Thus, noise must be taken into account when processing the antenna array signals for the desired application.

A common belief is that addition of noise to a system always degrades the quality of the response. However, this is not the case always. It has been shown that the sensitivity of muscle spindle receptors to a weak movement signal would be enhanced by adding noise through the tendon of the parent muscle [3]. In hearing systems noise enhanced peripheral sensory response has been demonstrated experimentally and theoretically [4–7]. Certain nonlinear systems have shown that there is an optimal nonzero noise intensity which can be added to a system to improve the response. This phenomenon is known as *stochastic resonance* [8–11]. Another kind of noise-induced resonance is the *coherence resonance* [12-15] and it has many practical applications.

Random numbers are useful to study various aspects of stochastic resonance in the theoretical model equations. Study of stochastic resonance is quite interesting for biological systems, especially in neurobiological systems since it provides a mechanism for such systems to detect and process weak signals. Stochastic resonance has been found to play a relevant role in several problems in biology, mammalian sensory systems, increment of tactile capacity, visual perception, low-frequency effects and low-amplitude electromagnetic fields. The presence of noise in nonlinear systems can give rise to many other fascinating phenomena like – noise-enhanced stability, noise-induced intermittency, spiral dynamics, phase transition, synchronization, pulse and pattern formation and noise-delayed extinction, a few to mention.

19.3 Uniform Random Numbers

An abstract representation for a trial and its possible outcome is known as the sample space denoted as S. A typical random number generator S consists of all integers between 0 and $2^{31} - 1$, a very large number. A point in space is called an *event* (or *outcome*), E. For every E in S a non-negative number called the *probability of E* denoted as $P(E)$, can be assigned with $0 \leq P(E) \leq 1$. A random variable X is a function on S with the following two properties:

1. The values of it are real numbers denoted as x.

2. For every x the probability that the value of the function is less than or equal to x (denoted as $P(X \leq x)$) can be calculated.

19.3.1 Mean and Variance

Consider the distribution of a single random variable x assuming real values in the interval say $[\alpha, \beta]$. Let us denote the probability for x taking values between x and $x+\mathrm{d}x$ as $p(x)\mathrm{d}x$. $p(x)$ is called *probability density* of the random variable x. The total probability is

$$P(\alpha \leq x \leq \beta) = \int_{\alpha}^{\beta} p(x)\mathrm{d}x = 1. \tag{19.6}$$

Divide the interval $[\alpha, \beta]$ into N subintervals with $N \to \infty$. Assume that the random numbers are distributed with equal probability in each interval. Such random numbers are called *uniform random numbers*. In this case the probability density $p(x) = p' = $ a constant. *What is the value of p'?* Since $P(\alpha \leq x \leq \beta) = 1$ Eq. (19.6) gives

$$1 = \int_{\alpha}^{\beta} p' \,\mathrm{d}x = p' \int_{\alpha}^{\beta} \mathrm{d}x = p'(\beta - \alpha)$$

or

$$p' = \frac{1}{(\beta - \alpha)}. \tag{19.7}$$

The probability for the random variable taking values between x and $x + \mathrm{d}x$ is $\mathrm{d}x/(\beta - \alpha)$ and $P(X \leq x) = (x - \alpha)/(\beta - \alpha)$.

For uniform distribution

$$\bar{x} = \int_{\alpha}^{\beta} xp(x)\,\mathrm{d}x = \frac{1}{(\beta - \alpha)} \int_{\alpha}^{\beta} x\,\mathrm{d}x = (\alpha + \beta)/2 \tag{19.8}$$

and

$$\sigma^2 = \int_\alpha^\beta (x - \bar{x})^2 p(x)\, dx$$

$$= \frac{1}{(\beta - \alpha)} \int_\alpha^\beta \left(x - \frac{\alpha + \beta}{2} \right)^2 dx$$

$$= \frac{1}{12}(\beta - \alpha)^2. \tag{19.9}$$

For a uniform random numbers in the interval $[0, 1]$ the results are $\bar{x} = 1/2$ and $\sigma^2 = 1/12$.

19.3.2   Generation of Uniform Random Numbers – Early History

How does one obtain a sequence of numbers which have the appearance of random? One can think of the following:

1. Take a telephone directory, open a page and pick a number on that page. Then, open another page and pick a number on that page and so on.

2. Write the numbers in pieces of paper, mix them in a box and pick one by one from the box.

These are not of practical use.

The history of generating random numbers is interesting. In 1927, L.H.C. Tippetti published a table of about $40,000$ random numbers (digits) taken at random from census reports. In 1939 M.G. Kendall and B. Babington-Smith built a mathematical device to produce 10^5 random digits. Then, computers were introduced and the search for efficient algorithms to generate random numbers began. In 1946 John von Neumann proposed a method of producing random numbers using arithmetic operations of a computer. In 1955, the RAND corporation published a table of a million random digits [16]. *Is it possible to generate random numbers by simply iterating an equation?* This is interesting because in such a case it is not necessary to store the data.

19.3.3   General Iterative Process

Computers themselves have a library routine that often has the name like *ran* or *rand* or *rnd* to generate uniformly distributed random numbers in the interval $[0, 1]$. Such system-supplied random numbers are generated by the recurrence relation

$$I_{j+1} = aI_j + c, \quad \mathrm{mod}(m), \quad j \geq 1 \tag{19.10}$$

with the four magic numbers

$$m \quad - \quad \text{the modulus with } m > 0$$
$$a \quad - \quad \text{the multiplier with } 0 < a < m$$
$$c \quad - \quad \text{the increment with } 0 \leq c \leq m$$
$$I_1 \quad - \quad \text{the starting value with } 0 \leq I_1 < m.$$

Start with an integer I_1. Calculate $aI_1 + c$ which is also an integer. Divide it by m and obtain the remainder and call it I_2. Repeating the above procedure a sequence of integers $\{I_1, I_2, \ldots\}$ can be computed. This is a sequence of random integers and is called a *linear congruential sequence*. The sequence obtained when $m = 1000$, $I_1 = a = c = 89$ is

$$89, 10, 979, 220, 669, 630, 159, \ldots .$$

TABLE 19.1

A few commonly used values of the parameters a, c and m in random number generators [17].

a	c	m	period
7^5	0	$2^{31} - 1$	$2^{31} - 2$
69069	0	2^{32}	2^{30}
1664524	1013904223	2^{32}	2^{32}

Equation (19.10) will repeat itself, with a period less than m. The sequence obtained when $m = 10$, $I_1 = a - c - 7$ is 7, 6, 9, 0, 7, 6, 9, 0, ... The above sequence is periodic with period 4. The desire is to find appropriate values for the parameters m, a, c and I_1 to get a good linear congruential sequence. *How does one choose good values for these four parameters?* In the following this aspect [17] is discussed.

19.3.4 Choice of the Modulus m

Let us consider the number m. The factors that influence the choice of m are

1. periodicity of the sequence and
2. speed of generation.

First, note that a periodic sequence cannot have more than m different numbers. So, m has to be chosen as very large. Secondly, the value of m should be such that the calculation of I_{j+1} in Eq. (19.10) with mod m operation is fast. Essentially, the avoid of mod operation would improve the speed of the calculation. For this purpose, set the value of m as the word size of the computer.

19.3.5 Choice of the Multiplier a

The value of a is such that the period of the sequence is maximum. Remember that a desirable criterion for the randomness of a sequence is the long period. As an example, consider the choice $a = c = 1$, $m = 6$ and $I_1 = 5$. Then, the random sequence is 5, 0, 1, 2, 3, 4, 5, 0, 1, ... Its period is 6, that is, the value of m. The period cannot be longer than m since the number of possible values are only m.

What are the possible values of a, c and I_1 that give a period of length m? The answer is given by the following theorem [17].

Theorem:

The period of the linear congruential sequence defined by the parameters m, a, c and the initial value of I, I_1, is m if and only if the following three conditions are satisfied:

1. *c is relatively prime to m,*
2. *$b = a - 1$ is a multiple of p, for every prime p dividing m and*
3. *b is a multiple of 4, if m is a multiple of 4.*

Table 19.1 shows a few commonly used values of the parameters in random number generators.

19.3.6   A General Simple Generator for Random Numbers

For generating a sequence of random numbers using a computer the following procedure gives a simple generator.

1. Choose an arbitrary value for the starting value of I, I_1.

2. Choose the value of m as large as possible, say at least 2^{30}. The choice of m being the computer word size makes the computation of $aI_j + c \bmod(m)$ as an efficient one. More over round-off error should not be introduced in the computation of $(aI_j + c)\bmod(m)$.

3. If m is a power of 2 then choose the value of a in such a way that $a \bmod(8) = 5$. This choice of a along with the choice of c given in step 5 is to make sure that the produced sequence will have m different values of I before the generator begins to repeat the numbers.

4. The value of a should be between $0.01\,m$ and $0.99\,m$ and further it should not have a regular pattern.

5. If a is a good multiplier then the value of c is immaterial. However, c should not have any factor in common with m. It is better to choose $c = 1$ or $c = a$.

Instead of considering a random integer between 0 and $m - 1$ one may consider I as a random fraction between 0 and 1. Then, the numbers generated are all in the interval $[0, 1]$. It can be converted into any suitable range $[-\alpha, \beta]$. Before using the sequence for applications apply all existing tests to make sure that the sequence is sufficiently random.

19.3.7   Park and Miller's Generator

Park and Miller proposed a generator based on the choices [18]

$$a = 7^5, \quad c = 0, \quad m = 2^{31} - 1 = 2147483647. \tag{19.11}$$

The maximum possible value of I is $m - 1$. When the product of a and I_j is greater than the maximum value of a 32-bit integer it is not possible to implement the Eqs. (19.10)–(19.11) in a computer. However, the following suggestion of Schrage is useful. From the approximate factorization of m given by

$$m = aq + r \tag{19.12}$$

one can write

$$q = \text{int}\,[m/a]\,, \quad r = m\,(\bmod\ a), \tag{19.13}$$

where int denotes integer part. For $r < q$ and $0 < I < m - 1$ both $aI\,(\bmod\ q)$ and $r.\text{int}[I/q]$ lie in the range $0, 1, \dots, m - 1$. Write

$$aI\,(\bmod\ m) = \begin{cases} aI\,(\bmod\ q) - r.\text{int}[I/q], & \text{if it is} > 0 \\[2mm] aI\,(\bmod\ q) - r.\text{int}[I/q] + m, & \text{otherwise.} \end{cases} \tag{19.14}$$

For simplicity, define

$$P_1 = I\,(\bmod\ q), \quad P_2 = \text{int}[I/q]\,. \tag{19.15}$$

Then,

$$aI\,(\bmod\ a) = \begin{cases} aP_1 - rP_2, & \text{if it is} > 0 \\[2mm] aP_1 - rP_2 + m, & \text{otherwise.} \end{cases} \tag{19.16}$$

TABLE 19.2

First five random numbers generated using Eqs. (19.10) and (19.16). Here, $m = 2147483647$, $a = 7^5$, $r = 2836$, $q = 127773$ and $c = 0$.

j	I_j	P_1	P_2	I_{j+1}	ran
0	3	3	0	50421	0.0000235
1	50421	50421	0	847425747	0.3946134
2	847425747	35211	6632	572982925	0.2668160
3	572982925	48793	4484	807347327	0.3759504
4	807347327	77513	6318	1284843143	0.5983017

Some useful choices of a, q and r are given below:

(i) $a = 7^5$, $q = 127773$, $r = 2836$.

(ii) $a = 48271$, $q = 44488$, $r = 3399$.

(iii) $a = 69621$, $q = 30845$, $r = 23702$.

The period of the numbers generated using the above procedure is $2^{31} - 2 = 2147483646 \approx 2.1 \times 10^9$. The above procedure generates numbers in the interval $[0, m]$. Note that the random numbers are now integers. The numbers generated can be converted into the range $[0, 1]$ by dividing them by m.

The following steps will generate a uniform random number in the interval $[0, 1]$:

$m = 2147483647$

$a = 7^5$

$r = 2836$

$q = 127773$

$I = 3$

$P_1 = \text{mod}(I, q)$

$P_2 = \text{int}(I/q)$

$I = aP_1 - rP_2$

if $(I < 0)$ then $I = I + m$

ran $= \text{float}(I)/\text{float}(m)$.

The value of ran is between 0 and 1. Repeating the lines $6 - 10$ generates a sequence of uniform random numbers. Table 19.2 gives the first few random numbers generated. Figure 19.1a depicts first 100 random numbers of the sequence generated. The mean and variance obtained with 10^4 numbers are 0.5017 and 0.0833, respectively, whereas the exact values are 0.5 and $0.0833\ldots$, respectively.

19.3.8 Numerical Calculation of Probability Distribution

To calculate the probability distribution of the random numbers, the numbers in the interval $[0, 1]$ are scaled into the range $[0, L]$ with $L = 50$. This range is then divided into L-equal subintervals. This simplifies the procedure to count the number of points occurring in each subinterval.

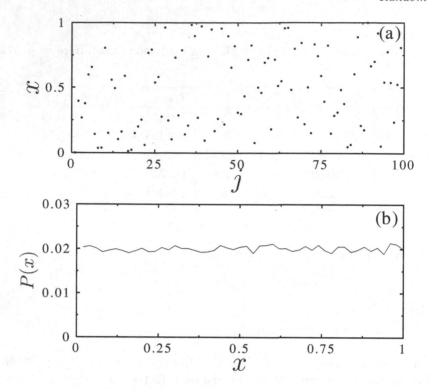

FIGURE 19.1
(a) Plot of first 100 uniformly distributed random numbers x generated using the Park and Miller algorithm. (b) The numerical probability distribution $P(x)$ of 5×10^4 numbers generated. The interval $[0, 1]$ is divided into 50 equal subintervals. $P(x)$ is almost constant.

For computational purpose, first the minimum, $xmin$, and the maximum, $xmax$, values of generated numbers $\{x_i\}$ are calculated. For uniform random numbers in the interval $[0, 1]$ these values are 0 and 1, respectively. Next, the numbers in the interval $[xmin, xmax]$ are converted into the interval $[0, L]$. Then, the numbers are changed into integers. Now, if the value of a number is K then it falls in the Kth subinterval. In this way, the number of x's falling in each subinterval is counted. Figure 19.1b shows the computed probability distribution of the numbers where 10^4 numbers are used. For uniformly distributed numbers the kth moments m_k are finite. First few m_k's are given by

$$m_0 = 1, \quad m_1 = \bar{x}, \quad m_2 = \sigma^2 + m_1^2 \tag{19.17}$$

which can also be verified numerically.

Before going to study how to generate other types of random numbers let us present how to analyze randomness of a set of random numbers in the next section.

19.4   Tests for Randomness

In the previous section how to generate a sequence of uniformly distributed random numbers with a long period is described. Absence of periodicity alone does not guarantee that the sequence generated is useful for practical applications. Before using it for a practical purpose it is necessary to know whether the sequence is sufficiently random. *What are the quantitative measures for randomness?* Some unbiased tests are needed to know the randomness of a sequence.

Randomness of a set of random numbers can be analyzed using the following tests:

1. Uniformity test.

2. Autocorrelation test.

3. Periodicity test.

4. Comparison of computed mean value with the expected one.

Let us consider the distribution of numbers in the interval $[\alpha, \beta]$ and divide the interval into L equally spaced subintervals. For perfectly distributed numbers, the probability of each interval is obtained from the expected or assumed probability density function. For uniformly distributed numbers this is $1/L$. Call this as expected probability E_i. For a given set of numbers numerically calculate the probability in each subinterval and denote them as O_i (observed). Then, the *chi-square*, χ^2, quantity is given by

$$\chi^2 = \sum_{i=1}^{L} \frac{(E_i - O_i)^2}{E_i + O_i} . \tag{19.18}$$

If the given set is perfectly distributed according to the assumed distribution then $\chi^2 = 0$. Deviation from zero can be used to describe nonuniformity.

The *autocorrelation function* for a set of N numbers is defined as

$$C(k) = \frac{\sum_{i=1}^{N} x_i\, x_{i+k} - N\bar{x}^2}{\sum_{i=1}^{N} x_i^2 - N\bar{x}^2} , \quad k = 0, 1, 2, \ldots, N-1 \tag{19.19}$$

where $\bar{x}$ is given by Eq. (19.1). For a sequence of truly random numbers $C(0) = 1$ and $C(k) = 0$ for $k > 0$.

The third test is the checking of whether the sequence has a long period. The fourth test is to compare the mean value of the sequence with the expected one. For a sequence of N truly uniform random numbers in the interval $[0, 1]$ the mean value $\bar{x}$ of it is expected to lie between $0.5 - \epsilon$ and $0.5 + \epsilon$. For large N, from the central-limit theorem, $\bar{x}$ is Gaussian with mean 0.5 and $\sigma^2 = 1/(12N)$. Suppose that $p(\epsilon)$ is the probability for the numbers in the interval $[0.5 - \epsilon, 0.5 + \epsilon]$. If a sequence of N random numbers has an average that falls inside (outside) the interval $[0.5 - \epsilon, 0.5 + \epsilon]$ then the conclusion is that it passes (fails) the test at 5% level.

In practice, it is better to apply all the known tests to make sure that the sequence is sufficiently random.

19.5 Gaussian (Normal) Distribution

A probability density function which is observed in many scientific processes is the *Gaussian*, also called *normal*. It is characterized by the density function

$$G(x) = p(x) = \frac{1}{\sqrt{2\pi}\,\sigma} e^{-(x-\mu)^2/(2\sigma^2)} , \quad -\infty < x < \infty \tag{19.20}$$

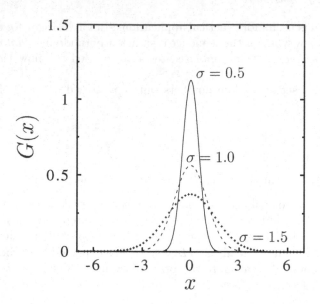

FIGURE 19.2
The Gaussian distribution, Eq. (19.20), with $\mu = 0$, and $\sigma = 0.5$, 1, 1.5.

where μ is the mean of x and σ^2 is the variance of x. Figure 19.2 shows $G(x)$ for $\mu = 0$ and $\sigma = 0.5$, 1 and 1.5. The Gaussian function is of fundamental importance in many physical and mathematical applications. The Gaussian random variable is utilized in digital communication (particularly, to analyze transceiver systems), machine learning, statistical modelling, data analysis, computer simulation, statistical sampling, cryptography and so on. Gaussian noise is employed in the numerical investigation of the effect of fluctuation of parameters and interaction of environments in the mathematical models of physical, chemical, biological and engineering systems. For these, it is necessary to have an efficient Gaussian random number generator. A very large number of techniques have been proposed to generate Gaussian random numbers. In the following, two simple methods are described.

19.5.1   Box–Muller Algorithm

Box and Muller proposed a simple algorithm to generate Gaussian random numbers. Their method is quite popular and called *Box–Muller algorithm*. The Box–Muller formula is [19]

$$x_1 = \mu + \sigma(-2\ln y_1)^{1/2}\cos(2\pi y_2) ,\qquad\qquad (19.21)$$

where y_1 and y_2 are uniform random numbers in the interval $[0, 1]$, μ and σ are, respectively, the mean and the standard deviation of the Gaussian random numbers to be generated and x_1 is the generated number.

Let us derive the Box–Muller formula (19.21). Consider a generation of a pair of Gaussian random numbers x and z. For simplicity set $\mu = 0$ and $\sigma = 1$. The joint distribution of x and z is

$$G(x, z) = p(x, z) = \frac{1}{2\pi}\,e^{-(x^2+z^2)/2} , \quad -\infty < x, z < \infty. \qquad (19.22)$$

Introducing the polar coordinates $r = \sqrt{x^2 + z^2}$, $\theta = \tan^{-1}(z/x)$ write

$$p(x, z)\,\mathrm{d}x\mathrm{d}z = \frac{1}{2\pi}\,e^{-r^2/2}\,r\,\mathrm{d}r\mathrm{d}\theta . \qquad (19.23)$$

Here, r is distributed according to $r\,e^{-r^2/2}$ whereas θ is uniformly distributed between 0 and 2π. Then, the distribution function for r is given by

$$F(r) = \int_0^r r\,e^{-r^2/2}\,\mathrm{d}r = 1 - e^{-r^2/2} = y_1\,, \tag{19.24}$$

where y_1's are uniformly distributed random numbers in the interval $[0, 1]$. The above equation gives

$$r = \sqrt{-2\ln(1 - y_1)}\,. \tag{19.25}$$

Since the replacement of $1 - y_1$ by y_1 does not change the probability distribution Eq. (19.25) can also be written as

$$r = \sqrt{-2\ln y_1}\,. \tag{19.26}$$

One can generate the random variable $\theta = 2\pi y_2$, where y_2's are uniformly distributed random numbers in the interval $[0, 1]$. Hence, x and z can be generated using the formula

$$x = r\cos\theta = \sqrt{-2\ln y_1}\,\cos 2\pi y_2\,, \tag{19.27}$$

$$z = r\sin\theta = \sqrt{-2\ln y_1}\,\sin 2\pi y_2\,. \tag{19.28}$$

To get a Gaussian distribution with mean μ and variance σ^2 rescale x as $\sigma x + \mu$ and obtain

$$x = \mu + \sigma\sqrt{-2\ln y_1}\,\cos 2\pi y_2 \tag{19.29}$$

or

$$x = \mu + \sigma\sqrt{-2\ln y_1}\,\sin 2\pi y_2\,. \tag{19.30}$$

Figure 19.3a shows the plot of first 100 Gaussian random numbers generated using the above formula (19.29) with $\mu = 0$ and $\sigma = 1$. Figure 19.3b depicts the probability distribution $P(x)$ of the generated numbers. 5×10^4 numbers are used to obtain this plot. The interval $[-5, 5]$ is divided into 50 equal subintervals in the calculation of $P(x)$. The moments, given by Eq. (19.3), of the Gaussian random numbers diverge with increase in N.

19.5.2 Fernańdez–Criado Algorithm

Fernańdez and Criado [20] proposed an algorithm which is based on an N-particles closed system interacting two at a time and conserving energy.

Consider N-particles each with same unit velocity: $\{v_i = 1,\ i = 1, 2, \ldots, N\}$. Randomly choose two particles and denote them as i and j. Redefine their velocities as per the iteration rule

$$v_i(\text{new}) = \frac{1}{\sqrt{2}}\left[v_i(\text{old}) + v_j(\text{old})\right]\,, \tag{19.31a}$$

$$v_j(\text{new}) = \frac{1}{\sqrt{2}}\left[v_j(\text{old}) - v_i(\text{old})\right]\,. \tag{19.31b}$$

Leave, for example, first $10N$ iterations or so as transients. Then, the velocities of the pair of particles chosen in all further iterations are the required pair of Gaussian random numbers with mean zero and variance unity. Gaussian random numbers of desired mean and variance σ are then obtained by the transformation

$$x = \mu + \sigma v\,. \tag{19.32}$$

This algorithm is found to be ten times faster than the Box–Muller algorithm.

How does one generate two-dimensional (that is, a sequence of pair of numbers (x, y)) random numbers with Gaussian distribution? (see Subsection 19.6.2 and Problem 19.29).

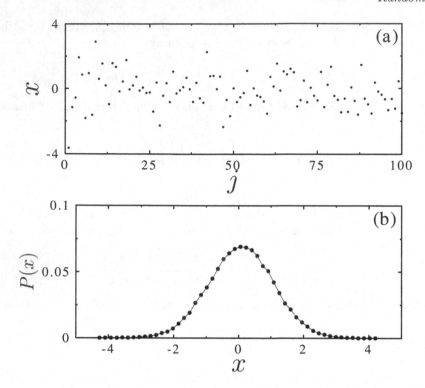

FIGURE 19.3
(a) Plot of first 100 Gaussian random numbers generated using the Box–Muller method. Here, $\mu = 0$ and $\sigma = 1$. (b) The numerically computed probability distribution of 5×10^4 Gaussian random numbers. The interval $[-5, 5]$ is divided into 50 equal subintervals. $P(x)$ is maximum at the mean $\bar{x} = \mu = 0$.

19.6 Generation of Random Numbers with Desired Distribution

It is easy to generate a sequence of random numbers other than uniformly distributed numbers and Gaussian numbers, for example, numbers with exponential distribution, circular probability density distribution, binomial distribution, etc. This can be achieved by employing the methods called *random sampling techniques*. Some of such methods are direct inversion, rejection method and Metropolis. In the following, these methods are described to obtain numbers with a chosen distribution. These methods can be easily applied to generate an arbitrary distribution. For other methods of generating exponential distribution, readers may refer to the refs. [21–26].

19.6.1 Method of Inversion

The method of inversion is based on the direct inversion of the cumulative probability distribution of the random variable x. Therefore, first introduce it. Let a random variable take values in the interval $-\infty$ to ∞ and the probability density is $p(x)$. The probability that the value of x is less than or equal to a value x_0 is

$$P(x \leq x_0) = F(x_0) = \int_{-\infty}^{x_0} p(x)\,\mathrm{d}x \tag{19.33}$$

with $0 \leq F(x_0) \leq 1$. The distribution function $F(x_0)$ is monotonic and nondecreasing function of its argument. $F(x_0)$ is called a *cumulative probability distribution* or simply a *probability distribution function*. The method of inversion is stated by the following theorem.

Theorem:

A random variable with probability density $p(x)$ can be generated from the formula

$$x = F^{-1}(y), \tag{19.34}$$

where $F^{-1}(y)$ is the inverse function of $F(x)$ and y is a uniformly distributed random number.

Proof:

Let us prove that x given by Eq. (19.34) is distributed as per $p(x)$. There is a one-to-one correspondence between x and y. Particularly, the probability for x assuming values between x and $x + \mathrm{d}x$ is the same for y and $y + \mathrm{d}y$. The probability in this interval is $\mathrm{d}y$ (*why?*). As $y = F(x)$ the probability for x between x and $x + \mathrm{d}x$ is given by

$$\mathrm{d}y = \mathrm{d}F(x) = F'(x)\,\mathrm{d}x = p(x)\,\mathrm{d}x. \tag{19.35}$$

Exponential Distribution

Choose the desired distribution as exponential of the form

$$p(x) = \begin{cases} \alpha\,\mathrm{e}^{-\alpha x}, & x \geq 0 \\ 0, & x < 0. \end{cases} \tag{19.36}$$

Let us mention a few applications of the exponential distribution. This distribution and its variants are used in queuing models, to model time duration between successive events such as request arrivals to a device and failure of a device, the length of phone calls and sales totals for customers, life span of electronic gadgets, shoppers at a shopping market, etc. [24-25]. It is used in the Monte Carlo simulation of electronic distribution of a hydrogen atom in various quantum states (see Chapter 20 in this book).

Consider the cumulative probability distribution defined as

$$F(x) = \int_{-\infty}^{x} p(x)\,\mathrm{d}x. \tag{19.37}$$

$F(x)$ is a monotonic nondecreasing function of x with $F(-\infty) = 0$ and $F(\infty) = 1$. For the function $p(x)$ given by Eq. (19.36)

$$F(x) = \int_{0}^{x} \alpha\,\mathrm{e}^{-\alpha x}\,\mathrm{d}x = 1 - \mathrm{e}^{-\alpha x}, \quad x > 0. \tag{19.38}$$

For the choice $\alpha = 1$ the function $F(x) = 1 - \mathrm{e}^{-x}$. Given a random number y_i, one has $x_i = F^{-1}(y_i) = -\ln(1 - y_i)$.

Figures 19.4a and 19.4b depict x versus $p(x)$ and x versus $F(x)$, respectively. Random numbers with the distribution given by Eq. (19.36) are obtained as follows.

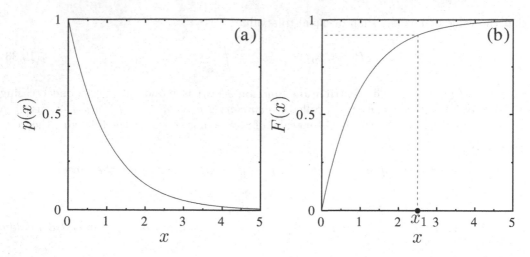

FIGURE 19.4
Plots of (a) x versus $p(x)$ (Eq. (19.36)) with $\alpha = 1$ and (b) x versus $F(x)$ (Eq. (19.38)).

(1)   Choose a point on the y-axis of Fig. 19.4b randomly between 0 and 1.

(2)   Draw a line parallel to x-axis and pass through the curve (dashed line in Fig. 19.4b). Obtain the intersecting point on the curve and call the x-component of it as x_1.

(3)   Repeating the steps (1) and (2) generates a sequence of numbers $\{x_i, \ i = 1, 2, \ldots\}$. These numbers have the chosen exponential distribution.

The above graphical scheme can be performed analytically as follows. Let $\{y_i, \ i = 1, 2, \ldots\}$ be a sequence of numbers $F(x)$ and are uniformly distributed between 0 and 1. The probability that it falls between $F(x)$ and $F(x) + \mathrm{d}F(x)$ is $\mathrm{d}F(x)$ which is equal to $p(x)\mathrm{d}x$. Hence, $x = F^{-1}(y)$ is distributed as per $p(x)$. For exponential distribution $F(x) = 1 - \mathrm{e}^{-x}$ and therefore $x = -\ln y$. As both y and $1 - y$ are uniformly distributed in the interval $[0, 1]$ one can set $x = -\ln y$.

The above scheme can be easily implemented in computers. Figure 19.5a shows a plot of first 100 numbers obtained. The numbers are usually generated in the interval $[0, \infty]$. For 5×10^4 numbers the numerically computed mean is 0.99 (the exact mean is $1/\alpha = 1$) and the variance is 0.98 (the exact variance is $(1/\alpha)^2 = 1$). Figure 19.5b shows the numerically computed probability distribution of the generated numbers.

The inversion technique can be used to generate random numbers with discrete distributions also. For some details see the ref. [27].

19.6.2   Rejection Technique

Another useful random sampling method is the von Neumann accept-reject or simply *rejection technique*. It is a powerful technique to generate random numbers whose distribution function is known and computable. This method does not rely on the analytic expression of inverse distribution but it is less efficient.

Let $f(x)$ be the desired distribution defined in the interval $[\alpha, \beta]$. Select a suitable bounding function $g(x)$ such that $cg(x) \geq f(x)$ for all values of $\alpha \leq x \leq \beta$. For the

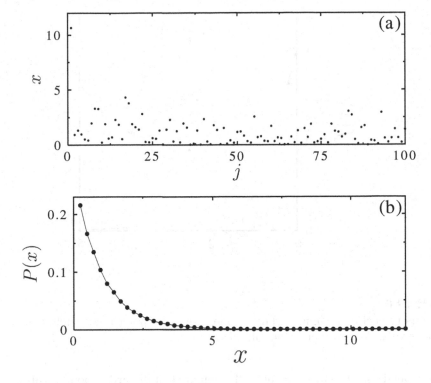

FIGURE 19.5
(a) Plot of first 100 random numbers with exponential distribution generated by the method of inversion. (b) The probability distribution of 5×10^5 numbers, where the interval $[0, 12]$ is divided into 50 equal subintervals.

exponential function

$$f(x) = \begin{cases} e^{-x}, & x \geq 0 \\ 0, & x < 0 \end{cases} \qquad (19.39)$$

one may choose the range $[\alpha, \beta]$ as $[0, 10]$ or a suitable interval for practical purpose. $cg(x) \geq 1$ because $f(x) \leq 1$ for $x \geq 0$. Set $c = 1$ then $g(x) \in [0, 1]$. Choose $g(x)$ and x as uniformly distributed random numbers in the intervals $[0, 1]$ and $[0, 10]$, respectively. The scheme is the following:

(1) Choose a value of x randomly and compute $f(x)$.

(2) Choose a value of cg and call it y.

(3) If $y > f(x)$ then go to step (1) otherwise x is a member of the chosen distribution number.

Repeating the above procedure, number of random numbers with the distribution $f(x)$ can be obtained. The procedure is simply the following. Consider Fig. 19.6 where the dots are the points with coordinate values (x_i, y_i), $x_i \in [0, 10]$ and $y_i \in [0, cg]$ $(= [0, 1])$. Discard the points falling above the curve $f(x)$. The x values of the points lying below the curve form random numbers with exponential distribution. Here, $cg(x)$ need not necessarily be defined in $[0, 1]$ and can be $[0, 1.5]$ or $[0, 2]$ or $[0, > 1]$. The number of points discarded to generate N

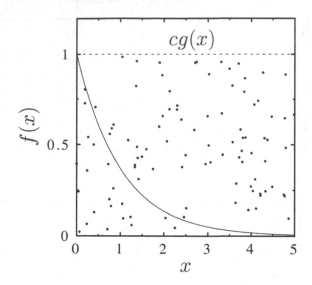

FIGURE 19.6
Description of rejection technique. The x-component of a point falling below the curve $f(x)$ is a number with the chosen exponential distribution.

numbers depends on the choice of $cg(x)$. The number of discarded points will be minimum if the maximum value of $cg(x)$ is chosen as the maximum value of $f(x)$.

Rejection technique is extendable to generate n-dimensional random numbers with a chosen distribution. As an example, consider a two-dimensional distribution function $f(x, y)$. Define x and y as uniformly random numbers in the interval $[a, b]$. $cg(z)$ is a suitable bounding function which is a set of random numbers in the interval, say, $[0, cgM]$ where cgM is a number such that $cgM \geq f(x, y)$. Choose the values of x, y and $cg(z)$. If $cg(z) < f(x, y)$ then accept the pair (x, y) otherwise leave. Repeating the above, a sequence of pair of numbers (x, y) can be obtained and will obey the distribution $f(x, y)$.

19.6.3   Metropolis Algorithm

Another simple technique of generating random numbers of a given distribution using uniform random numbers is the *Metropolis sampling* suggested by Metropolis and his collaborators [28]. This algorithm is widely used in Monte Carlo simulation of models in statistical physics.

The method is based on the concept of a *random walk*. In a typical one-dimensional random walk, a walker or a point is started at a reference point, say, the origin. At any instant of time, the point is allowed to move one step to the left or right. In the limit $t \to \infty$, the point will have visited all parts on a line. The point can spend much time in the region where the value of the probability density function sampled is large. In this way the set of positions of the points give a realization of the probability density function. This is essentially because the probability of finding a point in a given region is same as the chosen probability density function. Consequently, if all the positions of the points are collected then the probability would follow the probability density function.

First, describe the technique for an arbitrary distribution $f(x)$. Starting from an initial arbitrarily chosen value x_0 belonging to the value Ω of $f(x)$ the Metropolis technique generates a *Markov chain* (a stochastic process in which *past* history has no influence on the future if its *present* is specified) T of values x_1, x_2, ..., x_n. All the x_i's are values Ω of $f(x)$ and for $n \to \infty$, x_{n+1}, x_{n+2}, ... will have the given distribution $f(x)$. Call $\{x_{n+1}, x_{n+2}, \ldots\}$ as the desired states or values. Here, x_1 is based on x_0, x_2 is based on x_1 and so on. States are generated through the transition probability $W(x_i \to x_{i+1})$. For an asymptotic convergence to the desired values, the sufficient (but not necessary) condition is that

$$f(x_i) W(x_i \to x_j) = f(x_j) W(x_j \to x_i). \tag{19.40}$$

Equation (19.40) is called the *detailed balance*. This condition does not specify the transition probabilities from one value to another. Metropolis algorithm refers to a particular choice of W given by

$$W(x_i \to x_j) = T(x_i \to x_j) \min[1, f(x_j)/f(x_i)], \quad x_i \neq x_j \tag{19.41}$$

where $T(x_i \to x_j) = T(x_j \to x_i)$. T must be a symmetric stochastic matrix with positive elements but can be any distribution. Choose

$$T(x_i \to x_j) = \begin{cases} \text{constant,} & \text{for } x_j \text{ inside a certain} \\ & \quad \text{region around } x_i \\ 0, & \text{otherwise.} \end{cases} \tag{19.42}$$

Let us describe the implementation of the Metropolis sampling procedure for generating Gaussian random numbers obeying the probability density given by Eq. (19.20).

(1) Set the parameters of the required distribution. For Gaussian distribution, for simplicity, fix $\sigma = 1$ and $\mu = 0$.

(2) Set the index $i = 0$ and specify the initial value of $x_i (= x_0)$. A good choice of initial value is that value for which the probability is maximum. For Gaussian distribution, this is the case for $x = 0$ and hence choose $x_i - 0$.

(3) Select a trial state x_t randomly and with equal probability among the values Ω of the required distribution. Define

$$x_t = x_i - \epsilon + 2\epsilon\xi, \tag{19.43}$$

where ξ is a uniformly distributed random number in the interval $[0, 1]$ and ϵ is an adjustable parameter, for example, say $\epsilon = 3$. x_t is distributed uniformly in the interval $[x_i - \epsilon, x_i + \epsilon]$ with x_i as the centre.

(4) Calculate the ratio $\omega = f(x_t)/f(x_i)$.

(5) If $\omega \geq 1$ then set $x_{i+1} = x_t$. When $\omega < 1$, generate a (new) random number ξ and if $\xi \leq \omega$ set $x_{i+1} = x_t$ otherwise $x_{i+1} = x_i$.

(6) Set $i = i + 1$ and go to the step (3).

It may be required to generate several values of x, starting from x_0, before the sequence $x_1, x_2, \ldots$ follows the desired distribution. Therefore, leave, for example, 10^3 random numbers generated and store the further obtained numbers.

A characteristic measure of the Metropolis algorithm is the acceptance ratio given by

$$A_{\mathrm{R}} = \frac{\text{Number of accepted moves}}{\text{Number of trials}}. \tag{19.44}$$

A_{R} can be adjusted by changing ϵ. When A_{R} is too small then the value (of x) space or region will be explored much slowly because only very small number of trials are accepted.

TABLE 19.3
The Gaussian random numbers are generated by the Metropolis algorithm.

i	x_i	ξ	x_t	ω	ξ(new)	x_{i+1}
0	0.00	0.650	0.90	0.667	0.350	0.90
1	0.90	0.400	0.30	0.698	0.560	0.30
2	0.30	0.170	−1.68	0.255	0.910	0.30
3	0.30	0.610	0.96	0.659	0.390	0.96
4	0.96	0.890	3.30	0.002	0.001	3.30
5	3.30	0.250	0.55	199.100	——	0.55

On the other hand, when A_R is high, the trial moves are too small and in this case diffusion through the space will be very slow. An optimal choice of ϵ is the one which roughly accepts 50% of the trials.

Example:

The aim is to generate five Gaussian random numbers with mean zero and variance 1 using the Metropolis technique. Choose $\epsilon = 3$ and assume that the following sequence of uniformly distributed random numbers in the interval $[0, 1]$ are given:

$$0.650, \ 0.350, \ 0.400, \ 0.560, \ 0.170, \ 0.910,$$
$$0.610, \ 0.390, \ 0.890, \ 0.001, \ 0.250, \ 0.440.$$

Since the Gaussian distribution has maximum probability about $x = 0$ choose $x_0 = 0$. Table 19.3 presents the x_{i+1} for first five iterations.

19.7 Some Other Types of Random Numbers

Apart from uniform and Gaussian random noise, there are other types of noises such as Cauchy, Lévy and coin-toss square-wave dichotomous are realized in certain physical systems. These noises are also used in the study of certain physical and engineering problems. This section discusses the algorithms to generate them.

19.7.1 Cauchy Random Numbers

The *Cauchy probability distribution* is given by

$$P_C(x) = \frac{a}{\pi \left(a^2 + x^2\right)}, \tag{19.45}$$

where a is a constant. It has applications in electrical and mechanical theories, measurement and calibration problems and in representing the points of impact of straight-line of particles emitted from a source [29,30]. It is used in the analysis of quality of service in IP networks [31] and price fluctuation in research and development projects investments of governments [32].

Consider $P_C(x)\mathrm{d}x = P_y(u)\mathrm{d}y$ where y is a set of uniformly distributed random numbers in the interval $[0, 1]$ and obtain

$$\frac{a\,\mathrm{d}x}{\pi\,(a^2 + x^2)} = \mathrm{d}y. \tag{19.46}$$

Integration of the above equation gives

$$\frac{a}{\pi} \int_{-\infty}^{x} \frac{1}{a^2 + x^2}\,\mathrm{d}x = y. \tag{19.47}$$

That is,

$$\frac{1}{\pi}\left[\tan^{-1}\frac{x}{a} + \frac{\pi}{2}\right] = y. \tag{19.48}$$

From the above equation write

$$x = a\tan\pi\left(y - \frac{1}{2}\right). \tag{19.49}$$

Equation (19.49) can be used to generate random numbers x with the probability distribution $P_C(x)$ given by Eq. (19.45). Hundred random numbers generated using Eq. (19.49) with $a = 0.1$ are plotted in Fig. 19.7a. The probability distribution of 5×10^5 Cauchy random numbers is shown in Fig. 19.7b.

19.7.2   Lévy Random Numbers

Lévy [33] shown that the sum of n independent stochastic variable z with a probability distribution characterized by power-law tails $P(z > u) \propto |z|^{-\alpha}$, $0 < \alpha \leq 2$ converges to a stable process characterized by a probability density, which is now called a *Lévy distribution*. The Lévy stable processes are characterized by the probability density with diverging moments. The analytical form of the symmetrical Lévy stable distribution is known only for a few special values of α. The choices $\alpha = 1$ and 2 lead to Cauchy and Gaussian distributions, respectively.

A Lévy noise can increase the mutual information or bit count of several feedbacks neuron models that obey a general stochastic differential equation. Use of Lévy noise can benefit subthreshold neuronal signal detection [34], enhance subthreshold synchronization of the superparamagnetic tunnel junctions (spintronic nanodevices) [35], induce inverse stochastic resonance [36] and stochastic resonance [37] in neuron models, leads to synchronization of dynamical systems [38] and induce transport in multiple-well potential systems [39]. It has also been used in stochastic Norovirus epidemic model [40], stochastic COVID-19 epidemic models [41] and epidemic systems with quarantine strategy [42] and so on. In certain nonlinear systems the anomalous diffusion is found to show Lévy distribution.

The Lévy distribution is given by

$$L_{\alpha,\gamma}(z) = L_{\alpha,\gamma}(-z) = \frac{1}{\pi} \int_0^{\infty} \cos qz\, e^{-\gamma q^\alpha}\,\mathrm{d}q \tag{19.50}$$

or

$$P(x) = P(-x) = \frac{1}{2\pi} \int_{-\infty}^{\infty} e^{iqz}\, e^{-\gamma|q|^\alpha}\,\mathrm{d}q, \tag{19.51}$$

where α and γ are two parameters characterizing the distribution. α defines the index of the distribution and controls the scale properties of the stochastic process $\{x\}$ and γ selects the scale unit of the process [43].

Now, describe the method of Rosario Nunzio Mantegna [43] to generate Lévy random numbers. The method of Mantegna consists of the following four steps.

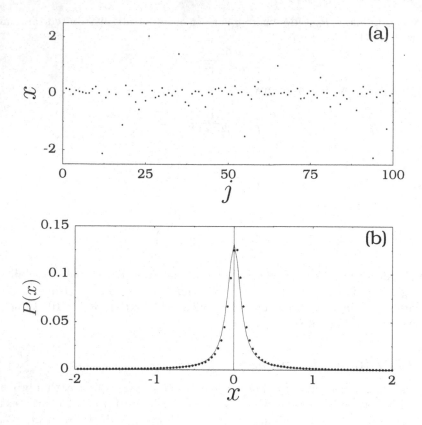

FIGURE 19.7

(a) Hundred Cauchy random numbers with $a = 0.1$. (b) The numerically computed probability distribution of 5×10^5 Cauchy random numbers (marked by dots) and the exact probability distribution (continuous curve).

Step 1:

With x and y being Gaussian random variables with their standard deviation given by

$$\sigma_x(\alpha) = \left[\frac{\Gamma(1 + \alpha) \sin(\pi\alpha/2)}{\Gamma((1 + \alpha)/2)\, \alpha\, 2^{(\alpha - 1)/2}} \right]^{1/\alpha}, \quad \sigma_y = 1, \tag{19.52}$$

respectively, define a random variable v,

$$v = \frac{x}{|y|^{1/\alpha}}. \tag{19.53}$$

For $|v| \gg 0$ the distribution of v is similar to the Lévy distribution (for a proof see the ref. [43]). One can show that the probability distribution of v is

$$P(v) = \frac{1}{\pi \sigma_x \sigma_y} \int_0^\infty y^{1/\alpha} \exp\left[-\frac{y^2}{2\sigma_y^2} - \frac{v^2 y^{2/\alpha}}{2\sigma_x^2} \right] \, dy. \tag{19.54}$$

Figure 19.8 shows the plots of $L_{1.5,1}(v)$ and $P(v)$ with $\sigma_y = 1$, $\sigma_x = 0.696575$. Both $L_{1.5,1}(v)$ and $P(v)$ are calculated numerically by integrating the integrals in Eqs. (19.50) and (19.54). $L_{1.5,1}(v)$ and $P(v)$ are different near the origin, however, they coincide for $|v| \gg 0$.

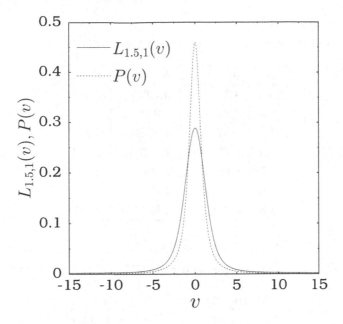

FIGURE 19.8
Plots of $L_{\alpha,\gamma}(v)$ and $P(v)$ (Eq. (19.54)) where $\alpha = 1.5$, $\gamma = 1$, $\sigma_x = 0.696575$ (Eq. (19.52)) and $\sigma_y = 1$. The two functions are almost coincident for $|v| \geq 10$.

Step 2:

The second step is to ensure that the $P(v)$ of the numerically generated $\{v\}$ coincides all over the range with the Lévy stable distribution of the same index α and the scale factor γ. For this purpose consider

$$z_n = \frac{1}{n^{1/\alpha}} \sum_{m=1}^{n} v_m . \tag{19.55}$$

The convergence of $\{z_n\}$ is quite slow. For a faster convergence, Mantegna proposed the transformation from v to ω as

$$\omega = \left\{ 1 + [K(\alpha) - 1] e^{-v/C(\alpha)} \right\} v , \tag{19.56}$$

where $C(\alpha)$ and $K(\alpha)$ are two parameters to be determined. Next, introduce, z_{cn}, weighted average of n independent stochastic variables ω, as

$$z_{cn} = \frac{1}{n^{1/\alpha}} \sum_{m=1}^{n} \omega_m . \tag{19.57}$$

z_{cn} quickly converges to a Lévy stable distribution.

Step 3:

The optimal value of $K(\alpha)$ can be determined analytically by requiring $P(\omega = 0) = L_{\alpha,1}(0)$. $\omega = K(\alpha)v$ for ω close to the origin. Then, $P(\omega = 0) = L_{\alpha,1}(0)$ is satisfied if

$$K(\alpha) = \frac{P(v = 0)}{L_{\alpha,1}(0)} . \tag{19.58}$$

TABLE 19.4
Values of σ_x, K and C for some values of α with $\gamma = 1$ [43].

α	$\sigma_x(\alpha)$	$K(\alpha)$	$C(\alpha)$
0.8	1.139900	0.795112	2.4830
1.0	1.0	1.0	
1.2	0.878829	1.205190	2.9410
1.4	0.759679	1.446470	2.8315
1.5	0.696575	1.59922	2.7370
1.6	0.628231	1.793610	2.6125
1.8	0.458638	2.501470	2.2060

Substitution of $L_{\alpha,1}(0) = \Gamma(1/\alpha)/(\pi\alpha)$ and Eq. (19.54) in the above equation yields

$$K(\alpha) = \frac{\alpha\Gamma((\alpha+1)/2\alpha)}{\Gamma(1/\alpha)} \left[\frac{\alpha\Gamma((\alpha+1)/2)}{\Gamma(1+\alpha)\,\sin(\pi\alpha/2)} \right]^{1/\alpha}. \qquad (19.59)$$

$C(\alpha)$ is the result of a polynomial fit of the values tabulated in the ref. [43]. The required variable is $z = C^{1/\alpha} z_{cn}$. Table 19.4 gives some of the values of the control parameters.

Step 4:

In the above, the value of the scale factor γ is set as 1. Lévy random numbers with $\gamma \neq 1$ can be obtained from z_{cn} using the transformation

$$z = \gamma^{1/\alpha} z_{cn}. \qquad (19.60)$$

Now, compare the numerically generated Lévy random numbers with the exact Lévy stochastic variable. Figure 19.9a is a plot of hundred Lévy random numbers generated using the above algorithm where $\alpha = 1.5$, $\gamma = 1$, $\sigma_x = 0.696575$ (Eq. (19.52)), $\sigma_y = 1$, $K(\alpha) = 1.59922$ (Eq. (19.59)), $C(\alpha) = 2.737$ (from the Table 19.4) and $n = 1$. Very few generated numbers have very large magnitude say greater than 20. They can be discarded. Probability distribution of the numerically generated random numbers is computed by dividing the interval $[-20, 20]$ into 120 bins. Then, the probability density, P_L is obtained by dividing the probability distribution by the bin size used. Figure 19.9b shows P_L of the numerically generated 5×10^5 Lévy random numbers and the exact Lévy distribution $L_{\alpha,\gamma}$. Notice that the P_L is in very good agreement with the $L_{\alpha,\gamma}$ all over the range of z. A small discrepancy observed near the origin can be eliminated by using large values of n. Figure 19.10 shows two probability densities calculated for $\gamma = 0.01$ and 100. In both cases, numerical simulation is in very good agreement with the actual Lévy distribution.

For two other methods of generating Lévy random numbers, one may refer to the refs. [18,44,45].

19.7.3 Dichotomous Noise

Dichotomous noise is characterized by whether it is *on* or *off* or whether it is *up* or *down* [46–48]. Rich and complex phenomena are realized due to an interplay of the dichotomous single-molecule gene noise with architecture of genetic networks. Bursting intermittency is

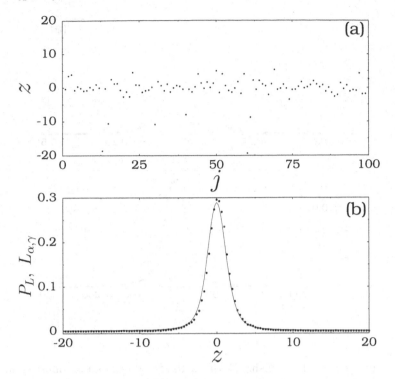

FIGURE 19.9
(a) Hundred Lévy random numbers with the control parameters $\alpha = 1.5$, $\gamma = 1$, $\sigma_x = 0.696575$ (Eq. (19.52)), $\sigma_y = 1$, $K(\alpha) = 1.59922$ (Eq. (19.59)), $C(\alpha) - 2.737$ (from the Table 19.4) and $n = 1$. (b) The numerically computed probability density P_L of 5×10^5 Lévy random numbers (marked by dots) and the exact probability density $L_{\alpha,\gamma}$ (continuous curve) given by Eq. (19.45).

found to occur in the generation of mRNAs [49] and in coupled nonlinear oscillators [50] due to dichotomous noise. Induced transitions in ZnSe interference filter with dichotomous fluctuations in the incident power are experimentally detected [51]. Due to dichotomous noise phase synchronization [52,53], hysteresis [54], patterning [55,56], improved resonance [57] and hypersensitivity [58,59] are reported.

A dichotomous coin-toss square-wave noise can be expressed as

$$G(t) = Da_n, \quad [\alpha + (n-1)]t_0 < t \le (\alpha + n)t_0, \tag{19.61}$$

where $n = \ldots, -2, -1, 0, 1, 2, \ldots$ is a set of integers, α is a random variable uniformly distributed between 0 and 1, a_n are independent random variables that take on the values -1 and $+1$ with equal probability 0.5, t_0 is a parameter of the process $G(t)$ and D is the amplitude of the noise. A rectangular pulse-wave of amplitude a_n and length t_0 centred at the coordinates $t_n = (\alpha + n - 0.5)t_0$ has a Fourier transform

$$G_n(\omega) = Da_n \left[(2/\omega) \sin(\omega t_0/2) \, e^{-i\omega t_n} \right]. \tag{19.62}$$

The pulse itself can be expressed as a sum of harmonic terms approximating as closely as desired the inverse Fourier transform of $G_n(\omega)$. Each realization of the coin-toss dichotomous square-wave can be approximated arbitrarily closely by a superposition of such sums, which

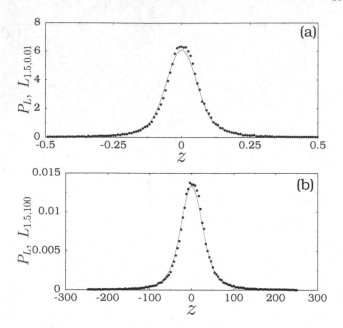

FIGURE 19.10

(a) The computed probability density P_L of 5×10^5 Lévy random numbers (marked by dots) and the exact probability density $L_{\alpha,\gamma}$ (continuous curve) given by Eq. (19.45). (a) $\gamma = 0.01$ and (b) $\gamma = 100$. The values of the parameters are $\alpha = 1.5$, $\sigma_x = 0.696575$ (Eq. (19.52)), $\sigma_y = 1$, $K(\alpha) = 1.59922$ (Eq. (19.59)), $C(\alpha) = 2.737$ (from the Table 19.4) and $n = 1$.

is itself a sum of harmonics, that is, a quasiperiodic function with parameters a_n. Figure 19.11 shows the numerically generated dichotomous coin-toss noise for $t < 1000$ with $t_0 = 100$ and t is incremented in units of 1 and $D = 1$.

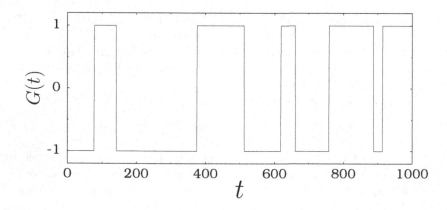

FIGURE 19.11

The numerically generated dichotomous noise for $t < 1000$ with $t_0 = 100$ and t is incremented in units of 1 and $D = 1$.

19.8 Skewness, Kurtosis and Students t Tests

In this section, quantities useful to characterize an asymmetry of a distribution and comparison of two different distributions are defined.

19.8.1 Skewness

An asymmetry of distribution of random numbers around its mean can be characterized by the skewness. It is defined as

$$S(x_1, x_2, \ldots, x_N) = \frac{1}{N} \sum_{i=1}^{N} \left(\frac{x_i - \bar{x}}{\sigma}\right)^3, \quad \bar{x} = \frac{1}{N} \sum_{i=1}^{N} x_i, \quad (19.63)$$

where σ is the standard deviation of the given set of random numbers. It is a dimensionless quantity. A distribution with a longer tail to the right of central maximum than to the left said to be *skewed to the right*. For this distribution S is positive. A negative value of S corresponds to a distribution with a longer tail to the left of central maximum than the right. Such a distribution is called *skewed to the left*. For a perfectly symmetrical curve S is zero.

19.8.2 Kurtosis

Kurtosis measures the degree of peakedness or flatness of a distribution relative to normal distribution. Like skewness, the kurtosis is also a dimensionless quantity and is defined as

$$k(x_1, x_2, \ldots, x_N) = \frac{1}{N} \sum_{i=1}^{N} \left[\frac{x_i - \bar{x}}{\sigma}\right]^4. \quad (19.64)$$

A distribution with a relatively high peak is called *leptokurtic*. In this case kurtosis (k) is > 0. A flat-topped distribution is called *platykurtic* for which $k < 0$. For normal distribution $k = 3$.

19.8.3 Students t-Test

When two distributions have same variance but different mean values then the significance of different mean values can be quantified by student's t-test. It is a number between 0 and 1. Let A and B represent two distributions with x_A and y_B are the set of N_A and N_B numbers, respectively. Denote $\bar{x}_A$ and $\bar{y}_B$ as mean values of the set numbers x_A and y_B, respectively. Then, the student's t is defined as

$$t = \frac{\bar{x}_A - \bar{y}_B}{s_D}, \quad (19.65)$$

where the standard error s_D of the difference of mean values is given by

$$s_D = \left[\left(\frac{1}{N_A} + \frac{1}{N_B}\right) \frac{\sum_{i=1}^{N_A}(x_i - \bar{x}_A)^2 + \sum_{i=1}^{N_B}(y_i - \bar{y}_B)^2}{(N_A + N_B - 2)}\right]^{1/2}. \quad (19.66)$$

t can be computed as follows:

1. Compute $\bar{x}_A$ and $\bar{y}_B$.

2.   Compute $\sum(x_i - \bar{x}_A)^2$ and $\sum(y_i - \bar{y}_B)^2$.

3.   Compute s_D given by Eq. (19.66).

4.   Compute t through Eq. (19.65).

Student's is a pseudonym for W.S. Gosset, *Biometrika* 6(1908)1.

Whether two given distributions are different or same can be determined by computing the chi-square–χ^2 test quantity and is already discussed in Section 19.4.

19.9   Concluding Remarks

In this chapter, numerical algorithms to generate different types of random numbers are discussed. The method of generating uniform random numbers is very simple. The methods of generating other types of random numbers require uniform random numbers. The random numbers play a key role in the Monte Carlo simulation of physical problems. Real physical, mechanical and biological systems are subjected to some kind of noise. In nonlinear systems the presence of even a noise of small strength can highly alter the response of the systems. In some cases the presence of noise is beneficial. The influence of noise in a system can be studied by adding appropriate noise in the mathematical model of the system. Now, one may ask how to solve, for example, a differential equation, in the presence of a noise. This is considered in Section 11.10.

19.10   Bibliography

[1]   W. Horsthemke and R. Lefever, *Noise-Induced Transitions*. Springer, Berlin, 1984.

[2]   J. Garcia-Ojalvo and J.M. Sancho, *Noise in Spatially Extended Systems*. Springer, Berlin, 1999.

[3]   P. Cordo, J.T. Inglis, S. Verschueren, J.J. Collins, D.M. Merfeld, S. Rosenblum, S. Buckley and F. Moss, *Nature* 383:769, 1996.

[4]   F. Jamarillo and K. Wiesenfeld, *Chaos, Solitons & Fractals* 11:1869, 2000.

[5]   I.C. Gebeshuber, *Chaos, Solitons & Fractals* 11:1855, 2000.

[6]   M. Chatterjee and M.E. Robert, *Proc. SPIE* 5110:348, 2000.

[7]   Z.C. Long, F. Shao, Y.P. Zhang and Y.S. Qin, *Phys. Lett. A* 323:434, 2004.

[8]   B. McNamara and K. Wiesenfeld, *Phys. Rev. A* 39:4854, 1989.

[9]   P. Jung, *Phys. Rep.* 234:175, 1993.

[10]   L. Gammaitoni, P. Hanggi, P. Jung and F. Marchesoni, *Rev. Mod. Phys.* 70:223, 1998.

[11]   S. Rajasekar and M.A.F. Sanjuan, *Nonlinear Resonances*. Springer, New York, 2016.

[12]   H. Gang, T. Ditzinger, C.Z. Ning and H. Haken, *Phys. Rev. Lett.* 71:807, 1993.

[13]   A.S. Pikovsky and J. Kurths, *Phys. Rev. Lett.* 78:775, 1997.

[14] B. Linder and L. Schimanwky-Geier, *Phys. Rev. E* 61:6103, 2000.

[15] C. Palenzuela, R. Toral, C.R. Mirasso, O. Calvo and J.D. Gunton, *Europhys. Lett.* 56:347, 2001.

[16] RAND Corporation, *A Million Random Digits with* 100000 *Normal Deviates*. The Free Press, Santa Monica, 1955.

[17] W. Jian-Sheng, C. Kan and Z. Fei, *Computational Techniques in Theoretical Physics*. Lecture Notes, 1998.

[18] W.H. Press, S.A. Teukolsky, W.T. Vetterling and B.P. Flannery, *Numerical Recipes in Fortran*. Foundation Books, New Delhi, 1993. Indian edition.

[19] G.E.P. Box and M.E. Muller, *Ann. Math. Stat.* 29:610, 1958.

[20] J.F. Fernańdez and C. Criado, *Phys. Rev. E* 60:3361, 1999.

[21] J.F. Fernańdez and R. Rivero, *Comp. Phys.* 10:83, 1996.

[22] C.S. Wallac, *ACM Trans. Math. Software* 22:119, 1996.

[23] J.H. Ahren and U. Dieter, *Comm. Assoc. Mach.* 15:873, 1972.

[24] Exponential Distribution Examples in Real Life; https://studiousguy.com/exponential-distribution-examples/ (accessed on 10 January 2022).

[25] J. Frost, *Exponential distribution: Uses, parameters & examples*; https://statistics byjim.com/probability/exponential-distribution/ (accessed on 10 January 2022).

[26] G. Marsaglia, *Ann. Math. Stat.* 32:899, 1961.

[27] K.P.N. Murthy, *Monte Carlo Basics*. Indian Society for Radiation Physics, Kalpakkam, 2000.

[28] N. Metropolis, A.W. Rosenbluth, M.N. Rosenbluth, A.H. Teller and E. Teller, *J. Chem. Phys.* 21:1087, 1953.

[29] A. Alzaatreh, C. Lee, F. Famoye and I. Ghosh, *Stat. Distri. Appl.* 3:12, 2016.

[30] N.L. Johnson, S. Kotz and N. Balakrishnan, *Continuous Univariate Distributions*. Volume-1. Wiley, New York, 1994.

[31] L. Rizo, *Cauchy distributions for Jitter in IP networks*; https://www.academia. edu/5044387/Cauchy_Distribution_for_Jitter_in_IP_Networks (accessed on 10 January 2022).

[32] S. Casault, A.J. Groena and J.D. Linton, *On the use of the Cauchy distribution to describe price fluctuations in R&D and other forms of real assets*; https://idus.us.es/bitstream/handle/11441/58827/On%20the%20use%20of%20-the%20Cauchy%20distribution.pdf?sequence=4&isAllowed=y (accessed on 10 January 2022).

[33] P. Lévy, *Theorie de l'Addition des Variables Aleatories*. Gauthier–Villars, Paris, 1937.

[34] A. Patel and B. Kosko, *IEEE International Conference on Acoustics, Speech and Signal Processing* 3:1413, 2007.

[35] Z. Liao, K. Ma, S. Tang, Md. Shamum Sarker, H. Yamahara and H. Tabala, *Res. Phys.* 27:104475, 2021.

[36] Y. Zhao and D. Li, *Mod. Phys. Lett.* 33:1950252, 2019.

[37] A. Patil and B. Kosko, *IEEE Trans. Neural Netw.* 19:1993, 2008.

[38] H. Zhou, Y. Li, W. Li and J. Feng, *Appl. Anal.* 101:2535, 2022.

[39] Y. Li, Y. Xu, J. Kurths and X. Yue, *Phys. Rev. E* 94:042222, 2016.

[40] T. Cul, A. Din, P. Liu and A. Khan, *Comp. Meth. Biomech. Biomed. Eng.*, 2022. DOI: 10.1080/10255842.2022.2106784

[41] P. Liu, L. Huang, A. Din and X. Huang, *J. Biol. Dyn.* 16:236, 2022.

[42] Y. Sabbar, D. Kiouach and S.P. Rajasekar, *Int. J. Dyn. Control* 11:122, 2023.

[43] R.N. Mantegna, *Phys. Rev. E* 49:4677, 1994.

[44] J.M. Chambers, C.L. Mallows and B.W. Stuck, *J. Am. Stat. Assoc.* 71:340, 1976.

[45] M. Leccardi, *Comparison of three algorithms for Lévy noise generation*; http://www.unipa.it/ocs/sito-strategico/relazioni/publicazioni_secondo_anno/-AL1.pdf (accessed on 10 January 2022).

[46] R. Kapral and S.J. Fraser, *J. Stat. Phys.* 71:61, 1993.

[47] J.P. Porra, J. Masoliver and K. Lindenberg, *Phys. Rev. E* 48:951, 1993; 50:1985, 1994.

[48] Y.R. Sivathanu, C. Hagwood and E. Simiu, *Phys. Rev. E* 52:4669, 1995.

[49] D.A. Potoyana and P.G. Wolynes, *J. Chem. Phys.* 143:195101, 2015.

[50] S. Rajasekar, M.C. Valsakumar and S. Paulraj, *Physica A* 261:417, 1998.

[51] I. Broussell, I. L'Heureux and E. Fortin, *Phys. Lett. A* 225:85, 1997.

[52] M.R. Roussel and J. Wang, *J. Phys. Chem. A* 105:7371, 2001.

[53] T. Yamada, T. Horita, K. Ouchi and H. Fujisaka, *Prog. Theor. Phys.* 116:819, 2006.

[54] R. Mankin, A. Sauga, A. Ainsaar, A. Haljas and K. Paunel, *Phys. Rev. E* 69:061106, 2004.

[55] D. Das and D.S. Ray, *Phys. Rev. E* 87:062924, 2013.

[56] J. Buceta and K. Lindenberg, *Phys. Rev. E* 68:011103, 2003.

[57] R. Rozenfeld, A. Neiman and L. Schimansky-Geier, *Phys. Rev. E* 62:R3031, 2000.

[58] M.B. Tarlie and R.D. Astumian, *Proc. Natl. Acad. Sci. U.S.A.* 95:2039, 1998.

[59] S.L. Ginzburg and M.A. Pustovoit, *Phys. Rev. Lett.* 80:4840, 1998.

19.11   Problems

A. *Uniform Random Numbers*

19.1 Generate uniformly distributed random numbers in the intervals a) $[-0.5, 0.5]$, b) $[1, 2]$ and c) $[-1, -2]$. For each sequence compute mean and variance and compare them with the exact values.

19.2 For a sequence of random numbers generated by the Park–Miller method verify that a) $\chi^2 \approx 0$, b) $C(k) \approx 0$, for $k > 0$, and c) the mean value passes the randomness test at 5% level.

19.3 For a random number generator available in your computer check whether it passes the randomness tests.

19.4 For a sequence of uniformly distributed numbers in the interval $[0, 1]$ verify that kth moments m_k are finite. Also, verify that $m_0 = 1$, $m_1 = \bar{x} \approx 0.5$ and $m_2 = \sigma^2 + m_1^2$, where $\sigma^2 = 1/12$.

19.5 Let $y(N, x) = (1/N) \sum_{i=1}^{N} x_i$ where x_i are uniform random numbers in the interval $[0, 1]$. Generate a large number of y's with $N = 2$. Compute the probability distribution of y, mean and variance. Repeat the above calculation for $N = 5$, 10, 50 and 100. Write a short note on your observation.

19.6 Write a single Python program which do all the following:

   (a) Generate $N = 5 \times 10^4$ uniformly distributed random numbers.

   (b) Calculate mean and variance of the sequence.

   (c) Calculate the probability distribution of the numbers.

19.7 Computer-generated random numbers are not really random. Why?

B. Gaussian Random Numbers

19.8 In the Box–Muller method instead of cosine function one may use sine function (Eq. (19.30)) also. Generate Gaussian random numbers with sine function. Sketch the probability distribution of the obtained numbers. Compare its mean and variance with the assumed values.

19.9 Numerically, shows that the moments of Gaussian random numbers diverge with N (the number of random numbers). (Compute kth moment as a function of N and show that it diverges with N).

19.10 Verify that

   (a) the sum $Y_M = \sum_{i=1}^{M} x_i$, where $\{x_i\}$ are Gaussian random numbers, is also a Gaussian with mean $M\mu$ and variance $M\sigma^2$,
   (b) the mean and variance of the sum Y_M scaled by M are μ and σ^2/M, respectively, and
   (c) the mean and variance of $Y_M/\sqrt{M}$ are $\mu\sqrt{M}$ and σ^2, respectively.

19.11 Compare the Box–Muller and the Fernańdez–Criado algorithms with special emphasis on mean, variance, probability distribution curve and CPU time taken. Form a table showing these values for the number of random numbers $N = 10^4, 2 \times 10^4, \ldots, 1 \times 10^5$.

19.12 A method of generating Gaussian random numbers using the central-limit theorem is the following: Generate N uniform random numbers in the interval $[0, 1]$. Consider the sum

$$x = \frac{\left(\sum_{i=1}^{N} y_i\right) - (N/2)}{\sqrt{N/12}}.$$

Generate a large number of x's. For large N the x's form a sequence of Gaussian random numbers with mean 0 and variance unity. Take $N = 2, 3, \ldots$ and demonstrate the approach of x to Gaussian. Did you observe tails in the probability distribution?

19.13 Write a single program which do all the following.

    (a) Generates $N = 5 \times 10^4$ normally distributed random numbers with mean = 0 and $\sigma^2 = 1$,

    (b) calculates the mean and variance of the obtained numbers and

    (c) computes the probability distribution of the numbers.

19.14 Can you apply the method of inversion to generate Gaussian random numbers? Why?

C. Random Sampling Techniques

19.15 Numerically, show that the moments

$$m_k = \frac{1}{N} \sum_{i=1}^{N} (x_i - \bar{x})^k , \quad k = 1, 2, \ldots, 10$$

for the exponentially distributed numbers diverge.

19.16 Calculate the mean and variance of exponentially distributed 5×10^4 numbers with $[\alpha, \beta]$ as a) $[0, 4]$, b) $[0, 5]$, c) $[0, 6]$, d) $[0, 10]$, e) $[0, 20]$ and f) $[0, 50]$. What did you observe?

19.17 Count the number of discarded points in generating 10^4 numbers with exponential distribution by rejection technique. Do this for the following bounding function $cg(x) = [\alpha, \beta] = [0, 1], [0, 2], [0, 3], [0, 4]$ and $[0, 5]$. Study the dependence of the discarded points on the parameter β.

19.18 Generate a sequence of random numbers with the distribution

$$f(x) = \begin{cases} \sqrt{2}\,e^{-\sqrt{2}\,x}, & x \geq 0 \\ \\ 0, & x < 0. \end{cases}$$

Also, compute its mean, variance and probability distribution.

19.19 Generate a sequence of numbers with the distribution $(1/\sqrt{2})e^{-\sqrt{2}\,x}$, $-\infty < x < \infty$. What are the values of its mean and variance?

19.20 Generate a sequence of N random numbers with the exponential distribution, Eq. (19.36). Sum them and divide by $\sqrt{N}$. Call it as y. Generate a large number of y. Then, compute their probability distribution. Compare it with the Gaussian distribution of the same mean and variance.

19.21 Obtain a sequence of random numbers with the distribution

$$f(x) = \begin{cases} \sin^2 x, & 0 \leq x \leq \pi \\ \\ 0, & \text{elsewhere} \end{cases}$$

using the expression $x = 2\sin^{-1}\sqrt{y}$ where y's are uniform random numbers in the interval $[0, 1]$. Verify that its mean is $\pi/2$ while the variance is ≈ 0.468. Draw the probability distribution curve of the generated numbers.

19.22 Generate random numbers with the Gaussian distribution

$$f(x) = \frac{1}{\sigma \sqrt{2\pi}}\, e^{-(x-\mu)^2/(2\sigma^2)} ,$$

where σ^2 and μ are the variance and mean, respectively, with fixed mean and variance. Fix the range of the Gaussian numbers as $[-5, 5]$. Compute the mean and variance and sketch the probability distribution of the obtained numbers.

19.23 Obtain a sequence of random numbers with the circular probability distribution density function

$$f(x) = \frac{4}{\pi}\sqrt{1 - x^2}, \quad 0 \le x \le 1.$$

(A convenient choice of the bounding function in the rejection technique is $g(x) = 1$, $0 \le x \le 1$ and c is $4/\pi$.) Verify that its mean is ≈ 0.43 and the variance is ≈ 0.074.

19.24 The Cauchy distribution is given by

$$f(x) = \frac{1}{\pi}\frac{a}{a^2 + x^2}, \quad -\infty < x < \infty$$

where $a > 0$ is a scale factor. Sketch the graph of $f(x)$ for $a = 1$, $-10 < x < 10$. Employing the rejection technique, generate random numbers with the Cauchy distribution. Compare the numerically obtained distribution with the exact distribution.

19.25 Consider the Metropolis algorithm for generating Gaussian random numbers with mean zero and variance 1. For each value of ϵ in the interval 0 to 4 with step size 0.01 compute the acceptance ratio. Find the optimum value of ϵ for which the acceptance ratio is 0.5. Repeat the above for several values of variance.

19.26 Generate a sequence of random numbers with the following probability distributions:

(a) $f(x) = x\,e^{-x}$.

(b) $f(x) = x^N\,e^{-x}/N!$ (Poisson distribution).

(c) $f(x) = {}^M C_N\, x^N (1-x)^{M-N}$ (Binomial distribution).

19.27 The gamma density function is given by

$$f(x) = \frac{(\sqrt{N})^N}{(N-1)!}\, x^{N-1}\, e^{-\sqrt{N}\,x}, \quad 0 \le x < \infty.$$

Its mean is $\sqrt{N}$ and variance is 1. For $N = 2, 3$ and 10 generate random numbers with the above distribution in the interval $[0, 10]$. (For larger values of N the above distribution is a bell-shaped form with a peak at $x = N$).

19.28 Write a short note on comparison of inverse and rejection methods.

19.29 Employing the rejection technique, generate a sequence containing 2000 pairs of random numbers $\{(x_i, y_i)\}$ with the following distributions and also plot the numbers in the $x - y$ plane.

(a) $f(x, y) = e^{-(x+y)}$, $x, y \in [0, 10]$.

(b) $f(x, y) = e^{-(x^2+y^2)}$, $x, y \in [-5, 5]$.

(c) $f(x, y) = \frac{1}{1+(x^2+y^2)}$, $x, y \in [-100, 100]$.

(d) $f(x, y) = \frac{4}{\pi}\sqrt{2 - (x^2 + y^2)}$, $x, y \in [-1, 1]$.

D. Skewness

19.30 Write a Python program to read N numbers x_i from an external data file and then to compute the skewness of the distribution of x_i.

19.31 Verify that for a set of Gaussian random numbers $S \approx 0$.

19.32 Generate random numbers with the distribution xe^{-ax} for several values of a. For each value of a sketch the distribution curve. Calculate S as a function of a.

E. Kurtosis

19.33 Write a Python program to calculate k of a given set of N random numbers. Read the data from an external data file.

19.34 Calculate k for Gaussian random numbers as a function of N. Choose $N = 10$, 20, 50, 100, 200, 500, 1000, 2000, 5000 and 10000. Verify that $k \to 0$ as $N \to \infty$.

19.35 Consider the two-dimensional map

$$x_{n+1} = y_n, \quad y_{n+1} = -x_n - r \sin y_n + \Gamma \quad \text{mod } 2\pi.$$

Fix $r = 6.34$ and $(x_0, y_0) = (5, 3)$. For $\Gamma = 0.3$ generate $N = 10^5$ points. Verify that the distribution of x is normal. Calculate $k(N)$ and show that it becomes 0 for large N. For $\Gamma = 0$ verify that the distribution of x deviates from normal and k diverges with N.

F. Students t-Test

19.36 Write a Python program to compute t for a given two sets of N random numbers.

19.37 Generate two sets of normal distributions each containing 10^5 numbers with the same variance but with different means. Define $d = \bar{x}_A - \bar{y}_B$. For several values of d compute t and sketch the graph of t versus d.

G. Chi-Square Test

19.38 Write a Python program to compute χ^2 of two given sets of random numbers. For verifying the program see the next two problems.

19.39 Collect two sets of 10^5 numbers from the following distributions. Divide their range into, say, 50 bins. Compute χ^2 and verify that in all the cases $\chi^2 \approx 0$.

(a) Uniform distribution.

(b) Gaussian distribution.

(c) $x e^{-x}$ distribution.

(d) e^{-x} distribution.

19.40 Calculate χ^2 for two sets of numbers with normal distribution but with different mean and variance. Verify that χ^2 is nonzero.

20

Monte Carlo Technique

20.1 Introduction

Computer simulations are methods that try to model a physical process rather than solving the equation that governs the physical process. Particularly, problems that are stochastic in nature are most suitable for computer simulation. Monte Carlo method is of such a type. The method has applications in traffic flow, stellar evolution, nuclear reactor design, radiation cancer therapy, oil-well exploration and so on. The method uses random numbers in a calculation that has the structure of stochastic process, a sequence of states whose evolution is determined by random events. Generally, any problem analyzed or studied with the use of random numbers and probability statistics is called a *Monte Carlo method* or an *experiment* or a *simulation*. The Monte Carlo simulation can be applied to study systems whose time evolution is not described by fully deterministic manner but behaves stochastically.

Nicholas Metropolis gave the name Monte Carlo to random sampling techniques. The term Monte Carlo is due to the name of the city of Monte Carlo (in Monaco which is located along the French Riviera between Mediterranean Sea and France) famous for its casino. The point is that the roulette is one of the simplest mechanical devices for generation of random numbers. The Monte Carlo method was introduced by the great mathematical physicists John von Neumann, Enrico Fermi and Stanislaw Ulam during 1949. There were some isolated instances earlier where Monte Carlo method has been used in some form or the other. The name Monte Carlo was initially applied to a class of mathematical methods used for the development of nuclear weapons in Los Alamos in the 1940s. The earliest documented use of random sampling to find solution of an integral is that of Comte de Buffon [see Problem 12 at the end of the present chapter]. In India, P.C. Mahalanobis [1,2] exploited random sampling technique to solve a variety of problems like the choice of optimum size and shape of plots in experimental works, etc. Description of several Monte Carlo techniques appeared in a paper by Kelvin [3]. He used random sampling to aid in evaluating certain integrals in the kinetic theory of gases. In 1930s, Fermi made some numerical experiments and are now called *Monte Carlo calculations*.

To understand the basic idea of the Monte Carlo method, assume that it is desired to find the area of the plane S shown in Fig. 20.1. It may be an arbitrary figure specified graphically or analytically. The area of S can be computed employing the Monte Carlo method. For simplicity choose the square $ABCD$ in Fig. 20.1 as a unit square. Generate a set of uniformly distributed N random numbers (points) inside the square $ABCD$. Count the number of points falling within the plane S and, say, it is N_S. Geometrically the area of S is approximately N_S/N. Greater the value of N higher the accuracy of the estimate. Essentially, if the geometry of the shape S is known then the ratio of the *hits* in the area S and *throws* is the area of the shape. The above procedure is also called *hit and miss* method [4].

DOI: 10.1201/9781032649931-20

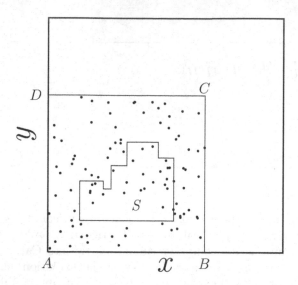

FIGURE 20.1
An arbitrary plane S within a square $ABCD$.

There are three distinct features of the Monte Carlo method:

1. The structure of computation algorithm is simple. A program is developed to perform only one random trial. This trial is repeated larger number of times with each being independent of all the others. The results of all the trials are then averaged.

2. The error is proportional to $\sqrt{D/N}$ where D is a constant and N is the number of trials.

3. The simulation with two different sequences of random numbers, drawn from same distribution with same characteristic properties such as mean and variance, will not produce identical results but will give values which are close to each other within some statistical error.

One should note that high accuracy cannot be obtained with the Monte Carlo approach. It never gives an exact answer, rather its conclusions indicate that *the answer is so and so, within such and such an error, with such and such probability*. In other words, it provides an estimate of the value of the numbers sought in a given problem. It is computationally effective, compared with deterministic methods for higher-dimensional problems. Generally, a Monte Carlo simulation application has the following primary components:

1. *A random number generator* – An appropriate uniform random number generator such as the Park and Miller (see Subsection 19.3.7) method is necessary.

2. *Probability distribution function* – The given physical process or phenomenon must be described by one or more appropriate probability distribution functions.

3. *Sampling rule* – A description for sampling from the given probability distribution function must be specified.

4. *Error estimation* – It is necessary to determine the variance of resulting data as a function of number of trials.

In this chapter, some simple applications of Monte Carlo method are described. Particularly, the following problems are considered:

1. Evaluation of definite integrals.

2. Square-root of real positive numbers.

3. Estimation of value of π.

4. Estimation of value of e.

5. Electronic distribution of hydrogen atoms.

6. Radioactive decay.

7. Diffusion of neutrons in a moderating material.

8. Percolation.

20.2 Evaluation of Definite Integrals

It is desired to compute the definite integral of the form

$$I = \int_a^b f(x)\,dx. \tag{20.1}$$

A. Monte Carlo Estimator

Consider the integral

$$I = \int_a^b f(x)p(x)\,dx, \tag{20.2}$$

where $p(x)$ is a probability distribution function and

$$\int_a^b p(x)\,dx = 1. \tag{20.3}$$

Then, the integral given by Eq. (20.2) can be thought of as the solution of the probabilistic problem of computing the average of $f(x)$, $\bar{f}$, over the function $p(x)$. To compute $\bar{f}$, draw a large number (N) of random numbers for x from the distribution function $p(x)$. Then,

$$\bar{f} = \langle f \rangle \approx \frac{1}{N}\sum_{i=1}^{N} f(x_i). \tag{20.4}$$

$\bar{f}$ is called the *Monte Carlo estimator* for the integral in Eq. (20.1). Write

$$\int_0^1 f(x)\,dx \approx \frac{1}{N}\sum_{i=1}^{N} f(x_i) = \bar{f}, \tag{20.5}$$

where x_i are uniformly distributed random numbers in the interval $[0, 1]$. Note that the $p(x)$ which appeared in the integral in Eq. (20.2) is absent in Eq. (20.5). *What is the reason for this?* x_i is distributed according to $p(x)$. Obviously, there would be more points in the region where $p(x_i)$ is large. But, if $p(x_i)$ is chosen as a uniform distribution then $p(x_i) = $ a constant.

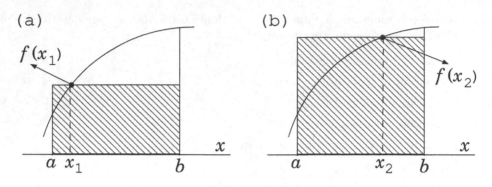

FIGURE 20.2
Approximate area under the curve $f(x)$ in the interval $[a, b]$ with (a) $x = x_1$ and (b) $x = x_2$.

B. Connection Between $\bar{f}$ and the Value of the Integral in Eq. (20.1)

Consider the Figs. 20.2a and b. $\int_a^b f(x) dx$ represents the area under the curve $f(x)$ in the interval $[a, b]$. For $x = x_1$ the value of the integral can be approximated by the area of the shaded rectangle in Fig. 20.2a. For another value of x, say, $x = x_2$ the value of the integral can be approximated by the area of the rectangle in Fig. 20.2b. The area of the rectangle in Fig. 20.2a underestimates the area below the curve (value of the integral) while the area of the rectangle in Fig. 20.2b overestimates the area below the curve (value of the integral). Averaging the area of rectangles formed by considering a large number of values of x in the interval $[a, b]$ gives an approximate value of the integral. The areas of the rectangles for the case of $x = x_1$ and $x = x_2$ are $(b - a)f(x_1)$ and $(b - a)f(x_2)$, respectively. Then,

$$I_N = \int_a^b f(x)\, dx = (b - a)\bar{f}, \tag{20.6}$$

where $\bar{f}$ is given by Eq. (20.4) with x_i are uniformly distributed random numbers in the range $[a, b]$.

If the values of $|a|$ and $|b|$ are sufficiently small then x_i can be chosen as uniformly distributed random number in the interval $[a, b]$. That is, $p(x)$ in Eq. (20.2) is a uniform distribution. In the case of $a = 0$ and $b = \infty$, practically, it is difficult to generate random numbers in the interval $[0, \infty]$. In this case choose $p(x)$ as some other suitable function. For example, in the integral $\int_0^\infty e^{-x} \cos x\, dx$ the function $p(x)$ can be e^{-x} whereas in $\int_0^\infty e^{-x^2} dx$, it can be e^{-x^2}.

C. The Monte Carlo Algorithm

The Monte Carlo algorithm of calculating the value of the integral (20.1) using (20.6) is the following:

1. Generate N uniformly distributed random numbers x_i, $i = 1, 2, \ldots, N$ in the interval $[a, b]$.

2. For each x_i calculate $f(x_i)$.

3. Calculate the average value of $f(x_i)$, that is $\bar{f}$, using Eq. (20.4).

4. $(b - a)\bar{f}$ is the value of the given integral I.

D. Convergence of the Monte Carlo Estimate

In the limit $N \to \infty$ the probability of getting $I_N = I$ is 1. That is,

$$P\left\{\lim_{N\to\infty} I_N = I\right\} = 1. \tag{20.7}$$

The above theorem guarantees that in the limit of very large number of points the Monte Carlo estimates converge to the exact answer. But it does not give the rate of convergence. However, much more information is obtained from the central-limit theorem.

Define

$$\sigma^2 = \int_a^b (f(x) - I)^2 p(x) \, dx \tag{20.8}$$

and obtain

$$\begin{aligned}
\sigma^2 &= \int_a^b \left(f^2 p - 2fIp + I^2 p\right) dx \\
&= \langle f^2 \rangle - 2I\langle f \rangle + I^2 \\
&= \langle f^2 \rangle - 2I^2 + I^2 \\
&= \langle f^2 \rangle - I^2 \\
&= \langle f^2 \rangle - \langle f \rangle^2.
\end{aligned} \tag{20.9}$$

Also, define a random variable

$$y_N = \frac{\sqrt{N}}{\sigma}(I_N - 1). \tag{20.10}$$

Further,

$$P\{a \le y_N < b\} = \frac{1}{\sqrt{2\pi}} \int_a^b e^{-y^2/2} \, dy. \tag{20.11}$$

Equation (20.11) implies that for very large N, I_N is a Gaussian distribution with mean I and variance σ^2/N. In the limit $N \to \infty$, I_N becomes very narrow near I.

E. Error in the Monte Carlo Estimation

Let us calculate the error in the Monte Carlo estimation. Write

$$I = I_N + \epsilon, \tag{20.12}$$

where ϵ is the error. From Eq. (20.10)

$$|\epsilon| = \frac{\sigma}{\sqrt{N}} \tag{20.13}$$

and the variance σ^2 is given by

$$\sigma^2 = \frac{N}{N-1}\left[\frac{1}{N}\sum_{i=1}^{N}\left((f(x_i))^2 - \bar{f}^2\right)\right]. \tag{20.14}$$

For large N, drop -1 in $N - 1$ in the above equation and then obtain

$$|\epsilon| = \frac{\sigma}{\sqrt{N}} = \frac{1}{\sqrt{N}}\left[\frac{1}{N}\sum_{i=1}^{N}\left((f(x_i))^2 - \bar{f}^2\right)\right]^{1/2}. \tag{20.15}$$

TABLE 20.1

The Monte Carlo estimation of the integral $\int_0^2 e^x dx$. The exact value of the integral is 6.38906.... σ is the standard deviation error.

N value	Integral value	% of error	σ
1000	6.16404	3.521	0.09727
2000	6.27201	1.832	0.06992
10000	6.41480	0.403	0.03187
50000	6.40922	0.316	0.01432
100000	6.38577	0.051	0.01003
200000	6.39625	0.113	0.00713
300000	6.39273	0.057	0.00583
400000	6.38591	0.049	0.00504
500000	6.38653	0.040	0.00451

The error in the Monte Carlo method decreases $1/\sqrt{N}$ as N increases.

F. Examples

   Now, present two examples of numerical computation of definite integrals.

Example 1:

Evaluate the integral $\int_0^2 e^x dx$ by employing the Monte Carlo method.

Since the integration is from 0 to 2 one has to use random numbers x_i, $x \in [0, 2]$. A Python or C++ program can be developed to calculate the value of I, the percentage of error in the value of the integral and the standard deviation for various values of N. Table 20.1 presents the computed value of the integral as a function of N. For a sufficiently large value of N the computed value of the integral is close to the exact value 6.38906.....

Example 2:

Evaluate the integral $\int_0^1 \frac{1}{1+x} dx$ by employing the Monte Carlo method using the 5 random numbers 0.501, 0.762, 0.243, 0.892, 0.123 (for the illustrative purpose only a few random numbers are chosen). The exact value of the integral is $\ln(1+x)|_0^1 = 0.69315....$

The value of the integral $I = (b - a)\bar{f} = \bar{f}$ is obtained as

$$
\begin{aligned}
\bar{f} &= \frac{1}{5}\left[f(0.501) + f(0.762) + f(0.243) + f(0.892) + f(0.123)\right] \\
&= \frac{1}{5}\left[0.66622 + 0.56754 + 0.80451 + 0.52854 + 0.89047\right] \\
&= \frac{3.45728}{5} \\
&= 0.69146.
\end{aligned}
$$

TABLE 20.2
The Monte Carlo computation of $\sqrt{10}$. Here, $N = 2^k$. The exact value of $\sqrt{10}$ is $3.16228\ldots$.

k	N	$\sqrt{10}$	% of error
14	16384	3.15924	0.09604
15	32768	3.16064	0.05165
16	65536	3.16116	0.03524
17	131072	3.16080	0.04682
18	262144	3.16074	0.04875
19	524288	3.16171	0.01799
20	1048576	3.16167	0.01917
21	2097152	3.16223	0.00140

20.3 nth Root of a Real Positive Number

How does one estimate $x^{1/n}$ using Monte Carlo method? First, find two numbers between which the root lies. Let these two numbers being N_1 and N_2 with $N_1 < N_2$. For simplicity, these two numbers may be chosen as integers. The value of N_2 is chosen such that $(N_2-1)^n \leq x \leq N_2^n$. Then, $N_1 = N_2 - 1$. The procedure is summarized below.

1. Read the value of x and N.
2. Find N_2 and then N_1.
3. Generate N (sufficiently large) uniformly distributed random numbers y in the interval $[N_1, N_2]$.
4. Count the number of random numbers y satisfying the condition $y^n < x$. Let this number being N_c.
5. Calculate $N_1 + (N_c/N)$. This number is the value of $x^{1/n}$.

Example:

Compute $\sqrt{10}$ by the Monte Carlo method.

For $\sqrt{10}$ the values of N_1 and N_2 are 3 and 4, respectively. Table 20.2 displays the computed value of $\sqrt{10}$ as a function of N. If the nth root of a number is perfectly an integer then the computed result would exactly coincide with the true value even for a small number of samplings.

20.4 Estimation of π

Another application of Monte Carlo method is the calculation of π. The early history of the calculation of π is interesting [5]. The value of π is known with high accuracy for a long time. An Egyptian text known as the *Rhind Papyrus* dating to 1650 BC contains the

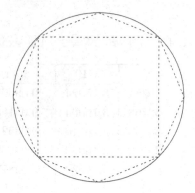

FIGURE 20.3

Archimedes idea of successive approximation of value of π. At nth stage, the value of π is given by the formula (20.19).

statement that

$$\text{area of a circle} \quad = \quad \text{area of the square with side}$$
$$\text{8/9 times the diameter of the circle}. \tag{20.16}$$

That is,

$$\pi \left(\frac{d}{2}\right)^2 = \left(\frac{8}{9}d\right)^2 \tag{20.17}$$

which gives

$$\pi = \frac{256}{81} = 3.16049. \tag{20.18}$$

A. Archimedes Method of Computation of π

Later Archimedes proposed a method which could give the value of π to any desired accuracy. His idea was to take a circle and inscribe in it a series of regular polygons of more and more sides as shown in Fig. 20.3. Remember that all sides of a regular polygon are equal in length. The perimeter of the polygon is slightly less than the circumference of a circle. However, the polygons approach the circle closer and closer as the number of sides increases. Then, write

$$2\pi r = na, \tag{20.19}$$

where r is the radius of the circle and a is the length of the side of the inscribed polygon with n sides. Calculate π from the approximate formula

$$\pi = \frac{na}{2r}. \tag{20.20}$$

Archimedes repeated the process with circumscribing polygons as shown in Fig. 20.3. Using the inscribed and circumscribing polygons of 6 sides to 96 sides Archimedes calculated the value of π in the range 3.1408 and 3.1428.

B. Monte Carlo Method of Computation of π

A Monte Carlo method of calculation of π is as follows, which is based on the relation that area of a circle of radius r is πr^2. Consider a square with size unity and its inscribed

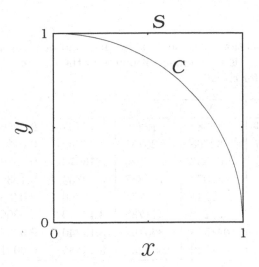

FIGURE 20.4
S-square part of a C-circular curve.

circle of radius, r, unity as shown in Fig. 20.4. The ratio of the area of the $(1/4)$th of the circle lying inside the square and the area of the square is $\pi/4$. One can also consider a square with a full circle embedded in it. In this case, the above ratio is π. Generate N pairs of uniform random numbers in the range $[0,1]$ and designate the ith pair as (x_i, y_i). x_i and y_i are the random numbers for the coordinates x and y, respectively. If these points are placed inside the square, some will fall within the circle and some will not. Count the number of points that fall inside the quarter circle and call it as N_c. Whether a point (x_i, y_i) falls inside the quarter circle can be easily identified. For a pair (x_i, y_i) calculate $d_i = \sqrt{x_i^2 + y_i^2}$. If $d_i < r(=1)$ then it lies inside the quarter circle. Then, N_c/N is the ratio of the areas of the quarter circle and the square, that is,

$$\frac{\pi r^2/4}{r^2} = \frac{N_c}{N}. \tag{20.21}$$

Thus,

$$\pi = 4\frac{N_c}{N}. \tag{20.22}$$

Using the above formula the value of π is computed for $N = 2^k$, $k = 14, 15, \ldots, 21$. The result is presented in the Table 20.3. The statistical convergence to the correct answer as the sample size increases is seen. For some other methods for the estimation of π see Problems 3, 4 and 12 at the end of the present chapter.

C. A Simple Exercise

Let us present a quick illustration of the Monte Carlo estimation of the value of π. Consider a quarter circle of radius $r = 1$ unit lying inside a square with sides 1 unit in the $x - y$ plane. Compute the value of π employing the Monte Carlo method using 10 pair of random points (x_i, y_i) generated using the Ran function in a calculator.

By the Monte Carlo method, the value of π can be computed from $\pi = 4N_c/N$ where N is total number of pair of points (x_i, y_i) and N_c is the number of pair of points lying inside the quarter circle of radius $r = 1$ unit. For a pair (x_i, y_i) if $d_i^2 = x_i^2 + y_i^2 < 1$ then it lies

TABLE 20.3
The Monte Carlo estimation of value of π by the ratio of area of a quarter circle and a square. Here, $N = 2^k$. N is the number of points in the square while N_c is the number of points within the quarter circle.

k	N	N_c	π value	% of error
14	16384	12839	3.13452	0.22509
15	32768	25731	3.14099	0.01915
16	65536	51425	3.13873	0.09103
17	131072	102837	3.13834	0.10366
18	262144	205977	3.14296	0.04351
19	524288	411735	3.14129	0.00968
20	1048576	823762	3.14240	0.02578
21	2097152	1647233	3.14185	0.00811

inside the quarter circle. Table 20.4 is obtained with the set of random numbers generated using a calculator. As $N_c = 8$ the value of $\pi \approx 4 \times 8/10 = 3.2$. Note that different realization of set of ten pairs of random numbers will give different values π and in some cases the result can be far from the actual value.

D. Standard Error

For a fixed value of k, if the computation of π (as well as e in the next section) is repeated M times with N different sequences of uniformly distributed random numbers one would get slightly different M values for π. The spread of the final result is specified by the *standard error* of the final result. It is given by

$$S_e^2 = \frac{1}{M} \sum_{i=1}^{M} \left(R_i - \bar{R} \right)^2 , \quad \bar{R} = \frac{1}{N} \sum_{i=1}^{N} R_i , \tag{20.23}$$

where R_i is the value of the quantity calculated in the ith repetition of the numerical scheme. The standard error is similar to the standard deviation. *What is the difference between these two?* The standard deviation describes the spread of the numerically computed values of the quantity (for example, value of π or e or mean of a uniform distribution) with the true value. The standard error, on the other hand, represents the spread of numerically computed values with the mean value.

TABLE 20.4
The Monte Carlo estimation of value of π using only 10 pairs of uniformly distributed random numbers.

x_i	y_i	d_i^2	N_c	x_i	y_i	d_i^2	N_c
0.218	0.706	0.545	1	0.094	0.050	0.011	5
0.958	0.945	1.810	1	0.881	0.119	0.790	6
0.664	0.725	0.966	2	0.885	0.116	0.796	7
0.422	0.622	0.564	3	0.782	0.700	1.101	7
0.330	0.417	0.282	4	0.041	0.439	0.194	8

20.5   Estimation of Value of e

The previous section presented an estimation of π by a Monte Carlo method. Another naturally occurring irrational number in mathematics is the base of the natural logarithm e. Mohazzabi [6] described three Monte Carlo algorithms for the estimation of e. Here, two of them are described.

20.5.1   Dart Method

Let us consider a dart board, divide it into R equal size regions and N darts are thrown onto it. The probability for a dart to strike a region is $p = 1/R$. *What is the probability $P(n)$ of finding n darts in a given region out of N throws?* It is given by the binomial distribution

$$P(n) = C_n^N p^n q^{N-n}, \tag{20.24}$$

where $q = 1 - p$ and C_n^N are the binomial coefficients $N!/(n!(N - n)!)$. Therefore, the probability of finding an empty region is

$$P(0) = q^N = (1 - p)^N. \tag{20.25}$$

When R is set to N then

$$P(0) = (1 - 1/N)^N. \tag{20.26}$$

For large N, $1/N \ll 1$ and the series expansion of $(1 - 1/N)^N$ is approximated to the series expansion of e^{-1} (verify). Further, for $p \ll 1$, $N \gg 1$ and $R = N$ the binomial distribution is approximated by the Poisson distribution:

$$P(n) - e^{-1}/n!. \tag{20.27}$$

Therefore, $P(0) = P(1) = e^{-1}$.

From the above, the following algorithm to estimate e is obtained.

1.   Throw a large number of darts, N, at a board which has been divided into N equal size regions.

2.   Count the empty cells. Denote this number as N_0.

3.   The ratio N_0/N is e^{-1}, that is, e $= N/N_0$.

What is the expression for e if the number of cells (N_1) with occupancy 1 is counted?

In a computer algorithm throw of a dart N times can be replaced by N uniformly distributed random numbers in the interval $[1, N]$. In this case, the algorithm is the following:

1.   Divide the range $[1, N]$ into N equal sizes.

2.   Treat the N uniformly distributed random numbers in the interval $[1, N]$ as N points on a line.

3.   Note down the number of points falling in each interval such as $1 - 2$, $2 - 3$, $\ldots$, $(N - 1) - N$.

4.   Count the number of empty intervals N_0.

5.   The ratio N/N_0 is e.

It is easy to develop a program code to compute e as a function of N where $N = 2^k$, $k = 14, 15, \ldots, 21$. Table 20.5 displays the obtained result.

TABLE 20.5
The Monte Carlo estimation of the value of e by a dart method. Here, $N = 2^k$. The exact value of e is $2.71828\ldots$.

k	N	N_0	e value	% of error
14	16384	6029	2.71853	0.02752
15	32768	12070	2.71483	0.12691
16	65536	24061	2.72374	0.20100
17	131072	48484	2.70341	0.54714
18	262144	96481	2.71705	0.04514
19	524288	193122	2.71480	0.12795
20	1048576	386234	2.71487	0.12536
21	2097152	771695	2.71759	0.02532

20.5.2   Derangement Method

Another simple method for estimation of e is a derangement method. Consider a permutation of N objects labelled as say $\{1, 2, 3, \ldots, N\}$. The objects are shuffled once. Now, check whether all the objects are moved from its original place. Such a permutation is called a *derangement* of the objects. The number of derangement of N objects is given by

$$N_{\mathrm{d}} = N! \left[\frac{1}{2!} - \frac{1}{3!} + \frac{1}{4!} - \cdots + (-1)^N \frac{1}{N!} \right]. \tag{20.28}$$

Since $e^x = 1 + x + x^2/2! + x^3/3! + \cdots$ the square bracket in the right-side of Eq. (20.28) is approximated as e^{-1} so that $N_{\mathrm{d}} = N!/e$. The number of permutations is $N!$. For a reasonably large N the probability of finding the derangement is e^{-1}. The following is a systematic algorithm of this method.

(1)   Consider an array of say $N = 10$ numbers. For simplicity and convenience choose them as $IX(i) = i$, $i = 1, 2, \ldots, N$. Set $N_{\mathrm{d}} = 0$.

(2)   Generate N uniformly distributed random numbers in the interval $[1, 10]$. Make them as integer numbers again in the interval $[1, 10]$. Using these numbers the values of $IX(i)$ are exchanged as per the Step (3).

(3)   First, change the content in $IX(1)$. If the first random number is, for example, 6 then exchange the contents in $IX(1)$ and $IX(6)$. If the second random number is supposed, say, 5 then exchange $IX(2)$ and $IX(5)$. Repeat this for $IX(3), \ldots, IX(N)$. Now, N number of IX's are shuffled.

(4)   Next, check for derangement. That is, check for $IX(i) \neq i$, $i = 1, 2, \ldots, N$. If this is realized then increment the value of N_{d} by one. If $IX(i) = i$ for at least one of the N values of i then no increment of N_{d}.

(5)   Repeat the above procedure for a large number of times, say, $N_N = 10^5$ times and record the total derangement N_{d}.

Then, $e = N_N/N_{\mathrm{d}}$. Let this is a value of e in one trial. An average value of e can be obtained over many trials. Table 20.6 displays the value of e in 10 trials. The average value of e is 2.71006 whereas the true value is $2.71828\ldots$. The average value of e obtained in 20, 50, 100

TABLE 20.6

The Monte Carlo estimation of e by the derangement method. N_d is the number of derangements observed in $N_N = 10^5$ shuffling. The average value of e is 2.71006.

Trial	N_d	e value	Trial	N_d	e value
1	36991	2.70336	6	36961	2.70555
2	36943	2.70687	7	36919	2.70863
3	36894	2.71047	8	36872	2.71209
4	36568	2.73463	9	36937	2.70731
5	36901	2.70995	10	37014	2.70168

and 500 trails are 2.7125, 2.7150, 2.7169 and 2.7180, respectively. Mohazzabi [6] computed e with 10^5 shuffling and 10^3 trials. The obtained e value is 2.7181 ± 0.0002.

20.6 Electronic Distribution of Hydrogen Atom

In the previous sections, the features of Monte Carlo technique are illustrated in the calculation of values of nth root of positive numbers, π, e and evaluation of definite integrals. This technique can also be applied to real physical problems. For example, it helps us visualize electronic distribution of harmonic oscillator, scattering of particles from a potential, phase transition and radiation transport studies and computer simulation of certain physical and engineering problems. In this section, the application of Monte Carlo technique to the representation of orbitals of hydrogen atom [7] is discussed. A few other interesting applications of Monte Carlo technique are presented in the subsequent sections.

20.6.1 Hydrogen Atom and Its Eigenfunctions

The hydrogen atom is a two-particle system, consisting of the atomic nucleus of charge Ze and an electron of charge $-e$. Let us use spherical polar coordinates system. If the nucleus is assumed to remain in static then the potential energy of the system is $-Ze^2/r$ where r is the distance between the electron and the nucleus. This potential is spherically symmetric because V does not depend upon θ and ϕ and is only a function of r.

The Schrödinger equation of the system is

$$\left[-\frac{\hbar^2}{2\mu} \nabla^2 + V \right] \psi = E\psi, \qquad (20.29)$$

where $\mu = m_e m_n/(m_e + m_n)$, m_e-mass of the electron and m_n-mass of the nucleus. For the construction of solution of Eq. (20.29), the meaning of eigenfunction, eigenstate and quantum numbers refer refs. [8-10]. Its eigenfunctions are denoted as ψ_{nlm} where n, l and m are quantum numbers and are given by

$$n = 1, 2, \ldots, \quad l = 0, 1, \ldots, n-1, \quad m = -l, -l+1, \ldots, 0, 1, 2, \ldots, l-1, l. \qquad (20.30)$$

The position probability density for the electron is given by $P_{nlm} = |\psi_{nlm}|^2$. The first few quantum energy eigenfunctions are given below where the Bohr radius a_0 and Z are set to unity:

$$nlm = 100: \quad \psi_{100} = \frac{1}{\sqrt{\pi}}\, e^{-r}\,.$$

$$nlm = 200: \quad \psi_{200} = \frac{1}{\sqrt{32\pi}}\, (2 - r)\, e^{-r/2}\,.$$

$$nlm = 210: \quad \psi_{210} = \frac{1}{\sqrt{32\pi}}\, z\, e^{-r/2}\,.$$

$$nlm = 211: \quad \psi_{211} = -\frac{1}{8\sqrt{\pi}}\, (x + iy)\, e^{-r/2}\,.$$

$$nlm = 300: \quad \psi_{300} = \frac{1}{9\sqrt{3\pi}}\, \left(27 - 18r + 2r^2\right)\, e^{-r/3}\,.$$

$$nlm = 310: \quad \psi_{310} = \frac{1}{162\sqrt{2\pi}}\, (6 - r)\, z\, e^{-r/3}\,.$$

$$nlm = 311: \quad \psi_{311} = -\frac{1}{81\sqrt{\pi}}\, (6 - r)(x + iy)\, e^{-r/3}\,.$$

In the above eigenfunctions x, y, z are Cartesian coordinates. They can be replaced in terms of r, θ, ϕ using spherical polar coordinates transformation. Here, $r = \sqrt{x^2 + y^2 + z^2}$.

20.6.2   Electronic Distribution in Various States

Consider the ground state eigenfunction ψ_{100}. The probability distribution of electron in this state is

$$P_{100} = |\psi_{100}|^2 = \frac{1}{\pi}\, e^{-2r}\,. \tag{20.31}$$

For the present discussion, P is chosen as e^{-2r} and the factor $1/\pi$ is dropped. The multiplicative factors in other P_{nlm}'s are also dropped. Now, generate points (x, y, z) obeying the distribution P_{100}. The plot of the generated points in the $x - y - z$ plane represents the electronic distribution in the ground state $nlm = 100$. It is convenient to choose $x - z$ plane with $y = 0$. A sequence of pair of points x, z with the distribution $P_{100} = e^{-2r}$ can be generated by the rejection technique described in Subsection 19.6.2. This is summarized below:

(1) Set $y = 0$ and let x and z are a set of uniformly distributed random numbers in the interval $[a, b]$.

(2) Define a bounding function $cg(x, z)$ such that the maximum of $cg(x, z)$ is greater than or equal to the maximum of $P_{100}(r)$. For $P_{100} = e^{-2r}$, the maximum of $cg(x, z)$ denoted as cgm can be chosen as 1.

(3) Treat $cg(x, z)$ as a sequence of random numbers in the interval $[0, cgm]$.

(4) Select a number for each of the variables x, z and $cg(x, z)$.

(5) Compute P_{100}. If $cg(x, z) < P_{100}$ then accept the pair (x, z). Otherwise reject it.

(6) Repeat Steps (4) and (5) and obtain large number, say, $N = 2000$ pair of points (x, z). This sequence obeys the distribution P_{100}.

The above procedure can also be used to obtain other distributions P_{nlm}. For each distribution function the values of a, b, cgm have to be chosen properly. For the ground state the values of a, b and cgm are set to -10, 10 and 1, respectively. Figure 20.5 displays the electronic distribution of hydrogen atom in its several states. The distributions corresponding

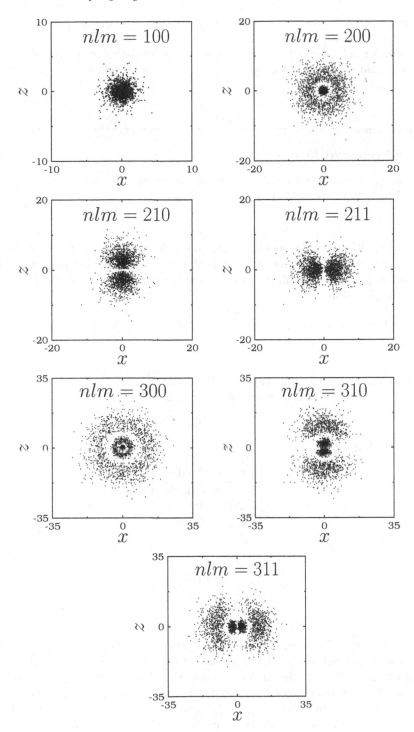

FIGURE 20.5

The electronic distribution of the hydrogen atom in some of its states produced by 2000 points in $x - z$ plane with $y = 0$.

to $nlm = 100, 200, 300$ exhibits spherical symmetry around their centre. The distributions for $nlm = 210, 211, 310, 311$ show rotational symmetry around the z−axis. The influence of applied uniform electric field along z−axis on the probability distribution called *Stark effect* [8-10] can also be visualized by the Monte Carlo method.

20.7   Radioactive Decay

Another physical process that can be simulated by the Monte Carlo method is the radioactive decay. This process is a purely random process.

A. Basic Idea

Let us start with a sample containing N_0 unstable nuclei at $t = 0$ and assume that the nuclei decay with a rate λ per second. The probability with which a nucleus decay in time dt is p. That is, $p = \lambda dt$ and $\lambda dt \ll 1$. In nuclear physics, the rate of decay is governed by the differential equation

$$\frac{dN}{dt} = -\lambda N.\tag{20.32}$$

Its solution is

$$N(t) = N_0 e^{-\lambda t},\tag{20.33}$$

where $N(t)$ is the number of undecayed nuclei at time t.

At time $t + dt$ the number of undecayed nuclei is

$$N(t + dt) = N_0 e^{-\lambda(t+dt)}.\tag{20.34}$$

Then, the probability for a nucleus to decay in time dt, P, is

$$
\begin{aligned}
P &= \frac{\text{Number of nuclei decayed in time } dt}{\text{Number of nuclei at time } t}\\
&= \frac{N(t) - N(t + dt)}{N(t)}\\
&= 1 - \frac{N(t + dt)}{N(t)}\\
&= 1 - \frac{N_0 e^{-\lambda(t+dt)}}{N_0 e^{-\lambda t}}\\
&= 1 - e^{-\lambda dt}\tag{20.35}
\end{aligned}
$$

$P = \lambda dt$ for $\lambda dt \ll 1$. That is, the probability for a nucleus to decay in time dt is $P = \lambda dt$. The time varies uniformly. $P = \lambda dt$ varies uniformly with the time dt. As the total probability of an event is 1, the probability for a nucleus to decay in time dt is $\in [0, 1]$. Therefore, to find whether a nucleus is decayed in time dt one can generate a uniform random number in the interval $[0, 1]$. If the number is $< P = \lambda dt$ then assume that a nucleus has decayed in time dt. If there are N nuclei then one can generate N uniform random numbers in the interval $[0, 1]$. If N' numbers are $< \lambda dt$ then N' nuclei are decayed. *Why are uniform random numbers used in this simulation process?*

B. Simulation Process

In the Monte Carlo approach, time t is divided into number of subintervals with an increment, say, dt. The steps involved in the Monte Carlo simulation are given below.

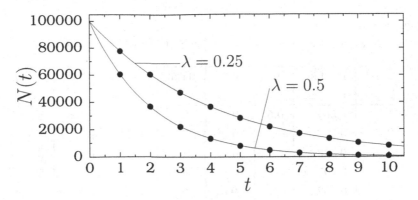

FIGURE 20.6

The number of undecayed nuclei $N(t)$ versus t (in sec) for $N(0) = N_0 = 10^5$, $\lambda = 0.25 \, \text{sec}^{-1}$ and $\lambda = 0.5 \, \text{sec}^{-1}$. The solid circles are the Monte Carlo simulation results while the continuous curves are given Eq. (20.33).

(1) Set, for example, $t = 0 \, \text{sec}$, $N(t = 0) = N_0 = 10^5$, $\lambda = 0.25 \, \text{sec}^{-1}$, $dt = 0.01 \, \text{sec}$ and $t_{\text{max}} = 10 \, \text{sec}$.

(2) Increment the time from t to $t + dt$.

(3) Set $N(t) = N(t - dt)$.

(4) Generate a uniform random number r in the interval $[0, 1]$. If $r < \lambda dt$ then decrement $N(t)$ by 1, that is, $N(t) \rightarrow N(t) - 1$.

(5) Repeat the Step (4) for $N(t - dt)$ random numbers. The final $N(t)$ is the number of undecayed nuclei at time $t \, \text{sec}$.

(6) Repeating the Steps (2)–(5) one can compute $N(t)$ for successive discrete values of t.

Figure 20.6 shows $N(t)$ as a function of t (in sec) for $\lambda = 0.25 \, \text{sec}^{-1}$ and $\lambda = 0.5 \, \text{sec}^{-1}$. The time step size used is $dt = 0.01$. For each value of λ continuous curve represents theoretically computed $N(t)$ (Eq. (20.33)) while solid circles are the Monte Carlo simulation predictions.

C. A Simple Exercise

Applying the Monte Carlo method compute the number of undecayed nuclei at time $t = 1$ and 2 units given that the number of undecayed nuclei at time $t = 0$ units is 10 and the decay constant $\lambda = 0.5$ units. Use the uniform random numbers generated in a calculator.

At $t = 0$ units, $N(0) = 10$, $\lambda = 0.5$ units. To find $N(0 + dt)$ generate 10 ($N(0)$) uniform random numbers in the interval $[0, 1]$. If a number is $< \lambda dt = 0.5 \times 1 = 0.5$ then a nucleus is decayed so decrease the number of undecayed nuclei by 1 from 10. Repeat this process for the other random numbers also. The final number is the number of undecayed nuclei at time $= 1$ unit. Table 20.7 is obtained at time $t = 1$ unit. From this table $N(1)$ by the Monte Carlo simulation is 5 while by the theory is 6. Proceed to compute $N(2)$ at $t = 2$ units with $N(1) = 5$ and by considering 5 random numbers. Table 20.8 is obtained at time $t = 2$ units. $N(2)$ by the Monte Carlo simulation is 4 while by the theory is 3.

TABLE 20.7
Number of undecayed nuclei at $t = 1$ units with $N(0) = 10$ and $\lambda = 0.5$ units.

Trial	random number	λdt	$N(1)$	Trial	random number	λdt	$N(1)$
1	0.561	0.5	10	6	0.441	0.5	8
2	0.796		10	7	0.355		7
3	0.573		10	8	0.262		6
4	0.497		9	9	0.291		5
5	0.668		9	10	0.631		5

TABLE 20.8
Number of undecayed nuclei at $t = 2$ units with $N(0) = 10$, $N(1) = 5$ and $\lambda = 0.5$ units.

Trial	random number	λdt	$N(2)$	Trial	random number	λdt	$N(2)$
1	0.593	0.5	5	4	0.376	0.5	4
2	0.845		5	5	0.699		4
3	0.592		5				

20.8 Diffusion of Neutrons by a Moderating Material Slab

This section presents the study of diffusion of neutrons through a slab of a moderating material by the Monte Carlo method [11]. Assume that slowing down of neutrons is due to elastic collisions only and the nucleus is at rest in the laboratory frame. In the centre-of-mass (CoM) system, a neutron with an energy E_0, after scattering will have an energy E and deflected by an angle θ. The ratio E/E_0 is given by [11]

$$\frac{E}{E_0} = \frac{\left(1 + 2A\cos\theta + A^2\right)}{\left(1 + A\right)^2}, \tag{20.36}$$

where A is the mass number of the nucleus. The scattering angle θ in the CoM system and the θ_{L} in the laboratory system are connected through the relation

$$\cos\theta_{\mathrm{L}} = \frac{1 + A\cos\theta}{\left(1 + 2A\cos\theta + A^2\right)^{1/2}}. \tag{20.37}$$

The total mean free path is

$$\frac{1}{\Lambda} = \frac{1}{\lambda_{\mathrm{c}}} + \frac{1}{\lambda_{\mathrm{a}}}, \tag{20.38}$$

where λ_{c} and λ_{a} are average elastic scattering free path and absorption free path, respectively.

In the Monte Carlo simulation consider that a flux of neutrons with energy, say, E_0 is allowed to incident normally onto a homogeneous infinite plate of width d. After a number of scattering events the neutron may cross the plate or captured inside the plate or reflected. Denote P_T, P_A and P_R as the transmission, absorption and reflection probabilities, respectively. They are defined as

$$P_T = \frac{N_T}{N}, \quad P_A = \frac{N_A}{N}, \quad P_R = \frac{N_R}{N}, \tag{20.39}$$

where N_T, N_A, N_R and N are the number of neutrons crossing the plate, number of absorbed neutrons, number of reflected neutrons and total number of neutrons, respectively. N_T, N_A and N_R can be computed using the Monte Carlo method.

The free path l, the diffusion angle ϕ and the scattering angle θ for each neutron can be determined stochastically. If γ is a uniform random variable between 0 and 1 then

$$l = -\Lambda \ln \gamma, \tag{20.40a}$$
$$\phi = 2\pi\gamma, \tag{20.40b}$$
$$\cos\theta = 2\gamma - 1. \tag{20.40c}$$

The neutron motion can be simulated. Denote j as the trajectory number and k as the collision number. For each neutron fix $\theta_L = 0$ and assign a value for the initial coordinate x_0. Determine stochastically l_k between two collisions and θ_k according to Eqs. (20.40). θ_k is then expressed in terms of θ_L using Eq. (20.37). The new coordinate x_{k+1} is given by

$$x_{k+1} = x_k + l_k \cos\theta_L. \tag{20.41}$$

After every collision calculate the position of the neutron and check the following possibilities:

1. $x_{k+1} > d$: Terminate the calculation of the trajectory for the neutron and update N_T as $N_T = N_T + 1$.

2. $x_{k+1} < 0$: Terminate the calculation of the trajectory and update N_R as $N_R = N_R + 1$.

3. $0 \le x_{k+1} \le d$: In this case in order to find whether the absorption of neutron will take place or further elastic scattering will occur, generate another random number, say, γ. If $\gamma < \Lambda/\lambda_a$ then absorption of the neutron is happened and so update N_A as $N_A = N_A + 1$. Otherwise, the neutron is assumed to be scattered.

Then, use Eq. (20.40c) to compute another θ and then θ_L from Eq. (20.37) and l from Eq. (20.40a). Repeat the above process number of times.

Now, present the simulation result for water. For water in the ref. [12] the values of λ_c and λ_a are available as 1.1 cm and 170 cm, respectively. Further, $A = 18.015$ g/mol, $N = 10^5$ and $E_0 = 1$ MeV. Figure 20.7 presents the numerically computed P_A, P_R and P_T as a function of thickness d (in cm). P_T decreases nonlinearly with d. $P_T \approx 0$ for sufficiently large d. P_R exhibits sigmoid type variation.

20.9   Percolation

Percolation is a phenomenon where a system exhibits a sudden change in its behaviour when a parameter of the system attains a threshold value. This provides a model for phase transitions and first-passage problems occurring in different areas of basic sciences and engineering. The name originates from the mathematical similarity of such problems with passage of coffee through a percolator.

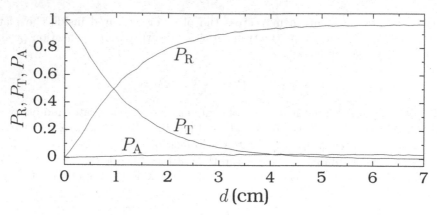

FIGURE 20.7

P_A, P_R and P_T of neutrons computed through the Monte Carlo simulation as a function of d for water.

20.9.1   Computer Simulation

Consider a two-dimensional square lattice. Each lattice site is termed as *participating* or *nonparticipating* with a probability p or $1 - p$, respectively. Algorithmically, one may set up a two-dimensional array in a, say, C_{++} or Python program. Initially, all elements are set to zero. Fix a value for p. The program now visits all the sites going successively through all rows (or columns) and assigns a uniformly distributed random number $R \in [0, 1]$. If $R \leq p$ then the site is set to 1. After having visited all sites, one realization or configuration is generated. Since, only lattice sites are involved, this is called a *site percolation model*. If two nearest (row- as well as column-wise) sites have the value 1, that is, they are participating, connect them by a line (bond). This is done for the entire lattice. This is called a *bond percolation* where adjacent sites are connected by bonds. A collection of participating sites connected by nearest neighbour distance is defined as a *cluster*. That is, in general, all sites r_j which are connected to r_i form a cluster containing r_i.

The meaning and significance of the above can be understood by considering the following examples.

1. Consider a lattice made up of a nonconducting material. Certain sites are replaced by a conducting material like graphite. Then, the nearest conducting sites are connected or made into contact.

2. Another simple configuration can be viewed by replacing participating sites by magnetic atoms and nonparticipating sites by nonmagnetic atoms.

3. Consider a fluid trying to pass through a porous medium, for example, a rock. The medium may have a number of small but random channels. It is desired to know whether the fluid can flow through the medium. Fluid flow obviously depends on the concentration and nature of the channels. When there are more channels the fluid can pass through it. The particular concentration or situation at which fluid flow becomes possible is called *percolation threshold*.

Figure 20.8a shows a configuration for $p = 0.4$. '+' symbol represents participating sites and dots represent nonparticipating sites. The nearest neighbour participating sites are connected by lines denoting clusters. In Fig. 20.8a, one can clearly observe the following:

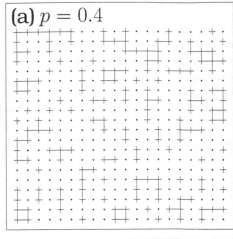

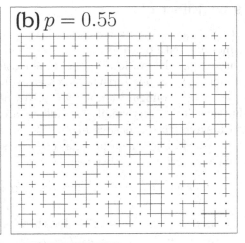

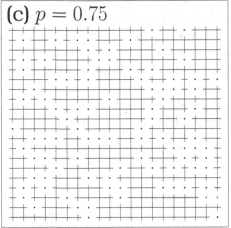

FIGURE 20.8

Percolation configuration for three values of p. The + symbol and the dots represent the participating and the nonparticipating sites, respectively. The nearest neighbour participating sites are connected by lines.

1. Some sites are participating.
2. Some sites are nonparticipating.
3. Some participating sites form a cluster.
4. Certain participating sites are isolated from the nearest neighbours.

Let us ask: *Will a person starting from a site on the first column reach the last column continuously visiting participating sites without crossing the nonparticipating sites?* In Fig. 20.8a such a reach is not possible. This is because in this figure, even though there are many clusters but all are scattered and disconnected. There is no path from first column to last column. In a physical experiment the sites can be replaced by conducting materials and in this case, one may ask whether conduction from one end to another occurs so that a deflection in the ohmmeter is observed. Deflection does not occur for the configuration shown in Fig. 20.8a.

20.9.2   Effect of p

What does happen if one increase p? As p increases, larger clusters will appear with small ones merging with one another. Figure 20.8b depicts a configuration obtained for $p = 0.55$. An interesting observation is that there exists a percolation, that is first and last columns are connected. As p is further increased one can find more number of percolation. Figure 20.8c depicts a configuration for $p = 0.75$ where one can clearly see the effect of p.

Further, if p is very small, the clusters are small. The probability $c(r_i, r_j)$ that two sites are in a same cluster, decays like $e^{-\alpha|r_i-r_j|}$ as $|r_i - r_j| \to \infty$. Denote p_c as the threshold value at which first time a percolation occurs in all the realizations. From a detailed study the following results are obtained:

1. The average size of the clusters increases with p. There are many small ones and a few large ones.

2. As $p \to p_c$ from a small value the average cluster size diverges. The mean cluster size scales $(p_c - p)^{-\nu}$ as $p \to p_c$.

3. There is a broad distribution of cluster size at $p = p_c$.

4. For $p > p_c$ there is an infinite cluster. The change of the lattice site from finite clusters to an infinite cluster is the geometric analogue of a phase transition such as when water changes from liquid to gas at a specific temperature and pressure.

5. Consider the root-mean-square distance ξ between pairs of sites belonging to the same finite clusters. ξ is also called *mean connectedness length*. ξ is maximum at p_c and decays exponentially above and below it.

For more details about percolation the reader may read the refs. [13-19] and also see the Problems 13–15 at the end of the present chapter.

20.10   Concluding Remarks

This last chapter introduced the Monte Carlo technique. The basic idea of this technique is illustrated by considering a few simple mathematical problems. Then, the method is applied to certain physical problems. The Monte Carlo methods play an important role in computational science and engineering [20]. An interesting development during the past few years is the use of Monte Carlo techniques to evaluate path integrals associated with field theories, variational and diffusion Monte Carlo.

Monte Carlo methods have been employed to analyze security pricing [21], materials studio applications for molecular simulation [22,23], radiological sciences [24], project management [25], medical physics [26], atmospheric optics [27], social sciences [28], finance engineering [29], semiconductor device simulation [30] and simulations of x-ray image devices [31]. For recent developments of the Monte Carlo methods refer to the refs. [32-34].

20.11   Bibliography

[1]   P.C. Mahalanobis, *Dialectia* 8:95, 1954.

[2]   C. Radhakrishna Rao, *Statistics and Truth: Putting Chance To Work*. World Scientific, Singapore, 1997.

[3]   L. Kelvin, *Phil. Mag.* 2:1, 1901.

[4]   T.A. Joy Woller, *The Basics of Monte Carlo Simulations*. Springer, Berlin, 1996.

[5]   Eli Maor, e: *The Story of a Number*. University Press, Hyderabad, 1994.

[6]   P. Mohazzabi, *Am. J. Phys.* 66:137, 1998.

[7]   V.M. de Aquino, V.C. Aguilera-Navano, M. Goto and H. Itwamoto, *Am. J. Phys.* 69:788, 2001.

[8]   J.L. Powell and B. Crasemann, *Quantum Mechanics*. Oxford and IBH Publi. Co., New Delhi, 1961.

[9]   P.M. Mathews and K. Venkatesan, *Quantum Mechanics*. Tata McGraw–Hill, New Delhi, 1976.

[10] S. Rajasekar and R. Velusamy, *Quantum Mechanics I: The Fundamentals*. CRC Press, Boca Raton, 2022.

[11] M.C. Capizzo, S. Nuzzo, R.M. Sperandeo Mineo and M. Zarcone, *Eur. J. Phys.* 26:85, 2005.

[12] E. Segre, *Nuclei and Particles*. Benjamin, New York, 1977.

[13] C. Vijayan and A. Arulgnanam, *Phys. Edu.* October–December, 1991. pp.229.

[14] D. Stauffer, *Introduction to Percolation Theory*. Taylor and Francis, London, 1985.

[15] G. Grimmet, *Percolation*. Springer, Berlin, 1999.

[16] G.A. Schwartz and S.J. Luduena, *Am. J. Phys.* 72:364, 2004.

[17] W. Jian Sheng, C. Kan and Z. Fei, *Computational Techniques in Theoretical Physics*. Lecture Notes, 1998.

[18] M. Barma and R. Ramaswamy, *J. Phys. A. Math. Gen.* 19:L605, 1986.

[19] M.A. Dubson and J.C. Garland, *Phys. Rev. B* 32:7621, 1985.

[20] J.G. Amar, *Comput. Sci. & Eng.* March-April 2006. pp.9-19.

[21] P. Boyle, B. Broadie and P. Glasserman, *J. Econ. Dyn. Control* 21:1267, 1997.

[22] R.L.C. Akkermans, N.A. Spenley and S.H. Robertson, *Molec. Simul.* 39:1153, 2013.

[23] A. Rahbari, R. Hens, M. Ramdin, O.A. Moultos, D. Dubbeldam and T.J.H. Vlugt, *Mol. Simul.* 47:804, 2021.

[24] R.L. Morin, *Monte Carlo Simulation in the Radiological Sciences*. CRC Press, Boca Raton, 1988. Reissued in 2019.

[25] Y.H. Kwak and L. Ingall, *Risk Manag.* 9:44, 2007.

[26] P. Andreo, *Phys. Med. Biol.* 36:861, 1991.

[27] G.I. Marchuk, G.A. Mikhailov, M.A. Nazaraliev, R.A. Darbinjan, B.A. Kargin and B.S. Elepov, *The Monte Carlo Methods in Atmospheric Optics*. Springer, Berlin, 1980.

[28] T.M. Carsey and J.J. Harden, *Monte Carlo Simulation and Resampling Methods for Social Sciences*. Sage, Los Angeles, 2014.

[29] P. Glasserman, *Monte Carlo Methods in Financial Engineering*. Springer, Berlin, 2003.

[30]   C. Jacobon and P. Lugli, *The Monte Carlo Methods for Semiconductor Device Simulation*. Springer, New York, 1989.

[31]   H. Fuchs, L. Zimmermann, N. Reisz, M. Zeilinger, A. Ableitinger, D. Greorg and P. Kuess, *Z. Med. Phys.* S0939-3889(22):00061-7, 2022.

[32]   Special issue on *Monte Carlo Simulation of Soft Matter Systems*, *Front. Phys.* 9, 2021.

[33]   A.A. Jaoude, *The Monte Carlo Methods: Recent Advances, New Perspectives and Applications*. IntechOpen, London, 2022.

[34]   A. Senova, A. Tobisova and R. Rozenberg, *Sustainability* 15:1006, 2023.

20.12   Problems

20.1   Applying a Monte Carlo method compute the area under the following curves:

a) $f(x) = e^{-x^2}$, $-5 \le x \le 5$.

b) $f(x) = e^{-x}$, $0 \le x \le 1$.

c) $f(x) = e^x$, $0 \le x \le 1$.

d) $f(x) = \sin^2 x$, $0 \le x \le 2\pi$.

e) $f(x) = xe^{-x}$, $0 \le x \le 10$.

20.2   The logarithm of a number $p > 0$ is not known. Employing a Monte Carlo method $\ln p$ can be determined. For example, consider the integral

$$\int_0^1 \left(\frac{1}{1-x} - \frac{p\,x^{p-1}}{1-x^p} \right) \mathrm{d}x .$$

Its value is $\ln p$. Evaluate this integral by the Monte Carlo method and make an estimate of $\ln p$ for $p = 2, 3, \ldots, 20$.

20.3   Compute the value of π by evaluating the integral in the identity

$$4 \int_0^1 \frac{1}{1+x^2} \, \mathrm{d}x = \pi$$

by the Monte Carlo method.

20.4   Evaluate the integral in the relation

$$\int_0^{\ln 2} \frac{x}{1-e^{-x}} \, \mathrm{d}x = \frac{\pi^2}{12}$$

by the Monte Carlo method and compute the value of π.

20.5   The error function is defined as

$$\mathrm{erf}(x) = \frac{2}{\sqrt{\pi}} \int_0^x e^{-u^2} \, \mathrm{d}u.$$

Compute erf(1) and erf(2) by evaluating the integral by the Monte Carlo method.

20.6 Apply the Monte Carlo method to calculate the value of the integral

$$\int_0^\infty x^2 e^{-x^2} \, dx$$

(exact value is $\sqrt{\pi}/4$).

20.7 Prepare a comparative study of the dart and the derangement methods. Out of these two methods which is efficient? Why?

20.8 Write a Python program to compute all n roots of a negative number.

20.9 Write a Python program to compute all n roots of a positive number.

20.10 Obtain the image representation of an electronic distribution in $x - z$ plane with $y = 0$ of hydrogen atom in the states

(a) $(\psi_{200} + \psi_{210})/\sqrt{2}$ and (b) $(\psi_{200} - \psi_{210})/\sqrt{2}$.

20.11 A three-dimensional harmonic oscillator potential is given by

$$V(r) = \frac{1}{2}r^2 = \frac{1}{2}\left(x^2 + y^2 + z^2\right).$$

In spherical polar coordinates its eigenfunctions are given by

$$\psi_{nlm} = N e^{-r^2/2} r^l L_k^{l+1/2}\left(r^2\right) Y_l^m(\theta, \phi).$$

The ψ_{nlm} for a few values of n, l, m are

$$\psi_{000} \propto e^{-r^2/2}, \quad \psi_{110} \propto z e^{-r^2/2}, \quad \psi_{200} \propto \left(\frac{3}{2} - r^2\right) e^{-r^2/2},$$

$$\psi_{220} \propto \left(3z^2 - r^2\right) e^{-r^2/2}, \quad \psi_{310} \propto z \left(\frac{5}{2} - r^2\right) e^{-r^2/2},$$

$$\psi_{400} \propto \left(\frac{15}{8} - \frac{5}{2}r^2 + \frac{1}{2}r^4\right) e^{-r^2/2}.$$

Obtain the probability distribution of a particle in the $x - z$ plane with $y = 0$ for the states represented by the above functions. Write a short note on your observation.

20.12 One of the oldest documented applications of Monte Carlo technique suggested by Comte de Buffon in 1777 [George Louis Leclere Comte de Buffon, *Essai d'arithmatique morale*, in Supplement á l'Histoire Naturella (de L'Imprimerie Royale, Paris 1777) Vol.4] was Buffon's needle. In this problem, a needle of length l is thrown randomly onto a horizontal plane ruled with straight parallel lines a distance d $(d > l)$ apart. Buffon shown that the probability p of a needle intersecting a line is given by

$$p = \int_0^\pi \int_0^{l\sin\theta} \frac{1}{d\pi} \, dA d\theta = \frac{2l}{d\pi}.$$

Laplace [Marquis Pierre-Simon de Laplace, *Theorie Analytique des Probabilities* in Oeuvres completes de Laplace (de L'Academie des Sciences, Paris, 1886), Vol.7, part 2, pp.365-366] suggested that this technique could be used to evaluate π: $\pi = 2l/(dp_n)$, where p_n is the numerically computed probability for a needle to intersect a line. Fix the values of l and d as 0.5 unit and 1 unit, respectively.

Consider a two-dimensional plane consisting of 10^5 lines drawn parallel to y-axis with x-values $1, 2, \ldots, 10^5$. Write a program to compute p_n taking the number of throws as $N = 2^{12}$. Then, using the formula compute π. Repeat the calculation for $N = 2^{13}, 2^{14}, \ldots, 2^{24}$.

[Hint: Generate a random number x_1 between 0 to 10^5. Treat this number x_1 as one end of the needle. The other end of the needle must be on the circumference of a circle of radius l with the centre whose x value is simply x_1. Now, generate a number for the angle θ between 0 to 360 (or 0 to 2π). Then, the x-component of the other end of the needle is $x_2 = x_1 + l \cos \theta$.]

20.13 For a percolation problem with 20×20 lattice counts, the number of

   a) nonparticipating sites,

   b) participating sites but not a member of a cluster and

   c) isolated clusters

as a function of p, for $p \in [0, 1]$ with $\Delta p = 0.05$. Describe the variation of these quantities with p.

20.14 Consider a 20×20 lattice. Increase the value of p from 0 to 1 in steps of 0.01 and find the threshold value p_c at which first time a percolation occurs. Then, for $p \geq p_c$, each value of p count the number of percolation N_p. Draw a graph between p versus N_p. Describe the trend in the figure. (The percolation configuration is sensitive to the sequence of random numbers. If you got a percolation for chosen p then repeat the experiment for several, say M, times for different random number sequences. If you got at least one percolation in each trial then call the p as p_c. Otherwise do the above for $p + \Delta p$.).

20.15 The threshold value p_c is sensitive to the size $n \times n$ of the lattice. However, as the value of n increases p_c approach a constant value. Compute p_c for $n = 20, 40, \ldots, 200$ and sketch a graph between n versus p_c. How does it approach a constant value? What is that constant value?

Answers to Some Selected Problems

Chapter 1 Preliminaries

1.1 $(382.382)_{10} = 3 \times 10^2 + 8 \times 10^1 + 2 \times 10^0 + 3 \times 10^{-1} + 8 \times 10^{-2} + 2 \times 10^{-3}$.

1.2 $11101.11101 = 29.90625$.

1.3 $1.9 = 1.1110011001\ldots$ and is a never ending sequence.

1.4 $(11010.11)_2 = (26.75)_{10} = (32.6)_8 = (1A.C)_{16}$.

1.5 $e^{0.1}_{\text{Taylor}} = 1.1$ and the percentage of relative error is 0.47. $e^{0.9}_{\text{Taylor}} = 1.9$ and the percentage of relative error is 22.75.

1.6 $f(0.8) = 2.049180328$, $f(0.80005) = 2.04958355$, $g(f(0.8)) = 6.248320476$, $g(f(0.80005)) = 6.250376278$. The error in f of the order of 10^{-4} gives an error in g of the order of 10^{-2}. That is, the error propagates.

1.7 $f(0.1) = 100$, $f(0.10005) - 99.90007$ and $|f(0.10005) - f(0.1)| = 0.09993$. That is, an approximation of the order of 10^{-5} in the value of x causes a change of the order of 10^{-2} in the function. Here the divisor is a small number. Next, $f(10) = 0.01$, $f(10.00005) = 0.99999 \times 10^{-2}$ and $|f(10.00005) - f(10)| = 10^{-7}$. When the divisor is a large number an approximation of the order of 10^{-5} in the value of x causes an error of the order of 10^{-7} in the value of the function.

1.8 For $a = 3$, $x_n \to x^* = 0.66666\ldots$. For $a = 3.3$ after few iterations $\{x_n\}$ becomes a period-2 sequence. For $a = 3.5$ after a few iterations $\{x_n\}$ becomes a period-4 sequence. For $a = 3.9$ the sequence is nonperiodic.

Chapter 2 Solutions of Polynomial and Reciprocal Equations

2.2 (a) $x = 0$, 2, 3.

    (b) $x = 2$, $\pm i$.

    (c) $x = 1.27816$, $-1.13908 \pm i1.61690$.

2.4 $\alpha = 1$.

2.5 $x = -2/9$, $2/3$, -2.

2.8 (a) $x = 1$, $(-3 \pm \sqrt{5})/2$.

    (b) $x = -2$, $-1/2$, -1, -1.

    (c) $x = -1$, $(1 \pm i\sqrt{3})/2$, $(-3 \pm \sqrt{5})/2$.

2.9 $x = -3$, -2, -1, 1, 2.

Chapter 3 Solution of General Nonlinear Equations

3.1 (a) $N = 4$, $x^* = -0.00313$.

    (b) $N = 3$, $x^* = 0.56875$.

3.3 It does not cross the x-axis.

3.4   (a) After 4 iterations $x^* = 0$.
       (b) After 3 iterations $x^* = 0.56714$.

3.8   After 4 iterations the solution is $\sqrt{11} = 3.31662$.

3.9   After 4 iterations $(50)^{1/3} = 3.68403$.

3.10   Consider the equation $\pi x - 1 = 0$. With $x_0 = 0.2$ after 2 iterations $1/\pi = 0.31831$.

3.11   Denote the length of AC as x and the length of CB as $1 - x$. Then $AB/AC = AC/CB$ is $x^2 + x - 1 = 0$. Its positive root is 0.61803.

3.12   After 4 iterations $t = 3.46574\,\text{sec}$.

3.13   After 4 iterations $x = 1.73937$.

3.14   With $x_0 = 0$ units a) after 4 iterations $x = -0.23873$ units and b) after 4 iterations $x = -0.15878$ units.

3.15 $E = -0.52507$ units.

3.16   After 4 iterations the result is 1.11010.

3.17   After 2 iterations $V_T = 1.44269\,\text{V}$.

3.18   After 4 iterations $R = 1\,\Omega$.

3.19   Define $f = x^5/(e^x - 1)$ then the problem is to find the value of x at which $\mathrm{d}f/\mathrm{d}x = 0$. The exact value of x is 4.965.

3.20   No. Because $f' = 0$ when $x_0 = 0$.

3.22 $x_{n+1} = (-x_n^3 + x_n^2)/(-x_n^2 + 2x_n + e^{x_n})$.

3.23 $T = 2.6180345\,\text{sec}$.

3.24   (a) After 4 iterations the roots are $x^* = (0.0 \pm i\,1.57080)$.
       (b) After 4 iterations the roots are $x^* = (0.5 \pm i\,0.86603)$.

3.26   (a) After 4 iterations the root is $(x^*, y^*) = (1, -2)$.
       (b) After 1 iteration the root is $(x^*, y^*) = (1, 3)$.

Chapter 4 Solution of Linear Systems $AX = B$

4.1   a) $(x_1, x_2) = (1, -2)$.
       b) $(x_1, x_2) = (2, 3)$.

4.3 $(x_1, x_2, x_3) = (2, 3, -1)$.

4.4 $(x_1, x_2, x_3) = (2, 1, -3)$.

4.5 $x = 1, y = 1$.

4.6   a) $(x_1, x_2, x_3) = (1, 0, 1)$.
       b) $(x_1, x_2, x_3) = (1, -3, -4)$.

4.9   a) $\begin{pmatrix} 0.875 & -0.6875 & 0.0625 \\ -0.125 & 0.3125 & 0.0625 \\ 0.375 & 0.0625 & -0.1875 \end{pmatrix}$.

       b) $\begin{pmatrix} -0.30769 & 0.03846 & 0.19231 \\ 0.53846 & -0.19231 & 0.03846 \\ 0.46154 & 0.19231 & -0.03846 \end{pmatrix}$.

4.11 $X = (4, 0, 11)$.

4.16   a) Exact solution is $(x_1, x_2) = (1, -2)$.
<u>Solution by Jacobi method:</u>
Using computer – After 9 iterations $(x_1, x_2) = (1, -2)$.
By hand calculation – After 2 iterations $(x_1, x_2) = (1.03, -2.06)$.
<u>Solution by Gauss–Seidel method:</u>
Using computer – After 5 iterations $(x_1, x_2) = (1, -2)$.
By hand calculation – After 2 iterations $(x_1, x_2) = (1.01, -2.0)$.

   b) Exact solution is $(x_1, x_2, x_3) = (1, 2, -3)$.
<u>Solution by Jacobi method:</u>
Using computer – After 10 iterations $(x_1, x_2, x_3) = (1, 2, -3)$.
By hand calculation – After 2 iterations $(x_1, x_2, x_3) = (1.06, 2.03, -3.05)$.
<u>Solution by Gauss–Seidel method:</u>
Using computer – After 7 iterations $(x_1, x_2, x_3) = (1, 2, -3)$.
By hand calculation – After 2 iterations $(x_1, x_2, x_3) = (1.02, 1.97, -3)$.

4.17   a) $(x_1, x_2, x_3) = (0, 1, 2)$.
   b) $(x_1, x_2, x_3) = (2, 1, -1)$.

4.18   a) $(x_1, x_2, x_3) = \alpha(1, 1, -1)$.
   b) $(x_1, x_2, x_3) = \alpha(3, 1, -2)$.

4.19   a) $(x_1, x_2) = (-1, 2)$.
   b) $(x_1, x_2) = (-2, 2)$.

4.20   a) $(x_1, x_2, x_3) = (-0.5, 1, 1)$.
   b) $(x_1, x_2, x_3) = (1, 1, \ 2)$.

Chapter 5 Curve Fitting

5.1 $a = 0.5\,\text{V}^{-1}$, $b = 1$.
5.2 $V_0 = 2\,\text{m}^3$, $a = 0.0006\,\text{C}^{-1}$.
5.3 $a = -3.25\,\text{J/kg}$, $b = 1052\,\text{J/(kg.C)}$.
5.4 $R = 100$ ohms.
5.5 $a = 0.00255\,\text{KJ/mol.K}^2$, $b = 32.2307\,\text{KJ/mol.K}$.
5.6 $N_0 = 1000$, $\lambda = 0.75072/\text{m}$.
5.7 $a = 1.99884/\text{eV}$, $b = 0.49959$.
5.8 $q_0 = 1 \times 10^{-4}$ coulomb, $\tau = 0.24711\,/\text{sec}$.
5.9 $N_0 = 800$, $\mu = -25.03827/\text{m}$.
5.10 $N_0 = 9990$, $\lambda = 0.00995/\text{hour}$.
5.11 $a = 2.55779$, $b = -0.95522$.
5.12 $a = 0.52995$, $b = 1.92825$.
5.13 $a = -0.496$ units, $b = 9.81659$ units.
5.14 $a = -1.50127$ units, $b = 0.0641$ units.
5.15 $a = -1.75851$, $b = 0.43337$.
5.16 $a = 0.4999$, $b = 0.7$.

5.17 $K = 0.80084$ volt, $a = 0.50749$ m.

5.18 $a = 2.99865$, $b = 0.50285$.

5.19 $a = -1.38279$, $b = 1.60641$.

5.20 $a = 1.12564$, $b = 0.47436$.

5.21 $a = 0.8$, $b = 1.175$.

5.22 $a = -0.09965$, $b = 1.04421$.

5.23 $a = 3.00564$, $b = -1.99818$.

5.24 $a = 2$, $b = 0.5$.

5.25 $a = 3$, $b = 0.25$.

5.26 $a = -1$, $b = 3$.

5.27 $a = -0.5$, $b = 2$.

5.28 $a = 0.51105$ MeV, $b = 1$.

5.29 $a = -1$, $b = 1$.

5.30   (a) $A = 90$ units, $B = 0.5$ units, $C = 0.003$ units.

5.31   (a) $C_1 = 5.30057$ units, $C_2 = -0.00011$ units, $C_3 = 0.00002$ units.

5.32 $C_1 = 0.49851$ units, $C_2 = -0.00217$ units, $C_3 = 0.00002$ units.

5.33 $A = 3335.79992$ units, $B = 1.022$ units.

5.34 $C_1 = 0$, $C_2 = -7$, $C_3 = 9$, $C_4 = 4$.

5.35   (a) $y = 0.5 \sin x$.

       (b) $y = 1.0 + 0.5 \sin x - 0.3 \cos x$.

Chapter 6 Interpolation and Extrapolation

6.1   a) $f(-1.5) = 0.25162$, $|E_1| = 0.04599$.

     b) $f(2.5) = 0.29167$, $|E_1| = 0.00926$.

6.2   a) $a_1 = 1$, $a_2 = 0.63212$, $a_3 = 0.21332$, $a_4 = 04450$, $P(-1.5) = 0.22015$, $P(-3.1) = 0.03803$.

     b) $a_1 = 0$, $a_2 = 0.89256$, $a_3 = -0.32648$, $a_4 = 0.13472$, $P(0.3) = 0.26247$, $P(-0.1) = -0.10351$.

6.5 $h^2 \leq (8/e^1)10^{-5}$, that is, $h < 0.00542$.

6.7   By Newton interpolation polynomial: $a_1 = 0.1$, $a_2 = 0.1$, $a_3 = 0.0$, $a_4 = 0.0$, $P(1.65) = 0.165$ ampere.

     By Gregory–Newton forward-difference method: $P(1.65) = 0.165$ ampere.

     By Gregory–Newton backward-difference method: $P(1.65) = 0.165$ ampere.

6.8   By Newton interpolation polynomial: $a_1 = 0.0$, $a_2 = 4.4$, $a_3 = -0.8$, $a_4 = 0.66667$, $P(0.9) = 3.648$ m.

     By Gregory–Newton forward-difference method: $P(0.9) = 3.648$ m.

     By Gregory–Newton backward-difference method: $P(0.9) = 3.648$ m.

6.9   By Newton interpolation polynomial: $a_1 = 2.108$, $a_2 = 0.00110$, $a_3 = 0.00001$, $a_4 = 0.0$, $P(100) = 2.11781$ m^3.

     By Gregory–Newton forward-difference method: $P(100) = 2.11781$ m^3.

     By Gregory–Newton backward-difference method: $P(100) = 2.11781$ m^3.

6.10 By Newton interpolation polynomial: $a_1 = 0.184$, $a_2 = -0.305$, $a_3 = 0.2625$, $a_4 = -0.25$, $P(0.6) = 0.15013$.

By Gregory–Newton forward-difference method: $P(0.6) = 0.15013$.

By Gregory–Newton backward-difference method: $P(0.6) = 0.15013$.

6.13 For 6.7: At $V = 1.65$V; $L_1 = -0.05950$, $L_2 = 0.77350$, $L_3 = 0.33150$, $L_4 = -0.04550$, $P(1.65) = 0.165$ ampere.

For 6.8: At $t = 0.9$ minute; $L_1 = -0.03200$, $L_2 = 0.21600$, $L_3 = 0.86400$, $L_4 = -0.04800$, $P(0.9) = 3.648$ m.

For 6.9: At $T = 100°$C; $L_1 = 0.3125$, $L_2 = 0.9375$, $L_3 = -0.3125$, $L_4 = 0.06250$, $P(100) = 2.11781$ m^3.

For 6.10: At $E = 0.6$; $L_1 = 0.3125$, $L_2 = 0.9375$, $L_3 = -0.3125$, $L_4 = -0.0625$, $P(0.6) = 0.15012$.

For 6.11: At 12 days; $L_1 = -0.056$, $L_2 = -0.288$, $L_3 = 1.008$, $L_4 = 0.224$, $P(12) = 0.12432$ mg.

Chapter 7 Eigenvalues and Eigenvectors

7.1 a) i) After 3 iterations $\lambda_1 = 12.03478$, $\mathbf{X} = (0.00506, 0.08309, 1)$.

After 17 iterations $\lambda_1 = 10$, $\mathbf{X} = (0.01, 0.1, 1)$.

ii) After 3 iterations $\lambda_{small} = 0.6625$, $\mathbf{X} = (1, 0.56698, 0.3)$.

After 17 iterations $\lambda_{small} = 0.5$, $\mathbf{X} = (1, 0.5, 0.25)$.

b) i) After 3 iterations $\lambda_1 = 1$, $\mathbf{X} = (1, 1, 1)$.

After 2 iterations $\lambda_1 = 1$, $\mathbf{X} = (1, 1, 1)$.

ii) After 2 iterations $\lambda_{small} = 1$, $\mathbf{X} = (1, 1, 1)$.

After 2 iterations $\lambda_{small} = 1$, $\mathbf{X} = (1, 1, 1)$.

7.2 a) After 3 iterations $\lambda_1 = 2.80000$, $\mathbf{X} = (1, 0.57143, 0.14286)$.

b) After 3 iterations $\lambda_1 = 3.83333$, $\mathbf{X} = (0.65217, 1, 0.82609)$.

7.3 After 3 iterations $\lambda = 2$ and $\mathbf{X} = (1/8, 1, 1/8)$. $\mathbf{X}_1 = (1/2, 1, 1/2)$, $\mathbf{X}_2 = (1/4, 1, 1/4)$ and $\mathbf{X}_3 = (1/8, 1, 1/8)$. Hence, $\mathbf{X}_n = (1/2^n, 1, 1/2^n)$ giving $\mathbf{X} = (0, 1, 0)$.

7.4 a) After 1 iteration $\lambda = -1, 3$ and $\mathbf{X} = \begin{pmatrix} 1/\sqrt{2} & 1/\sqrt{2} \\ -1/\sqrt{2} & 1/\sqrt{2} \end{pmatrix}$.

b) After 1 iteration $\lambda = \sqrt{2} \pm 1$ and $\mathbf{X} = \begin{pmatrix} 1/\sqrt{2} & 1/\sqrt{2} \\ -1/\sqrt{2} & 1/\sqrt{2} \end{pmatrix}$.

7.5 a) After 2 iterations $\lambda = -1, 1, 5$ and

$$\mathbf{X} = \begin{pmatrix} 1/\sqrt{2} & -1/2 & 1/2 \\ 0 & 1/\sqrt{2} & 1/\sqrt{2} \\ -1/\sqrt{2} & -1/2 & 1/2 \end{pmatrix}.$$

b) After 1 iteration $\lambda = -1, 1, 3$ and

$$\mathbf{X} = \begin{pmatrix} 1/\sqrt{2} & 0 & 1/\sqrt{2} \\ 0 & 1 & 0 \\ -1/\sqrt{2} & 0 & 1/\sqrt{2} \end{pmatrix}.$$

7.8   a) $\lambda = 1, 3.$

     b) $\lambda = -2.72545, 1.60147, 4.12398.$

7.9   a) $\lambda = 1.58579, 3, 4.41421.$

     b) $\lambda = 1.58579, 3, 4.41421.$

7.10   a) $\lambda = 0.25761, 1.85228, 3.96828, 7.92184.$

     b) $\lambda = -1.23607, 0.76393, 3.23607, 5.23607.$

Chapter 8 Numerical Differentiation

8.4.   Exact value of $f_x(0.2, 0.2) = -0.25761457.$

     $f_x(0.2, 0.2, h = 0.05) = -0.25698847.$ $f_x(0.2, 0.2, h = 0.1) = -0.25511928.$

     Exact value of $f_y(0.2, 0.2) = -1.28807283.$

     $f_y(0.2, 0.2, h = 0.05) = -1.29022069.$ $f_y(0.2, 0.2, h = 0.1) = -1.29667718.$

8.5.   Exact value of $f_x(0, 0) = 1.0.$

     $f_x(0, 0, h = 0.05) = 1.0.$ $f_x(0, 0, h = 0.1) = 1.0.$

     Exact value of $f_y(0, 0) = 0.$

     $f_y(0, 0, h = 0.05) = 0.$ $f_y(0, 0, h = 0.1) = 0.$

8.6. $x' = 11.28000004 \, \text{m/sec}$, $v = 0.49999936 \, \text{m/sec}$ and the exact $v = 0.5 \, \text{m/sec}.$

8.7. $x'' = -3.72500011$ and $\omega_0^2 = 4.04451693.$

8.8. $x' = -0.200833483 \, \text{m/sec}$ and $E = 0.98056708 \, \text{J}.$

8.9. $v''(1.2) = -0.37249860$ and $L = 0.97288956.$

8.10. $f' = (3/4h) \left[f(x_0 + h) - f(x_0 - h) \right] - (3/20h) \left[f(x_0 + 2h) - f(x_0 - 2h) \right]$
        $+(1/60h) \left(f(x_0 + 3h) - f(x_0 - 3h) \right) - (4/21)h^6 f^{(7)}(x_0) .$

8.11. $D^{0.5}(t^2 - t)$ values at $t = 0.5$ for $h = 0.01, 0.001$ and 0.0001 are $-0.26642,$
        -0.26598 and -0.26596, respectively.

8.12.   At $t = 0.1, 0.2, 0.3, 0.4$ and 0.5 the values of $D^{0.5}t^2$ are $0.04758, 0.13457, 0.24722,$
        0.38061 and 0.53192, respectively.

8.13.   At $t = 0.1, 0.2, 0.3, 0.4$ and 0.5 the values of $D^{0.5}t$ are $0.35682, 0.50463, 0.61804,$
        0.71365 and 0.79788, respectively.

8.14.   The values of $D^{1.5}t^2$ at $t = 0.1, 0.2, 0.3, 0.4$ and 0.5 are $0.71365, 1.00925, 1.23608,$
        1.42730 and 1.59577, respectively.

Chapter 9 Numerical Minimization of Functions

9.1.   a) After 3 iterations the root of $f'(x) = 0$ is 2.45667. $f(x)$ is locally minimum at
        this point.

     b) After 3 iterations the root of $f'(x) = 0$ is 1.01484. $f(x)$ is locally maximum at
        this point.

9.2. $f(x)$ has a local minimum at $x = -0.70705$ while a local maximum at $x = 0.70705.$

9.3.   The given function has local minima at $x = \pm 1.23411$ and a maximum at $x = 0.$

9.4.   After 3 iterations $x^* = 0.01949$ and $f'' > 0$ at this point.

9.6. The 1st iteration gives $x_0 = 0.1$, $x_1 = -0.025$, $x_2 = -0.15$ and $x_1^* = -0.00016$. In the 2nd iteration Δx less than 0.001 is realized therefore $x_1^* = -0.00016$ is the point of minimum. The exact value of x^* is 0.

9.7. The first iteration gives $x_0 = 0.1$, $x_1 = -0.9$, $x_2 = -1.9$ and $x_1^* = -0.98864$. The second iteration gives $x_0 = -0.98864$, $x_1 = -0.48864$, $x_2 = 0.01136$ and $x_2^* = -0.72294$. The third iteration gives $x_0 = -0.72294$, $x_1 = -0.69169$, $x_2 = -0.66044$ and $x_2^* = -0.70717$. In the fourth iteration $\Delta x < 0.001$ is realized. Therefore, $x_1^* = -0.70717$ is the point of minimum.

9.8. a) After 5 iterations $(x^*, y^*) = (0, 3.14159)$, $\Delta^2 = 2$ and $f_{xx} = 2$.

   b) After 5 iterations $(x^*, y^*) = (0, 0)$, $\Delta^2 = 4$ and $f_{xx} = 2$.

9.9. For $(x_0, y_0) = (0.8, 0.8)$ after 6 iterations $(x^*, y^*) = (1, 1)$ and $\Delta^2 = 48$, $f_{xx} = 8$. (x^*, y^*) is a minimum.

   For $(x_0, y_0) = (0.8, -0.8)$ after 12 iterations $(x^*, y^*) = (1, -1)$ and $\Delta^2 = 80$, $f_{xx} = 8$. (x^*, y^*) is a minimum.

   For $(x_0, y_0) = (-0.8, 0.8)$ after 13 iterations $(x^*, y^*) = (-1, 1)$ and $\Delta^2 = 80$, $f_{xx} = 12$. (x^*, y^*) is a minimum.

   For $(x_0, y_0) = (-0.8, -0.8)$ after 15 iterations $(x^*, y^*) = (-1, -1)$ and $\Delta^2 = 128$, $f_{xx} = 12$. (x^*, y^*) is a minimum.

   For $(x_0, y_0) = (0.2, 0.2)$ after 4 iterations $(x^*, y^*) = (0, 0)$ and $\Delta^2 = 4$, $f_{xx} = -2$. (x^*, y^*) is a maximum.

   For $(x_0, y_0) = (0.0, 0.7)$ after 3 iterations $(x^*, y^*) = (0, 0.707107)$ and $\Delta^2 = -12$, $f_{xx} = -3$. Therefore, f does not have a local maximum at (x^*, y^*).

9.10. a) After 17 iterations $(x^*, y^*) = (0.000309, 3.141850)$.

   b) After 18 iterations $(x^*, y^*) = (0.000717, 0.000276)$.

Chapter 10 Numerical Integration

In the following TR, $S_{1/3}$, $S_{3/8}$ and GL denote trapezoidal rule, Simpson's 1/3-rule, Simpson's 3/8-rule and two-point Gauss–Legendre rule, respectively. CTR, $CS_{1/3}$ and $CS_{3/8}$ denote composite trapezoidal rule, composite Simpson's 1/3-rule and composite Simpson's 3/8-rule, respectively. Further, exact value is the value of the given integral obtained by direct integration or by composite Simpson's 3/8-rule with large n.

10.2 (a) Exact value of the integral $= \sin 1 = 0.8414710$.

   TR: 0.7701512. $S_{1/3}$: 0.8417721. $S_{3/8}$: 0.8416044.   GL: 0.8412698.

   CTR: 0.8336651. $CS_{1/3}$: 0.8414746. $CS_{3/8}$: 0.8414726.

   (b) Exact value of the integral $= 0.7468241$.

   TR: 0.6839397. $S_{1/3}$: 0.7471804. $S_{3/8}$: 0.7469923.   GL: 0.7465947.

   CTR: 0.7399865. $CS_{1/3}$: 0.7468304. $CS_{3/8}$: 0.7468269.

10.3 (a) Exact value of the integral $= 0.9460831$.

   TR: 0.9207355. $S_{1/3}$: 0.9432914. $S_{3/8}$: 0.9461109.   GL: 0.9460411.

   CTR: 0.9432914. $CS_{1/3}$: 0.9460838. $CS_{3/8}$: 0.9460834.

   (b) Exact value of the integral $= 1/12 = 0.0833333$.

   TR: 0.0000000. $S_{1/3}$: 0.1308997. $S_{3/8}$: 0.0956496.   GL: 0.0459383.

   CTR: 0.0850218. $CS_{1/3}$: 0.0828822. $CS_{3/8}$: 0.0831402.

10.4 Exact value of the integral $= 0.25$.   TR: 0.125. $S_{1/3}$: 0.25.

10.8 a) TR: 0.1666667. $S_{1/3}$: 0.0083333. $S_{3/8}$: 0.0037723.   GL: 0.0055555.

b) TR: 0.1666667. $S_{1/3}$: 0.0. $S_{3/8}$: 0.0.   GL: 0.0.

10.9 a) CTR: $n_{opt} = 129$. $CS_{1/3}$: $n_{opt} = 5$. $CS_{3/8}$: $n_{opt} = 2$.

b) CTR: $n_{opt} = 223$. $CS_{1/3}$: $n_{opt} = 0$. $CS_{3/8}$: $n_{opt} = 0$.

10.10 The exact value of the integral is 2.1779795.

CTR: $n_{opt} = 12$. Therefore, $n = 13$. With this value of n the value of the integral is 2.1786000.

$CS_{1/3}$: $n_{opt} = 1$. Therefore, $n = 2$. The value of the integral is 2.1780266.

$CS_{3/8}$: $n_{opt} = 0$. Therefore, $n = 1$. The value of the integral is 2.1783107.

10.14 Exact value of the integral $= -0.0900633$.

TR: 0.0. $S_{1/3}$: 0.0. $S_{3/8}$: -0.0937500.   GL: -0.1726528.

CTR: -0.0833333. $CS_{1/3}$: -0.0919278. $CS_{3/8}$: -0.0908453.

10.15 Exact value of the integral $= 3.6275987$.

TR: 3.1415927. $S_{1/3}$: 3.1415927. $S_{3/8}$: 3.6853298.   GL: 4.1094810.

CTR: 3.6249146. $CS_{1/3}$: 3.6249146. $CS_{3/8}$: 3.6279342.

10.16 Exact value of the integral $= 2.6220576$.

TR: 2.6815171. $S_{1/3}$: 2.6039055. $S_{3/8}$: 2.6146593.   GL: 2.6349313.

CTR: 2.6220879. $CS_{1/3}$: 2.6220474. $CS_{3/8}$: 2.6220538.

10.17 Exact value of the integral $= 0.7357589$.

TR: 2.3504024. $S_{1/3}$: 0.7834675. $S_{3/8}$: 0.7573709.   GL: 0.7043259.

CTR: 0.9343744. $CS_{1/3}$: 0.7364409. $CS_{3/8}$: 0.7360627.

10.18 Exact value of the integral $= 1.7627472$.

TR: 1.4142136. $S_{1/3}$: 1.8047379. $S_{3/8}$: 1.7765783.   GL: 1.7320508.

CTR: 1.7363156. $CS_{1/3}$: 1.7628166. $CS_{3/8}$: 1.7627755.

10.19 Exact value of the integral $= 0.5204999$.

TR: 0.5017904. $S_{1/3}$: 0.5206015. $S_{3/8}$: 0.5205447.   GL: 0.5204318.

CTR: 0.5184609. $CS_{1/3}$: 0.5205011. $CS_{3/8}$: 0.5205004.

10.20 Exact value of the integral $= 0.1666667$.

TR: 0.6666667. $S_{1/3}$: 0.2222222. $S_{3/8}$: 0.1978850.   GL: 0.1384280.

CTR: 0.2499718. $CS_{1/3}$: 0.1702858. $CS_{3/8}$: 0.1684600.

10.21 Exact value of the integral $= 2.4221121$.

TR: 2.3561945. $S_{1/3}$: 2.4411629. $S_{3/8}$: 2.4302092.   GL: 2.4088172.

CTR: 2.4219854. $CS_{1/3}$: 2.4221542. $CS_{3/8}$: 2.4221279.

10.22 Exact value of the integral $= 0.5 \ln 5 = 0.8047190$.

TR: 0.4000000. $S_{1/3}$: 0.8000000. $S_{3/8}$: 0.8061538.   GL: 0.8108108.

CTR: 0.7610256. $CS_{1/3}$: 0.8053092. $CS_{3/8}$: 0.8049718.

10.23 Exact value of the integral $= 0.0750000$.

TR: 0.0000000. $S_{1/3}$: 0.0750000. $S_{3/8}$: 0.0750000.   GL: 0.0750000.

CTR: 0.0666667. $CS_{1/3}$: 0.0750000. $CS_{3/8}$: 0.0750000.

10.24 Exact value of the integral $= 3.1415927$.

TR: 2.5132741. $S_{1/3}$: 2.5132741. $S_{3/8}$: 3.1657588.   GL: 3.9444714.

CTR: 3.0932605. $CS_{1/3}$: 3.0932605. $CS_{3/8}$: 3.1476207.

10.25 Exact value of the integral $= 0.2000000$.

TR: 0.3000000. $S_{1/3}$: 0.2000000. $S_{3/8}$: 0.2000000.   GL: 0.2000000.

CTR: 0.2111111. $CS_{1/3}$: 0.2000000. $CS_{3/8}$: 0.2000000.

10.26 a) Exact value $= 0.7468241$. By GL 3-point method: 0.7468146.

b) Exact value $= 0.6931472$. By GL 3-point method: 0.6931217.

c) Exact value $= 0.9460831$. By GL 3-point method: 0.9460832.

10.27 The formula is

$$0.347858451 \left[f(0.8611363116) + f(-0.8611363116) \right]$$
$$+ 0.6521451549 \left[f(0.3399810436) + f(-0.3399810436) \right]$$
$$+ \frac{1}{3472875} f^{(8)}(x').$$

a) Exact value $= 0.6931472$. By the 4-point GL method: 0.6931490.

b) Exact value $= 0.4352099$. By the 4-point GL method: 0.4352117.

10.28 a) Exact value $= 0.7468241$.

By the 3-point Lobatto method: 0.7471805.

b) Exact value $= 0.6931472$.

By the 3-point Lobatto method: 0.6944445. 0.9461459.

10.29 a) Exact value $= 0.7468241$.

By the 3-point Radan method: 0.7472275.

b) Exact value $= 0.6931472$.

By the 3-point Radan method: 0.6933333.

10.30 $I = 0.526745$.

10.32 3.97807, 3.99927 and 3.99998.

10.33 0.02777, 0.02689 and 0.02688.

10.34 1.02436, 1.00784 and 1.00248.

Chapter 11 Ordinary Differential Equations – Initial-Value Problems

For simplicity, the units of the variables are not mentioned in the answers.

11.5 At $x = 0.1$ $\text{erf}_{\text{Eu}} = 1.2412171$, $\text{erf}_{\text{RK2}} = 1.2406557$, $\text{erf}_{\text{RK4}} = 1.2408421$.

At $x = 1$ $\text{erf}_{\text{Eu}} = 2.0060514$, $\text{erf}_{\text{RK2}} = 1.9703879$, $\text{erf}_{\text{RK4}} = 1.9710800$.

11.6 At $x = 0.1$ $C_{\text{Eu}} = 1.1000000$, $C_{\text{RK2}} = 1.0999938$, $C_{\text{RK4}} = 1.0999974$.

At $x = 1$ $C_{\text{Eu}} = 1.8272711$, $C_{\text{RK2}} = 1.7772711$, $C_{\text{RK4}} = 1.7798945$.

11.7 At $t = 0.1$ $x_{Ex} = 0.1003347$, $x_{Eu} = 0.1000000$, $x_{RK2} = 0.0995000$,
$x_{RK4} = 0.0996679$.

11.8 At $t = 0.1$ $x_{Ex} = 1.0996680$, $x_{Eu} = 1.1000000$, $x_{RK2} = 1.0995000$,
$x_{RK4} = 1.0996679$.

11.9 At $t = 0.1$ $x_{Ex} = 0.9128709$, $x_{Eu} = 0.9000000$, $x_{RK2} = 0.9135500$,
$x_{RK4} = 0.9128709$.

11.10 At $t = 0.1$ $x_{Ex} = 0.5378994$, $x_{Eu} = 0.5375000$, $x_{RK2} = 0.5378606$,
$x_{RK4} = 0.5378994$.

11.11 At $t = 0.1$ $x_{Ex} = 0.9003320$, $x_{Eu} = 0.9000000$, $x_{RK2} = 0.9005000$,
$x_{RK4} = 0.9003321$.

11.12 At $t = 0.1$ $x_{Ex} = 0.4534796$, $x_{Eu} = 0.4500000$, $x_{RK2} = 0.4535125$,
$x_{RK4} = 0.4534796$.

11.13 At $t = 0.1$ $x_{Eu} = 4.7500000$, $x_{RK2} = 4.7562500$, $x_{RK4} = 4.7561471$.

11.14 At $t = 0.1$ $x_{Eu} = 0.1900000$, $x_{RK2} = 0.1852502$, $x_{RK4} = 0.1854837$.

11.15 At $t = 0.1$ $x_{Ex} = 2.7145240$, $x_{Eu} = 2.7000000$, $x_{RK2} = 2.7150171$,
$x_{RK4} = 2.7145236$.

11.16 At $t = 0.1$ $v_{Eu} = 1.4550000$, $v_{RK2} = 1.3616488$, $v_{RK4} = 1.3821320$.

11.17 At $t = 0.1$ $x_{Ex} = 0.5124974$, $x_{Eu} = 0.5125000$, $x_{RK2} = 0.5124961$,
$x_{RK4} = 0.5124974$.

11.18 At $t = 0.1$ $q_{Ex} = 0.0380650$, $q_{Eu} = 0.0400000$, $q_{RK2} = 0.0380000$,
$q_{RK4} = 0.0380650$.

11.19 At $t = 0.1$ $i_{Ex} = 0.0195082$, $i_{Eu} = 0.0200000$, $i_{RK2} = 0.0195000$,
$i_{RK4} = 0.0195082$.

11.20 At $t = 0.1$ $x_{Ex} = 4.5717684$, $x_{Eu} = 4.5500000$, $x_{RK2} = 4.5725000$,
$x_{RK4} = 4.5717687$.

11.21 At $x = 0.1$ $I_{Ex} = 2.7145123$, $I_{Eu} = 2.7000000$, $I_{RK2} = 2.7150000$,
$I_{RK4} = 2.7145125$.

11.22 At $t = 0.1$ $T_{Ex} = 57.1451225$, $T_{Eu} = 57.0000000$, $T_{RK2} = 57.1500000$,
$T_{RK4} = 57.1451250$.

11.23 At $t = 0.1$ $x_{Ex} = 0.2000000$, $x_{Eu} = 0.2000000$, $x_{RK2} = 0.2000000$,
$x_{RK4} = 0.2000000$.
$x'_{Ex} = 2.0000000$, $x'_{Eu} = 2.0000000$, $x'_{RK2} = 2.0000000$, $x'_{RK4} = 2.0000000$.

11.24 At $t = 0.1$ $x_{Ex} = 0.1490000$, $x'_{Ex} = 1.9800000$,
$x_{Eu} = 0.1000000$, $x'_{Eu} = 1.9800000$,
$x_{RK2} = 0.1490000$, $x'_{RK2} = 1.9800000$,
$x_{RK4} = 0.1490000$, $x'_{RK4} = 1.9800000$.

11.25 At $t = 0.1$ $N_{1,Ex} = 93.2393820$, $N_{2,Ex} = 6.7129655$,
$N_{1,Eu} = 93.0000000$, $N_{2,Eu} = 7.0000000$,
$N_{1,RK2} = 93.2450000$, $N_{2,RK2} = 6.7060000$,
$N_{1,RK4} = 93.2393833$, $N_{2,RK4} = 6.7129639$.

Chapter 12 Symplectic Integrators for Hamiltonian Systems

12.1 $\mathbf{p}_{n+1} = \left(1 + \frac{1}{2}\right)\mathbf{p}_n + \frac{1}{2}h\mathbf{F}\left(\mathbf{q}_n + \frac{1}{2}h\mathbf{p}_n\right)$.

$\mathbf{q}_{n+1} = \mathbf{q}_n + \frac{1}{2}h\mathbf{p}_n + h\mathbf{p}_{n+1}$.

12.2 $a_1 + a_2 + a_3 = 1$, $\quad b_1 + b_2 + b_3 = 1$, $\quad a_1 b_2 + (a_1 + a_2)\,b_3 = 1/2$,

$a_1 b_1^2 + a_2\,(b_1 + b_2)^2 + a_3 = 1/3$, $\quad a_1^2 b_2 + (a_1 + a_2)^2\,b_3 = 1/3$.

$(a_1,\,a_2,\,a_3,\,b_1,\,b_2,\,b_3) = (2/3,\,-2/3,\,1,\,7/24,\,3/4,\,-1/24)$.

12.4 The exact solution at $t = 0.1$ is $q(0.1) = 0.0998334$ and $p(0.1) = 0.9950042$.

By the explicit Euler method: $q(0.1) = 0.10$, $p(0.1) = 1$, $D = 0.0049986$ and $H(0.1) = 0.505$.

By the symplectic Euler method (updated q is used): $q(0.1) = 0.1$, $p(0.1) = 0.99$, $D = 0.0050069$ and $H(0.1) = 0.49505$.

By the symplectic Euler method (updated p is used): $q(0.1) = 0.1$, $p(0.1) = 1.0$, $D = 0.0049986$ and $H(0.1) = 0.505$.

By the Störmer–Verlet method: $q(0.1) = 0.1$, $p(0.1) = 0.995$, $D = 0.0001666$ and $H(0.1) = 0.5000125$.

By the Candy–Rozmus method: $q(0.1) = 0.0998331$, $p(0.1) = 0.9950042$, $D = 2.84466 \times 10^{-7}$ and $H(0.1) = 0.5$.

12.6 By the explicit Euler method: $q(0.1) = 0.10, p(0.1) = 1.0$ and $H(0.1) = 0.5050250$.

By the symplectic Euler method (updated q is used): $q(0.1) = 0.10$, $p(0.1) = 0.98990$ and $H(0.1) = 0.4949760$.

By the symplectic Euler method (updated p is used): $q(0.1) = 0.10$, $p(0.1) = 1.0$ and $H(0.1) = 0.5050250$.

By the Störmer–Verlet method: $q(0.1) = 0.10$, $p(0.1) = 0.994950$ and $H(0.1) = 0.4999878$.

By the Candy–Rozmus method: $q(0.1) = 0.0998325$, $p(0.1) = 0.9949793$ and $H(0.1) = 0.50$.

12.8 By the explicit Euler method: $q(0.1) = 2.0$, $p(0.1) = -0.0909297$ and $H(0.1) = 0.4202809$.

By the symplectic Euler method (updated q is used): $q(0.1) = 2.0$, $p(0.1) = -0.0909297$ and $H(0.1) = 0.4202809$.

By the symplectic Euler method (updated p is used): $q(0.1) = 1.9909070$, $p(0.1) = -0.0909297$ and $H(0.1) = 0.4161469$.

By the Störmer–Verlet method: $q(0.1) = 1.9954535$, $p(0.1) = -0.0910239$ and $H(0.1) = 0.4161511$.

By the Candy–Rozmus method: $q(0.1) = 1.9954519$, $p(0.1) = -0.0909929$ and $H(0.1) = 0.4161469$.

12.10 $q(0.1) = 0.3484456$, $p(0.1) = 0.4681761$.

Chapter 16 Fractional Order Ordinary Differential Equations

16.1 Replace α by $-\alpha$ in Eq. (16.3).

16.5 For $\alpha = 0.5$, $x(0.001) = 0.03159$ and $x(0.002) = 0.04735$. For $\alpha = 0.75$, $x(0.001) = 0.00561$ and $x(0.002) = 0.00982$.

16.6 $x(0.001) = 0.45007$ and $x(0.002) = 0.47413$.

16.9 $x(0.001) = -0.0112$ and $x(0.0002) = -0.01572$.

16.10 $x(0.001) = 1.03656$ and $x(0.002) = 1.05300$.

16.11 $x(0.01) = 1.00075$ and $x(0.02) = 1.00213$.

16.12 $x(0.01) = 0.74849$ and $x(0.02) = 0.70298$.

16.14 $x(0.001) = 1.0$ and $x(0.002) = 1.00004$.

16.15 $x(0.001) = 0.98216$ and $x(0.002) = 0.96804$.

Index

Printed in the United States
by Baker & Taylor Publisher Services

Printed in the United States
by Baker & Taylor Publisher Services